Wireless Networks

Series editor

Xuemin Sherman Shen
University of Waterloo, Waterloo, Ontario, Canada

More information about this series at http://www.springer.com/series/14180

Yang Yang • Jing Xu • Guang Shi
Cheng-Xiang Wang

5G Wireless Systems

Simulation and Evaluation Techniques

 Springer

Yang Yang
CAS Key Lab of Wireless Sensor
  Network and Communication
Shanghai Institute of Microsystem
  and Information Technology
Shanghai, China

Guang Shi
China Institute of Communications
Beijing, Beijing
China

Jing Xu
Shanghai Research Center for Wireless
  Communications
Shanghai, China

Cheng-Xiang Wang
School of Engineering & Physical Sciences
Heriot-Watt University
Edinburgh, United Kingdom

ISSN 2366-1186   ISSN 2366-1445   (electronic)
Wireless Networks
ISBN 978-3-319-61868-5   ISBN 978-3-319-61869-2   (eBook)
DOI 10.1007/978-3-319-61869-2

Library of Congress Control Number: 2017947041

Printed on acid-free paper

This Springer imprint is published by Springer Nature
The registered company is Springer International Publishing AG
The registered company address is: Gewerbestrasse 11, 6330 Cham, Switzerland

Preface

It is anticipated that the Fifth Generation (5G) mobile communication systems will start to be commercialized and deployed in 2018 for new mobile services in three key scenarios, i.e., enhanced Mobile Broadband (eMBB), massive Machine Type Communications (mMTC), and Ultra-Reliable and Low Latency Communications (uRLLC). Based on the preliminary research for 5G standardization by telecom industry leaders since 2012, the International Telecommunication Union (ITU) has identified and announced the 5G vision and Key Performance Indicators (KPI) on spectrum efficiency, energy efficiency, peak data rate, traffic density, device connectivity, radio latency and reliability for achieving more comprehensive and better service provisioning and user experience. At present, the telecom industry is actively developing a variety of enabling technologies, such as massive MIMO, Ultra-Dense Network (UDN), and mmWAVE to accelerate the ongoing 5G standardization and precommercial trial processes. On the other side, new 5G technologies drive the design and development of corresponding simulation platforms, evaluation methods, field trails, and application scenarios to compare, screen, and improve 5G candidate technologies for the ongoing 5G standardization, system development, and performance enhancement. In view of these 5G technological trends and testing requirements, this book aims at addressing the technical challenges and sharing our practical experiences in the simulation and evaluation of a series of 5G candidate technologies. In particular, it reviews the latest research and development activities of 5G candidate technologies in the literature, analyzes the real challenges in testing and evaluating these technologies, proposes different technical approaches by combining advanced software and hardware capabilities, and presents the convincing evaluation results in realistic mobile environments and application scenarios, which are based on our long-term dedicated research and practical experiences in wireless technology evaluation and testbed development. This book reviews and provides key testing and evaluation methods for 5G candidate technologies from the perspective of technical R&D engineers. In order to facilitate readers with different backgrounds to better understand important concepts and methods, we include in this book many examples of various simulation and testing cases, which can be used as technical references by 5G researchers, engineers, and postgraduate students in wireless mobile communications.

This book is organized as follows. Chapter 1 reviews the vision and technical requirements of 5G mobile systems. Some 5G candidate technologies and the challenges in technical evaluation and verification methods are then discussed. Chapter 2 introduces the evolution history of a variety of testing and evaluation technologies, thus offering a comprehensive and systematic understanding of the importance and difficulties of testing technologies. Since wireless channel characteristics play a key role in system design and performance evaluation, Chapter 3 introduces state-of-the-art wireless channel measurement, modeling, and simulation methods in detail. Chapter 4 is focused on the development of a large-scale system-level software simulation platform by using parallel computing technology. Based on Hardware-in-the Loop (HIL) technology, Chapter 5 develops a hardware and software co-simulation platform with a real-time channel emulator in the loop, thus balancing the simulation speed and system flexibility. Further, Chapter 6 presents a truly real-time hardware testing and evaluation platform with a set of advanced measuring equipments and instruments for small-scale 5G technology evaluation. Finally, Chapter 7 gives the layout and design of a real 5G testbed with six macro-cells for field trials and future mobile services.

Last but not least, we would like to acknowledge the generous financial supports over the last ten years from the National Science and Technology Major Projects "New Generation Mobile Wireless Broadband Communication Networks," National Natural Science Foundation of China, Ministry of Science and Technology of China, Ministry of Industry and Information Technology of China, Chinese Academy of Sciences, Science and Technology Commission of Shanghai Municipality, and Shanghai Municipal Commission of Economy and Informatization. In addition, we are very grateful to the following colleagues for their kind help and constructive comments on earlier versions of this book, i.e., Xin Yang, Haowen Wang, Hui Xu, Kai Li, Xiumei Yang, Jian Sun, Jinling Du, Yuanping Zhu, Chenping, Panrongwei, Guannan Song, Guowei Zhang, and Kaili Wang.

Shanghai, China   Yang Yang
Shanghai, China   Xu Jing
Beijing, China   Shi Guang
Edinburgh, UK   Wang Cheng-Xiang

Contents

Abbreviations

5G	The Fifth Generation mobile communications system
2D	Two-Dimensional
2G	2-Generation wireless telephone technology
3D	Three-Dimensional
3G	the 3rd Generation mobile communications system
3GPP	the 3rd Generation Partnership Project
4G	Fourth Generation mobile communications system
5GIC	5G Innovation Center
A2GSM	Ad hoc assisted GSM
AAA	Authentication, Authorization, Accounting
AAS	Active Antenna System
A-BFT	Associated Beam Forming training Time
AC	Access Controller
ACK	ACKnowledgement
A-D	Anderson-Darling
AD/DA	Analogy Digital/Digital Analogy converter
ADC	Analog-to-Digital Converter
AFD	Average Fade Duration
AMC	Adaptive Modulation and Coding
A-MPDU	Aggregated MAC Protocol Data Unit
ANO	Automated Network Organization
AOA	Angle Of Arrival
AOD	Azimuth angle Of Departure
AP	Access Point
APDP	Averaged PDP
ARM	Advanced RISC Machines
ARS	Ad hoc Relay network Stations
AS	Angle Spread
ASA	Azimuth angular Spread at Arrival
ASIC	Application-Specific Integrated Circuit

AT	Announcement Time
AUT	UT Austin
AWGN	Additive White Gaussian Noise
AWRI	Advanced Wireless Research Initiative
AXI	Advanced eXtensible Interface
AXIe	Advanced eXtensible Interface express
B/S	Browser/Server
BBU	BaseBand Unit
BDS	BeiDou Navigation Satellite System
BER	Bit Error Rate
BFS	Breadth-First Search algorithm
BI	Beacon Intervals
B-ISDN	Broadband Integrated Services Digital Network
BLER	BLock Error Rate
BRP	Beam Refinement Phase
BS	Base Station
BSC	Base Station Controller
BT	Beacon Time
CA	Carrier Aggregation
CAPEX	CAPital EXpenditure
CAPWAP	Control and Provision of Wireless Access Points
CB	Coherence Bandwidth
CBP	Competition-Based access Period
CC	Cooperative Communications
CDD	Cyclic Delay Diversity
CDD-OFDM	Cyclic Delay Diversity Orthogonal Frequency Division Multiplexing
CDF	Cumulative Distribution Function
CDL	Cluster Delay Line
CDM	Cyclic Delay Modulation
CDMA	Code-Division Multiple Address
CDMA2000	Code Division Multiple Access 2000
CDN	Content Delivery Network
CI	Close-In the reference short
CIF	CI with Frequency
CIR	Channel Impulse Response
CMD	Correlation Matrix Distance
CoMP	Coordinated Multiple Points
COST	European Cooperation in Science and Technology
COTS	Commercial Off The Shelf
COW	Cluster of Workstation/PC
CP	Cyclic Prefix
CPCI	Compact PCI
CPRI	Common Public Radio Interface

CPU	Central Processing Unit
CQI	Channel Quality Indicator
C-RAN	Cloud Radio Access Network
CRC	Cyclic Redundancy Check
CRLBs	Cramer Rao Lower Bounds
CRS	Cell Reference Signal
CS	Compressive Sensing
CS/CBF	Coordinated Scheduling/BeamForming
CSCM	Correlation-based Stochastic Channel Model
CSF	Cluster Shadow Fading
CSI	Channel State Information
CTD-ML decoder	Circular Trap Detection based ML decoder
CTIA	Cellular Telecommunications and Internet Association
CTS	Clear To Send
CTTL	China Telecommunication Technology Labs
CUDA	Compute Unified Device Architecture
CVA	Circular Viterbi Algorithm
CW	Continuous Wave
D2D	Device to Device
DAB	Digital Audio Broadcasting
DAC	Digital-to-Analog Converter
DC	Direct Current
DCT-CDA	Discrete Cosine Transform-CDA
DDC	Direct Digital Controller
DDP	Delay-Doppler Profile
DFT	Discrete Fourier Transform
DL	DownLink
DMA	Direct Memory Access
DMC	Diffuse scattering Multipath Components
DoD	Department of Defense
DOF	Degrees Of Freedom
DS	Delay Spread
DSD	Doppler power Spectral Density
DSM	Distributed Shared Memory multiprocessor
DSP	Digital Signal Processors
DTT	Data Transmission Time
DUT	Device Under Test
DVB	Digital Video Broadcasting
DVB-H	Digital Video Broadcasting Handheld
EB	Elektrobit
ECU	Electronic Control Unit
EDA	Electronics Design Automation
EKF	Extended Kalman Filter

EM	Expectation-Maximization
EOA	Elevation angle Of Arrival
EOD	Elevation angle Of Departure
EPC	Evolved Packet Core
ER	Effective Roughness
ESNR	Estimated Signal-to-Noise Ratio
ESPRIT	Estimation of Signal Parameter via Rotational Invariance Techniques
EU	European Union
EuQoS	End-to-end Quality Of Service Support over Heterogeneous Networks
EUT	Equipment Under Test
EVA	Extended Vehicular A model
EVM	Error Vector Magnitude
FB	Flexible Backhauling
FBMC	Filter Bank based Multi-Carrier
FCC	Federal Communications Commission
FDD	Frequency Division Duplex
FDMA	Frequency Division Multiple Address
FD-SAGE	Frequency Domain SAGE
FER	Frame Error Rate
FFR	Fractional Frequency Reuse
FFT	Fast Fourier Transform
FIFO	First Input First Output
F-OFDM	Filtered OFDM
FPGA	Field Programmable Gate Array
FR	Fraunhofer Region
FST	Fast Session Transfer
FT	France Telecom
GBC	General purpose Baseband Computing
GBCM	Geometric-Based stochastic Channel Model
GCS	Global Coordinate System
GIS	Geographic Information System
GO	Geometrical Optics
GoF	Goodness of Fit
GPIB	General Purpose Interface Bus
GPP	General Purpose Processor
GPS	Global Positioning System
GPU	Graphics Processing Unit
GSCM	Geometric-based Stochastic Channel Model
GSM	Global System for Mobile communication
GSM-R	Global System for Mobile Communications-Railway
GTD	Geometrical Theory of Diffraction
HARQ	Hybrid Automatic Repeat reQuest

H-CRAN	Heterogeneous Cloud Radio Access Network
HD	High Definition
HDL	Hardware Description Language
HeNB	Home eNodeB
HetNet	Heterogeneous Network
HF	High Frequency
HFB	High Frequencies Band
HIL	Hardware-In-the-Loop
HSDPA	High-Speed Downlink Packet Access
HSPA	High-Speed Packet Access
HSR	High-Speed Rail
HSS	Home Subscriber Server
HTRDP	High Technology Research and Development Program
I/O	Input/Output
I2I	Indoor-to-Indoor
IBM	International Business Machines Corporation
IBSS	Independent Basic Service Set
iCAR	integrated Cellular and Ad hoc Relaying System
ICI	Inter-Cell Interference
ICIC	Inter-Cell Interference Coordination
IDFT	Inverse Discrete Fourier Transform
IDMA	Interleave Division Multiple Access
IETF	Internet Engineering Task Force
IF	Intermediate Frequency
IFDMA	Interleaved FDMA
IFFT	Inverse Fast Fourier Transform
IID	Independent and Identically Distributed
IMC	Intel Mobile Communications
IMS	IP Multimedia Subsystem
IMT-2020	International Mobile Telecommunications-2020
IMT-A	International Mobile Telecommunications Advanced
InH	Indoor Hotspot
INNL	Intel Nizhny Novgorod Lab
IOI	Inter-Operator Interference
IoT	Internet of Things
IoV	Internet of Vehicles
IP	Internet Protocol
IPP	Integrated Performance Primitives
IPSec	Internet Protocol Security
ISA	Industry Standard Architecture
ISD	Inter-Site Distance
ISDN	Integrated Services Digital Network
ISI-SAGE	Initialization and Search Improved SAGE
ISM	Industrial Scientific Medical

IT	Information Technology
ITU	International Telecommunications Union
JP	Joint Processing
KBSM	Kronecker product based CBSM
KEST	Kalman Enhanced Super resolution Tracking algorithm
KF	K-Factor
KPI	Key Performance Indicators
K-S	Kolmogorov-Smirnov
LAA	LTE Assisted Access
LAN	Local Area Network
LCR	Level Crossing Rate
LCS	Local Coordinate System
LDPC	Low Density Parity Check
LDSMA	Low Density Spreading Multiple Access
LFDMA	Localized FDMA
LMDS	Local Multipoint Distribution Services
LO	Local Oscillator
LOS	Light Of Sight
Low Power SCPHY	Low Power Single Carrier PHYsical layer
LS	Least Square
LSF	Local Scattering Function
LSPs	Large-Scale Parameters
LTE	Long-Term Evolution
LTE-A	LTE-Advanced
LTE-U	LTE use Unlicensed band
LUT	Look Up Tables
LXI	LAN eXtension for Instrumentation
M2 M	Machine to Machine
MAC	Media Access Control
MAGNET	My personal Adaptive Global NET
MAN	Metropolitan Area Network
MCD	Multipath Component Distance
MCN	Multi-hop Cellular Network
MCS	Modulation and Coding Scheme
MED	Maximum Excess Delay
METIS	Mobile and wireless communications Enablers for Twenty-twenty (2020) Information Society
MIC	Many Integrated Cores
MIMO	Multiple Input and Multiple Output
MiWEBA	Millimetre-Wave Evolution for Backhaul and Access
ML	Maximum Likelihood
MLE	Maximum Likelihood Estimation
MME	Mobility Management Entity

MMSE	Minimum Mean Squared Error
MOBYDICK	MOBilitY and DIfferentiated serviCes in a future IP networK
MoM	Method of Moments
MPCs	Multi-Path Components
MPDU	Message Protocol Data Unit
MPP	Massively Parallel Processor
MSDU	MAC Service Data Unit
MT	Mobile Terminal
MUI	MultiUser Interference
MU-MIMO	MultiUser MIMO
MUSIC	MUltiple SIgnal Classification
NACK	Negative ACKnowledgement
NFV	Network Functions Virtualization
NGN	Next Generation Network
NI	National Instrument
NICT	National Institute of Information and Communications Technology
NIST	National Institute of Standards and Technology
NLOS	Non-Light Of Sight
NOMA	Non-Orthogonal Multiple Access
NSF	National Science Foundation
NTB	Non-Transparent Bridge
NYU	New York University
O2I	Outdoor-to-Indoor
O2O	Outdoor-to-Outdoor
OAI	Open Air Interface
OCXO	Oven Controlled Crystal Oscillator
OEW	Open-Ended Waveguide
OFDM	Orthogonal Frequency Division Multiplexing
OFDMA	Orthogonal Frequency Division Multiple Access
OFDMPHY	OFDM PHYsical layer
OLAP	OnLine Analysis & Process
OLTP	OnLine Traffic Process
OPEX	OPerating EXpense
OSI	Open System Interconnection
OTA	Over The Air
OTT	Over The Top
P2P	Peer to Peer
PAPR	Peak-to-Average Power Ratio
PAS	Power Angle Spectrum
PBSS	Personal Basic Service Set
PCA	Principal Component Analysis
PCA-CDA	Principal Component Analysis-CDA
PCC	Policy Control and Charging

PCFICH	Physical Control Format Indicator CHannel
PCI	Peripheral Component Interconnect
PCIe	Peripheral Component Interface express
PCP	PBSS Control Point
PCRF	Policy and Charging Rules Function
PDCCH	Physical Downlink Control CHannel
PDCP	Packet Data Convergence Protocol
PDF	Probability Density Function
PDMA	Power Domain Multiple Access
PDN	Public Data Network
PDP	Power Delay Profile
PDSCH	Physical Downlink Shared CHannel
PF	Particle Filter
PGW	Packet data network GateWay
PHICH	Physical HARQ Indicator CHannel
PHY	Physical Layer
PHY Control	PHYsical Control layer
PL	Path Loss
PLI	Programming Language Interface
PLL	Phase-Locked Loop
PMC	Production Material Control
PN	Pseudo-random Noise
PPS	PCP Power Save
PRB	Physical Resource Block
P-SAGE	Parallel sounding SAGE
PSS	Primary Synchronization Signal
PUCCH	Physical Uplink Control CHannels
PUSCH	Physical Uplink Shared CHannel
PVP	Parallel Vector Processor
PXI	PCI eXtensions for Instrumentation
PXIe	PCI eXtensions for Instrumentation express
PXImc	PXI multi-computing
PXISA	PXI Systems Alliance
QAM	Quadrature Amplitude Modulation
Q-D	Quasi-Deterministic
QoS	Quality of Service
QPSK	Quadrature Phase Shift Keying
RAM	Random Access Memory
RB	Resource Block
RE	Resource Element
RF	Radio Frequency
RI	Rank Indication
RIMAX	Richter's Maximum Likelihood Framework For Parameter Estimation

Rma	Rural Macro
RMSE	Root Mean Squared Error
RN	Relay Node
RNC	Radio Network Controller
ROM	Read Only Memory
RRC	Radio Resource Control
RRH	Remote Radio Head
RRU	Remote Radio Unit
RS	Reference Signal
RSRP	Reference Signal Receiving Power
RSRQ	Reference Signal Receiving Quality
RT	Ray Tracing
RTS	Request To Send
RVs	Random Variables
SAE	System Architecture Evolution
SAGE	Space-Alternating Generalized Expectation-maximization
SBR	Shooting and Bouncing Ray
sBS	small Base Stations
SC	Specular Component
SC-FDE	Single-Carrier Frequency Domain Equalization
SC-FDMA	Single-Carrier Frequency-Division Multiple Access
SCM	Spatial Channel Model
SCMA	Sparse Code Multiple Access
SCME	Spatial Channel Model Extension
SCPHY	Single Carrier PHYsical layer
SCTD	Simplified CTD
SD	Spectral Divergence
SDMA	Space Division Multiple Access
SDN	Software Defined Network
SDP Based	Semi-Definite Programming Based algorithm
SDR	Software Defined Radio
SDRAM	Synchronous Dynamic Random Access Memory
SF	Shadow Fading
SFC	Shadow Fading Correlation
SFE	Spatial Fading Emulator
SFG	Sweep Frequency Generator
SGi	Short Guard interval
SGW	Serving GateWay
SI	Self Interference
SIMD	Single Instruction Multiple Data
SIMO	Single Input Multiple Output
SINR	Signal to Interference plus Noise Ratio
SIR	Signal Interference Ratio
SISO	Single Input Single Output

SLS	Sector Level Sweep
Sma	Suburban Macro
SMP	Symmetric MultiProcessor
SNR	Signal-to-Noise Ratio
SON	Self-Organizing Network
SOPRANO	Self-Organizing Packet Radio Ad hoc Networks with Overlay
SoS	Sum of Sinusoids
SP	Service Period
SPM	Standard Propagation Model
SR	Scheduling Request
SS	Small Scale
SSPs	Small-Scale Parameters
SSS	Secondary Synchronization Signal
STA	STAtion
STBC	Space-Time Block Coding
STRC	Shared Trigger Reference Clock
STTC	Space-Time Trellis Codes
SV	Saleh-Valenzuela
SVB-SAGE	Sparse Variational Bayesian SAGE
SWR	Standing Wave Ratio
T-CDA	Treelet-based Compressive Data Aggregation
TCP/IP	Transmission Control Protocol/Internet Protocol
TDD	Time Division Duplex
TDL	Tapped Delay Line
TDM	Time Division Multiplexing
TDMA	Time Division Multiple Address
TDMS	Technical Data Management Streaming
TDP	Time-Delay Profile
TD-SCDMA	Time Division-Synchronous Code Division Multiple
TR	Technical Report
Tracking	Tracking phase
TRP	Total Radiated Power
TRS	Total Radiated Sensitivity
TTI	Transmission Time Interval
TUI	Technical University of Ilmenau
UCA	Uniformly Cylindrical Array
UCAN	Unified Cellular and Ad hoc Network architecture
UC-LMC	User-Centric Local Mobile Cloud
UDN	Ultra Dense Network
UDP	User Datagram Protocol
UFMC	Universal Filtered Multi-Carrier
UL	UpLink
ULA	Uniform Linear Array
UMa	Urban Macro

UMass	University of Massachusetts
UMB	Ultra Mobile Broadband
UMi	Urban Micro
UMTS	Universal Mobile Telecommunications
UMVUE	Uniformly Minimum Variance Unbiased Estimation
USB	Universal Serial Bus
USRP	Universal Software Radio Peripheral
UT	User Terminal
UTD	Uniform Theory of Diffraction
UWB	Ultra Wide Band
V2 V	Vehicle to Vehicle
VHDL	Very-High-Speed Integrated Circuit Hardware Description Language
VISA	Virtual Instrumentation Software Architecture
VME	Versa Module Eurocard
V-MIMO	Virtual MIMO
VNA	Vector Network Analyzer
VoIP	Voice over Internet Protocol
VR	Visibility Region
VSA	Vector Signal Analyzer
VSG	Vector Signal Generator
VT	Virginia Tech
VXI	VMEbus eXtensions for Instrumentation
WCDMA	Wideband Code Division Multiple Access
WEB	World Wide Web
WG1	Working Group 1
WiMAX	Worldwide Interoperability for Microwave Access
WINNER	Wireless World Initiative New Radio
WLAN	Wireless Local Area Network
WTP	Wireless Termination Point
XPR	cross-polarization Power Ratio
ZF	Zero Forcing
ZOA	Zenith angle Of Arrival
ZOD	Zenith angle Of Departure
ZSA	Zimuth angular Spread at Arrival
ZSD	Zenith angle Spread of Departure

Chapter 1
Candidate Technologies and Evaluation Challenges for 5G

With several decades of booming development, mobile communications technology has penetrated into many related fields of our daily life. Led by various emerging applications, users have increasing higher requirements for wireless services, posing almost stringent requirements for the technical indicators of network. Therefore, the Fifth Generation mobile communications system (5G) emerges at a historic moment, devotes itself to open curtain of comprehensive informational era and provide excellent user experience. Recently, researches on key technologies for 5G are springing up in the industry, and the corresponding testing, evaluation and verification system need to be established and improved. This chapter will briefly introduce the features of candidate technologies for 5G, and then analyze the challenges faced by 5G evaluation system.

1.1 5G Requirements Analysis

Mobile communications industry has been developing at an amazing speed, and at present it has become one of the important pillar industries in the global economic development. There is no doubt that mobile communications technology is changing people's life and work, and will continue to have an important impact on social development [1, 2]. Correspondingly, people have ever-increasing dependence and demand for mobile communications. In recent years, diversified businesses are emerging in mobile communications, which has led to the rapid development of storing and processing technology of massive data. Meanwhile, many breakthroughs have been made on the research and development of artificial intelligence processor and real-time equipment. The emergence of these new technologies has brought great convenience to people's life. Meanwhile, it has also put forward a greater challenge to the modern mobile communications technologies [3].

Therefore, 5G is facing both opportunities and challenges. For users, 5G's vision is "information comes as you wish, and everything is in touch" [4]. We are going to

© Springer International Publishing AG 2018
Y. Yang et al., *5G Wireless Systems*, Wireless Networks,
DOI 10.1007/978-3-319-61869-2_1

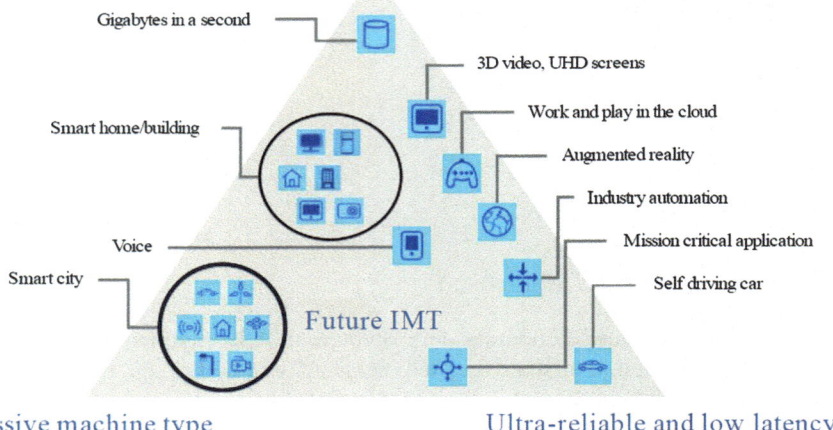

Fig. 1.1 Usage scenarios of IMT for 2020 and beyond

personally feel a gluttonous feast of the information age. The purpose of 5G age is to construct a stable, convenient and economic information ecosystem for human beings. As shown in Fig. 1.1, various features of the information age will be included in the development of 5G, and users can enjoy more convenient intelligent life [5]. With the popularity of wearable devices, the types and number of mobile terminals will experience explosive growth. Predictably, in the future, demand for virtual reality and augmented reality experience, demand for cloudization of massive office data, wireless control of industrial manufacturing or production processes, remote medical surgery, automation in a smart grid, transportation safety and other aspects, not only require 5G network data transmission rate to reach a very high level, but also require real-time experience with almost zero latency. In addition, cost reduction and energy saving should also be considered.

1.1.1 5G Application Types

The birth of 5G largely benefited from the large scale growth of the mobile Internet and the Internet of Things (IoT), and the application of 5G also mainly lies in the development of these two networks [6–9]. In recent years, mobile Internet, as the carrier of main businesses of mobile data communications, has greatly promoted the development of various fields of information service. Various service providers made full use of the advantages of their resources and services and developed numerous refreshing applications, reaching an effect of "flowers blooming together" in market. By 2020, users will see more convenient functions realized via mobile Internet, and demand for virtual reality and augmented reality in game experience will be met. With the further development of mobile Internet, information transmission rate will grow by thousands of times.

IoT is an ideal example of the full application of new generation technology. In all walks of life, human beings are able to manage production and live in a more precise and dynamic manner via IoT, achieving an "intelligent" state and improving resource utilization and productivity level [10]. IoT, as a main tool to realize intelligent life, work and production, extends the communications from person-to-person to person-to-thing and thing-to-thing [11]. The application of IoT is extremely wide, including environmental protection, intelligent transport, public safety, government work, home safety, intelligent firefighting, environmental monitoring, lighting control, health care, food traceability, floriculture, water system monitoring, enemy spy, information collection and many other fields. We can say that IoT will be the next "important productive power" promoting society progress [12]. Therefore, it is of prime importance to foster the development of IoT, and the development of IoT depends on the development of communications technology. It is conceivable that when IoT is everywhere in our life, which means the "everything connected" is realized, information transmission will be very frequent among thing-to-thing, person-to-thing and person-to-person, and this change not only brings vitality and opportunities but also poses a big challenge to mobile communications [13].

1.1.2   5G Application Scenarios and Requirements Indicators

Application scenarios of 5G are related to every aspect in people's daily life, work, entertainment and transportation, and wireless communications will show different characteristics in all kinds of different scenarios. For example, the crowed or dense mobile devices areas such as residential areas, stadiums and marketplaces, wireless communications will have the characteristics of high traffic volume density and high number of connections, while on transport tools such as subway and high-speed railways, the high mobility feature of wireless communications will be prominent. At present, the Fourth Generation mobile communications system (4G) is not able to satisfy the requirements of some special scenarios featuring high traffic volume density, high number of connections and high mobility [14].

In crowded scenarios such as stadiums which need ultrahigh traffic volume density and ultrahigh connection density, we need the wireless communications' transmission rate as high as that of the optical fiber so that it can carry the businesses like photo transmission, video transmission, live broadcast and other services. In high-speed mobility scenarios, e.g., High-Speed Rail (HSR), the traffic volume density and connection are relatively lower than those of stadiums. Since HSR's speed is usually above 200 km/h, it has high requirements for the wireless communication systems to support high-speed mobility.

Table 1.1 5G performance requirements

	Value
Peak data rate	>10 Gbps
User experienced data rate	>0. 1 Gbps
Connection density	Million connections/square kilometers
Service density	10 Gbps/square kilometers
End-to-end latency	Millisecond order

Although it's very convenient for us to access the Internet now, half of our world is still beyond the coverage of Internet after several decades when mobile terminals came into being. With the development and change of Internet, the denotation of Internet is expanded accordingly and more and more devices are connected with each other. Cisco System forecasts that by the year of 2019, the whole world will have 11.5 billion devices to be connected [15], including some hard-to-connect devices which are under water or beyond the coverage of satellite, so it becomes more and more important to meet the requirements for wide coverage in future.

It is predicted that in a long period in the future, mobile data traffic will continue to show explosive growth [4]: from 2010 to 2020, global mobile data traffic growth will increase by more than 200 times while from 2010 to 2030, more than 20,000 times. Meanwhile, the growth of Chinese mobile data traffic is higher than the global average level. It is predicted that from 2010 to 2020, it will grow by more than 300 times, while from 2010 to 2030, more than 40,000 times.

Therefore, based on the requirements above, 5G's overall goals are: much faster, more efficient, and more intelligent. Specific 5G performance requirements are shown in Table 1.1 [16].

To satisfy users' experience in multiple dimensions, combination of different technologies is needed to reach the 5G performance requirements in the above table. For example, ultra-dense wireless communications technology can make contributions to improving performance indicators of user experienced data rate, connection density and service density through increasing the base station deployment density. Massive antenna technology can effectively improve the spectrum utilization efficiency through increasing antenna's number and it has an important significance in improving peak data rate, user experienced data rate, connection density and service density. Millimeter wave communications technology can increase usable spectrum in a large scale, which is good for improving performance indicators of peak data rate, user experienced data rate and service density.

1.2 5G Research and Development Process

From 2009, Long Term Evolution (LTE) set off a boom across the globe. At the end of 2012, European Union (EU) invested 27 million euros and launched the first 5G research project in the world Mobile and wireless communications Enablers for

Twenty-twenty (2020) Information Society (METIS). There were 29 project members participating in this project, including main equipment manufacturers, operators, car manufacturers and academic institutions, and the research contents covered 5G scenarios and requirements, network architecture and various types of new wireless technology [17, 18]. In 2013, China's 863 Plan invested 300 million RMB to research and development of 5G system, and the research contents included wireless network architecture, massive antenna, ultra-dense wireless network, soft base station platform, wireless network virtualization, millimeter wave indoor wireless access, as well as evaluation, test and verification. In order to promote the development of 5G system, governments worldwide have established technology communications platforms such as China's "International Mobile Telecommunications-2020 (IMT-2020) (5G) advance group", Japan's "2020 and beyond ad hoc"', Korea has "5G forum" and so on.

At present, the International Telecommunications Union (ITU) has defined a clear 5G work timetable [16]: ITU completed the formulation of 5G vision and key technical indicators in 2015. In 2016, it presented 5G technology performance requirements and evaluation methods. In 2017, it begun to collect 5G international standards. And by the end of 2020, it will complete technical specifications of 5G. The 3rd Generation Partnership Project (3GPP), as an international mobile communications standardization organization, has identified 5G research plan and initiated the research on requirements for 5G technology in 2015, and started 5G technology specialized seminar in the second half of 2015. It plans to complete the study stage of 5G technology scheme from 2016 to 2017, and complete the formulation of 5G technology specification "Release 14" from 2018 to 2019. IEEE 802.11 standardization group, in order to meet the massive business in hot spot and indoor areas in the future, intends to support 5GHz unlicensed band transmission in 802.11ax, meanwhile, use orthogonal frequency division multiple access and interference compensation technology, which can make the peak transmission rate up to 10Gbps, and Wireless Local Area Network (WLAN) performance more robust. According to the IEEE 802.11 working group's plan, it is expected to complete 802.11ax standard specification in 2018.

1.3 5G Candidate Technologies

In order to meet the wireless service growth needs of 1000 times in the next ten years, the capacity of the wireless networks can be expanded from three directions in rising the spectrum utilization, enhancing the spatial multiplexing and bandwidth expansion. For example, it can effectively improve spectrum efficiency, throughput per unit area and power efficiency through deploying ultra-dense small base station, and at the same time shorting the distance between wireless access networks and terminals. It can greatly increase 5G system's available frequency bandwidth through expanding the use of unlicensed spectrum, high frequency band and millimeter wave band. And through massive MIMO it can further tap the potentials of the space, and greatly improve spectrum utilization. The characteristics and performance of various wireless candidate technologies vary greatly, and we will briefly introduce the several typical 5G candidate technologies in the following.

1.3.1 Ultra-Dense Wireless Network

How can we effectively increase the capacity of wireless network? A simple and effective way is to decrease the coverage area of each cell, and to deploy the cellular network in a dense way to improve the spatial multiplexing of frequency resources and accordingly enhance the throughput per unit area. Meanwhile, due to the shortened distance between base station and terminals, transmission loss will be reduced, thus greatly improving the power efficiency.

Statistics show that in wireless networks, more than 70% wireless data generated from indoors and hotspots [19]. If we want to significantly improve the network capacity through utilizing the outdoor macro cells, there will be the two limitations. On the one hand, due to the scarcity of macro cell site resources, the deployment density could not be further increased. On the other hand, when the macro cell base stations realize the indoor coverage, wireless signals will experience severe penetration loss, leading to very poor indoor coverage performance and thus making it very difficult to meet the business needs of indoor and hotspots. Therefore, a large number of various types of small access points will be densely deployed, forming ultra-dense wireless network. These small access points include Home eNodeB (HeNB), WLAN Access Point (AP), Relay Node (RN), Micro Cell, Pico Cell and etc. In particular, both Home eNodeB and WLAN access points can be connected to the operators' core network through the cable broadband in a plug-and-play way, and their deployments are very flexible and convenient.

Since the ultra-dense wireless network access points may be randomly deployed by the end users according to their own needs, and business needs change more frequently, the traditional network planning will have to be faced with huge challenges. According to the deployment scenarios and application requirements, ultra-dense wireless networks will have multiple features:

• Ultra-dense wireless network sites need to have the capability of ad hoc networking, such as neighbor cell discovery by itself, physical cell identifier configuration, access point adaptive activation, carrier adaptive selection and re-selections.
• Because it is very difficult to receive wireless positioning satellite signal of Global Positioning System (GPS) and BeiDou Navigation Satellite System (BDS), ultra-dense wireless networks need to realize wireless network synchronization through air interfaces.
• Due to access points' limited coverage, ultra-dense wireless networks need to have the capacity to distinguish users' moving speed, adaptively connect fast-

moving users and quasi-stationary users respectively with macro coverage and ultra-dense wireless networks.

- In order to enhance the mobile robustness, we can separate the user plane and control plane, and keep the control plane in the macro cell to reduce the handover latency and ensure the consistency of user experience.
- With relatively small coverage of every access point, high line-of-sight transmission probability, and less path transmission loss, ultra-dense wireless networks can make full use of high-frequency band and millimeter wave transmission, whose spectrum resources are still sufficient.
- Since the hotspots and indoor business requirements change frequently and meanwhile we need to reduce the deployment cost, the ultra-dense wireless backhaul network can use wireless link, and the spectrum resource of wireless backhaul and access links can be shared.

In order to better measure and evaluate the cost of ultra-dense wireless networks, J. G. Andrews et al. [3] defined base station density gain ρ ($\rho > 0$). ρ is the gain of data transmission rate with the increasing density of the network. If we define the original data transmission rate as R_1, the density of base station is $\lambda_1 BSs/km^2$. Through the ultra density networks, the data transmission rate is defined as R_2, the density of base station is $\lambda_2 BSs/km^2$. Then the corresponding density gain is:

$$\rho = \frac{R_2 \lambda_1}{R_1 \lambda_2} \tag{1.1}$$

According to this formula, if the network density increases twice and at the same time, the data transmission rate increases twice, then the density gain ρ is 1.

1.3.2 Large Scale Antenna Technology

In practical scenarios, the point-to-point multiple antenna technology needs multiple antennas to be deployed on terminals with limited physical size, making it hard to effectively improve the system spectrum utilization. Therefore, downlink Multi-User MIMO (MU-MIMO) and uplink Virtual MIMO (V-MIMO) are introduced to overcome the limitations of multiple antennas, which can effectively improve the system spectrum utilization. On the other hand, multiple antennas technology has evolved from passive to active, from Two-Dimensional (2D) to Three-Dimensional (3D), from high order MIMO to the development of large-scale array, and the spectrum efficiency will be expected to increase by dozens of times or even higher. Due to the introduction of the active antenna array, the base station side can support 128 cooperative antennas, or even a larger array of antennas. In addition, the original 2D antenna array expanded to 3D antenna array, the formation of a novel 3D–MIMO technology can support multi-user smart beamforming for reducing user interference. It can be predicted that massive MIMO, combined with technologies of high frequency band and millimeter wave, will further improve the performance of wireless signal coverage [21].

Since the proposal of massive MIMO, it got immediate attention from academia and industry. Operators, equipment manufacturers and research institutions all showed great interest and made a series of achievements [21–24]. From 2010 to 2013, led by Bell Lab, Lund University and Linkoping University in Sweden, Rice University in the U. S. and many other research institutions, the international academia has made extensive exploration of massive MIMO channel capacity, transmission, Channel State Information (CSI) acquisition and testing.

Massive MIMO has certain advantages in technology, for example:

- Assuming that channels are independent of each other, since the base station uses a large number of transmit and receive antennas, the transmitter or receiver based on filter matching can effectively suppress multiuser interference. Therefore, the optimal performance can be achieved via linear complexity greatly improving system spectrum utilization.
- The narrow beam formed by massive MIMO can greatly improve the energy efficiency.

Correspondingly, the research on massive MIMO is also faced with some challenges:

- The pairing freedom has greatly increased between downlink multiuser MIMO and uplink virtual MIMO user, and it needs to design a realizable and high performance radio resource scheduler;
- In the uplink transmission of large scale antenna system, a large number of users will be multiplexed. Under the limited condition of bandwidth and orthogonal code sequence, pilot channel will be polluted, which will affect the performance of uplink receiver.

1.3.3 Millimeter Wave Communications Technology

According to the latest spectrum requirements researches [20], in 2020 the world's incremental spectrum requirements will be 1000–2000 MHz while the low frequency resource has been largely depleted. Compared with the deployed low frequency band, the available frequency resources of millimeter wave band (30–300 GHz) are quite abundant, which is about 200 times of the low frequency band. Therefore, the industry began to explore how to use the millimeter wave band (30–300 GHz) in wireless communications. Millimeter wave band has always been considered to be not suitable for wireless transmission, due to relatively large path loss, absorption by atmosphere and rain, relatively poor capacity of diffraction, big phase noise, high cost of measurement equipment. However, with increasingly developed semiconductor technology, the cost of equipment has been declining rapidly. At present, in satellite communications, Local Multipoint Distribution

Services (LMDS), cellular communications backhaul link and wireless transmission standard 802.11ad, the millimeter wave band has been used.

Because millimeter wave has relatively small wave length, at least more than 10 times of antennas can be deployed in the same region, which is very suitable for building massive Multiple Input and Multiple Output (MIMO), greatly improving the channel capacity. However, due to the propagation characteristics of millimeter wave band, its application in cellular wireless communications has several challenges:

- The transmission loss of millimeter wave is proportional to the square of the frequency, and it needs to be compensated by the high gain beamforming.
- Millimeter wave transmission is very sensitive to the object occlusion, and the transmission loss is very large.
- When line-of-sight transmission path is existed, the transmission quality of millimeter wave is very high, but the transmission path is longer, the probability of line-of-sight transmission is lower.
- Channel fading coherence time is proportional to the carrier frequency, and the millimeter wave frequency is very high, which results in that the small scale fading of channel is very fast in the same moving speed conditions compared with the low frequency.

The latest channel measurement results of 28 and 73 GHz show that the millimeter wave transmission can be effectively in non-line-of-sight environments [25, 26]. Due to the special propagation characteristics, in addition to line-of-sight models and non-line-of-sight models, a third state is proposed to explicitly model the possibility of outages. That is to say, the three-state millimeter wave transmission, including line-of-sight, non-line-of-sight and outage, can objectively reflect the path loss rule of the millimeter wave [26, 27]. Outage state is a good description of the characteristics of the object occlusion and penetration of millimeter wave. In [26], the measurement data analysis shows that the probability of line-of-sight transmission and non-line-of-sight transmission is closely related to the transmission distance, and the probability functions for the three states are given as Eq. (1.2).

$$
\begin{aligned}
P_{out}(d) &= \max\left(0, 1 - e^{-a_{out}d + b_{out}}\right) \\
P_{LOS}(d) &= (1 - p_{out}(d))e^{-a_{los}d} \\
P_{NLOS}(d) &= 1 - p_{out}(d) - p_{LOS}(d)
\end{aligned}
\tag{1.2}
$$

where a_{LOS}, a_{out} and b_{out} are the parameters decided by the measurement data of typical network deploying scenarios. When the terminal is close to the base station, millimeter wave transmission has close-to-zero probability of outage state, and it mainly relies on line-of-sight transmission. When the terminal is moving away from the base station, the millimeter wave transmission is in the state of line-of-sight transmission or non-line-of-sight transmission. When the terminal is far away from the base station, millimeter wave transmission is in non-line-of-sight transmission or outage states. Of course, line-of-sight and non-line-of-sight transmissions have different corresponding path loss models. The model is very similar to the relay path transmission model, so it can be considered as an extension of relay path transmission model.

1.3.4   Flexible Spectrum Usage

In the process of the rapid development of mobile communications, one of the bottlenecks is the shortage of spectrum resources. Yet the requirements for radio frequency spectrum resources in wireless communications have been expanding. The competition between different radio technologies and applications is becoming more and more intense, highlighting the increasing shortage of spectrum resource. Under this circumstance, on the one hand, we need to expand the spectrum resources like millimeter wave which is abundant but hasn't yet been used in a large scale. On the other hand, we need to try to improve the spectrum utilization of existing resources. The flexible use of spectrum resources is an effective way to improve the efficiency of spectrum utilization. Flexible spectrum usage includes the spectrum refarming in operator itself [28], LTE use Unlicensed band (LTE-U) [29], peer spectrum sharing between operators [30] and other potential ways.

Firstly, the spectrum refarming in operator itself is mainly based on the same operator with different modes to share frequency resource. For example, with the extensive deployment of LTE network sites, 2-Generation wireless telephone technology (2G) and the 3rd Generation mobile communications system (3G) users will gradually migrate into the LTE network, and the spectrum utilization of 2G and 3G networks will have lower efficiency. It can be predicted that when 5G network deployment is completed in the future, there will be more users migrating into 5G network. Then 5G network can exchange the information through 2G, 3G and LTE networks and obtain user distribution and spectrum usage in 2G, 3G and LTE, and then determine how to use the spectrum resources strategy of 2G, 3G and LTE network, and actively control interference between networks.

Secondly, the sharing of LTE-U spectrum resources mainly refers to how the cellular network use unlicensed band. WLAN plays a very important role in LTE-U. Sharing of unlicensed frequency band between cellular networks and WLAN has two possible ways: (1) LTE Assisted Access (LAA) [31], i.e., cellular network configures the carrier of LTE-U as auxiliary carrier though Carrier Aggregation (CA), cooperating with main carrier of licensed band to provide service for users. (2) Independent LTE-U communications, which is that cellular network only uses unlicensed bands for communications. Compared with WLAN, since cellular network uses technique of turbo codes, Hybrid Automatic Repeat reQuest (HARQ), adaptive modulation and coding and frequency selective scheduling, it can achieve greater coverage and higher network throughput in unlicensed band.

Finally, in peer spectrum sharing between operators, operators can jointly purchase spectrum, and then fairly share spectrum resources, or composite spectrum sharing resource pool through holding out part of licensed spectrum which is owned by a plurality frequency license holders. Multiple operators need to follow a certain negotiated spectrum sharing strategy, exchange spectrum using information and form the spectrum use database so that the spectrum can be used in a dynamic way in space and time dimensions. This sharing way can not only improve the spectrum utilization, but also ease the operators' financial pressure in purchasing spectrum.

1.3.5   Waveform and Multiple Access

Although 4G's physical layer uses the core technology based on Orthogonal Frequency Division Multiplexing (OFDM), but the OFDM waveform itself has some defects, for example: (1) Slow rolling-off leads to high out-of-band leaking. Therefore, larger spectrum guard interval is needed. (2) To avoid interference between carriers, (coarse) synchronization is needed between nodes transmission. In the hierarchical network, base stations with different coverage should be synchronized. Because the application scenarios of 5G is far more complex than 4G, and its latency and the number of access requirements are very strict. However, OFDM's slow rolling-down characteristic and strict synchronization requirement cannot adapt to real-time business asynchronous fast access and efficient use of non-continuous spectrum.

Targeting at the disadvantages of OFDM, a lot of alternative waveforms have been proposed, such as Filter Bank based Multi-Carrier (FBMC), Filtered OFDM (F-OFDM) and Universal Filtered Multi-Carrier (UFMC) [3], etc. By filtering the sub-band or sub-carrier, the rate of spectrum rolling-off is increased and the frequency spectrum leakage is reduced, so that the time frequency synchronization requirement is reduced, the frequency protection band and the time domain protection interval is removed. The above new waveform technologies can be well combined with OFDM and MIMO, and increase the flexibility of 5G air interface design, so as to match different traffic latency and data rate requirements.

Mobile Internet and IoT are the driving force of the development of 5G, and the requirements for various applications greatly vary. For example, various types of real-time traffic transmission in the 5G network have set down requirements for millisecond order end-to-end latency. Undoubtedly, such a strict traffic latency requirement will pose very high requirements on the physical layer design (including symbol duration, synchronization process, random access and frame structure, etc.). In addition, the traditional cellular communications establish and complete synchronization based on connection, and they schedule with the form of the network resource. That is to say, firstly the end-to-end connection is established, and then data is transmitted. In IoT applications led by machine communications, wireless sensor nodes are mainly energy constrained, and transmission data packets are usually quite small. Hence, the requirements are very high for traffic latency and energy efficiency. If we still use the traditional cellular communications method based on the connection, it will bring a lot of unnecessary network expenditure and lead to too long activation time of wireless sensor node, which is not good for reducing energy consumption. In order to support fast access to businesses and simplify the channel accessing process and signaling process, 5G has proposed non orthogonal multiple access method, which is to use the modulation, spread spectrum, power and space dimension for joint mapping so that users may transmit data without being scheduled, and the spectrum utilization can be effectively improved.

1.3.6 Device to Device

Device to device (D2D) means that two terminal devices directly communicate without base station and the core network. Its typical applications include cellular assisted D2D communications and Vehicle to Vehicle (V2V) direct communications. Cellular assisted D2D communications is a new technology that allows terminals to multiplex cell resource for direct communications under the control of a cellular system.

D2D communications can increase the spectrum efficiency of the cellular communications system, reduce the terminal transmitting power, and to a certain extent, solve the scarcity of spectrum resources in the wireless communications system. In addition, it can bring additional benefits including: reducing the burden of cellular network, reducing the mobile terminal battery power consumption, increasing the transmission rate and improving the robustness of network infrastructure failures, what's more, supporting the new way of point-to-point data services in a small area.

D2D communications technology is also faced with the following problems:

- Since the D2D communications will multiplex cellular resources, the mutual interference between D2D communications and cellular communications arises. When D2D communications multiplex uplink resources, in the system it is the base station that is interfered by the D2D communications while the base station can control the interference by adjusting the transmitting power of D2D communications and the multiplexed resources. When D2D communications multiplex downlink resources, in the system it is any downlink user that is interfered by the D2D communications. The interference is not controllable, which may result in the link failure.
- The synchronous mode of D2D communications can improve energy efficiency. When the terminal is in the coverage of base station, the terminal can make the base station as a time synchronization source, so as to realize the synchronous D2D communications mode. When the D2D communications terminals are not in the coverage of the base station, we can select a terminal as the cluster head for D2D communications to transmit synchronous signals. When there is a multi-hop in D2D communications, it will lead to multi-hop transmission of

synchronization time, resulting in multiple synchronization sources. Alternatively, when multiple D2D communications groups coexist with each other, it can lead to multiple synchronization sources. The problem of multiple synchronization sources also needs to be solved in the research of D2D communications technology.

1.4 Challenges in 5G Testing and Evaluation

1.4.1 Challenges Posed by New Technologies to Testing and Evaluation

According to the network requirements and technical indicators defined in 5G white paper [16], 5G network system will achieve huge growth in network capacity and connected user number. Meanwhile it put forward the new network vision and design principles based on network maintenance, energy efficiency, user experience and many other aspects. In order to reach all kinds of technical indicators of the 5G network, the research on new network architecture and several types of candidate technologies is going on intensively. Compared with 4G network, 5G is not only more flexible in the network architecture, but also more complex in signal processing mechanism used in transmission system, and more multi-dimensional in performance evaluation indicators. There is no doubt that both the proposed new network architecture and the emergence of various types of transmission technology will pose new challenges to 5G air interface technology standardization, program design and simulation.

- For physical layer transmission technology, 5G system will introduce new waveform and non-orthogonal multiple access at the physical layer to achieve the required traffic latency in air interface. But in order to overcome the interference caused by the non-orthogonal transmission technology, we need to introduce nonlinear receiver, and in the performance evaluation we need to consider the effect of user pairing and multi-dimensional parameters, which will greatly increase the difficulty of physical layer technology link simulation and evaluation. Also in order to simulate and evaluate the system of new wave and non-orthogonal multiple access, we also need to establish mapping method of the nonlinear receiver link in system.
- In order to further explore the spatial freedom and improve the network throughput, 5G will introduce massive MIMO technology, which will greatly enhance the baseband signal dimensions, and make the link simulation evaluation complexity increase by thousand times. And in simulation evaluation system, massive MIMO and MU-MIMO technology will greatly increase computational interference complexity. And multi-channel synchronous, isolated, multi-channel data storage pose severe challenges to the test implementation.
- The new channel propagation model will be introduced based on high frequency band transmission technology, D2D technology and massive MIMO technology.

On the one hand, how to build a more accurate channel propagation model for the 5G candidate technologies will pose challenges for corresponding testing and verification. On the other hand, the new channel model based on expanding the traditional channel model will bring the increased types and dimensions of parameters, which will lead to increased computation complexity and storage space for the link simulation and system simulation evaluation of the above technologies.

- In terms of the wireless network technology, the user plane and control plane are separated, and the wireless network resources are controlled and optimized in a centralized way, which will lead to a sharp increase in the space of the feasible solution of control and optimization. Through multi-core Central Processing Unit (CPU), Graphics Processing Unit (GPU) and Field Programmable Gate Array (FPGA) computing resources, we can build a heterogeneous computing platform, and distribute the above computing resources for the baseband signal processing operations which have higher calculation complexity. We need to design scheduling algorithm for heterogeneous computing resources, accurately estimate the consumed time of heterogeneous computing and interface data transmission, and meanwhile design the synchronized mechanism for computing tasks to make full use of the heterogeneous computing platform.

1.4.2   Four Elements of Testing and Evaluation Requirements

All in all, to meet the requirements for realness, comprehensiveness, rapidity and flexibility is an important challenge for the evaluation of 5G testing and evaluation.

Realness

In the 5G testing and evaluation, the requirement for realness is reflected in the 5G wireless channel model, verification methods, user experience, and other aspects.

From the point of establishing channel model, describing the channel characteristics in a numerical way with testing and measurement methods, and providing a close-to-objective physical channel simulation environment in a real and reproducible way, are of self-evident importance for 5G communications technology verification. Compared with 3G/4G system, new application scenarios are added in 5G, which have the various characteristics of ultra high traffic volume density, ultra high connection density, ultra high mobility, etc. From the point of the network topology, various link types in 5G communications system will coexist in a same region, extending from the traditional macrocell and microcell, to picocell, femtocell and nomadic base station, and support D2D, Machine to Machine (M2M), V2V and the fully-connected network. In the complex and diversified network architecture, traditional channel model does not give sufficient consideration to the consistency of small and medium scale parameter space, which may lead to the exaggerated simulation performance of in technological evaluation (for example, MU-MIMO) [32]. In addition, two-way mobility of D2D/V2V will introduce Doppler model which is different from the traditional model, and massively intensive scattering exists in both transmitter and the receiver and the stationary cycle is short, all of which need to be considered in the channel model. Under the condition of high mobility, the features of fast time-varying, more severe fading and mixed propagation scenarios make it difficult for the current channel to be well applied in 5G high speed mobile scenarios. Therefore, wide-range propagation scenarios and diversified network topologies have posed challenges to the 5G channel model. At the technological level, a variety of networks and transmission technologies emerge in the 5G research, which pose the following challenges to the channel model: unique transmission characteristics of radio waves in higher frequency and bandwidth, including non-line-of-sight path loss, high resolution angle characteristics and outdoor mobile channel; in the greater antenna array (including antenna unit number and the physical dimensions), the original plane wave propagation model assumption is no longer applicable, and scattered clusters' non-stationary feature is reflected not only in the time axis, but it' s also changing along the array. In the third chapter, we will describe how to build a real and reliable 5G channel model from the aspects of the characteristics of the channel model, measurement methods and parameters extraction and data analysis.

In terms of the verification methods, according to the traditional methods, the system design of wireless link in the mobile communications field is mainly based on the software simulation methods to verify and evaluate the algorithms. However, there is a certain gap between the system environment under the pure software simulation and the real situation. First of all, the software simulation assumes that the hardware design can perfectly realize the software algorithms, and thus we cannot introduce the impact of hardware conditions on the communications system. However, in reality algorithm design is often restricted by the hardware conditions, and the software algorithms are often greatly reduced when realized in hardware. For example, in the actual hardware environment, the algorithms with particularly large amount of computation and extremely high calculation accuracy requirements are often unable to obtain the optimal performance. And the precisely designed algorithms are often vulnerable to the outside interference. The algorithms at the front of transmitter and receiver have to consider power, the impact on other hardware components and many other factors. If we only use software simulation for system verification and evaluation, it's easy to ignore the above problems, which will make the designed algorithm design stay in the theoretical stage and cannot meet the needs of the real situation. Secondly, in the former software simulation systems, simplified methods are often introduced in network and channel modeling, simulation business abstraction, design and realization and other processes for convenience of processing. These simplified methods usually reduce the simulation computational cost at the cost of authenticity. Thirdly, accurate math modeling of the complex time-varying nonlinear system is very difficult. No matter how good the model is, it cannot replace the real scenarios.

In order to more accurately reflect the impact of the actual link process on the system simulation, we can use a combination of hardware and software, using real hardware platform, deploying the link to be simulated in the hardware (such as FPGA, channel simulator) for real-time calculation, and thus improving the calculation accuracy through hardware simulation; in addition, introducing hardware in the loop technology in 5G testing, can not only get closer to the real environment, increase the reliability of the simulation test, but also test the communications equipment under extreme conditions, which is difficult to achieve in the laboratory environment. For the corresponding contents, please refer to Chaps. 5 and 6 of this book. Furthermore, in terms of environmental authenticity, in the process of research and development of 5G key technologies, in addition to all kinds of simulation and laboratory evaluation, we also have to verify the technological feasibility through outfield experiments, and realize the verification of the tested technology or prototype equipment in real scenario through "algorithm realization → data acquisition → system optimization → outfield verification", providing field basis for standardization. Therefore, in Chap. 8 of this book, we will analyze the planning for test outfield from the perspective of the diversity of application scenarios, integration of heterogeneous networks and large scale users.

Comprehensiveness

The demand for testing and evaluating comprehensiveness mainly involves two aspects: on the one hand, it lies in the comprehensive support for the evaluation of 5G performance indicators; on the other hand, it lies in the comprehensive support for the diversification of candidate technologies. 5G evaluation indicators system will include the objective indicators like transmission and network as well as the user experience indicators. Transmission is mainly reflected in the physical layer's transceiver performance, link layer performance indicators, while the network layer is reflected in the evaluation indicators for system performance. In evaluating the performance of 5G system, we need to not only comprehensively measure all kinds of KPI, but also consider the relationship between each indicator, and guarantee users' consistent experience in pace, time, network, business, and other dimensions. 5G candidate technologies can be divided into two categories of air interface technology and network technology. The air interface technology includes massive MIMO, full duplex, novel multiple access and waveform, high frequency range communications and spectrum sharing and flexible using. Network technologies include Cloud Radio Access Network (C-RAN), Software Defined Network/Network Functions Virtualization (SDN/NFV), ultra-dense network, multi-RAT and terminal direct communications, etc. As for 5G software simulation technology, this book will, from the architecture level, element level and module level, give a detailed analysis on various 5G candidate technologies, propose and implement the corresponding resource model and design scheme, and build a comprehensive performance evaluation system, which will be shown in detail in Chap. 4.

Rapidity

Requirement for rapid test performance is mainly derived from the huge growth in 5G network traffic and a variety of performance. In order to meet the KPI s in 5G white paper, the computing performance of the simulation system must grow by more than 1000 times before meeting the requirements of the timely evaluation of simulation task. With such rapid growth in computational performance, a systematic and brand new design and realization are needed in simulation system's hardware platform, software platform, and simulation application. It can be predicted that within a period of time, multi-core and Many Integrated Cores (MIC) parallel computing mode will become the main technical means to enhance computational efficiency, and cloud computing, super computer will gradually become the calculation means of simulation system. For the simulation system, the key problem is how to complete the software system concurrent design and coding implementation on the new and powerful hardware platform with powerful computational capabilities.

Flexibility

Affected by flexible network architecture, network resource virtualization management, parallel computing needs and other factors, simulation and validation system needs to have enough flexibility. Architecture design, module design, and interface design have the characteristics of coupling, modularization, interface expansion, easy integration and so on, which have posed higher requirements for the network resources model design, software function module design, network element and the inter module interface design and other aspects.

In this book, Chap. 4 will introduce, from the perspective of the application of multi-core simulation technology and parallel processing, elaborate on how to deal with the challenges in "rapidity" and "flexibility" software simulation testing verification requirements, so as to implement 5G link-level simulation and system-level simulation. Furthermore, Chaps. 5 and 6 of this book, from the perspective of software and hardware co-simulation and hardware platform of software defined radio, will describe how to use these two technologies to enhance 5G key technology verification and evaluation.

From the point of system comprehensiveness, this book presents the 5G test system covering 5G channel model (Chap. 3), software simulation system (Chap. 4), software and hardware co-simulation technology (Chap. 5), hardware platform of software defined radio (Chap. 6), and the actual outfield testing and verification system (Chap. 8), displaying various technological means of the whole 5G testing and verification system from all-round angles. Relationship between chapters and the four elements of evaluation are shown as Fig. 1.2.

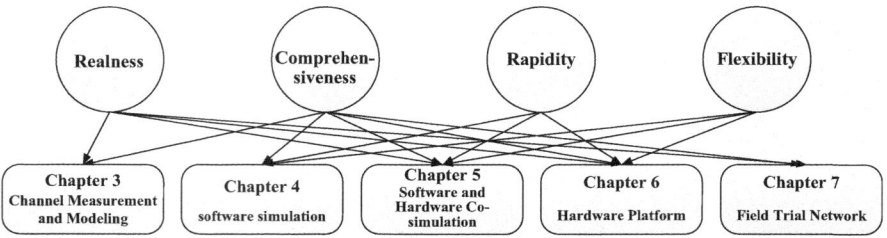

Fig. 1.2 Relationship between chapter setting and the four elements of evaluation

1.5   Summary

Starting from 5G application requirements, this chapter firstly introduces the typical 5G application, deployment scenarios, key technical indicators, as well as the plans and progress of global 5G research and development. After providing a brief overview of various 5G candidate technologies and their characteristics and analyzing the challenges posed by these candidate technologies to 5G testing and verification, we expounded the challenges brought by the four elements requirements for realness, comprehensiveness, rapidity and flexibility in testing and evaluation, as well as the relationship between chapter setting and these four elements of evaluation, playing a leading role in outlining the subsequent chapters.

References

1. C. Wang, F. Haider, X. Gao, et al. Cellular Architecture and Key Technologies for 5G Wireless Communication Networks. IEEE Communications Magazine, 2014, 52(2):122–130.
2. X. Chu, D. Lopez-Perez, Y. Yang, and F. Gunnarsson. Heterogeneous Cellular Networks: Theory, Simulation and Deployment. New York, USA: Cambridge University Press, 2013.
3. J. G. Andrews, S. Buzzi, W. Choi, et al. What Will 5G Be? IEEE Journal on Selected Areas in Communications, 2014, 32(6):1065–1082.
4. IMT-2020 (5G) Promotion Group. White Paper on 5G Vision and Requirements, 2014. http://www.imt-2020.cn/zh/documents/download/1.
5. ITU. (2015). Recomendation ITU-R M.2083–0: IMT Vision—Framework and overall objectives of the future development of IMT for 2020 and beyond. Technical report, ITU-R.
6. J. Yang, Y. Qiao, X. Zhang, et al. Characterizing User Behavior in Mobile Internet. IEEE Transactions on Emerging Topics in Computing, 2015, 3(1):95–106.
7. W. Huang, Z. Chen, W. Dong, et al. Mobile Internet big data platform in China Unicom. Tsinghua Science and Technology, 2014, 19(1):95–101.
8. L. Xu, W. He, S. Li. Internet of Things in Industries: A Survey. IEEE Transactions on Industrial Informatics, 2014, 10(4):2233–2243.
9. Z. Sheng, S. Yang, Y. Yu, et al. A survey on the IETF protocol suite for the internet of things: standards, challenges, and opportunities. IEEE Wireless Communications, 2013, 20(6):91–98.
10. C. Perera, A. Zaslavsky, P. Christen, D. Georgakopoulos. Context Aware Computing for The Internet of Things: A Survey. IEEE Communications Surveys & Tutorials, 2014, 16 (1):414–454.

11. C. Perera, C. H. Liu, S. Jayawardena, C. Min. A Survey on Internet of Things from Industrial Market Perspective. IEEE Access, 2014, 2:1660–1679.
12. A. Zanella, N. Bui, A. Castellani, et al. Internet of Things for Smart Cities. IEEE Internet of Things Journal, 2014, 1(1):22–32.
13. C. Tsai, C. Lai, M. Chiang, L. T. Yang. Data Mining for Internet of Things: A Survey. IEEE Communications Surveys & Tutorials, 2014, 16(1):77–97.
14. IMT-2020(5G)Promotion Group. White Paper on 5G Wireless Technology Architecture, 2015. http://www.imt-2020.cn/zh/documents/download/61.
15. Cisco, Cisco Visual Networking Index: Global Mobile Data Traffic Forecast Update, 2014–2019. White Paper, 2015.
16. IMT-2020(5G)Promotion Group. White Paper on 5G Concept, 2015. http://www.imt-2020.cn/zh/documents/download/23
17. A. Osseiran, V. Braun, T. Hidekazu, et al. The foundation of the Mobile and Wireless Communications System for 2020 and beyond Challenges, Enablers and Technology Solutions. IEEE Vehicular Technology Conference, 2013:1–5.
18. A. Osseiran, F. Boccardi, V. Braun, et al. Scenarios for 5G mobile and wireless communications: the vision of the METIS project. IEEE Communications Magazine, 2014, 52(5):26–35.
19. Qualcomm, The 1000x Mobile Data Challenge, 2013. https://www.qualcomm.com/documents/1000x–mobile-data-challenge.
20. T. Wang, B. Huang, J. Pang. Current Situation and Prospect of Spectrum Requirements Forecasting of the Future IMT System. Telecom science, 2013, 29(4):125–130.
21. G. Zhen, D. Ling, M. De, et al. MmWave Massive MIMO based Wireless Backhaul for the 5G Ultra-Dense Network. IEEE Wireless Communications, 2015, 22(5):13–21.
22. E. Larsson, O. Edfors, F. Tufvesson, and T. Marzetta. Massive MIMO for next generation wireless systems. IEEE Communications Magazine, 2014, 52(2):186–195.
23. J. Shen, J. Zhang, K. B. Letaief. Downlink User Capacity of Massive MIMO Under Pilot Contamination. IEEE Transactions on Wireless Communications, 2015, 14(6):3183–3193.
24. L. Lu, G. Y. Li, A. L. Swindlehurst, et al. An Overview of Massive MIMO: Benefits and Challenges. IEEE Journal of Selected Topics in Signal Processing, 2014, 8(5):742–758.
25. G. R. MacCartney, T. S. Rappaport. 73 GHz millimeter wave propagation measurements for outdoor urban mobile and backhaul communications in New York City. Proceedings of IEEE International Conference on Communications, 2014: 4862–4867.
26. M. R. Akdeniz, Y. Liu, M. K. Samimi, et al. Millimeter wave channel modeling and cellular capacity evaluation. IEEE Journal on Selected Areas in Communications, 2014, 32 (6):1164–1179.
27. A. Ghosh, T. A. Thomas, M. C. Cudak, et al. Millimeter-wave enhanced local area systems: a high-data-rate approach for future wireless networks. IEEE Journal on Selected Areas in Communications, 2014, 32(6):1152–1163.
28. Nokia, Optimizing mobile broadband performance by spectrum refarming. White paper, 2014., http://networks.nokia.com/file/35861/optimizing-mobile-broadband-performance-by-spectrum-refarming.
29. F. M. Abinader, E. P. L. Almeida, F. S. Chaves, et al. Enabling the coexistence of LTE and Wi-Fi in unlicensed bands. IEEE Communications Magazine, 2014, 52(11):54–61.
30. B. Singh, S. Hailu, K. Koufos, et al. Coordination protocol for inter-operator spectrum sharing in co-primary 5G small cell networks. IEEE Communications Magazine, 2015, 53(7):34–40.
31. H. Zhang, X. Chu, W. Guo, et al. Coexistence of Wi-Fi and heterogeneous small cell networks sharing unlicensed spectrum. IEEE Communications Magazine, 2015, 53(3):158–164.
32. METIS. Deliverable D1. 4: METIS Channel Models, 2015. https://www.metis2020.com/wp-content/uploads/METIS_D1.4_v3.pdf.

Chapter 2
Evolution of Testing Technology

Testing and evaluation are indispensable links of technology and product inspection. The final birth of every new technology or product must go through strict testing and evaluating process. In the 5G era, the testing tools and evaluation methods are faced with new challenges along with the continuous development of the new 5G technologies. In the section of testing technology evolution of this chapter, we will give a brief review of the history of testing technologies, and make analysis on the development trend of the testing technology ecosystem according to the existing technologies. In the section of testing technology challenges, based on the characteristics of wireless communications technology and 5G technology, we will sort out and analyze the challenges of wireless communications testing technologies in terms of overall performance, the number of RF channels, and high throughput, etc.

2.1 The Importance of Testing Technology

"Testing" means measurement, inspection and test. In different application fields, concepts of testing are different, but they all play important roles. For example, in the medical field, "testing" is an important step and indicator to determine whether a person is healthy or not; in the teaching area, "testing" is an important criterion to review people's knowledge and skills, etc. "Testing" in this book means to measure and test the performance and precision of science, technology and equipment, and lay more emphasis on the examination and verification of the new wireless communications technology.

In the field of wireless communications, testing technology plays a decisive role in the entire industry. When it comes to the new technological revolution, Academician Xuesen Qian said, "The key technology of new technology revolution is information technology. Information technology is composed of three parts, namely measurement technology, computer technology and communications technology,

© Springer International Publishing AG 2018
Y. Yang et al., *5G Wireless Systems*, Wireless Networks,
DOI 10.1007/978-3-319-61869-2_2

among which measurement technology is the key and basis." [1] Computer, communications and instruments are three internationally recognized information technologies and they are the industries that have achieved the fastest development since the twenty-first century. Technological development of different fields usually mutually supports and promotes each other. Thanks to the advance of computer and communications technology, measuring instruments have achieved rapid development and changes in the past 30 years, which will be described in subsequent chapters of this book. As one of the three key technologies of the information industry, testing and measuring technology has become the foundation and development guarantee for the electronic information industry. Measuring instrument is a country's strategic equipment, whose development level has become a sign of national scientific and technological level, comprehensive national strength and international competitiveness. In communications, radar, navigation, electronic countermeasure, space technology, measurement and control, aerospace and many other fields, the electronic measuring instrument is indispensable technical equipment. In modern manufacturing industry, the advanced instruments are needed in research and design, production process control, and product testing as a support to testing technology.

Instruments are the realization carrier of testing technology. Moreover, testing is an important step for checking the entire product's quality, and also an important basis to help the researchers to improve the product. Any new technology or product must be fully tested and evaluated before entering the industrial production and becoming a standard technology or qualified product in the market. Testing technology needs to run through the entire process, from the pre-design stage to design stage to production completion stage. Take the production of mobile phones as an example. Before the development and production, engineers need to test all kinds of components related to the phone. During the development and production, they need to take qualitative or quantitative tests of the machining parts, semi-finished mobile phone and all kinds of mobile phone parameters. And after the production completed, the comprehensive tests of mobile phones' functionality, performance and reliability are needed, such as mobile phones' radio frequency conformance test, WLAN test, GPS test and so on, so as to judge whether a phone meets various protocol standards and authority requirements. Different tests are taken to determine whether mobile phone is up to the standard or not. After organizing and analyzing the test data, if the testing shows that it does not meet the standards, then it will go back to optimize the performance and be tested again. The test data can be used to assist the researchers with performance optimization and problem solving. The product should not enter the market to become a qualified mobile phone until the test results finally meet the standards. The quality and efficiency of products are restricted by the quality and efficiency of testing technology. Only the accurate and efficient testing process can guarantee more efficient production of high performance products for the fast-updating terminal market nowadays. Take the software testing as another example. Engineers run or test software manually or automatically to find bugs or defects in the software, to help researchers in software design optimization and provide quality assurance for software. The process of compiling software is the process of testing, debugging and optimizing software, and the software testing is throughout the entire period of the software definition and development. Testing technology plays an important role in all stages of a product or technology, from idea budding to prototype to standardization. The testing and measurement solutions of each stage are important guarantees for the successful application of new technologies and new products.

The development of testing technology is mutually enhanced with research and application of new technologies and new products. It provides an important testing tool for the development of new equipment while testing instruments are indispensable tools for testing technology. Only when equipped with powerful instruments can we turn a lot of theories into the reality. Now the development of test instruments is more and more going toward the direction of multi-interface, multi-function, intelligence, high density, and high speed. Some instrument manufacturers even join equipment manufacturers in the standards development and equipment manufacturing, which greatly reflects the importance of testing and measurement technology [2].

2.2 Testing Technology Evolution

2.2.1 Development of Testing Instruments

Instruments are combinations of various sciences and technologies with many varieties and wide applications. By the intended purpose and usage, instruments can be divided into measuring instruments, radio test instruments, automobile instruments, aircraft instruments, navigation instruments, geological survey instruments, time measuring instruments, teaching instruments, medical instruments and many other types [3]. The testing instruments in this book are all related to the field of wireless technology, falling into the category of radio test instruments.

From the mobile phones with simple call function to the multimode smartphones and netbooks with various functions like GPS, Wi-Fi, Bluetooth, payment function, and digital television, personal mobile terminals have more and more functions and applications with more and more complex performances. Accordingly, to ensure more reliable performances and more powerful integrated functions, the mature testing equipment and advanced and reliable testing methods are needed. Powerful testing instruments are indispensable to the stable and reliable running of new wireless communications equipment [4].

Going through the history of testing instruments, we can take the beginning of automated testing technology and network-based remote services as critical points. By far, the development of test instruments can be roughly divided into three eras: "Instrument 1.0 Era", which means the times of traditional stand-alone and pure manual testing instruments; "Instrument 2.0 Era", which is the software radio era trigged by the technology progress of computer and communications; and

"Instrument 3.0 Era", which has introduced interconnected network testing and evolved from simple testing instruments to the integration of testing equipment and services. At present, the testing industry is still in Instrument 2.0 Era and the early stage of Instrument 3.0 Era. Whether changes will take place in the development of instruments in the future and whether the change will be fast or slow all need our technical workers' continued joint efforts. The development trend of testing technologies will be described in the subsequent chapters. It should be noted that the division of different eras is not absolute. Instruments of another era will not immediately disappear when new technology emerges. Instead, they are more likely to continue to progress with the technology development or be replaced and stop production after a certain length of time. The renewal of old and new instruments is a process of coexistence and substitution.

The development of testing technologies is actually the development process of the test instruments. This section will lead readers to review the development and evolution process of the testing instruments. The main realization carrier of testing technologies is the testing instruments. It's easier for the readers to understand the development of testing technologies in the corresponding periods with the introduction to the development of testing instruments.

Instrument 1.0 Era

Unlike the high-end testing instrument that the test engineers use now, the initial test instruments were simple and the measurement results were displayed by the pointer. They had single functions and low openness, and were known as the "analog instruments", such as the analog voltmeter and ammeter. Afterwards, with the emergence of electronic tube technology, electronic tube instruments came into being, such as the early oscilloscope, etc. Subsequently along with the emergence of transistors and integrated circuits, the instrument industry, combining with integrated circuit chips, produced the instruments based on integrated circuit chips, which were known as "digital instruments". The basic working principle of the digital instruments is that the analog signals are converted to digital signals in the measurement process, and the test results are finally displayed and output in the digital form. Compared with the analog instruments, digital instruments have more intuitive and clearer test results, faster response, and relatively higher accuracy, but the instruments then depended on manual operation. In the 1970s, the testing instruments began to use the microprocessor. Simple "intelligent instruments" began to appear in the field. For the first time, the engineers have tasted the sweetness of intelligence. The introduction of microprocessors has greatly improved the performance and automation degree of the instruments, facilitating automatic range conversion, automatic zero adjustment, trigger level automatic adjustment, automatic calibration, self-diagnosis and many other functions [5].

Instrument 2.0 Era

As time goes by, technologies in different fields are all developed. Owing to the development and progress of the computer technology, software, data processing, data bus, and reprogrammable chips, test instruments began to transform from the functional fixed discrete instruments to flexible instruments based on the software design. The instrument based on the software design allows users to redefine the software function according to their own needs. The development of network technology also experienced the user defined stage of Web2.0. Similarly, we call this stage of the instruments as "Instrument 2.0 Era". Therefore, the previous traditional instrument times can be called as "Instrument 1.0 Era".

In Instrument 1.0 Era, engineers totally relied on hardware to realize the testing and measurement. It was expert and manufacturing plant who took charge of design and manufacturing. Although users were allowed to put forward some opinions and demands, they could not be realized immediately. Moreover, users were unlikely to participate in the design and manufacture of products. The hardware itself and its analysis function were defined by the instrument suppliers, and self-definition function was not provided to users. Even if the instrument was connected to the computer, the transmitted information waste test results defined by equipment manufacturers. Users are unable to obtain the original measure data to do self-defined analysis. During the development of the testing instruments, with the help of the progress in computer technology and communications technology, the instrument industry had seen fast development and changes from the traditional hardware instruments. The emergence of bus technology made automated testing possible, and the rapid growth of the bus technology led to the fast development of the automated testing. The development of automated testing offers a guarantee for wide applications of wireless communications products. Traditional testing has complicated procedures, high operations difficulty and strict requirements on the test engineers, while automatic testing has overcome those disadvantages and brings many advantages, such as having testing specifications, high testing speed, avoidance of maloperations, and being able to guarantee the accuracy and authenticity of the testing results to the greatest extent [6]. Among them, the most representative one is the proposed concepts of the PCI eXtensions for Instrumentation (PXI) bus technology and Software Defined Radio (SDR), which changed the single direction of the traditional testing technology to the stage when users could redevelop the instruments according to their own needs, bringing us to Instrument 2.0 Era, namely, the software defined radio era. In this era, engineers had more control rights to the instruments. After getting access to the original real-time data, engineers could use software to design their user interface and define measurement tasks in order to obtain the desired results.

During this period, the instrument technology had experienced great improvement. The major improvement and change compared with the 1.0 Era mainly stemmed from and were reflected in the development and emergence of bus technology, instrument module technology, SDR, hybrid systems and many other representative technologies. The combination of multiple techniques helped to create a number of high-performance testing systems that increased the flexibility of testing and brought higher performance and lower cost.

Bus Technology

Since the birth of the first-generation test bus technology, i.e., General Purpose Interface Bus (GPIB) technology, in the 1970s, testing has gradually developed from a single manual operation to a large-scale automatic testing system. Afterwards, all measuring instruments with GPIB standard interfaces sprang up, and the unified standards of interconnected instruments were gradually formed. The standards allow testing engineers to assemble a variety of automatic measurement systems with powerful functions by very convenient means. Meanwhile, with the passing of time and the improvement of test requirements, the bus technology is not limited to GPIB anymore. Other bus technologies have been put forward and improved. The representative bus technologies include VMEbus eXtensions for Instrumentation (VXI), Peripheral Component Interconnect (PCI), Peripheral Component Interface expres (PCIe), PXI, PCI eXtensions for Instrumentation expres (PXIe), LAN eXtension for Instrumentation (LXI) and Advanced eXtensible Interface expres (AXIe), etc.

Through the test bus, the data communications between different units and modules in the testing instruments are realized. Thanks to the development of bus technology, testing is no longer limited to the manual operation of a single instrument and the automated testing becomes possible. This progress plays an important role in promoting the development of automated testing system, so the bus technology is an important technology of the 2.0 Era and the main changes in the testing instruments cannot be separated from the development of the bus technology.

The birth of the GPIB technology enables engineers to directly connect the test instruments with computers. Almost every device has a GPIB interface, which is firm, reliable, universal, simple and convenient. The simplest is to connect one computer with a testing instrument. However, it will be limited if the test needs to use multiple instruments. In that case, with low data transmission rate, high cost, limited bandwidth and poor reliability, the synchronization and trigger function of multiple instruments are unable to be provided, which will affect the test performance.

VXI has higher bandwidth and lower latency than GPIB. But it is difficult to be applied to other fields because of its high cost. It is mainly used in military and aerospace. In addition, VXI is based on the outdated VME bus and the modern computer does not support this kind of bus structure, which has affected the development of the bus technology.

Intel put forward the concept of PCI bus for the first time in 1990s, which was used to connect peripherals and computer backplane. The PCI bus is a parallel bus that can work in 33 MHz and 32 bit. The maximum theoretical bandwidth is 132 Mbytes/s, and it employs the shared bus topology.

After the PCI bus was used in the instrument field, under the combined effect of various factors, the PXI technology bus was obtained through the corresponding expansion of the PCI bus. It is a new bus standard launched by American National Instrument (NI) Company in 1997. PXI is based on the mature PCI bus technology, so it has faster bus transmission rate, smaller volume and better cost performance compared with VXI. In addition, PXI is capable of providing nanosecond timing and synchronization features as well as robust industrial characteristics. Finally, thanks to the flexibility of the software and the continuous updating of the modular hardware, users can upgrade the entire test system at any time with the least investment. Such an excellent scalability and flexible software architecture make the system integration based on PXI modular instrumentation platform more common.

In 2004, PCI Express was launched. It is an expansion on the PCI bus. The testing system with the Instrument 2.0 technology can enhance the performance of data stream transmission and data analysis when adopting this kind of bus. Comparing with the PCI bus, PCI Express uses the point-to-point bus topology, which provides a separate data transmission path for each device. All slots of PCI Express have dedicated bandwidth to connect the PC memory so that they do not have to share bandwidth like traditional PCI. Data will be transmitted and received in the packet form through the symmetric channels with the bandwidth of 250 Mbytes/s in each direction. Multiple channels gathering together can maximally form a 32 times channel width, significantly improving the data transmission bandwidth, minimizing the demand on memory and speeding up the transmission of data stream.

PXI Systems Alliance (PXISA) official organization officially launched the hardware and software standards for PXI Express in the third quarter of 2005. By using the technology of PCI Express on the backplane, PXI Express was able to increase the bandwidth by 45 times, from the original 132 MB/s of PXI to current 6 GB/s. At the same time, the compatibility of the software and hardware was retained with original PXI modules. The performance was enhanced so much that PXI Express could enter more application fields that were used to be controlled by dedicated instruments, such as Intermediate Frequency (IF) and Radio Frequency (RF) digital instruments, communications protocol verification and so on. Thanks to the software compatibility, PCI Express and PXI can share the same software architecture. In order to provide hardware compatibility, the latest Compact PCI Express standard has defined hybrid slots to support modules based on the architecture of PXI or PCI Express simultaneously [7]. PXI Express also offers a 100 MHz differential system clock, differential signals, and differential trigger time and synchronization characteristics. Prior to the launch of the technology, some fields had to only rely on expensive dedicated device to solve the problem. PCI Express technology solves the problem in these fields with high performance, for example, high bandwidth IF instruments for communications system testing, the protocol based on low voltage difference signals, interfaces of high-speed digital protocols such as the FireWire protocol and the optical fiber channel protocol, large-scale channel data identification systems for structural and incentive test, high-speed image recognition and data stream processing, etc.

The AXIe bus technology, an open standard for modular test equipment, has been proposed in recent years. It inherits the advantages of modular architecture of Advanced TCA and refers to the existing standards of VXI, PXI, LXI and IVI. It has larger circuit boards, higher data transmission rate, greater power and better heat dissipation than VXI or PXI. Moreover, a modular flexible platform with long life cycle, high performance and strong scalability are provided. Its goal is to create an ecosystem composed of various components, products and systems, promote the development of general instruments and semiconductor testing, provide maximum scalability and meet the needs of various platforms including the general rack stacking system, modular system, semiconductor AT, etc. [8].

Software Defined Radio Technology

Software Defined Radio, known as the third revolution in the information field, is the most representative technology of the Instrument 2.0 Era. The software radio is to take hardware as the basic platform of wireless communications and implement the maximal functions of wireless communications and personal communications with software.

In the 1990s, as a new concept and system of wireless communications, software radio began to receive attention at home and abroad. It gave communications system good universality and flexibility, and made it easy for system interconnection and upgrading. The technology turned out to be a major breakthrough in the field of radio and was primarily applied in military field. The basic idea is to let all the tactical radios in use be based on the same hardware platform, install different software to form different types of radio, complete functions of different natures, and get the software programmable capability [9, 10]. At the beginning of the twenty-first century, with the efforts of many companies, the application of software radio was transformed from military to civilian fields, such as multi-band multi-mode mobile terminals, multi-band multi-mode base stations, WLAN and universal gateways, etc. The software defined radio technology, as a new wireless communications system structure with strong flexibility and openness, has naturally become the strategic base of global communications [11].

Because of the increase in users' demand, the shortage of spectrum resources and the huge attraction of new business have brought a lot of stress to equipment manufacturers and operators and promoted the updating of wireless technology standards. As there are significant differences between various wireless technology standards and systems, the existing hardware-based wireless communications system is difficult to adapt to this situation, which has led to the concept of Software Defined Radio and the updated technology and equipment. The update of technology and equipment usually causes the waste of equipment and investment, while the software defined radio technology is able to save the inconvenience caused by technology updating through the self-defined software.

The architecture of software radio is different from that of the traditional software. In conventional wireless communications systems, the RF part, up/down frequency conversion, filtering and baseband processing are all in the analog form. A certain band or type of modulation of communications system corresponds to a certain specialized hardware structure. The low-frequency part of the digital radio system adopts the digital circuit, but the RF and IF parts are still inseparable from the analog circuit [10]. Compared with conventional radio systems, software radio realizes all of the communications functions by the software. The key idea of software radio is to construct a standardized, modular universal hardware platform, to realize various functions by software, and to make the broadband A/D and D/A convert to IF, near the RF side of the antenna, and strive to carry out the digital processing from IF. In addition, different from digital radio using digital circuits, software radio adopts high-speed DSP/CPU. The software radio system must work in parallel with a number of CPUs, and the digital signal processing data should be exchanged at high speed, which requires the system bus to have a very high data transmission rate. DSP devices are used to replace the dedicated digital circuit board so that the system hardware structure and functions are relatively independent. Modular design is also employed to give the platform openness, scalability and compatibility. Based on the relatively universal hardware platform, software radio realizes different communications functions by loading different software. It can rapidly change channel access methods or modulation modes and adapt to different standards by utilizing different software, thus forming the highly-flexible multi-mode terminals and multi-functional base stations to achieve interconnections.

In practical applications, software radio requires a very high speed of hardware and software processing. Due to the limitation of the technological level of hardware, the concept of pure software radio has not been widely used in practical products. The SDR technology based on the concept of software radio has attracted more and more attention. Software defined radio is a system, which must have the ability of reprogramming and reconstructing, so that the equipment can be used in various standards and multiple frequency bands and realize a variety of functions. It will not only use programmable devices to implement digital baseband signal processing, but also carry out programming and reconstruction on analog circuits of radio frequency and intermediate frequency, and have the ability of reprogramming, reset, providing and changing services, supporting multiple standards and the ability of intelligent spectrum utilization, etc.

The software defined radio technology is mainly realized based on single-chip FPGA or DSP. However, with the continuous increase in the volume of collected data, signal processing capabilities of software radio platform also need to be further enhanced. In order to improve modern signal processing capacity, the new technology is also proposed. For example software defined radio technology based on Compact PCI (CPCI) has attracted many teams and manufacturers to develop the software radio platform system in line with the CPCI standard [12].

Modular Instrument Technology

Different from simple length measurement, measurement in the wireless communications field has become more and more systematic. Many brand-new test methods and instrument concepts emerge one after another. High-speed communications test instruments featuring modular, software, and integration continue to spring up, which complement and get closely integrated with traditional methods, expanding the applications continuously and forming a technical highlight in the communication-centered application fields.

As mentioned in [7], VXI was the earliest bus which introduced the concept of modular instrument. It successfully reduced the size of the traditional instrument system and improved the level of system integration. It mainly used to meet the needs of high-end application of automated test, and had been successfully applied in the military aviation test, the manufacturing test and so on. Like the tested terminals, the test system is developing from the hardware-centric, single-purpose and limited-functions stage to the software-centric, multi-purposes and unlimited-functions stage. The modular instrument technology is the epitome of the technological progress.

Traditional desktop instruments generally have only a single function. They achieve instrument functions through the man-machine operation step by step. Modular instruments, on the other hand, highly integrated with computers and suitable for PCI and PXI platforms, are highly-flexible plug-in computer boards. The functions of modular instrument are similar to the traditional desktop instruments but better. The modular instrument is an important part of the SDR technology. Being defined by software enables it to define the measurement and analysis in real time, breaking the rule that the traditional instrument manufacturer defines the fixed functions. The users can realize the required test tasks within a short time through the functions of software-defined modular instruments. They can apply the customized data analysis algorithms, create a customized user interface, change the situation of purely displaying the test results of traditional instruments, and add more test engineers' ideas and testing intentions, thus giving engineers more initiatives. The difference of modular instruments with the traditional instruments mainly comes from the progress of the bus technology, with which different modular instruments share a power supply, a chassis and a controller. The bus can guarantee the data transmission channel between the modules. Through the bus control, different functions of the modular instruments can be integrated to reduce the volume of the instruments and simplify the testing complexity.

With the aid of modular instruments, engineers can choose different kinds of modular instruments according to their measurement needs and set up a test system. Due to the adoption of the software defined modular structure, system measurement can be realized through the corresponding software configuration operation. The service cycle of modular instruments can be increased through software upgrading. Different test requirements can be achieved through software programming. The life of the instruments can be ensured through repeatable and flexible use of modular instruments, relieving the pressure of increasingly complex equipment and technology on the test time. Especially, after the introduction of new wireless standards, the engineers using traditional instruments need to wait for suppliers to develop a corresponding desktop instrument before testing the standards. However, with the new software defined radio technology and modular instruments technology, the engineers are able to test the wireless standards in the standard-setting process with the universal module instruments and user-defined wireless protocols and algorithms. Due to the different characteristics from the traditional instruments, the modular instrument technology has become the representative technology of the "Instrument 2.0 Era".

Hybrid Test System

On account of the advance of the computer technology, the bus technology and their penetration in instrumentation, the hybrid systems composed of different bus technologies and instruments are gradually appearing in the testing field. Users of such kind of hybrid systems can not only enjoy the high speed and flexibility of modular instruments, but also use existing discrete instruments for some special measurements.

In a hybrid test system, different components of multiple automated test platforms are integrated in a system, including PXI, PCI, GPIB, VXI, Universal Serial Bus (USB), Local Area Network (LAN) and LXI, and other different buses. Its emergence is not a coincidence. Although the bus technology is developing forward and PXI has been widely recognized and used, other bus technologies cannot completely disappear. The instruments that include other bus technologies, such as capture card based on the Industry Standard Architecture (ISA) bus technology and GPIB control cards, are still trusted and being used by many engineers and manufacturers, and the bus cannot be completely replaced by PXI. Moreover, from the test engineers' point of view, when designing a test system, there are various factors to balance. Now the products are becoming more and more complex, the requirements for the mixed signal test are also getting higher and higher, so it is necessary to use the advantages of different bus test platforms and build a hybrid test system to meet the testing demands.

In the examples of the equipment system indicated in Fig. 2.1, the hardware employs the GPIB, PXI, LXI and many other test buses to form a hybrid system. What the test engineers should do is not just simply connect several instruments, i.e. to make the hybrid system that is composed of different products of different manufacturers and different buses work normally and smoothly, they also need to consider the software architecture of the hybrid system. Having unified software architecture can greatly simplify the complexity of system programming and avoid the problem of compatibility between different instruments.

For example, NI Company offers unified software architecture composed of the measurement and control service layer and the application development layer. The architecture provides software for the hybrid systems. If the hardware is the skeleton of the hybrid system, then the software is the soul to control the whole system. The measurement and control service layer includes a flexible device driver and is used to connect the software and hardware and simplify the test code of the hardware configuration part. The company also proposed Virtual Instrumentation Software Architecture (VISA) standards to provide the API with high-performance,

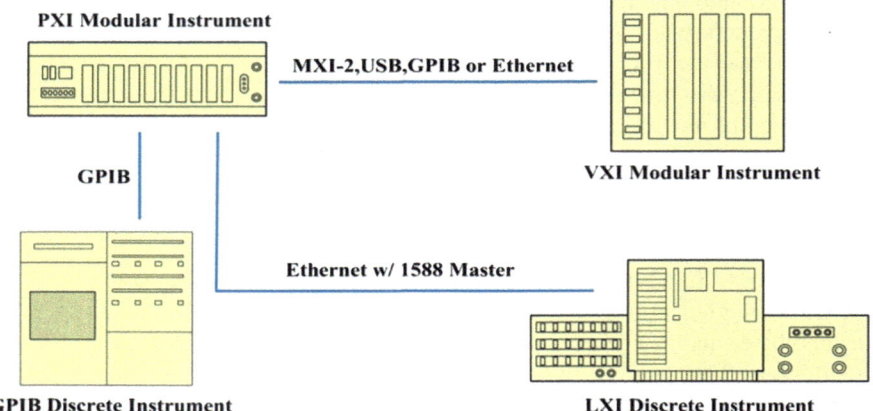

Fig. 2.1 A kind of hardware components of the hybrid bus test system [7]

programmability and continuous upgrading to help with the development. VISA, responsible for the communications with driver software, is independent of the instrumental bus and will help hardware seamlessly integrate into the software. Correspondingly, LabVIEW, the graphical programming language developed mainly by the company, programs in a graphical manner, helping users to efficiently and quickly develop and test applications.

The emergence and use of hybrid systems have brought tangible benefits to the engineers. For example, some test of engineers may need high throughput and excellent integration provided by the modular instrument bus like PXI and PCI Express. At the same time, they may also need discrete instruments based on USB or LAN (including LXI) for some specific test functions. In addition, if there are new testing needs, users can use hybrid systems. Engineers can easily upgrade or add new components in the existing system without redesigning the whole system, which ensures the continuity and reusability of the test system. The hybrid system is logically called a representative product of Instrument 2.0 Era, and with the development of technologies such as the bus and software technologies, it will continue to evolve.

Instrument 3.0 Era

Different from the evolution time from Instrument 1.0 to 2.0, after the Instrument 2.0, thanks to the rapid development of software technology and network technology, the 3.0 Era has arrived.

Instrument 3.0 Era reflects the characteristics of openness, union, and service. Generally, there are two kinds of definitions for the 3.0 Era. From the standpoint of test measurement industry, the Instrument 3.0 is defined as follows: in addition to the function and performance of the equipment itself, the technical service appears with the instrument. From the technical and academic point of view, the Instrument 3.0 can be defined as follows: in addition to the characteristics of Instrument 2.0, features like network-based combination, cross-layer test and remote test have emerged. Cloud testing, software and hardware joint multi-user system simulation test is a series of applications confirming the characteristics of the instrument 3.0 technology.

2.2.2 Development Trend of Testing Technology

Through the above description on the development, the key technologies and challenges of testing technology, it is not difficult to find that to cope with the changing challenges of new standards and technologies, engineers are facing many difficulties in testing wireless devices and more complex testing work. Although different testing techniques have developed towards software-based, modular and intelligent directions, it is still unable to meet the needs of the rapid technical development. In this section, we summarizes the main development trend of wireless testing technology based on the testing technology development status and the development needs of wireless communications,

Coexistence of New and Traditional Test Equipment

The instrument system for test and measurement has been developing towards the software-centered modular system. Accordingly, users can integrate test into the design process in a faster and more flexible way, shorten the development time, and improve the test efficiency. While the software defined instruments depending on software have become increasingly popular and been widely used, the synthesis of the instruments is also proposed. However, the traditional instruments are also advancing with the bus technology. Therefore, the emergence and widespread adoption of modular instruments and software defined radio technology cannot immediately replace the traditional test instruments. The renewal of the traditional test instruments will still be competitive, and the two will continue to coexist.

The Leading Trend of Many-Core/Parallel Technology and FPGA Real Time Technology

In the testing field, many-core parallel technology and FPGA real time technology will become important supports for the development of testing technology, and can lead the development trend of testing technology.

Many-Core Parallel Technology

Over the years, the microprocessor has been developed from single core into multiple core. Commercial equipment with dual-core or even eight cores are now everywhere. Owing to the software defined instruments, users can enjoy the huge performance improvement of automated test applications brought by the multi-core processor immediately. While multi-core is no longer satisfactory, many-core comes into being. Moore's law states that the number of transistors would be doubled every year, and processor manufacturers can use these increasing transistors to create more cores. Today, desktops, mobile and ultra-mobile computing commonly use dual-core and quad core processors, and servers usually use ten or more cores. More cores are being plugged into smaller, less-power encapsulations.

The number of cores in a server processor is increasing drastically to more than 10, and the processor is developing towards many-core. Super computers enable us to understand what the future processors will be like. The fastest computers in the world use millions of cores. Although these multi megawatt super computers are not suitable for the test stations in the production workshop, integrating more functions into smaller and smaller space means that many-core processors will be able to shrink to equal size of other processors due to reducing the volume of power supply. An example of this tendency is the coprocessors of Intel Xeon Phi class, which provide up to 61 cores and 244 threads of parallel execution capability.

Market demand is driving the improvement in the performance of graphics cards and augment of the number of cores. Although the test and measurement applications basically do not use the graphics function, the new processor with more kernel can provide higher performance to test applications specifically designed for a higher number of cores. The migration from single core, multi-core to the many-core processor has been very common in the field of high-performance computing. With more advantages of parallel processing being found, many-core processors will continue to be adopted by more routine applications.

The emergence and development of multi-core/many-core processors, is not only an inevitable choice of the development of the semiconductor industry, but also a very reasonable architecture. This architecture is well adapted to the current computing environment and application mode of Internet. Although there are still large quantities of problems to be solved, covering from the internal structure, such as network on chip and cache organization and coherence protocol, to the external environment, including programming model and key system software, there is no doubt that as the core component of future computing platform, multi-core/many-core processor is welcoming the best time for development [13]. The development of multi-core/many-core technology will surely promote the development of the test industry.

Real-Time FPGA Technology

The functions of the FPGA technology will become more and more prominent in the future. In many cases, when the test system is developed on a multi-level, it is always a trade-off between customized hardware and commercial general components to find the best balance point between the throughput and the test coverage. FPGA can realize the advantages of customized hardware tools through the flexible software system. The FPGA technology has many advantages. One key benefit that FPGA brings to the test application is to liberate the master processor from a large number of data streams or internal tasks. FPGA is suitable for uninterrupted filtering, modulation/demodulation, encryption/decryption or other data processing. Another key benefit is to ensure the inherent parallelism of data. Chip is able to allow any number of parallel data to process channels. Finally, FPGA can reconfigure traditional hardware. FPGA is defined by the software first and then downloaded to a FPGA chip for an actual execution. Only through downloading a new program to the FPGA chip can the hardware be reconfigured. Although in the past FPGA was applied in some forms in the testing field, there is no significant return on investment. It is mainly because the two key technical difficulties. Firstly, FPGA is a kind of discrete programmable chip. To apply this technology in practice, users must design peripheral circuits, including digital to analog converter, analog to digital converter, clock, power and some necessary components for FPGA programming. Likewise, the Equipment Under Test (EUT) that can communicate with the peripheral equipment must be provided. Secondly, the hardware description languages of FPGA, such as Verilog or Very-High-Speed Integrated Circuit Hardware Description Language (VHDL), have not been widely used in the industry. Only a few digital development engineers are able to use these languages. In the system of Instrument 2.0, the first problem has been solved by using standard commercial universal module combined with the FPGA technology. Many manufacturers provide FPGA based hardware systems, which have already embedded a lot of peripheral circuits in the development boards or platforms. In addition, the current trend shows that the hardware description languages have become more and more abstract, and been provided to all the engineering disciplines for application. At the same time, in order to support the design of FPGA, the important work has been completed in the C to VHDL language conversion and graphical programming technology, which ensures the development of the FPGA technology and reflects the importance of the FPGA technology.

As design and testing requirements become higher and higher, the FPGA technology is introduced into the FPGA based customized instruments. FPGA with high performance and reconfigurable characteristics provides the support for the application of FPGA. Especially, in the 5G wireless communications system in the future, performance requirements will be higher, and FPGA will be needed to ensure that the response is real time and the inflow and outflow speed of the data are high in the real-time system simulation and high-speed memory test applications. As the core of 5G communications test equipment, the digital signal processor based on FPGA has the following advantages [14]:

1. FPGA has an internal parallel mechanism, which can perform complex mathematical calculations simultaneously without occupying the host processor.
2. According to the ability of reprogram, FPGA can be used for testing applications of a variety of current or future communications standards.

3. FPGA's powerful functions can reduce the cost and size of the RF test equipment.
4. Compared with traditional firmware or software-led tests, using FPGA significantly shortens the test time, so that large-scale industrial applications become possible.

The advantages and application fundamentals of FPGA ensure its bright future in testing and application. The test industry has developed and applied the products combining with FPGA. For example, R-series data collection and FlexRIO products family provided by NI have integrated the high-performance FPGA into the readily available Input/Output (I/O) broad card for users to customize and repeat configurations according to the applications. At the same time, with the easy and intuitive graphical programming of LabVIEW FPGA and without the need to write the underlying VHDL code, users can rapidly configure and program FPGA functions for test automation and control applications [15], simplifying the application complexity of FPGA technology.

Test Ecosystem Becomes the Trend

The challenges of testing technology have depicted the future of test instrument times. Through the introduction to the evolution and challenges of test instruments, we can find that in addition to the above trends of future testing technology, an upstream and downstream test system will be formed, i.e. a software-centered test ecosystem.

How to keep up with the rapid development of many communications technologies and standards, how to improve the testing budget cost performance, how to flexibly redefine testing requirements and methods, how to effectively use the multi-core technology, how to use real-time processing technology to improve the test throughput -- the answers to all these problems will point to the solution of "software centric" [15].

FPGA Based IP to the PIN Technology

For decades, the electronic and communications industry has been pursuing the ideal state of mutual improvement of design and testing. In view of the difference between design and testing, this goal has not been reached yet. In the design phase, the latest Electronics Design Automation (EDA) software is applied to the system level design, while the testing area is slightly independent and lagging. Therefore, with the latest software-centered electronic communications equipment, a new test solution is often needed to be found.

Adopting "system-level approach, integrating the concepts of design and test, and expanding software architecture to FPGA" is one of the effective means to balance the development of the two areas and improve the efficiency of

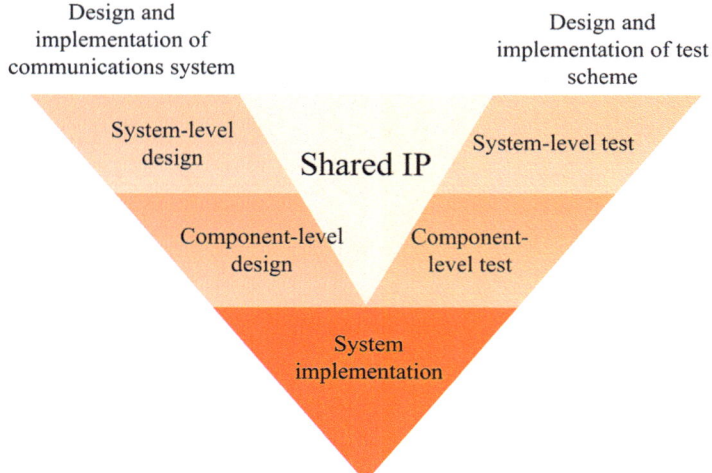

Fig. 2.2 "V Graph" to realize "IP to the Pin" [14]

communications testing. The way to integrate design and test is to deploy the designed IP cores to Device Under Test (DUT) and integrated test platform. This deployment process is called "IP to the Pin" [16], because it enables tester-defined software IP to be close to the hardware I/O pins of integrated test platform as much as possible. These software IP can include: data acquisition, signal generation, digital protocol, mathematical operation, RF and real-time signal processing, etc. Regardless of throughput or power consumption of a single device's original data processing, FPGA is better than digital signal processor, traditional processor, and even graphics processor [17].

The specific implementation of IP to the Pin technology can be expressed as the "V graph". Each phase of the design has a corresponding verification or testing phase. By sharing IP, the design and test team can respectively move along the two sides of the V graph, from modeling and design of the top layer to the implementation of the bottom layer, carrying out the corresponding test at each stage (Fig. 2.2).

Heterogeneous Computing Architecture Supporting Parallel Testing and Massive Signal Processing

The heterogeneous computing architecture is a system for assigning data processing and program execution tasks among different computing nodes, so that each node can handle the most appropriate test and calculation task. This technology can be used to deal with the increasingly complex testing calculations in future wireless mobile communications, to store and process massive amounts data of MIMO and massive MIMO in the RF back end, and to enable the multiple nodes to make

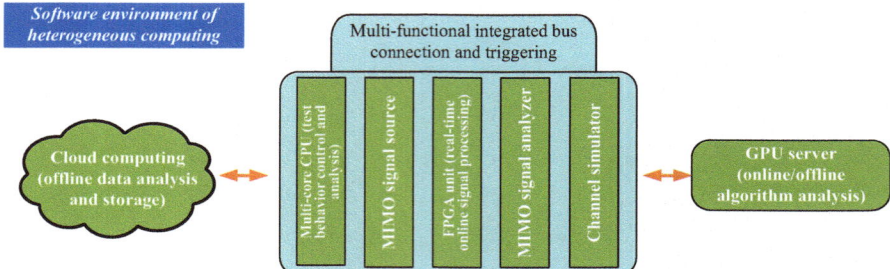

Fig. 2.3 Heterogeneous computing architecture for complex mobile communications testing

simultaneous spectrum sensing and detection, and collaborative test of the PHY layer and the Media Access Control (MAC) layer.

Take the MIMO RF test system using the heterogeneous computing architecture as an example. CPU can be used to control the program execution. FPGA is used for online demodulation, and the GPU is used for the calculation of multi-antenna test parameters. Finally all the processing results are stored on a remote server. The following figure shows a heterogeneous computing architecture that can be applied to the 5G communications test (Fig. 2.3).

Along with the rapid development of high bandwidth and high data rate of 5G mobile communications, the combination of the heterogeneous computing architecture and the multi-core parallel programming technology will be the indispensable main technology with which 5G test is able to deal with massive data processing and improve the parallel testing.

2.3   Challenges of Testing Technology

2.3.1   Challenges of Multi-Function and High-Performance

The demand for wireless communications is increasing, and the requirements of new technical standards are continuously improving. The development of digital simulation technology and the increase of the frequency band of RF test are urging the development of the testing technology. For example, the increasing bandwidth brings challenges to the testing technology. WLAN standards are generally applied in the bandwidth which is lower than 100 MHz, so it does not need special methods of the testing technology. However, the Ultra Wide Band (UWB) technology requires the bandwidth to be 500 MHz, which brings a great impact on the design of the testing instruments, because the test methods and techniques with different bandwidths are very different. In addition, with the development of technologies, the RF channel and the I/Q baseband channel both challenge the multi-channel instrument technologies. Moreover, the development of the digital simulation technology, speed improvement of digital interface and other growing technologies challenge the testing industry, and meanwhile promote the change of the testing technology.

The popular WLAN (802.11) protocol also drives the increase of RF channel number, and supports multiple antennas and multiple communications frequencies. In addition to the increasing RF channels, the IQ baseband channels of these transceivers are also increasing. All of these have put forward higher requests to the test ability of the baseband mixed signals in the test system. The independent development of the digital and the analog simulation technologies also promotes the change of testing, which brings the challenge to the test engineers.

Due to the various applications, the current devices usually integrate some parts of different technologies like the cellular wireless technology, the short distance wireless technology and the GPS technology. Therefore, it is not rare to have multiple RF channels on a single chip. The brand-new structure of new testing instruments should be able to handle the interface test efficiently. At the same time, the coexistence demand of LTE, Wideband Code Division Multiple Access (WCDMA)/ High-Speed Packet Access (HSPA), Global System for Mobile communication (GSM) and other networks promotes the development of multi-mode base stations and multi-mode terminals, and the corresponding test solutions of multi-mode measurements must be specially designed. Unlike the 2G and 3G tests, the 4G test not only tests the physical layer, but also comes up to the MAC layer measurement before getting a more comprehensive understanding of the product performance. It also presents new challenges for wireless communications testing, and the wireless testing technology needs to adapt to the needs of the forward progress. In addition, high-speed data transmission system has developed in the direction of high frequency. The high frequency testing is also an inevitable challenge in testing.

2.3.2 Challenges of Multi-channel

In order to promote its service support capacity, 5G will have new breakthrough in the wireless transmission technology and the network technology. In terms of the wireless transmission technology, technologies which can further explore the potential of spectrum efficiency improvement will be introduced, such as the advanced multiple access technology, multi-antenna technology, the modulation and coding technology, the new waveform design technology and so on. Among them the wireless transmission technology based on the large scale MIMO (or massive MIMO) will be likely to make the spectrum efficiency and power efficiency to upgrade an order of magnitude on the basis of 4G. This section introduces the development and application of the MIMO technology, and discusses the possible challenges of the testing technology brought by the massive MIMO technology in 5G.

As an effective means to improve the system spectrum efficiency and the transmission reliability, the multi-antenna technology has been applied to a variety of wireless communications systems, such as the 3G system, LTE, LTE-Advanced (LTE-A), WLAN, etc. According to information theory, the more number of antennas are, the more obvious increase in the spectrum efficiency and reliability is. Especially when the number of transmitting and receiving antennas is large, the MIMO channel capacity will have a close-to-linear increase along with the minimum number of transmitting and receiving antennas. Therefore, it provides an effective way to increase the capacity of the system by adopting a large number of antennas. Due to the limitation of technical conditions such as the multi-antenna occupied space, the complexity of the implementation and so on, the present wireless communications systems are equipped with only a few antennas at the transceivers. For example, there are at most four antennas in the LTE system and eight antennas in the LTE-A system. However, due to its huge capacity and reliability gain, the research on massive MIMO system has attracted great attentions, such as the research works on MU-MIMO where the base station is equipped with much larger numbers of antennas than that of mobile stations. In 2010, Marzetta in Bell Labs studied the MU-MIMO technology with an unlimited number of antennas were configured in each base station for multi cells with Time Division Duplexing (TDD). He proposed the concept of massive MIMO and found some different characteristics with those in a single cell with a limited number of antennas. Thereafter, many researchers studied the configuration of the base station with limited number of antennas on this basis. In massive MIMO, the number of antennas configured by a base station is very large (usually tens to hundreds of antennas, which is 1 to 2 orders of magnitude of the existing systems), serving several users on the same time-frequency resources simultaneously. With regard to the antenna configuration, these antennas can be centrally configured on one base station to form a centralized massive MIMO system, and can also be configured in multiple nodes to form a distributed massive MIMO system. It is worth mentioning that China's scholars have been at the forefront of the international academia in the study of the distributed MIMO. The benefits of massive MIMO are mainly reflected in the following aspects. Firstly, the spatial resolution of massive MIMO is significantly enhanced compared with the existing MIMO. It can further utilize the space dimension resources to make a plurality of users in the network communicate with the base station on the same time-frequency resource, so that the spectrum efficiency can be greatly improved without increasing base station density and bandwidth. Secondly, massive MIMO can concentrate the beam in a very narrow range, greatly reducing the interference. Thirdly, the transmission power can be reduced significantly, so as to improve the power efficiency. Fourthly, when the number of antennas is large enough, the simple linear precoding and the linear detector tends to be optimal, and noise and uncorrelated interference can be ignored.

The research on massive MIMO has been mainly concentrated in the channel model, capacity and transmission performance analysis, precoding, channel estimation and signal detection in the recent two years. But due to little work of the experimental model, the channel model has not been widely recognized. In addition, different from the traditional evaluation of the antenna radiation pattern, 5G tests also need to complete the evaluation and verification of the end-to-end

receiving performance of the wireless system equipment. All of the above requirements need a large number of Over The Air (OTA) tests for support. The fifth
generation mobile communications technology in China defined that the number of
cooperative antenna at the side of the future 5G base station should not be less than
128 [18]. The number of antennas on this massive MIMO technology has risen to
the hundreds, greatly larger than that of traditional techniques. The factors such as
the number of test channels, the synchronization and isolation among multi channels, the storage of multi-channel data, present severe challenge to the massive
MIMO test implementation.

When validating the characteristics of the MIMO device, or achieving some of
the non-product MIMO systems like RADAR and beam forming, the multi-channel
RF architecture is needed. From the previous 2×2MIMO system to the present
8×8MIMO system as well as the future massive MIMO, scalability has become the
key requirement for the next generation of the RF test system. Testing MIMO
wireless devices usually does not require a multi-channel architecture, because it
generally does not require the full spectrum characteristics test. But MIMO itself
needs to add a test channel. In order to the implement parallel test of multi-protocol
wireless devices, engineers need to upgrade the existing RF devices to more
channels at a low cost, and it should be flexible enough to measure multiple
frequency bands. According to the above requirements, it can be seen that the
new generation of RF devices should have a fully parallel architecture of hardware
and software, and have advanced synchronization. In addition, the new RF devices
must be "ready to be used", and provide a higher accuracy of synchronization, not
just to realize synchronization by using reference clock (usually 10 MHz) and
trigger. The reason is that although the traditional synchronization methods can
ensure the synchronous acquisition of the signal, it cannot ensure the signal phase
synchronization. The software part is even more important in the whole test
architecture, because it requires a lot of computations to deal with many kinds of
infinite standards. The modern software architectures can implement parallel data
streams, which make one or more processing units to be dedicated to a RF channel.
There are universal parallel processing architectures in the current market, including multi-core processors, multi-thread, multi-core and FPGA. There are also some
other promising technologies, such as the Intel Turbo Boost technology used in the
latest generation of the Intel Nehalem. It automatically allows the processors to
overspeed in excess of the basic operating frequency, as long as it runs below the
rated power and the limits of the current and temperature. In order to fully utilize
these processor technologies, engineers need to apply the parallel programming
technology on the algorithm level and the application software level, such as task
parallelism, data parallelism and pipeline operations. The multi-channel test architecture reduces the total test time, increases the test throughput, and improves the
performance of the instruments. However, the flexibility of the architecture is also
very important. For example, the MIMO configuration is a typical dynamic configuration, and the phase and amplitude of each transmitter can be used to optimize
the performance and direction of signals. Every addition of a MIMO transmitter
will make the software complexity grow in an exponential way.

Similarly, the emerging wireless technologies, for example, the MIMO antenna systems, have a great impact on the design of the transceiver, which will also affect the RF devices. The future multi-channel wireless system should be based on the low-cost architecture, in which the signals and the software bear the characteristics of parallelism.

Therefore, the application of the massive MIMO technology in 5G has brought unprecedented challenges to the traditional RF test technology at aspects of both testing methods and testing devices.

2.3.3 Challenges of High Throughput

The vigorous development of mobile Internet is the main driving force of 5G mobile communications. Mobile Internet will be the basic business platform of various future emerging businesses. The existing fixed Internet services will be increasingly provided to the users in a wireless way. The wide applications of cloud computing and backstage services will put forward higher requirements on the transmission quality and the system capacity to 5G mobile communications system. The main development goal of the 5G mobile communications system is to closely connect with other wireless mobile communications technologies, to provide the omnipresent fundamental business capacity to the rapid development of mobile Internet. According to the present preliminary estimates of the industry, future wireless mobile networks including 5G, will promote service ability in three dimensions at the same time: (1) improve the utilization rate of resources by more than 10 times on the basis of 4G by introducing new wireless transmission technologies; (2) increase the system throughput rate by about 25 times by introducing new system structures (such as ultra-dense cells, etc.) and deeper intelligence capabilities; (3) further explore new frequency resources (such as high frequency band, millimeter wave and visible light, etc.), so that the frequency resources of future wireless mobile communications can be expanded to around 4 times. Therefore, the amount of the related data from 5G terminals is bound to increase drastically. It puts forward higher requirements on the 5G test systems: how to capture, analyze, store and manage the massive data in real time.

As mentioned in the previous section, the applications of massive MIMO in 5G has brought the need for the multi-channel RF testing. Similarly, massive data has put forward new requirements on the RF test in both the method and the equipment. At present, advanced testing instruments have begun to take these requirements into account. For example, the Keithley's new-generation MIMO test platform makes it simpler and cheaper to increase new signal standards and MIMO options. The platform includes the Keithley 2920 RF vector signal generator, the 2820 vector signal analyzer, the 2895MIMO synchronization unit and the Signal Meister waveform generation software. However, when faced with 5G, such devices are still far from enough. More advanced 5G test equipments are essential. Besides the requirements on equipment, the more suitable testing method for massive data computing is also one of the main research focus. At present, the heterogeneous computing architecture is a 5G technology with much application potential. Along with the rapid development of high bandwidth and high data rate of 5G mobile communications, the combination of heterogeneous computing architecture and multi-core parallel programming technology will be the indispensable main technology with which 5G test is able to deal with massive data processing and improve the parallel testing.

2.4 Summary

This chapter focuses on the theme of "testing technology evolution" to explain the importance of the testing technology in the industry chain, and guides the readers to review the whole development process of the testing instrument industry. Due to the development of the computer technology and the micro processing technology, the traditional testing instruments are continuously developing towards the bus-based, intelligent, distributed, modular and network directions. The improved testing instruments are applied to the wireless communications, which facilitate the test in the communications field. There is usually a system level test in the wireless field. The development of the testing technology has avoided the stack-based test methods needed for the system level test, simplified the test volume, and reduced the human error. Although the test instruments have made great progress, we still need to consider various requirements from the wireless field especially the future wireless communications technology field. The constantly increasing requirements are posing ever higher requirements on the communications technology. The communications technology is also developing rapidly. All of these are challenging the testing technology. This chapter analyzes the possible challenges as well as the future development trends of the testing technology. With the constantly emerging challenges, the testing technology will surely make great progress accordingly. On the basis of the existing testing technology, it will grow toward the healthy direction of the ecological test system.

References

1. X. You, Z. Pan, X. Gao, S. Cao and H. Wu. 5G Mobile Communication Development Trend and Some Key Technologies. China Science, Information Science, 2014, 44(5):551–563.
2. C. Wang, F. Haider, X. Gao, et al. Cellular architecture and key technologies for 5G wireless communication networks, IEEE Communications Magazine, 2014, 52(2):122–130.
3. E. Larsson, O. Edfors, F. Tufvesson, T. Marzetta, Massive MIMO for next generation wireless systems. IEEE Communications Magazine, 2014, 52(2):186–195.
4. J. G. Andrews, S. Buzzi, W. Choi, and et al. What Will 5G Be? IEEE Journal on Selected Areas in Communications, 2014, 32(6):1065–1082.
5. IMT-2020 (5G) Promotion Group, 5G Network Technology Architecture, 2015.

6. A. Gohil, H. Modi, S. K. Patel. 5G technology of mobile communication: A survey. Intelligent Systems and Signal Processing (ISSP), 2013, 288–292.
7. X. Ge; H. Cheng, M. Guizani and H. Tao, 5G wireless backhaul networks: challenges and research advances. IEEE Network, 2014, 28(6):6–11.
8. A. Osseiran, F. Boccardi, V. Braun and et al. Scenarios for 5G mobile and wireless communications: the vision of the METIS project. IEEE Communications Magazine, 2014, 52 (5):26–35.
9. P. Prinen. A brief overview of 5G research activities. International Conference on 5g for Ubiquitous Connectivity. 2014:17–22.
10. X. Shen. Device-to-device communication in 5G cellular networks. IEEE Network, 2015, 29 (2):2–3.
11. S. Chen, J. Zhao. The requirements, challenges, and technologies for 5G of terrestrial mobile telecommunication. IEEE Communications Magazine, 2014, 52(5):36–43.
12. N. D. Han, Y. Chung, M. Jo. Green data centers for cloud-assisted mobile ad hoc networks in 5G. IEEE Network, 2015, 29(2):70–76.
13. E. Dahlman, G. Mildh, S. Parkvall and et al. 5G wireless access: requirements and realization. IEEE Communications Magazine, 2014, 52(12):42–47.
14. W. Nam, D. Bai, J. Lee and I. Kang. Advanced interference management for 5G cellular networks. IEEE Communications Magazine, 2014, 52(5):52–60.
15. M. Condoluci, M. Dohler, G. Araniti, A. Molinaro and K. Zheng. Toward 5G densenets: architectural advances for effective machine-type communications over femtocells. IEEE Communications Magazine, 2015, 53(1):143–141.
16. O. Galinina, A. Pyattaev, S. Andreev and et al. 5G Multi-RAT LTE-WiFi Ultra-Dense Small Cells: Performance Dynamics, Architecture, and Trends. IEEE Journal on Selected Areas in Communications, 2015, 33(6):1224–1240.
17. F. Boccardi, R. W. Heath, A. Lozano and et al. Five disruptive technology directions for 5G. IEEE Communications Magazine, 2014, 52(2):74–80.
18. P. Ameigeiras, J. J. Ramos-Munoz, L. Schumacher and et al. Link-level access cloud architecture design based on SDN for 5G networks. IEEE Network, 2014, 29(2):24–31.

Chapter 3
Channel Measurement and Modeling

The wireless channel model is a deterministric or stochastic numerical description of physical propagation enviroments based on channel measurement or analysis and simulation on the basis of propagation theories. It provides an effective and simple means to approximately express the channel characteristics of the wireless transmission. Channel models can be used for link and system-level performance evaluation, optimization, and comparative studies on wireless communications technologies, algorithms, products and system. Channel model is of great importance to all wireless communications (including 5G mobile communication) research. This chapter focuses on the progress in 5G channel measurement and channel modeling. Section 3.1 introduces the requirements for 5G wireless channel model. Section 3.2 introduces and compares five commonly used channel modeling methods. Section 3.3 introduces the knowledge on channel measurement, channel measurement system and the currently ongoing 5G wireless channel measurement activities. Section 3.4 describes several channel parameter extraction algorithms and statistical analysis of channel parameters. Section 3.5 introduces several existing channel models. Section 3.6 gives an example to show the generating process of stochastic channel models. And the last section is the summary of this chapter.

3.1 Requirements for 5G Wireless Channel Model

The 5G typical scenarios will touch many aspects of life in the future, such as residence, work, leisure, and transportation, mainly involving dense residential areas, office, stadiums, indoor shopping mall, open-air festivals, subways, highways and high-speed rails. Compared with 3G/4G system, 5G adds some new application scenarios, which have diversified features like ultra-high traffic volume density, ultra-high connection density and ultra-high mobility. A variety of technologies are used to serve the end users, such as massive MIMO, millimeter wave

© Springer International Publishing AG 2018
Y. Yang et al., *5G Wireless Systems*, Wireless Networks,
DOI 10.1007/978-3-319-61869-2_3

(mmWave) communications, ultra dense networks, and D2D, etc. The application of these new technologies and the requirements for 5G mobile communications system present new challenges to the wireless channel models. The 5G channel models should support wide range propagation scenarios, higher frequency and larger bandwidth, as well as larger antenna array and so on, and keep consistency in space, time, frequency and antenna, which are mainly showed in the following aspects [1]:

1. Support wide range propagation scenarios and diverse network topologies

The vision of 5G is that anybody and anything can get access to information and share data with each other whenever and wherever they are. IMT 2020 5G promotion group of China has interpreted this as "Information a finger away, everything in touch" [2]. Current cellular mobile communications networks serve the static and mobile users with fixed base station(BS), so traditional channel model developed to serve for this type of applications.While under the 5G architecture, network topology should support not only cellular networks, but also the communication of device to device (D2D), machine to machine (M2M) and vehicle-to-vehicle (V2V) as well as the whole interconnected network.Accordingly, the 5G channel models should support mobile to mobile links and networks.

In D2D/V2V communications, the user antenna is generally as low as 1-2.5 m, which limits its coverage to be smaller than the ordinary cellular system. Besides, massive dense scatterers exist around both the transmitter and receiver. The stationary interval of channel is short due to the mobility of both ends. All of those are the unique characteristics of D2D/V2V transmission. Many channel measurement activities have been carried out in D2D/V2V, and several preliminary models have been presented. For example, METIS (an EU 5G R&D projects) proposed a D2D/V2V model with several scenarios [1], 3GPP proposed a preliminary D2D channel model [3], and MiWEBA (an EU-Japan joint project) proposed a D2D models in 60GHz mmWave band [4].

2. Support higher communications frequency and larger bandwidth

5G may operate in a larger frequency range from 350 MHz to 100 GHz. Taking into account the available bandwidth and communications capacity demand, the bandwidth of 5G system may be higher than 500 MHz, even 1~2 GHz or higher. Thus, the 5G channel models should also meets such requirements. In 3GPP-HF channel model, if the bandwidth of a channel is beyond c/D(wherein c is the speed of light, D is the aperture size of antenna), then such bandwidth can be refered to as Big Bandwidth and special processing should be applied [6]. High frequency (HF) band is brought into keen focus for its large available communications bandwidth. For example, METIS defines its medium and high priority frequency bands to develop, including 10 GHz, 28~29 GHz, 32~33 GHz, 43 GHz, 46~50 GHz, 56~76 GHz and 81~86 GHz [5]. HF band (6~100 GHz) has many radio transmission characteristics that are different from the frequency band below 6 GHz (sub-6 GHz) [4]:

Firstly, according to the Friis Formula of free space propagation, propagation in HF band will undergo higher path loss (PL) due to short wavelength. Friis Formula also points out that the antenna gain is proportional to the square of frequency. Given a fixed physical aperture size and transmit power, higher receive power will be obtained when propagation in HF band than in low frequency band [7]. Therefore, high gain directional antennas or beamforming technology should be adopted for HF band to ensure communications over several hundred meters distance. Meanwhile, the antenna also needs to be able to adjust its beam directing to the UTs.

Secondly, diffraction in HF band weakens due to shorter wavelength, or to say the propagation has quasi-optical characteristics, which results in lack of rich scattering in HF band than the sub-6 GHz band. In the line of sight (LOS) conditions, the received signal is comprised of LOS path and a few low-order reflection paths, whereas in the non-line of sight (NLOS) conditions, signal propagation may mainly depend on reflection and diffraction. Thus the channel shows temporal and spatial sparsity. Blockage of human or vehicles will attenuate the signal vastly. Quasi-optical properties also make it possible to use ray tracing technique to assist channel modeling.

Despite the fact that the propagation characteristics of mmWave have been widely investigated, especially in the 60 GHz, many important characteristics, such as blockage loss, penetration loss, high resolution angle characteristics, frequency dependence and outdoor mobile channel, still need further measurement and exploration. In the aspect of channel model, several HF channel models (including 60 GHz) have been proposed [1, 4, 6–12].

3. Support largeassive antenna array or massive MIMO

The existing channel models [13, 14] assume the superposition of plane waves bouncing from the scatterers (i.e. scatterers are in the the far field of transmitting and receiving), and assume that the antenna array is in smaller size. In both ends of the antenna array, the arrival or departure direction of the radio waves only lead to little difference in the phase of the signal while the amplitude of the signal remains constant, i.e. the propagation characteristics of the both ends of the antenna array are similar. When applying larger antenna technologies, including massive MIMO and pencil beamforming, the antenna number can reach tens to hundreds, which will bring about two impacts, as shown in Fig. 3.1.

Firstly, planar wavefront assumption for the conventional MIMO channels is no longer applicable and should be replaced with spherical wavefront assumption. The radio waves propagating in the form of plane wave need conform to the far field assumption, i.e. the distance between the transmitter/receiver between scaterers should be larger than the Fraunhofer distance [16], $R_f = 2D^2/\lambda$, D is the aperture size of antenna array, and λ is wavelength. D and corresponding Fraunhofer distance of array will increase with the number of antennas, assuming antenna element spacing and λ remain unchanged, i.e. the far field of antenna system expands. The receiver may be in the near field zone (also known as the Fresnel zone) of transmitter, or the scatterers will be in the near field zone of transmitter and receiver. In this case, the wavefront should be modeled as spherical wavefront instead of phane wavefront.

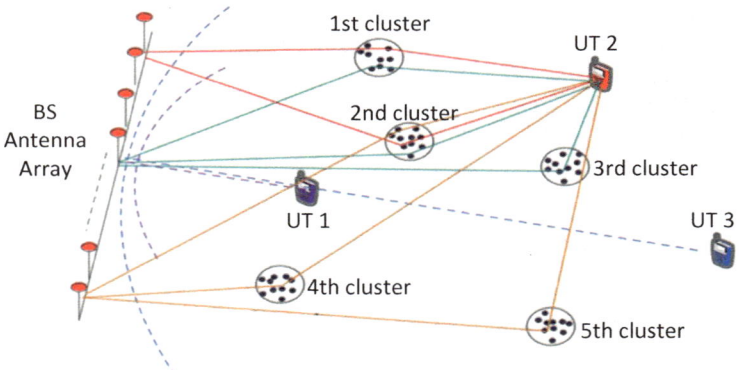

Fig. 3.1 Non-stationary and near field effect in Massive MIMO system [15]

Secondly, the researchers from the Lund University in Sweden have verified through measurement that any scattering cluster in the Massive MIMO systems can not affect all the antennas in the array. That is to say each antenna of the array has different scattering clusters which can affect its signal propagation. In traditional MIMO system, the scattering clusters that affect signal propagation only change with time due to the relative movement among the transmitter, receiver and scatterers Whereas in the massive MIMO system, scattering clusters also change along the axis of antenna array.

Additionally, as the large antenna array own high spatial resolution to distinguish closely located users in horizontal and perpendicular directions, the channel model must provide fine three-dimension (3D, includes both azimuth and elevation) angle information. The angular resolution should be at least one degree or less. The model should support a varity of antenna array, such as linear, planar, cylinder and spherical array. Therefore, for massive antenna array, we need to accurately model the azimuth and elevation angles at departure or arrival direction for each scattered multipath, as well as the distances or locations of first-bounce/last-bounce scatterer for spherical wavefront modeling. By far, several channel models have been proposed according to the aforementioned requirements [17, 18].

4. Support spatial consistency and dual mobility

Spatial consistency (continuity) has two meanings: one is that the channels of close links are of highly correlation, and the other is that the channel should evolve smoothly rather than discontinuity or interruption when transmitter/receiver moves or scenarios switch. The latter is commonly called the ability of dynamic simulation.5G communications network will contain a variety of link types, which will coexist in the same area, from the traditional macro cell, micro area, to picocell, and femtocell, as well as nomadic BS and D2D connection between UTs in the future. All of these ask 5G channel models to support spatial consistency.At present, most of the widely used channel models [13, 14] consider the spatial consistency of large scale parameters (LSPs), such as Path Loss, Shadow

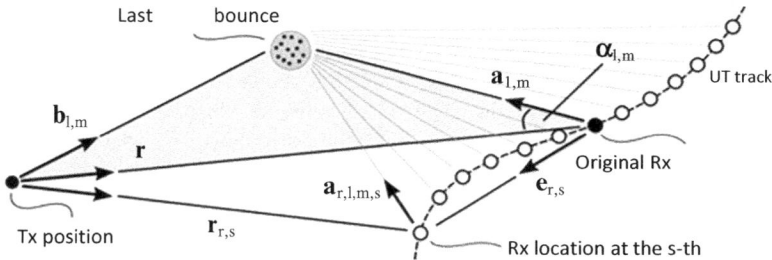

Fig. 3.2   Scatterer location and AOA update in dynamic simulation [19]

Fading (SF), Delay Spread (DS), Angular Spread (AS) and Ricean K-factor. However, the spatial consistency of small scale parameters (SSPs) is considered inadequately. Moreover, The LSPs is generated with snapshot randomly and assume the scattering environment of different snapshots are statistically independent and the scatterers seen from close mobile devices are not related, which leads to the exaggeration of the simulation performance of some multiple antenna technologies such as multiple-user MIMO (MU-MIMO) [1]. In addition, most traditional channel models consider the situation that one end of the link is fixed and the other end can move arbitrarily. New Doppler model should be introduced to support bidirectional mobility of D2D/V2V.

Withcoexistence and density increase of links, and with the application of D2D/V2V, to support spatial consistency becomes especially important for of wireless channel models. Space consistency model can be built through defining the scatterer's geometric position of the first bounce (the transmitter to scatterer) and the last bounce (scatterer to the receiver) of each scattering path. As shown in Fig. 3.2, QuaDriGa [19] channel model library supports dynamic evolution of the channel caused by the movement of user terminal (UT). The channel coefficients are calculated based on the location of the last bounce scatterer, the distance and relative angle of arrival (AOA) changes between it and MT, and the distance between mobile station and BS.

5. Support high mobility

Typical 5G high mobility scenarios involve mobile to mobile communication (e.g., V2V), and mobility to infrastructure communication (e.g., subway, high-speed rail, etc.). 5G should be able to support the mobility speed of above 500 km/h of high-speed rail, the ultra-high user density of 6 people /m^2 on subway, and the millisecond level end-to-end delay on highways. The channels in high mobility condition have the following characteristics.

Fast time-varying characteristics: Different from the low mobility channel, high mobility channels are fast time-varyingand show non-stationary characteristic. In other words, the channel can only be stationary in a shorter period of time. Large Doppler frequency shift or spectral spread: For either mobile to mobile channels, or mobile to fixed channels, either the transmitter or the receiver is in high-speed mobility.

More severe fading: The received signal is superposition of a number of multipath signals with different amplitudes, phases, frequency shifts and delays. The phase of each path is related to carrier wavelength, delay and Doppler shift. The small path delay will bring great phase change with the increase of carrier frequency. The signal phase caused by Doppler shift changes over time. Therefore, superposition of those multipath signals with opposite phases may lead to severe signal fading in some locations.

Mixed propagation scenarios: High-speed rail will encounter one or several scenarios when it runs, mainly including open area, cutting, viaduct, tunnel and station. Subway will be faced with two main scenarios: tunnel and platform. Automobile will meet a variety of scenarios in driving process, including city blocks, expressway, tunnels, bridges, and so on. These require the channel model to support smooth evolvtion in switching scenarios.

At present, the existing channel models cannot precisely describe the characteristics of the wireless channel in high-speed rail scenario. LTE-A system provides a channel model that can support the two high speed rail scenarios, i.e. open area and tunnel. However, the non-stationarity of the channel is not considered. WINNER II model considers the high speed rail at the speed of 350 km/h and the impacts caused by mobile relay, but it still assumes the channel is stationary. A few Some measurement have been done in this area, such as the high speed rail/subway measurement by State Key Lab of Rail Traffic Control & Safety in China, the vehicle measurement by Lund University in Sweden and the measurement of high-speed mobile scenarios by WINNER II project. As the high mobility increases the difficulty and cost of measurement, more and more channel modeling is carried out by means of ray tracing technology.

3.2 Channel Modeling Method

Channel modeling can be broadly split into deterministic and stochastic channel modeling or their combination. When detailed environment data (including man-made objects such as houses, buildings, bridges, roads, etc. as well as natural objects such as foliage, rocks, ground, etc.) is available to a sufficient degree, wireless propagation is a deterministic process that allows predicting its characteristics at every point in space. This is the basis of deterministic channel model and Ray Tracing is a such method.Stochastic channel modeling is one kind of method to obtained huge measurement data set that containing the underlying statistical properties of wireless channel by conducting channel measurements in a large variety of locations and environments, from which path parameters are extracted and used to statistically represent the channel parameters. Stochastic channel models can be roughly divided into several categories as Geometric-based Stochastic Channel Model (GSCM), Correlation-based Stochastic Channel Model (CSCM), extended Saleh-Valenzuela (SV) stochastic channel model, and Ray Tracing based Channel Model. GSCM is also known as Geometric-Based

Stochastic channel Model (GBCM) in some literature. There are two types of GSCM. One is constructed based on the channel measurement, and the other is based on the regular geometric shape of scatterers. Their model structure are same, but the former is based on the measured data to extract the path parameters, and obtain statistical characteristics through data analysis. The latter places emphasis on deriving the autocorrelation functions, cross-correlation functions and other statistical characteristics of the spatial, temporal and frequency domain of the channel.

Channel Parameters are usually divided into Path Loss (PL), large-scale parameters (LSPs, such as shadowing, delay spread, angular spread, etc.) and small-scale parameters (such as delay, angle of arrival and departure, etc.), which jointly reflect the channel fading characteristics. Path loss is usually expressed in one or two formulas and a set of numerical values of parameters, reflecting the relationships with transmission environment, distance and frequency and so on. Large-scale parameters can be regarded as a statistical average in a channel segment (or a quasi-static channel area), within which the LSP or probability distribution of LSPs do not change significantly. The length channel segment is dependent on the propagation environment. It is about dozens of wave-length. Small-scale parameter is used to describe statistical characteristics of path parameter in a wavelength range. Path Loss and Shadowing Model.

Friis formula gives the signal propagation model in free space. For the actual scenarios, a more general path loss model is constructed by introducing a path loss exponent dependent on environment. For LOS condition, a common path loss model can be presented as [12]

$$PL(dB) = 20\log_{10}(4\pi f/c) + 10n\log_{10}(d/1m) + X_\sigma \qquad (3.1)$$

This is called Close-in Reference short (CI) model [20], where f is the carrier frequency (Hz), c is the speed of light, d is the distance between transmitter and receiver (m). The first item at right side of the equation is the path loss in free space at the distance 1 m. Xs is shadowing. The relationships between path loss and frequency is same with Friis formula in free space. For NLOS, another model is usually been used,

$$PL(dB) = 20\alpha\log_{10}(d) + \beta + 10\gamma\log_{10}(f) + X_\sigma \qquad (3.2)$$

This equation is called ABG model [12] (named by three factors Alpha, Beta and Gamma). It is the extension of FI (Floating Intercept) model [20] to reflect frequency relevance.

3.2.1 Measurement-Based GSCM

MIMO channel is the superposition of rays with different power, delay and angular information which can be determined by the geometry among the transmitter (Tx), the receiver (Rx) and the scatterers. The rays with similar delays and angles are

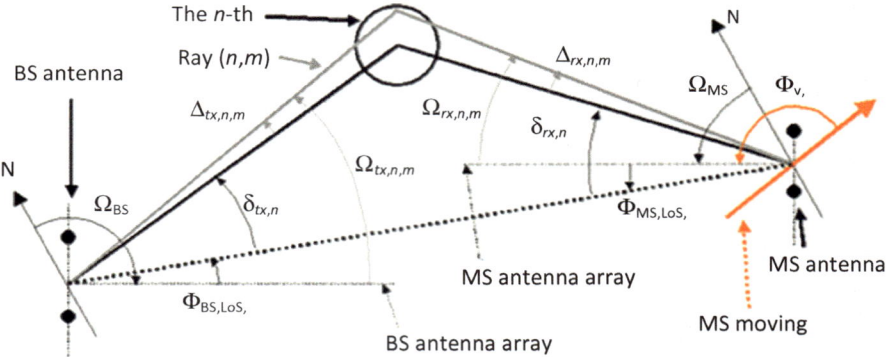

Fig. 3.3 Clusters and rays of GBSM channel model

grouped into one cluster [21], so that the channel can be described by a cluster-ray structure as shown in Fig. 3.3. The GSCM modeling method separates antennas and propagation channel. The user can combine arbitary type and layout of antennas with the propagation channel model to obtain transmission channel model. The whole 3D MIMO channel model from transmitting antenna array to receiving antenna array is composed of the sub-channel $H_{u,s}(t;\tau)$ from transmitting antenna element s to receiving antenna element u. $H_{u,s}(t;\tau)$ can be expressed as

$$H_{u,s}(t;\tau) = \sum_{n=1}^{N}\sum_{m=1}^{M_n} \begin{bmatrix} F_{rx,u,\theta}(\Omega_{rx,n,m}) \\ F_{rx,u,\varphi}(\Omega_{rx,n,m}) \end{bmatrix}^T \begin{bmatrix} \alpha_{n,m}^{\theta\theta} & \alpha_{n,m}^{\theta\varphi} \\ \alpha_{n,m}^{\varphi\theta} & \alpha_{n,m}^{\varphi\varphi} \end{bmatrix} \begin{bmatrix} F_{tx,s,\theta}(\Omega_{tx,n,m}) \\ F_{tx,s,\varphi}(\Omega_{tx,n,m}) \end{bmatrix}$$
$$\times e^{j2\pi\left(\overline{\Omega}_{rx,n,m}\cdot \bar{d}_{rx,u}\right)/\lambda} e^{j2\pi\left(\overline{\Omega}_{tx,n,m}\cdot \bar{d}_{tx,s}\right)/\lambda} e^{j2\pi v_{n,m}t}\delta(\tau - \tau_{n,m}) \quad (3.3)$$

Where N is the number of clusters between antenna element u and s. M_n is the number of ray in the n-th cluster. $F_{tx,s,\theta}$ and $F_{tx,s,\varphi}$ are the radiattion patterns of u for vertical and horizontal polarization respectively, $F_{tx,s,\theta}$ and $F_{tx,s,\varphi}$ are the elevation angle and azimuth radiation field patterns of s respectively. More commonly, vertical direction is called elevation (offset from horizontal plane) or zenith (offset from the vertical upwards z-axis) and horizontal direction is called azimuth. $[\alpha_{n,m}^{\theta\theta} \; \alpha_{n,m}^{\theta\varphi} \; \alpha_{n,m}^{\varphi\theta} \; \alpha_{n,m}^{\varphi\varphi}]$ are the complex gains of four polarization pairs of the m-th ray in cluster n (using $ray(n,m)$ for short), namely elevation-elevation ($\theta\theta$), elevation-azimuth ($\theta\varphi$), azimuth-elevation ($\varphi\theta$) and azimuth-azimuth ($\varphi\varphi$). The AOA ($\Omega_{rx,n,m}$) of ray (n,m) contains elevation angle of arrival (EOA) $\theta_{rx,n,m}$, and azimuth angle of arrival (AoA) $\varphi_{rx,n,m}$. AoD ($\Omega_{tx,n,m}$) contains elevation angle of departure (EOD) $\theta_{tx,n,m}$, and azimuth angle of departure (AOD) $\varphi_{tx,n,m}$. $\overline{\Omega}_{rx,n,m}$ and $\overline{\Omega}_{tx,n,m}$ are the unit vectors of AoA and AoD respectively. $\bar{d}_{tx,s}$ and $\bar{d}_{rx,u}$ are the location vector of s and u. λ is the wavelength of carrier. $v_{n,m}$ and $\tau_{n,m}$ are the Doppler shift and propagation delay of ray n,m. If the wireless channel is dynamic, all the channel parameters are time-varying, i.e. the function of t.

The clusters and rays in channel is parameterized by the path loss, shadowing and other parameters in large scale and small scale. In order to keep spatial consistency, the channel model considers the correlation of different LSPs in the same station and the correlation of same LSP in the different stations. Cluster parameters (including the number, arrival rate, power decay rate, angular spread) and ray parameters within clusters (arrival rate, the average time of arrival, power decay rate, etc.) are important parameters of the model, the existing GSCM usually assumes the numbers of the clusters and the number of rays within one cluster and are fixed, arrival interval of clusters follow exponential distribution, clusters power decrease exponentially with delay, and cluster angles follow wrapped Gaussian or Laplace distribution. The rays within one cluster have the same delay and power, and different angles.

GSCM can adapt to the different scenarios and different types of antennas easily by using rational parameters. For some new scenarios, like different high-speed rail scenarios, such as the plains, hills, etc., there's no need to build any new model. One only need to measure the channel of specific scenario and then update scenario-related parameters and parameters of clusters and rays in the model. The common GSCM modeling steps can are divided into three stages as illustrated in Fig. 3.4 and described below [13].

Stage 1: preparation and measurement. Determine the generic channel model formula and the parameters to be measured. Draw up a detailed measurement plan, take into account all aspects of measurement: first, determine the scenarios of wireless channel; select the measurement environment; determine antenna height, speed of transmitter and receiver, the placement of channel sounder, measurement routes and link budget, measurement time and other general requirements. Then make actual measurements and store measurement data onto a mass memory.

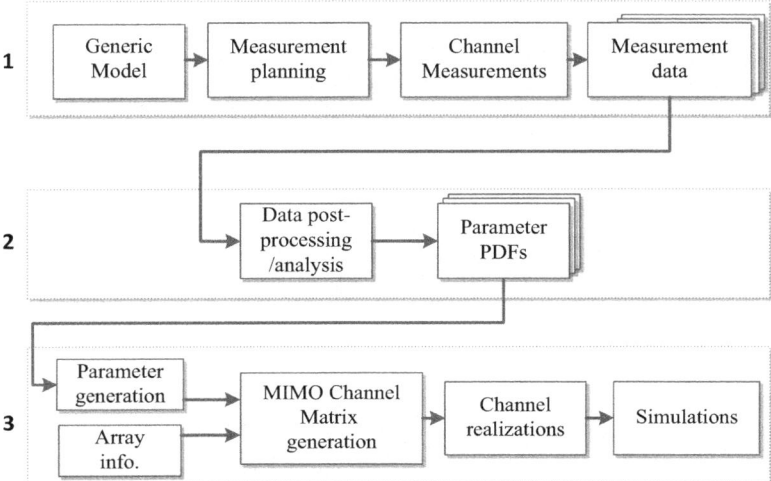

Fig. 3.4 GSCM basic modeling steps

Frequency response of the system RF chains and the radiation pattern of antenna system are required to be measured and used in system calibration, subsequent data processing and simulation.

Stage II: post-processing of the measured data. Different analysis methods are applied depending on the required parameters. Typically, high-resolution parameter estimation algorithm, such as Expectation-Maximization (EM), Space-Alternating Generalized EM (SAGE), RIMAX or other algorithms, is used to extract path parameters from the measured data. Output of data post-processing could be, e.g., a set of impulse responses, path-loss data, or multidimensional propagation parameters, such as gain or polarization gain, delay, AoA, AoD, as well as dense scattering multipath components (DMC) and so on. The rays with similar parameters are grouped into one cluster, so that the parameters are divided into inter-cluster ones and intra-cluster ones. Make statistical analysis on the post-processed data to obtain probability distribution functions (PDFs) and the corresponding numerical values of statistical parameters. The Goodness of Fit (GoF) tools may be used to select the optimal distribution function among several candidates.

Stage III: the generation of simulation model. At first, clusters and ray parameters are generated according to their PDFs and statistical parameters. Then MIMO transmission matrix is obtained by combining the generated parameters with antenna information. At last, the time-varying Channel Impulse Response (CIR) is generated to be used in simulations.

Measurement-based GSCM modeling method has been widely recognized and used, such as COST 259/273/2100, SCM, SCME, WINNER /I/II/+, IMT-Advanced, IEEE 802.16m, 3GPP 3D MIMO, 3GPP D2D, QuaDRiGa, mmMAGIC, 5GCM, METIS and so on. Many new 5G channel model is modified upon the basis model, e.g. by introducing correlation of cluster number N and Tx/Rx (u/s) position to model the visibility of scatterers by antennas, by introducing the correlation between $\Omega_{tx,n,m}$ ($\Omega_{rx,n,m}$) and $s(u)$ position to support the spherical wavefront propagation, by adding a another Doppler shift to characterize dual mobility and so on. QuaDRiGa model library supports the realization of dynamic simulation (time evolution) which depends on determing the locations of first-bounce and last-bounce scatterers. With the locations of scatters, it is easier to model spherical wavefront, spatial non-stationarity in Massive MIMO.

3.2.2 Regular-shape GSCM

The scatterers distribute randomly in the actual environment. Whereas, the regular-shape GSCM distributes scatterers with similar properties onto regular geometric shapes for simplifying characteristics analysis of the MIMO channel. The characteristics of the propagation channel are described according to the geometric relationships among the transmitter, receiver and scatterers. Only one-order or two-order reflections during the propagation process are taken into account, while high-order reflections are ignored because of higher attenuation. As shown in

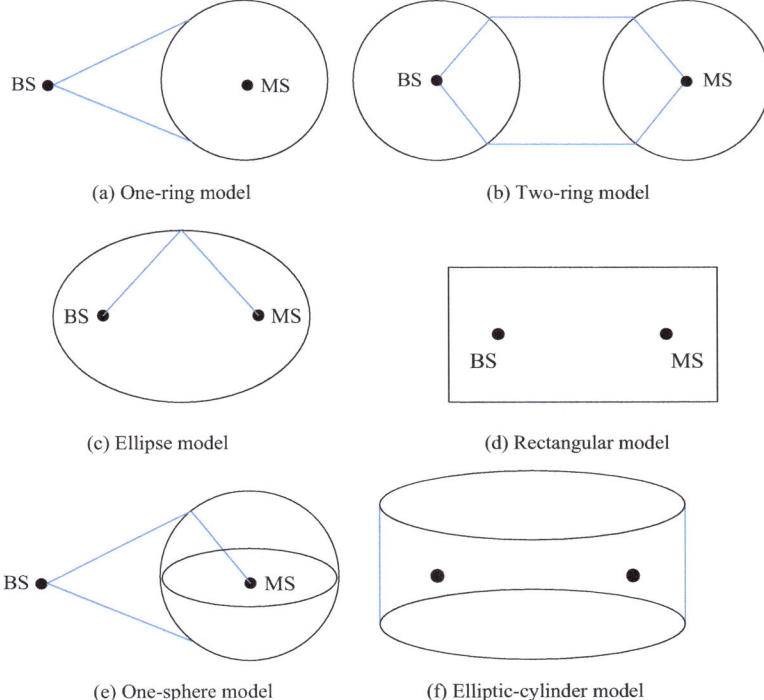

(a) One-ring model   (b) Two-ring model

(c) Ellipse model   (d) Rectangular model

(e) One-sphere model   (f) Elliptic-cylinder model

Fig. 3.5 Scatterer regular geometric sharpe-based model

Fig. 3.5, the basic shapes of 2D distribution of scatterers mainly includes: one-ring, two-ring, ellipse and rectangular [22]; the basic shapes of 3D distribution of scatterers mainly includes one-sphere, two-sphere, and elliptic-cylinder etc.. The geometric model in practical application can be a combination of these basic shapes, and it is also possible to add new geometric shapes.

The regular-shape GSCM is a parametric modeling method with high flexibility. Different channel characteristics can be described accurately by adopting combination of several geometry shapes and adjusting the related parameters. For some 5G typical scenarios and channel characteristics, Chengxiang Wang's team at the Heriot-Watt University has presented a series of channel models based on this method, mainly including: the modeling of Massive MIMO, M2M as well as High Speed Rail (HSR) [17, 23, 24]. For massive MIMO channel modeling, S. Wu et al. built a 2D massive MIMO channel model based on non-stationary twin clusters, from which space-time correlation function and power spectrum density are investigated [17]. Spherical wavefront was assumed and its impact on the statistical properties of the channel model was investigated. Birth-death process was incorporated to capture the dynamic properties of clusters on both the array and time axes. In channel modeling for V2V scenario, Yi Yuan et al. extended the geometric model represented as two circles and an eclipse in 2D scenario into two spheres and an ellipse cylinder in 3D scenario, which could be used to simulate the propagation environment of the V2V channel [23]. In addition, the Chanel non-stationarity and the influence of vehicle density on the channels were also taken into account. Ghazal et al. proposed a non-stationary elliptical channel model of the high speed rail and analyzed its related statistical characteristics [24]. The advantage of this model was that it could be adapted to different high-speed rail scenarios (except tunnel) by adjusting several parameters.

3.2.3 CSCM

CSCM describes the MIMO channel by the correlation matrix of antennas. It is favoured due to ease of use for analysis of MIMO capacity and system performance. Assuming that the number of receiving and transmitting antennas are N_r and N_t respectively, the MIMO channel matrix H is a N_r by N_t matrix, and the full correlation matrix of the channel R_H is a (N_r*N_t) by (N_r*N_t) matrix. Usually it is very hard to use R_H directly to generate channel. The reason is that it is too large and needs to specify $(N_r*N_t)^2$ values, and each elements in the R_H is difficult to correspond to the physical propagation channel, so that the channel correlation matrix model with low complexity is expected. The existing CSCM model mainly includes: independent and identically distributed (IID) channel model with minimum complexity, Kronecker product based CBSM (KBSM), VCR [25], and Weichselberger channel model [26].

IID model assumes that all Tx antennas and Rx antennas are independent. Obviously this assumption does not in line with the actual situation. This model is commonly used to show the performance advantage of MIMO than SISO. KBSM assumes that Tx and Rx have independent correlation of antennas, namely ignoring the coupling between Tx and Rx antennas. MIMO wideband access system (IEEE 802.11 n, 802.16 m/ac) using this model. Both VCR and Weichselberger model consider the coupling between Tx and Rx, but both require the calculation of eigenvalue decomposition. It is quite cumbersome to use, especially in the application of large-scale antenna array. Therefore some more simplified models have been proposed. For example, J. Hoydis et al., proposed a CBSM for the configuration that the BS is equipped with massive antennas and the user has a single antenna [27]. When the antennas number in MIMO Massive system increase extremely, the mutual coupling between antenna elements cannot be ignored. Masouros et al. propose a new model which considers the effect of antenna impedance, load impedance, and mutual coupling impedance matrix [28]. For more detailed analysis of antenna coupling, please refer to [29–31] and their references.

IEEE 802.16m document has given two specific calculation methods to convert GSCM parameters to antenna correlation matrix. One method is to use 20 sub-paths of each cluster to approximate the Power Angle Spectrum (PAS) with shape of Laplacian function, and to calculate the antenna correlation matrix by using AOD/AOA parameters and antenna geometric relations [32].

3.2.4   Extended SV Model

R. Valenzuela and A. M. Saleh of the AT&T Bell Labs proposed the SV model by analyzing the indoor channel measurement data [33]. The channel is composed of multiple clusters and rays. The number of clusters and the number of rays within a cluster obey Poisson distribution, namely the arrival of the cluster and the intra-cluster rays are two Poisson processes with different arrival rates. The arrival intervals of clusters or intra-cluster rays obey exponential distribution with different means. The average power of the cluster or intra-cluster ray decay exponentially with delay. The instaneous power of the clusters or intra-cluster rays obey Rayleigh distribution or lognormal distribution. The initial SV model can only describe the delay information of the channel, and its resolution is higher than 10ns subjected to the sounding equipments. After that, scholars continued to improve the time resolution of the model to a nanosecond or less with the help of many measurement in ultra wide band and mmWave band. Moreover, angle information and polarization characteristics are also integrated to develop the extended SV model. Hereinafter the IEEE 802.11ad channel model is taken as an example to show the detail of the extended SV model. The CIR, h, can be expressed as [9]

$$h(t, \varphi_{tx}, \theta_{tx}, \varphi_{rx}, \theta_{rx}) = \sum_i H^{(i)} C^{(i)} \left(t - T^{(i)}, \varphi_{tx} - \Psi_{tx}^{(i)}, \theta_{tx} - \Theta_{tx}^{(i)}, \varphi_{rx} - \Psi_{rx}^{(i)}, \theta_{rx} - \Theta_{rx}^{(i)} \right)$$

$$C^{(i)}(t, \varphi_{tx}, \theta_{tx}, \varphi_{rx}, \theta_{rx}) = \sum_k \alpha^{(i,k)} \delta \left(t - \tau^{(i,k)} \right) \delta \left(\varphi_{tx} - \varphi_{tx}^{(i,k)} \right)$$

$$\cdot \delta \left(\theta_{tx} - \theta_{tx}^{(i,k)} \right) \delta \left(\varphi_{rx} - \varphi_{rx}^{(i,k)} \right) \delta \left(\theta_{rx} - \theta_{rx}^{(i,k)} \right)$$

$$(3.4)$$

Where t, φ_{tx}, θ_{tx}, φ_{rx}, θ_{rx} are time, azimuth angle of departure, elevation angle of departure, azimuth angle of arrival, elevation angle of arrival, respectively. $H^{(i)}$ is a 2×2 polarization matrix. $C^{(i)}$ is the CIR of the i-th cluster. $\delta(\cdot)$ denotes Dirac impulse function. $T^{(i)}$, $\Psi_{tx}^{(i)}$, $\Theta_{tx}^{(i)}$, $\Psi_{rx}^{(i)}$, $\Theta_{rx}^{(i)}$ are the delay, arrival and departure angle of the i-th cluster respectively. $\alpha^{(i,k)}$ is the amplitude of the k-th path in the i-th cluster. $\tau^{(i,k)}$, $\varphi_{tx}^{(i,k)}$, $\theta_{tx}^{(i,k)}$, $\varphi_{rx}^{(i,k)}$, $\theta_{rx}^{(i,k)}$ are the delay and angle value of the k-th path in the i-th cluster relative to the central ray within the cluster respectively. The intra-cluster rays are divided into forward part and backward part, represented by subscript f and b respectively. The distributions of all of the parameters are as follows.

$$p\left(T^{(i)} | T^{(i-1)} \right) = \Lambda \left[-\Lambda \exp \left(T^{(i)} - T^{(i-1)} \right) \right], l > 0$$

$$p\left(\tau_{\substack{f \\ b}}^{(i,k)} \middle| \tau_{\substack{f \\ b}}^{(i,k-1)} \right) = \lambda_{\substack{f \\ b}} \left[-\lambda_{\substack{f \\ b}} \exp \left(\left| \tau_{\substack{f \\ b}}^{(i,k)} - \tau_{\substack{f \\ b}}^{(i,k-1)} \right| \right) \right], k > 0 \qquad (3.5)$$

$$\overline{\left|\alpha_{f/b}^{(i,k)}\right|^2} = \overline{\left|\alpha^{(0,0)}\right|^2} K_{f/b} \exp\left(-T^{(i)}/\Gamma\right) \exp\left(-\tau_{f/b}^{(i,k)}/\gamma_{f/b}\right)$$

The arrival of clusters and intra-cluster rays obey the Poisson distribution, where Λ and $\lambda_{f/b}$, respectively are the arrival rate of clusters and arrival rate of intra-cluster forward/backward rays. Γ and $\gamma_{f/b}$ respectively are the power decay rate of the clusters and the forward/backward rays. The average power of the k-th path in the i-th cluster is denoted as $\overline{\left|\alpha_{f/b}^{(i,k)}\right|^2}$, which is also the mean of instantaneous rays' power that obey Rayleigh distribution or lognormal distribution. $K_{f/b}$ denotes the Ricean K-factor of forward/backward rays. For the angle of clusters and intra-cluster rays, different models have different assumptions. IEEE 802.15.3c deems the angles of clusters are uniformly distributed between 0 and 2π; the azimuth angles of rays within a cluster follow Gaussian distribution or Laplacian distribution [8]. IEEE 802.11ad uses ray tracing method to determine the azimuth and zenith angle of clusters. The azimuth and zenith angles of the intra-cluster rays are Gaussian distributed with zero mean and variance of 5 degrees [9]. The extended SV model is also independent of the antenna just as GSCM, so the MIMO channel model can be obtained by applying specific antenna information.

3.2.5 Ray Tracing based Model

The propagation prediction of electromagnetic wave can be solve by analytical method based on electromagnetic theory or by corresponding computational electromagnetics (MOM, FDTD, etc.). However, both methods are hard due to high complexity under the actual propagation environment. Ray Tracing (RT) method which based on relatively simple Geometrical Optics (GO), Geometric Theory of Diffraction (GTD) and Uniform Theory of Diffraction (UTD) is therefore used to predict the propagation properties of radio wave. RT based channel modeling have high flexibility which channel measurements cannot achieve, and it can support modelling for D2D, massive MIMO, arbitrary frequencies, wide frequency band, and spatial consistency. Its basic principle is to use rays to simulate the electromagnetic wave propagation process, such as direct ray, reflection, scattering, and diffraction. By collecting the geometric dimension and electromagnetic properties including complex relative permittivity and conductivity of materials, the propagation environment can be constructed into a 2D or 3D digitized map composed of geometric structures and objects. Searching the propagation paths from the transmitter antenna to receiver antenna, RT method calculates path gain and shadowing using classical electromagnetic theory, and further obtain other channel parameters including path delays, angle, and CIR using antenna radiation pattern [34]. Calculation the field strength of rays and path tracking of rays are two main aspects of Ray tracing technique.

RT treats the propagation of electromagnetic wave by means of rays propagation. Based on three well-built propagation theory of optics (GO, GTD, and UTD), RT can describe the basic propagation behavior of rays, such as line of sight, reflection, transmission, diffraction and diffuse scattering., and calculate the electric field intensity contribution of each ray. GO is only suitable for direct ray, reflection and transmission of electromagnetic waves. It fails for diffraction arising when electromagnetic waves encountering discontinuous surfaces of objects. Keller presented the GTD theory by classifying the diffraction as edge diffraction, wedge diffraction and curved surface diffraction, and gives the formula to calculate diffraction coefficients under the three conditions. The field strength of diffraction wave can be obtained through multiplying the field strength of incident wave by the diffraction coefficient at the diffraction point. While GTD theory has the discontinuity problem of electromagnetic field in the transition region of diffraction, namely between the bright and dark region of diffraction objects. The Uniform Theory of Diffraction (UTD) proposed by Pathak and KouyoumJian overcomes the disadvantage [35].

Ray tracing can be divided into forward and backward algorithms. The forward algorithms include test ray method and Shooting and Bouncing Ray (SBR). The backward algorithms include mirror method, minimal light path method, etc. The most widely used SBR method needs to emit a large number of dense three-dimensional cone ray tubes into different directions, covering the entire three-dimensional spherical space with the transmitting source as its center. Each ray tube is ergodic of all obstacles in scenario, and undergoes reflection, diffraction propagation until the reception point is contained in the ray tube. Mirror method bases on geometric optics. The source and the reception antennas are regarded as points in geometry, and the mirror point on the reflection plane that corresponds to the source point is found through the plane symmetry method, and then the propagation path of rays in space is obtained. The complexity of the mirror method can increase exponentially with the increase of the complexity of the scenario or the order of mirror images, so it can only be applied to the scenarios with a few obstacles and simple geometric shapes. It cannot describe and calculate diffraction and scattering.

Adding one-order or two-order diffuse scattering into RT, namely so called effective roughness (ER) approach, can model the dense multipath component (DMC) and obtain PDPs which matched well with measurements [36]. Since RT does not need high-cost channel measurement devices and hard measurement compaigns, it have gathered great interest. Several novel 5G channel model were proposed by combining the RT model and the statistical model based on measurement, e.g., the Q-D model proposed by MiWEBA [4]. In the IEEE 802.11ad/ay and MiWEBA model, RT modeling method is used to generate the cluster parameters of 60 GHz mmWave channel, and the inter-cluster parameters are obtained by channel measurements. METIS project first proposed a systematic channel model based on RT, which considered the reflection, diffraction, diffuse reflection, blockage and other propagation mechanisms, and claimed that the model could satisfy all of the requirements of 5G system channel models (see Table 3.1 or Table 4.1 [1]).

Table 3.1 Channel modeling methods adopted by a variety of channel models

GSCM	3GPP SCM/3D/D2D, SCME, WINNER(I, II and +), IMT Advanced, COST 259/273/2100/IC1004, IEEE 802.16m, QuaDRiGa, METIS(GSCM)
CSCM	IEEE802.11n/ac, IEEE 802.16m
Extended SV	IEEE802.11n/ac, IEEE 802.15.3c, IEEE 802.11ad/ay, MiWEBA
RT	IEEE 802.11ad/ay, METIS(Map), METIS(GSCM with Point cloud field prediction), MiWEBA

3.2.6 Comparison of Modeling Methods

There are pros and cons for GSCM, CSCM, Extended SV and RT modeling methods. The first three types could separate the antennas from the propagation channels, thus the transmission channel can be obtained by applying any type of antennas with the propagation channel. GSCMs and extended SV model use a cluster-ray structure to describe the propagation channel. Each cluster is composed of multiple rays, and the rays within one cluster have similar amplitude, phase, AOA and AoD. The RT channel model is expressed with multiple rays. In CSCM, each tap has its antenna correlation matrix, which may restrain the angle information of rays. Specific antenna correlation matrics are required to be calculated for different configurations and radiation pattern of the antennas, which make the model lack of flexibility.

By rational parameterization, GSCM can describe different scenarios accurately and flexibly. GSCM model does not directly output the antenna correlation matrix. But if needed, it can be obtained from channel path parameters [32]. The accuracy of RT based model is closely related to the quality of electronic maps built for the enviroments. Some notable scatterers (e.g., buildings, desk, etc.) are easy to be modeled, but small irregular objects (such as trees, shrubs, and street lamp) may be ignored or cannot be accurately modeled which leads to improper results. MiWEBA model need channel measurement to obtain the statistical features of small scale parameters. Extended SV is suitable for channel modeling with high bandwidth and high delay resolution. It could give intuitive and accurate description of densely arrival of broadband signals in the indoor and outdoor environment. However, SV model lacks the support for spatial consistency.

The advantage of CSCM model is that it can provide spatial correlation matrix which make analyzing MIMO theoretical performance easier. However, it has some disadvantages. It makes over-simplification on the true channel. There is big difference in channel correlation matrices for LOS and NLOS, which make smooth transiting from each other impossible. Correlations in time-domain and space-domain are independent, which make the possible existed temporal-spatial correlation be hard to describe. The characteristics of LSPs, such as DS and AS changing over time and locations, are difficult to be captured in the model, because these parameters are hidden in the channel correlation matrix. CSCM is difficult to model the spatial non-stationary phenomenon and the spherical wavefront in near field propagation.

In simulation, the computational complexity of a channel model is very important. When talking about the channel modeling method, people used to consider that CSCM has lower complexity and higher computational efficiency than GSCM because of its relatively simple concept. The complexity of WINNER model (GSCM) and CSCM were revealed and compared in [37]. The number of GSCM sub-paths was set to 10 or 20, and the tap with a certain Doppler spectrum were generated by 8-order IIR filters in CSCM. It was found that the complexity (represented as the number of real calculations at each tap) of them is the same order of magnitude when the number of MIMO antennas is small. When the number of MIMO antennas is larger (>16), the complexity of CSCM is surprisingly higher. RT calculation is too complicated to generate the CIRs in link-level and system-level simulation in real-time. Off-line calculations of CIR on each UT location are required to establish a CIRs database for simulation, which will ask for huge storage resources. In addition, the computational quantity of channel convolution in simulation is very large and usually several times of that of CIR generation.

In addition to the above several channel modeling methods, recently, some scholars presented propagation graph theory for channel modeling. It takes the positions of transmitter, receiver, and scatterers, and propagation coefficient between them as input, establishes three transmission matrics for expressing the propagation from transmitter to the scatterers, between scatterers, and from scatterers to receiver, and calculates the channel matrix analytically. The location of the scatterers can be gained by measurement or ray tracing technology and propagation coefficient can be obtained by theoretical calculation based on GO, GTD, UTD. ER method can be combined with propagation graph to model the channel and obtain the PDP conformly alined with the measurement [38]. Compared with the ray tracing, propagation graph method is less computational, so that it can generate random channel samples in real-time.

The channel modeling methods used by a variety of standardized organizations or project teams are shown in Table 3.1. From the table, we can find that GSCM was widely used in 3G / 4G channel modeling. With the progress of 5G R&D, more and more groups combine ray tracing and measurement to obtain the channel models, such as METIS, MiWEBA, 5GCM channel models and son on.

3.3   Channel Measurement

Channel measurement, also known as channel sounding, is a kind of research activity which aims at extracting the channel parameters for acquiring the knowledge of wireless channel. Channel measurement is a direct and most important mean to obtain channel information and understand the characteristics of the channel. The transmitter emits sounding signals to the receiver through the wireless channel to be investaged. By exploiting the received signals, multiple propagation paths can be identified. The parameters of every resolvable path are outputted, with which a variety of important channel statistical information can be obtained, such as the angular distribution, coherence time, DS, etc. Sometimes it is enough to use the measurement results for transmission performace analysis. But more often, it is better to make a general or representative description of a certain type of propagation scenario. In this case the channel models are needed, and channel measurement can provide measurement data for channel modeling and verify the effectivity of the model.

The differences between the channel measurement methods exist in many aspects, such as the form of the stimulating signals, the number of RF channels, the types and number of antennas, etc. This section is going to introduce various measurement methods in a unified framework, and several common channel sounding systems in academia and industry, as well as various channel measurement activities.

3.3.1 Channel Measurement Methods

Large bandwidth and multiple antennas are the main features of 5G, so in this subsection we take the broadband MIMO channel as the sounding target. According to the numbers of channels being measured simultaneously, channel measurement methods can be divided into Single Input Single Output (SISO) [39–43], Multiple Input Multiple Output (MIMO), and the compromising Single Input Multiple Output (SIMO) [44]. MIMO method simultaneously measures multiple channels in parallel. It has the highest measurement speed and can be applied to the high-speed mobility scenario. However, multiple parallel RF units are required for MIMO measurement method. Besides the complex structure and high cost, the inherent requirements of consistency and synchronization increase the difficulty of system implementation and calibration. In addition, multiple concurrent sounding signals can interfere with each other. Even when orthogonal sequences are used as the excitation signals, under multipath channel conditions, the orthogonality between these sequences with different delay is hard to be guaranteed. The interference will still be produced, so that a more complex processing algorithm is needed. SISO only needs one RF chains at the transmitter and the receiver respectively, and adopts the Time Division Multiplexing (TDM) mode to sounding the subchannel iteratively. There is one electronic switch box at both ends to synchronously switch different antenna element connecting to the RF chain. This method has relatively simple structure and low cost. The number of antennas at transmitter and receiver can be flexibly configured. Because there is only one RF chain, the calibration and consistency between the multiple channels can be achieved much easier. In addition, since only one probe signal is transmitted, there is no demand for the orthogonality. However, its measurement time is generally longer, which is not suitable for the measurement of high mobility scenarios. The switch boxes on transmitter and receiver act according to a designed sequence. For example, the

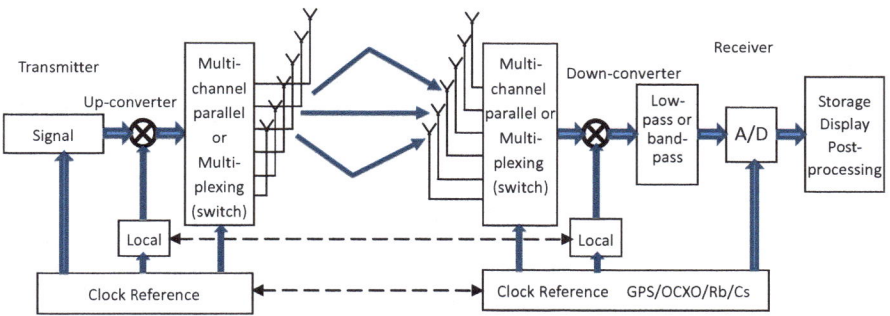

Fig. 3.6 Illustration of a unified modular MIMO channel sounder

possible switching sequence is as follows. The Tx box switches and keeps connection to one transmitting antenna when the Rx box switches to each of the receiving antennas and followed by subchannel sounding sequentially. After that the Rx box switches from the last antenna back to the first antenna when finishing sounding a group of subchannels, at the same time the Tx box switches to next antenna and begins sounding next group of subchannels. SIMO is a compromise between SISO and MIMO methods. It uses switched signal transmission and parallel signal reception, and obtain a certain balance among system complexity, cost, and measurement speed. Although a few systems [45, 46] have MIMO sounding ability in terms of the system structure and the principle of operation, they are SIMO in nature. It is possible to utilize the advantages of both MIMO and SISO, and extend a MIMO sounder to support larger arrays through a switch system.

A unified modular channel sounder, as shown in Fig. 3.6, is used as a framework to illustrate various methods, techniques, and devices in the channel measurement. We mainly focus on four aspects of the measurement system: antenna system (including type and switching of antenna), RF unit (local oscillator and up/down converter), sounding signal (baseband part), and transmitter/receiver synchronization method.

1. Antenna and antenna array

The transmitting and receiving antenna elements used in sounding have many types. In order to ensure the radiation and collection of radio signal in all the spaces and angles, omnidirectional or low gain antennas are adopted, such as half-wave dipole antenna, patch antenna, diskcone antenna, biconical antenna and open-ended waveguide (OEW), etc. In order to ensure the sufficient signal strength, directional antennas are adopted, such as standard gain horn antenna, parabolic antenna, lens antenna, etc. Directional antennas can only guarantee the signal sounding of certain angles. In order to achieve coverage of all space angles, multiple directional antennas splicing in different directions or rotating a directional antenna for scanning is needed. When the polarization characteristics of the channel need to be detected, the dual polarization antenna element is necessary.

Table 3.2 antenna array type set 1 [13]

Name	ODA_5G25	PLA_5G25	UCA_5G25	SPH_5
Manufacturer	Elektrobit	Elektrobit	Elektrobit	HUT(Aalto)
Array type	Omnidirectional array	Rectangular array	Uniform circular array	Hemisphere
Polarization	Dual polarization (+/- 45°)	Dual polarization (+/- 45°)	Perpendicularity	Dual polarization (H/V)
Center frequency [GHz]	5.25	5.25	5.25	5
Number of antenna elements	50 (25 dual)	32 (16 dual)	8	32 (16 dual)
Element type	Patch	Patch	Patch	Patch
Picture				

The resolution of direction or angle of the multipath channel is closely related to the structure of the antenna array and the number of the antenna elements. In order to obtain high angle resolution, large antenna array is usaually adopted in channel sounding. The antenna array can be formed in two types: virtual antenna array and real antenna array. The virtual antenna array shifts only one antenna element to the required locations (directions) through spatial translation or rotation by electromechanical device. CIR is measured independently at each location. For virtual antenna array, only serial sounding mode can be used. In existing channel measurement systems such as Elektrobit (EB) Propsound and Medav RUSK MIMO channel sounder, the real antenna arrays are widely used like linear array, area array, circular array, spherical array, quasi spherical array, etc., as shown in Table 3.2 and 3.3. In the mmWave sounding system, the antenna element rotation and translation are often used (e.g., Table 3.4) to form a virtual antenna array. Recently, the mmWave channel sounding is carried out by using the electronically controlled antenna array [4]. Such modular antenna array can be used to form the scanning beam with different directions in the space via phase control. In addition to the channel sounding, this antenna can quickly and adaptively change the direction of beam focusing, which can be applied to the actual mmWave communications system. Real antenna array can be used in SISO, SIMO or MIMO sounding. When it's used in SISO or SIMO sounding, the switching box is required to switch between different elements or polarization elements. Real antenna array system can achieve high speed sounding, which is suitable for sounding in multiple locations and mobile scenarios. Since the mechanical moving speed is much slower

Table 3.3 antenna array type set 2 [13]

Name	PULA8	UCA16	PUCPA24	SPUCPA4 × 24
Manufacturer	IRK Dresden	TU Ilmenau	IRK Dresden	IRK Dresden
Array structure	Uniform linear array	Uniform circular array	Uniform circular array	Multi uniform circular array
Polarization	Dual (V/H)	Perpendicularity	Dual (V/H)	Dual (V/H)
Center frequency [GHz]	5.2	5.2	5.2	5.2
Bandwidth [MHz]	120	120	120	120
Maximum power [dBm]	27 (40)	27	25	24
Number of elements	8	16	24	96
Array type	Patch	Diskcone	Patch	Patch
Size	Unit interval 0.4943λ	Diameter 10.85 cm	Diameter 19.5 cm	Diameter 19.5 cm, ring interval 0.4943λ
element direction				
Picture				

Table 3.4 Millimeter wave antennas

Name	LB-22-25	NICT [47]	MD249-AA	OEW 2650
Manufacturer	Ainfoinc	–	Flann	Satimo
Polarization	Linear	Linear	Linear	Linear
Frequency [GHz]	33~50 GHz	50~75 GHz	50~75 GHz	26.5~40 GHz
Antenna type	Pyramid standard gain horn antenna	Conical horn antenna	Monopole	OEW
Beam width (Azimuth/ zenith)	8.6°/7.5°	30°	360°/60°	64.5°/ 112.45°@26.5 GHz 53.3°/ 71.7°@40 GHz
Picture				

than the switching speed of the electric switch, the virtual antenna array is only suitable for the measurement with static or low speed. In addition, in the measurement system using real antenna array, when a new frequency band is measured, not only a new antenna array must be developed, but also a full set of complex calibration need to be done for it. In this aspect, the virtual antenna array only needs one antenna element, so it is much easier in terms of the cost and the calibration complexity.

2. RF unit

The important parameters of the RF unit include the number of RF channels, RF frequency range, and RF signal bandwidth. Parallel sounding mode requires multiple parallel RF units. It is necessary to consider the consistency and mutual interference of the RF units. Presently, most models have designated the applicable frequency range or sounding bandwidth, such as 3GPP SCM is 1~3 GHz/5 MHz, WINNER II/+ and IMT-A 0.45~6 GHz/100 MHz, 3GPP D2D and 3D 1~4 GHz/ 100 MHz. For sub-6 GHz bands, the existing channel models supporting 100 MHz signal bandwidth seems to be enough, but larger bandwidth requirement will continue to emerge. For example, the highest bandwidth in IEEE 802.11ac is 160 MHz. Therefore, the channel models have to support higher bandwidth, such as more than 500 MHz. In the 60 GHz mmWave band, the bandwidth of sounding system which was used for the development of IEEE 802.11ad is 800 MHz [48]. In the future mmWave applications, the bandwidth of the signal waveform may exceed 2 GHz, which requires that the RF unit of the mmWave channel sounder should be able to support bandwidth larger than 2 GHz.

In the mmWave channel sounding, due to the cable loss of high frequency signal is great, many sounding systems use an independent external mixer which is close to the antennas. Lower frequency excitation signal is transmitted to the mixer through a cable. The lower frequency receiving signal outputted by the downconverter is transmitted to the signal analysis system through the cable. The sounding system based on the Vector Network Analyzer (VNA) requires special considerations: (1) the local oscillator (LO) signals for the mixers at both transmitter and receiver should be from the same local oscillator source, so that the consistency of reference frequency and phase in each measurement frequency point can be guaranteed. (2) Sometimes in order to obtain a longer sounding distance, a long cable is used. Even the low frequency signal still has huge propagation loss, the signal strength need to be guaranteed by using an electric power amplifier or fiber optic [1, 49].

3. Sounding signal and detection technology

Different sounding signals have large impacts on the performance of sounding systems and correspond to different parameter extraction algorithms. How to choose sounding signals to achieve the optimal performance of the channel sounder is very worthwhile to explore. The pros and cons of a sounding signal can be evaluated from several aspects, such as, signal duration, signal bandwidth, time-bandwidth product, power spectral density, peak-to-average power ratio (PAPR) and correlation property. Belows are some commonly used channel sounding signals.

(a) Periodic pulse signal

Pulse signal (high bandwidth) is the most intuitive choice for channel sounding. Shortening the sounding signal pulse width can improve the path delay resolution. However, the narrow pulse width of sounding signal causes the high peak power and arsing serious nonlinear distortion in circuit. One has to reduce the signal power for lower distortion, as a result the sounding system has limited measurement range. For digital realized broadband sounding system, according to Shannon-Nyquist sampling theorem, the receiver should have an ADC with ultra high sampling rate, which will be highly costy expensive. Tamir et. al, suggest using compressed sensing technology to reduce the sampling frequency [50]. In the early stage of channel sounding, periodic pulse was used more extensively. Later, the spread spectrum sequence and signal with flat envelope in frequency domain have been more widely used.

(b) Time-domain flat signal

The time-domain flat signal is the signal whose envelope is relatively flat in the time domain, which is different from the signal whose envelope is flat in frequency domain. These signals include the sequences with good periodic autocorrelation properties, such as Pseudo-random Noise (PN) (e.g., M sequence), Chirp sequences, Zadoff-Chu sequences (also the frequency-domain flat signals), etc.. Systems applying this type of time-domain signal have been demonstrated excellent performance in delay resolution and system dynamic range, so it has been widely used, such as the Propsound channel sounder.

Two different processing methods can be used in the receiver when the PN sequences are sent in transmitter. One method is to use correlator or matched filter. Related operations are performed at each sampling time, and the CIR is directly obtained, or the signal is first recorded and then processed. In recent years, some high time and spatial resolution parameter extraction algorithms, such as RIMAX, SAGE, etc., have been widely used in channel measurement. This processing method requires the high-speed ADC to sample and perform analogue-to-digital conversion, as well as store a large number of data. Another method is sliding correlation technology. Transmitter and receiver use the same PN sequence, but with very small chip rate difference. Due to different chip rates (clock frequency), the PN sequence of transmitter and receiver will produce relative sliding. When the two sequence chips realign due to propagation delay, maximum correlation peak will be produced so as to achieve the despreading of the spread spectrum signal. The advantage of this technique is that it does not require high speed sampling, and Newhall et al. [43] use this technique to sound mmWave channels. However, since the single measurement time is determined by the clock frequency difference of sending and receiving PN sequences and the maximum DS length. If the frequency difference is small, i.e., the relative "sliding" of the two PN sequences is slower, a longer time is needed to complete a measurement. And hence this method is not suitable for the sounding of high mobility scenario.

(c) Frequency-domain flat signal

The frequency-domain flat signal is the signal whose envelope is relatively flat in the frequency domain, which is different from the signal whose envelope is flat in time domain. Early narrowband channel sounding techniques use sinusoidal signal as the sounding signal. After the signal goes through the propagation channel, the changes of signal amplitude and phase are the channel frequency response at this frequency. In the VNA-based broadband sounding system, the whole sounding frequency band is divided into a number of frequency points. At each frequency point, sinusoidal wave sounding signal is still used, in order to scan all the frequencies. This sweep frequency measurement method is generally slow, which is not suitable for measurement of dynamic scenarios. At different frequency points, the RF units at the transmitter and receiver must ensure the consistent frequency and phase. In order to overcome the limitation of the sweep frequency method, broadband frequency domain measurements commonly use the multi-carrier (OFDM) signal, which is used to detect the whole frequency band at a time. In addition, frequency domain excitation symbol $X(f)$ may be the PN sequence composed of $\{+1, -1\}$. And the frequency response can be obtained very easily without division. However, it should be noted that the multi-carrier signals need to be carefully designed to have a lower level of PAPR.

4. Synchronization method

The sounding system has reference clocks both at the transmitter and the receiver, from which the baseband clock, ADC/DAC sampling driving clock, and LO (may use a different reference clock), are obtained by means of frequency doubling, phase lock, etc. The transmitter and receiver need to ensure the consistency of baseband clock, sampling clock, LO frequency and phase. If they are inconsistent and cannot be accurately estimated and compensated, the difference parts will be regarded as a part of channel and thus the outputted channel parameters will be inaccurate. Therefore, the sounding system should minimize the error in reference clocks of transmitter and receiver, or it should try to use one reference clock. This requirement is easy to achieve in the short distance sounding (such as indoor sounding). For the VNA-based broadband sounding system, if one VNA is used for both transmitting and receiving, the clock sharing between channels should be achieved within the VNA. If both signal source and VNA are used for transmitting and receiving, the reference clocks of the two devices should be connected by cable. Due to the use of single frequency step sweeping, the consistency of the phases at both ends with different frequencies should be ensured, which requires that the RF signals of the signal source should be connected to the VNA as a reference with coupler and long cable. The coupling mode can be electrically amplified, or it can be realized by means of photoelectricity/electro-optical conversion and fiber optic extension [52]. In outdoor mobile environment, the shared reference clock usually uses GPS-tamed Oven Controlled Crystal Oscillator (OCXO) or the atomic clock with high precision and high stability (rubidium or cesium clock), or both. Even so, it still can't avoid the deviation of the clock. And the jitter or phase noise of each clock will affect the sounding performance.

In addition, when SISO or SIMO is used for MIMO sounding, the synchronous switching of the switch arrays at the receiver and the transmitter must be strictly controlled. The clock of the switching control circuit is also derived from the same reference clock.

5. Channel sounding system

   The following is a summary of some commonly used channel sounding systems, as shown in Table 3.5. The early systems that undertook channel sounding tasks for 4G or earlier communications systems only paid attention to frequency bands below 6 GHz. These systems include Propsound [39] (stop production), RUSK [40], the channel sounder of Aalto University (Aalto U., formerly known as the Helsinki University of Technology, HUT) [41] the channel sounder of NTT DoCoMo [42], etc. Many systems are applied in measurement tasks of various standard channel models, e.g., Propsound, Medav RUSK and the Aalto University system are widely used in the development of the channel modeling of COST, WINNER, IMT-Advanced, 3GPP 3D, etc. In recent years, these systems are also used in the channel sounding of massive MIMO. With the increase of the demand for the sounding of more high frequency band, the broadband mmWave sounding systems [1, 43, 48] and VNA-based measurement systems [1, 49, 51] were developed for a specific frequency (customization). Of course, the precious sounding system can also extend frequency to support the mmWave sounding. The VNA-based measurement system is often used in the sounding of massive MIMO. From the consideration of reducing the cost in hardware and calibration, most of the existing systems use a RF unit. But there are also some sounders with multiple receivers (such as Durham University sounder [45]) and some sounders with multiple receivers and multiple transmitters such as the Tokyo Institute of Technology (TIT) sounder [45, 46], etc.

   Figure 3.7 shows the MIMO channel sounding which is achieved using SISO method by sequential switching of the receiver and transmitter antenna (SISO). Here a simple analysis of the capability of the dynamic channel sounding of the system will be shown. Let's take that N_t denotes the number of transmitting antennas, N_r denotes the number of receiving antennas, and τ_{max} denotes the channel maximal excess delay. Take the PN sequence as the sounding sequence and SISO sounding as an example. A periodic sounding sequence is composed of K chips. The length of the chips is T_p, and the length of the sequence is $T_a=KT_p$. A longer sounding sequence T_a can get greater processing gain (see SAGE algorithm in Sect. 3.4), but it also limits the measurable Doppler shift. The time span T_o of one sounding should equal the sum of the length of sounding sequence T_a and the maximal DS τ_{max}, i.e., $T_o=T_a+\tau_{max}$. If we take into account the antenna switching time T_w, following the assumption in the RUSK, the measurement time and switching time are equal, i.e., $T_w=T_o$. The measurement time of a subchannel is $2T_o$. The time spent in a snapshot to iterate all of the sub-channels is $T_{snap}=2T_oN_t N_r$, i.e., the time in which a full channel sounding is carried out. For time-varying channel, the sounding period is denoted as T_{snap}. If the gap between snapshots is zero, i.e., $T_f=T_{snap}$, the maximum Doppler shift that can be measured by the system

Table 3.5 Channel sounding system parameters

Channel sounding system	RF units	Frequency range	Bandwidth (GHz)	Antenna units type	Antenna array type	Antenna units	Sounding signal type	Synchronization method
Propsound	Tx1 Rx1	<6 GHz, scalable	≤0.2	Arbitrary (monopole, dual polarized, diskcone, biconical, patch, horn antennas), often omnidirectional antennas	Fixed array, can be customized	Arbitrary	PN or OFDM	GPS, OCXO, Rubidium clock, cesium clock
RUSK			≤0.24			Limited by sounding speed		
Alto U.			≤0.2					
DOCOMO			0.05					
Durham U.	Tx 1 Rx 8		~0.06					
TIT U.	Tx≥1, Rx≥1 modular, scalable	11 GHz, can be customized	0.4	Arbitrary, can be customized		Arbitrary		
Broadband high frequency sounder	Tx 1 Rx 1~2	>6G	<1	Short distance, omnidirectional or directional; Long distance, often directional antennas	Rotate or translate 2D/3D to form virtual array	A few		
VNA-based sounding system	Tx 1 Rx 1	Arbitrary	>1			1	Frequency scanning	LO extension [1, 42] Transmitted signal extension [44–47]

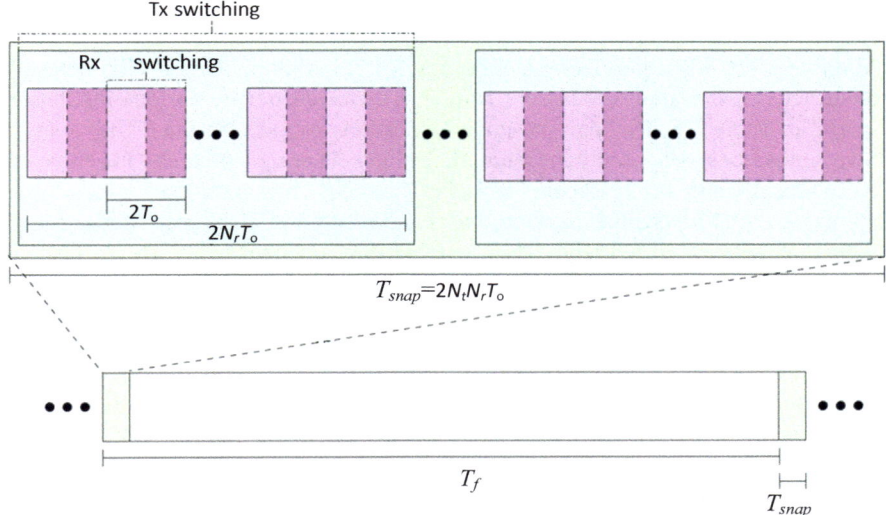

Fig. 3.7 Timing sequence in MIMO channel sounding via switching transmitting and receiving antennas

is $f_D = 1/T_{snap}/2 = 1/(4N_t N_r)/(KT_p + \tau_{max})$. Thus it is can be seen that, in the system sounding capabilities, large Doppler shift and more antennas/ large processing gain/ large DS are mutually exclusive. To obtain support of larger DS (larger coverage), the system has to sacrifice the mobility or the number of channels or the processing gain. To ensure support of mobility, the system has to sacrifice the sounding range or number of channels or the processing gain. Such contradictions mainly exist in the mobile environments in which either the transceivers or the scatterers move. It is essential to carefully balance various measurement objectives in the measurement process. In addition, the channel coherence time can been roughly estimated in terms of Doppler shift, $T_c = \sqrt{9/16\pi}/f_D$, and a snapshot should be completed in that time, i.e., $T_{snap} < T_c$

3.3.2   Channel Measurement Activities

In recent years, with the ongoing progress in 5G R&D, many communication enterprises, universities, and research institutes all over the world have carried out extensive channel measurement campaigns. These measurements are divided into three categories, namely the measurement of Massive MIMO, D2D/V2V/HSR and mmWave band. The measurement campaigns, parameter settings and results are briefly reviewed as follows.

3.3.2.1 Massive MIMO measurement

Massive MIMO is a promising technology for 5G communication. The measurement activities related to Massive MIMO are carried out by various institutions shown in Table 3.6. The main research institutions include the Lund University in Sweden, Aalborg University in Denmark, Beijing Jiaotong University in China, etc.

In recent years, the reseachers of Lund University had completed a lot of work on massive MIMO channel measurements and modeling. They used Medav RUSK with real antenna arrays and VNA with virtual antenna arrays to perform the measurements. Some new propagation phenomena were observed, which greatly promote the understanding of Massive MIMO channel and the development of channel model. Their works are outlined as follows002E

In [52], the BS was placed indoors and configured with a polarized uniform cylindrical array with 128 ports (4*16 patch antenna elements, each with two polarized input ports). Only one element in a polarized uniform antenna array with 32 ports (2*8 patch antenna units, each with two polarized input ports) was used as the user antenna. The channel correlation between two users was directly analyzed and found that it would decline rapidly (orthogonality will increase) when the number of BS antennas increased. When there were 20 antennas at the BS, the performance of linear pre-coding was close to the performance of optimal dirty paper coding.

In [53], measurements for controlled outdoor scenario in both LOS and NLOS conditions were carried out by using a VNA based system with 2.6 GHz carrier frequency, 50 MHz bandwidth and 1601 frequency points. An approximately omnidirectional and perpendicularly polarized antennas were used at both side, while at the receiver side the antenna was moved along a 7.3 m straight sliding trail with step of half wavelength to form a virtual antenna array with 128 elements. The path gain, Ricean K-factor, and PAS were analyzed and found obvious changes in the PAS of the incident wave along the linear array. The specific non-stationary characteristic and near-field effects of massive MIMO were demonstrated and both would helped to reduce the correlation between the users. Gao et al. extended the COTS 2100 channel model to support the newly discovered propagation characteristics [18]. The statistical model of cluster number as well as the visibility region and visibility gain of clusters at the BS side was obtained from the measured data.

In [54, 55], measurements were carried out outdoor with a larger space using two systems at in same position, and the results were compared. A uniformly cylindrical array (UCA) with 128 ports the same as [52] and a virtual uniformly linear array (ULA) with 128 position the same as [53] were used at the BS, while a perpendicularly polarized omnidirectional antenna was used at the UT. The measurement results of using the ULA are similar with the results in [53]. The power variation over the antenna arrays that is critical to the performance of massive MIMO were observed. For a ULA, the power variation came from different large scale fading that each element experienced. While for UCA, these large power variation were resulted from the polarization and directional radiation pattern of each patch

Table 3.6 Summary of massive MIMO measurement activities

Research institutes	Scenario	Antenna configuration	Channel sounder	Frequency/ Bandwidth	parameters
Lund University in Sweden	O2I [52]	Tx: 1 patch; Rx UCA(128 patch)	RUSK	2.6 GHz/50 MHz	channel correlation between users
	Outdoors [53]	Tx: 1 omnidirectional Rx: virtual ULA (128 omnidirectional), 7.3 m long	VNA	2.6 GHz/50 MHz	path gain, K- factor, PAS, eigenvalue, channel correlation between users
	Outdoors [54, 55]	Tx: 1 omnidirectional; Rx1: virtual ULA(128); Rx2: 128 UCA	VNA RUSK	2.6 GHz/50 MHz	Large-scale fading, angular resolution, singular value spread [55]
	Outdoors [56]	Tx: 8 omnidirectional Rx: UCA(128 elements)	RUSK	2.6 GHz/40 MHz	Singular value spread
Alcatel-Lucent Bell Labs in German	Outdoors [57]	Tx: UCA (112 patch) Rx: 2 omnidirectional (2 m spaced)	Customized	2.6 GHz/20 MHz (LTE)	Correlation, inverse condition number
Aalborg University in Denmark	Indoors [58]	Tx: 8*2 patch (6 orientations and combinations) Rx: 64 omnidirectional (3 array modes)	Customized	5.8 GHz/ 100 MHz	Correlation matrix, channel singular value, condition numbers, power variation over array
BJTU	Indoors, outdoors [59]	Tx: 128 virtual ULA(~12.9 m) & virtual UCA (biconical); Rx 1/2 omnidirectional	VSG +Mixer, VSA	1427~1518 MHz 4400~4500 MHz	Small scale parameters
	Outdoors [60]	Tx: 1 biconical; Rx: 64 virtual ULA(biconical)	VSG VSA	3.33 GHz/ 100 MHz	PADS, PDP, PAS, DS, AS

antenna elements. The ULA could provide better azimuth resolution than the UCA, however the latter could provide the resolution in both azimuth and zenith dimension. The analysis results showed that both arrays could approach the capacity that a independently idendifical channel (i.i.d) provided even for closely located users in the LOS case. When the number of BS antennas is 10 times of the number of users (the number of antennas), it could be considered as a massive MIMO array. The performance gain would be diminishing if the number of BS antennas continued to increase.

In [56], the UCA in [52] was used as the receiving antenna at the B. Eight users walked randomly in a circle with radius 5 m at about 0.5 m/s speed and the user antennas were connected to the transmitter of a RUSK channel sounder, which constitute a synchronous massive MIMO measurement system to investigate the spatial separability of adjacent users. As reported in [54, 55], even users located close to each other in the LOS case can be spatially separated in a massive MIMO system.

A massive MIMO measurement was also carried out by Alcatel-Lucent Bell Lab in Germany [57]. Seven antennas perpendicular spaced with half wavelength was rotated around a circle with radius 1 m and 3.5 degree steps pointing to 16 angular directions to form a virtual antenna array with 112 elements. A LTE-like system with a subcarrier interval of 15 kHz but only 400 subcarriers (every 3rd sub-carrier of total 1200 subcarriers) being valid was used to sound the channel and analyzed the channel characteristics. It mainly studied the orthogonality of the channel at different measurement positions with correlation coefficients and inverse condition number. The analysis results confirmed the theoretical advantages of Massive MIMO over convention MIMO. Using more antennas could improve the orthogonality of the channels between users, but the performance would not improve significantly when antenna number exceeds a certain amount.

As a part the METIS project, a massive MIMO measurement was completed by the Aalborg University in Denmark using self-developed sounding system with 5.8 GHz carrier frequency, 100 MHz bandwidth [58]. With 16 transmitting antennas and 64 receiving antennas, a 64×16 channel matrix was obtained. The 64 identical receiving antenna elements (seperated into 8 groups, and 8 antennas per group) formed three array shapes with different aperture, i.e. ultra large linear array (6 m long), large linear array (2 m long) and square array 8 users (two antennas per user). Six 6 schemes with different user locations were designed to achieve various measurement conditions, such as LOS or NLOS cases, the users being sparsely or densely distributed, and the user antennas being perpendicular or parallel to the array. The user/antenna correlation matrix, the singular value and condition number of the channel, and the power variation over the antenna array was studied. The measurement results showed that larger antenna aperture could produce more spatial degrees of freedom (DOF) which was close to the theoretical performance. The array with the largest aperture was able to approach the performance of an i.i.d channel since it enabled the channel discrimination among users and between antennas of a same device.

The researchers in Beijing JiaoTong University (BJTU) recently has launched indoor and outdoor measurement work. A commercial vector signal generator (VSG) solely or combined with mixer was used as the transmitter, a commercial signal analyzer (VSA) or a data recorder was used as the receiver, and there were two measurement configurations related to antenna and frequency band. In the 1st configuration, biconical antenna was used at the transmitter to form a virtual large-scale linear or a virtual uniformly circular array (UCA) both with 128 elements [59]. One and two omnidirectional antennas were used at the receiver respectively. The channel was explored at two frequency bands, i.e. 1427~1518 MHz and 4400~4500 MHz. In the 2nd configuration, a single biconical antenna was used at the transmitter and a virtual uniformly linear array with 64 points was used at the receiver. The channel was explored in a frequency band of 3.33 GHz with 100 MHz bandwidth [60]. The non-stationary phenomenon of massive MIMO, the SSPs such as PDP, PAS, DS and AS, as well as the propagation differences among different frequencies were analyzed and compared.

3.3.2.2 Measurement of D2D, V2V and HSR

D2D/V2V and high speed rail (HSR) communications are of importance for the future 5G wireless communications. IEEE 802.11p [61] has been committed to the standardization of vehicle safety communications and defines three link types such as V2V, V2I (I refers to Infrastructure) and I2V. IEEE 802.11p adopts 5.9 GHz frequency band for the V2V communication, and gives a simple SISO channel model with a tapped-delay line (TDL) structure. The envolope of channel coefficient of each tap is assumed to be Ricean distributed or Rayleigh distributed. In the future, MIMO technology will be widely used in vehicle communication system because it can significantly improve the system capacity and guarantee reliable data transmission. Therefore, it is essential to strengthen MIMO channel sounding for D2D, V2V, and HSR scenarios. Extensive channel measurement activities for D2D/V2V and HSR scenarios have been carried out [1, 62–82]. Part valueable measurement activities and the related details are listed in Table 3.7. The following is a brief review about these measurement activities.

In the METIS project, one channel measurement for D2D scenario was conducted by DOCOMO NTT in Japan using self-developed channel sounder with at 2.225 GHz carrier frequency [42]. A sleeve dipole antenna and a slotted cylinder antenna were used at the BS side to transmit vertically and horizontally polarized waves, respectively. A UCA composed of 48 dual polarized antennas (96 ports) was used at the receiver. The height of antennas at both ends were set to 1.45 m. The UT moved along five designated courses at about 1 m/s. The received power, PDP, azimuth angular spread at arrival (ASA) and zenith angular spread at arrival (ZSA) were analyzed. Measurements were carried out during the day and night time in order to observe the effects of the pedestrians on the channel. Measurement results showed that at night the received power and DS are larger because the pedestrians block long-delay path during the daytime, and the ZSA is

Table 3.7 Summary of D2D/V2V/HSR measurement activities

Research institutes	Scenario	Antenna configuration	Channel sounder	Carrier/ bandwidth (units: GHz)	Measurement parameters
NTT DOCOMO [1]	D2D urban,1 m/s	Tx: Sleeve, Rx 96 ports, dual polarization, 1.45 m high	DOCOMO	2.225/0.050	Received power, PDP,ASA, ZSA
University of Oulu [1, 62]	V2V, same/ opposite direction,20 km/h	Tx 1 omnidirectional, 1.6 m high; Rx:56 ports cylinder array (2.3 GHz), 50 ports cylinder array (5.25 GHz), 2.5 m high	Propsound	2.3/0.1 5.25/0.2	PL, DS, SF, K, correlation distance(DS, SF, K)
Lund University [63–66]	V2V, 2 scenarios, same/opposite direction	UCA(4 patch), 2.5 m high	RUSK	5.2/0.24	TDP(2D), DDP (2D), path power, antenna correlation matrix
	V2V, 3 scenarios	ULA(4 patch), 1.73 m high	RUSK	5.6/0.24, 0.020 analysis bandwidth	AOA,AOD,DS,Doppler, ASA
	V2V, 10 scenarios				LSF, time frequency varying PDP, DSD, DS, Doppler spread, K
Aalto University	V2V, 4 scenarios, two vehicles, same direction, 5–15 km/h [67]	Tx/Rx: SPH_5	Aalto sounder	5.3/0.06	PL and exponent, SC parameters, DMC parameters, SS fading gain
	V2V, 5 scenarios, <40 km/h [68]	Tx: ULA 4,Rx: SPH_5,~2 m high			PDP,APDP,ACF,CB,CMD, SD,SC
NEC [70]	V2V, 7 scenarios, <108 km/h	Single antenna 1.5 m~3 m high	Self-developed	5.9/0.020	Large- and small- scale power
German Aerospace Center [69]	V2V speed 30 km/h	Omnidirectional dipole	RUSK	5.2/0.12	DDP
University of Southern California [71]	V2V, 2 scenarios, <1 m/s	Single antenna Tx: 1.5 m/3 m high Rx:1.5 m high	Tx:WARP Rx: VSA	5.8/0.015	PL,SF, amplitude distribution, correlation of DS and SF

BJTU [74–77]	HSR,multi scenarios <350 km/h	Single antenna	Tx:GSM-R Rx: VSA	0.93/0.001	PL, SF, amplitude distribution, DS, K- factor, LCR, AFD
	HSR, cutting	Single antenna	Propsound	2.35/0.05	
BJTU, UPM [78–81]	Subway, <120 km/h	Single antenna	Tx customized source, Rx:VSA	2.4/ 0.002,0.92/- ,2.4/-,5.705/-	
WINNER + [164]	HSR 20/100/240 km/h	Tx: monopole, 5 m high; Rx:UCA (16 diskcone), 6 m high	RUSK	5.25/-	PL, DS, K, ASA
BUPT [83]	HSR 370 km/h	Tx: 2 vertically polarization directional 45°/135° layout, 40 m high, Rx 2 3 dBi, ~4 m high	VSG+VSA	2.6/0.02	Correlation Coefficients between subchannels,PL,SF, DS.K-factor

small because the pedestrians become the scatterers in the zenith direction). Whereas, the ASAs in the daytime and night didn't show significant difference. The statistical parameters obtained according to the measurement results are listed in details in Table A-14 [1].

Also in the METIS project, a V2V channel measurement was carried out using Propsound sounder at 2 frequency bands (100 MHz bandwidth in 2.3 GHz band and 200 MHz bandwidth in 5.25 GHz band) by University of Oulu in downtown of Oulu, Finland [62]. A single omni-directional antenna with 1.6 m high was used at the transmitter side, an array antenna with 28 dual-polarized elements (56 ports, 2.3 GHz) and an array antenna with 25 dual polarized elements (50 ports and 5.25 GHz) were fixed at 2.5 m high at the receiver side. Two types of V2V scenarios, moving in the same direction with low traffic density and moving in the opposite direction with high traffic density, were measured, in which the maximum relative speed was 20 km/h. The path loss, DS, Maximum Excess Delay (MED), SF standard deviation, K-factor, as well as their respective correlation distances were investigated. The results showed that the correlation distance is less than 11 m.

A series of 4-by-4 MIMO channel measurements were carried out by Lund University in Sweden in many scenarios using a RUSK channel sounder.

In [63], the system configuration for the measurement was 5.2 GHz carrier frequency and 240 MHz bandwidth. A circular array comprising four microstrip antennas was placed on the roof of cars, the antenna height was 2.5 m and the antenna radiation directions were 45°/135°/225°/315°, respectively. A sounding sequence of 3.2μs long was used and the snapshot (tranversing 16 subchannels) interval was 0.3072 ms which corresponds to maximum Doppler shift of 1.6 kHz (equivalent to to 338 km/h). The measurement scenarios included one-way rural expressway and one-way two-lane highway. The movement of vehicles could be in the same direction or in the opposite direction. Firstly, the Time-Delay Profile (TDP) and Delay-Doppler Profile (DDP) of the V2V channel were analyzed and the results showed that the propagation channel was composed of diffuse scattering and discrete specular scattering. The path loss model was a combination of attenuation with distance and a slowly varying stochastic processes. Based on the observations, a GSCM channel model suitable for V2V transmission was propased. The antenna correlation matrix of the model was verified to be consistent with the measurement results.

In [64], the system configuration for the measurement was 5.6 GHz carrier frequency and 240 MHz bandwidth. Four microstrip antennas were placed on the car roof and along the vehicle axle direction with half wavelength spaced, the antenna height was 1.73 m and the maximum radiation direction pointed toward front, back, left and right respectively. Three scenarios were designed as follows. The 1st scenario was at crossroads, where the transmitter standed statically near an intersection, while the receiver moved in the perpendicular direction of intersection with a speed of 30~40 km/h. The 2nd scenario is at expresway with two lanes, where the transmitting car was blocked in one lane and the receiving car approached to the transmitter in another lane with a speed of 70 km/h. In the 3rd scenario, two cars were driven at a speed of 110 km/h in one lane of a two-lane highway where both cars were obstructed by a tall van. The AOA, AOD, delay, and Doppler shift of the V2V channel were analyzed. It was found that in 2nd and 3rd scenarios the channel was mainly composed of one-order reflection paths and has smaller AS, so that beamforming technique was more suitable than diversity technique for both scenarios. In the 1st scenario, there were a lot of multiple-order reflections leading to a larger AS, so that diversity technique could be more suitable.

In [65], the same sounding system as [64] was adopted to measure in more scenarios including the following situations: (1) Road crossing (consisting of four sub-scenarios with respect to suburban or urban, with or without traffic, and single lane or multiple lanes) where both vehicles approach the crossing from perpendicular directions, driving at 10~50 km/h. (2) General LOS obstruction in highway where both vehicles drove in the same direction with similar speed of 70~110 km/h. (3) Ramp merging with partly obstructed main road in rural enviroments where both cars drived in the same direction at 80~90 km/h. (4) Traffic congestion including two cases:both vehicles were stuck in a traffic jam and slowly drived at 15~30 km/h; one vehicle was totally stuck in traffic while the other approached at a relative high-speed of 60~70 km/h. (5) In-tunnel with one way two lans where both vehicles were driving at 80~110 km/h. (6) On-bridge where both vehicles aparted from 150 m were drived at about 100 km/h. There were 3~15 measurements carried out for each scenario. The research showed that V2V channel is non-stationary or could be considered to be locally stationary. Therefore, the time-varying local scattering function (LSF) was calculated, based on which the time-varying power delay profile (PDP) and Doppler power spectral density (DSD) were obtained. The time varying mean square DS and Doppler spread were further obtained and these two statistical characteristics could be modeled as a bimodel Gaussian mixture distribution. Bernado et al. further analysed the measured data to conclulde that the power of the LOS path is Ricean distributed with with a time-varying K-factor, while the power of other paths follow Rayleigh distribution [66].

As early as in 2007, some V2V MIMO channel soundings for four scenarios (campus, six-lane highway, urban, suburban) were carried out by Aalto University in Finland using self-developed channel sounder with 5.2 GHz carrier frequency and 60 MHz bandwidth [67]. One snapshot took 8.4 ms and the snapshot repetition rate was 14.3 Hz allowing a maximum Doppler shift of 7.15 kHz. Two vehicles were drived in the same direction at 5~15 km/h. Either of the transmitter and receiver was equipped with one hemispheric antenna array comprised of 15 dual-polarized patch antennas (30 ports) to obtain 30-by-30 MIMO channel response data in one measurement. A simplified 2-D geometry of the scattering environment given the parameter of the scatters including densities, birth/death and locations was used to build a GSCM for V2V channel, in which the time-varing fading gain of the specular components (SC) was estimated and modelled as combination of path loss, large-scale fading being by a Gaussian random process and small-scale fading being Weibull distributed. Meanwhile, the DMCs were estimated and modeled as a stochastic process with large-scale power decaying exponentially (in dB) in the delay domain and small-scale fading being being Weibull distributed.

In another measurement, the same sounding system is used to study the quasi-stationary region of V2V channel [68]. The same receiving antenna arrys as above was used, while a four-element uniformly linear array was used at the transmitter side. One snapshot took 1.632 ms and the snapshot repetition rate was 66.7 Hz, which limit the maximum speed could not exceed 40 km/h for the five scenarios. The distance between two cars was between 10 m and 500 m according to the traffic conditions. The PDP, averaged PDP (APDPs), autocorrelation of small scale fading, coherence bandwidth (CB), correlation matrix distance (CMD), spectral divergence (SD) and shadow-fading correlation (SC) were analyzed for different scenarios and in the LOS condition. It was found that the quasi- stationary interval is around 3~80 m, which is dramatically affected by the existence of LOS component, vehicle speed, and antenna array size and configuration, etc.

One V2V channel measurement was carried out by the German Aerospace Center using a RUSK channel sounder configured with single antenna, 5.2 GHz carrier frequency and 120 MHz bandwidth [69]. This measurement mainly aimed to verify the delay-Doppler PDF proposed for a GSCM. Some measurements in the LOS and NLOS(due to vehicles or buildings/foliage) case for 7 scenarios were carried out by NEC laboratory in German using the self-developed IEEE 802.11p devices with single antenna, 5.9 GHz carrier frequency and 20 MHz bandwidth (larger than the standardized 10 MHz of IEEE 802.11p) [70]. The measurement mainly cared about the large-scale and small-scale signal variations. Another V2V channel sounding was carried out by University of Southern California [71]. - Software-defined radio platform WARP [72] and a commercial spectrum analyzer were used as the transmitter and receiver respectively, in which carrier frequency was set to 5.805 GHz quite close to standardized 5.9 GHz proposed by IEEE 802.11p, and bandwidth was 15 MHz. The channel that the link is blocked by other trucks was considered for two scenarios, open space and high-rise building areas. The path loss, SF, amplitude distribution (consistent with the Nakagami distribution) of small-scale fading, and the correlation of DS and SF was measured and analyzed.

Professor Bo Ai of BJTU divided the HSR scenarios into 12 categories including viaduct, cutting, tunnel, and station and further 18 subcategories [73]. Extensive measurements were done with the GSM-R system as the transmitter and a commercial signal analyzer as the receiver, or with Propsound channel sounder [74–77]. As the train reaches up to 350 km/h, the coherence distance can be reduced to 10 cm [74], which presents challenges to the measurements. Some subway channel measurement activities were also carried out in Madrid, Spain, and in Shanghai and Beijing, China, and other cities [78–81]. The SISO channel was concerned in these measurements and several large- and small- scale parameters such as path loss, SF, amplitude distributions, DS, K-factor, level crossing rate (LCR), average fade duration (AFD), etc, were mainly analyzed. The HSR channel measurement for the viaduct scenario were conducted at 2.6 GHz along the Harbin-Dalian HighSpeed-Railway (HD-HSR) in China by the researchers of Beijing University of Posts and Telecommunications (BUPT) to obtain 2×2 MIMO channel measurement data, from which the correlation coefficients between the subchannels, PL, SF, DS and K-factor were analyzed.

As a part of the WINNER II project, some fast-train channel measurements with SIMO setup on the express railway between Siegburg and Frankfurt in Germany was carried out the by Medav, Technical University of Karlsruhe (TUK) and Technical University of Ilmenau (TUI) in 2006 using RUSK channel sounder [82]. During the measurement, the train could run at different speeds of 20 km/h, 100 km/h and 240 km/h. Monopole antenna fixed on the top of the train carriage with 5 m high was used at the transmitter side, and the receiving antenna was a circular array composed of 16 diskcone antennas (UCA16) placed at 6 m high. Based on the measurement data, the path loss, K-factor, mean square DS, MED, ASA, and some other parameters were analyzed. Because the WINNER II D2a model only supports the LOS case, it is not suitable for the most HSR communications scenarios. Besides, it frequency band supported is 2 GHz, which is not suitable for the GSM-R or LTE-R system in nowadays HSR communications.

With the deployment of multi-antenna LTE-R system, it is necessary to carry out more channel measurements with multiple antennas in various scenarios, such as HSR, ordinary railway and subway, so as to analyze and better utilize the channel information in the angle domain [83].

3.3.2.3 Measurement of Millimeter Wave Band

The measurements and researches on high frequency (HF) band channel have been carried out over 20 years. The main concerns have changed from narrowband channel characteristics to wideband channel characteristics. The concerned small-scale characteristics have changed from simple multipath characteristics in delay domain to joint characteristics in both delay domain and angular domain. In recent years, with the the gradual progress of 5G R&D, some key frequency bands in 6 ~ 100 GHz, such as 11 GHz, 15 GHz, 28 GHz, 38 GHz, 60 GHz and E band, receive more and more attention, in which the path loss, blockage loss, penetration loss, SF, DS, AS, and other channel characteristics are being extensively studied. Many communication enterprises, universities, and research institutes, including Professor Rappaport's team, National Institute of Information and Communications Technology (NICT) in Japan, Intel Laboratory (in Russia and Germany), Fraunhofer HHI laboratory in Germany, Aalto University, Tokyo Institute of Technology, etc., have carried out extensive HF channel measurements in multiple frequency bands. They also formed several influential international project teams such as METIS, MiWEBA, mmMAGIC, 5GCM, etc, to do the channel measurements and modeling in cooperation. Domestically, HUAWEI, Southeast University, Beijing University of Posts and Telecommunications, Tongji University, Beijing Jiaotong University, Shandong University and other research organizations and institutes have begun the relevant researches and participated in the international cooperation. Some measurements and the related details are listed in Table 3.8.

Table 3.8 Summary of high frequency band measurement activities

Research institute	Frequency (GHz)/Bandwidth (GHz)	Platform type	Antenna and array	Scenario	Analysis items
NTT DOCOMO [1]	0.8,2,2.4,7,8,45,27,36	Customized narrowband, sync with cesium clock	Sleeve antenna	Outdoors: street canyon (<1 km), TxH:1.5,6,10 m RxH:1.5,2.5 m	PL
Nokia/Aalborg [12]	10,18	Commercial PSG +PSA	Omni-directional antenna	Street canyon, 6 subscenarios (LOS 10~120 m, NLOS 20~270 m); TxH:7 m, RxH: 1.85 m	PL
TIT [109]	11/0.4	Customized, sync with cesium clock	1. linear array:Tx*6/Rx*2 with 4 dBi dual polarized dipole 2. circular array, Tx*12/ Rx*12, with 6 dBi dual polarized patch	Indoors: same floor	PL, XPR, DS, coherent bandwidth, Ricean K- factor
NICT(IEEE 802.15.3c) [47]	62.5/3	VNA	Rectangular or conical horn antennas, rotating in azimuth with 5° step	Indoors: living room, office, desktop, library, corridor	All large-/small-scale polarized parameters, without zenith angle
INNL(IEEE 802.11ad) [9, 4, 48]	60/0.8 60/0.8	Customized VSG, VSA+ commercial up/down-converter	18 dBi horn antenna 3D rotating with 10° step	Indoors: conference room, office with cubicles, living room; H:0.7/0.9/1.5/2.9 m	All large-/small-scale polarized parameters
IMC (MiWEBA) [4]			Tx:19.8 dBi horn antenna / 34.5 dBi lens antenna, Rx:12.3 dBi conical horn antenna	Outdoors: campus access, TxH:6.2 m,RxH:1.5 m	All large/ small scale polarized parameters
Rappaport: VT,Austin U., NYU [90~99]	28,38,60,73.5/0.2~0.8	Customized PSG+ up/down-converter, sliding correlation receiving, sync with rubidium clock	horn antennas (13.3/15/19/ 24.5/25 dBi), OEW, 39 dBi parabolic, 2D/3D rotating	Indoors; outdoors: roof to ground, UMa TxH:8/23/35 m @38 GHz, UMi TxH: 7/17 m, RxH:2/ 4.06 m @28/73 GHz; distance 30~200 m	PL and exponent, DS, PAS, AOA, PDP, outage, penetration loss

Institution	Frequency/bandwidth	Equipment	Antenna	Environment	Parameters
Fraunhofer HHI	60/0.25 [4]	Customized baseband,commercial RF module	2 dBi omnidirectional antenna	Street canyon, TxH:3.5 m, RxH:1.5 m	All large-/small-scale polarized parameters,
	10/0.25, 60/0.25 [1]		Dipole antenna (10 GHz) 5 dBi horizontally omnidirectional (60 GHz)	Street canyon, distance 28,142 m; TxH:5 m, RxH:1.5 m	PL and exponent, DS
	Multi-frequency: 10.25/0.5, 28.5/1, 41.5/1.5, 82.5/1.5 [11]		omnidirectional	Street canyon; Rx 3 LOS routes(<310 m),4 NLOS routes(<110 m); TxH:5 m, RxH:1.5 m	-
Ericsson Ltd. [1] [11]	Multi-frequency: 2.44/0.08, 5.8/0.15,14.8/0.2,58.68/2	VNA+ up/down-converter	Tx: 7 dBi patch @2.44/5.8 GHz, OEW@14/58 GHz; Rx:2 dBi omni-directional; all vertically polarized	O2I(outdoor 60 m to indoor 2~15 m); TxH/RxH:1.5 m	Peneration loss, DS
				Indoor(<80 m),Street canyon (~130 m); TxH/RxH:1.5 m	PL, diffraction and diffuse scattering phenomena
	Multi-frequency: 5.8/0.15, 14.8/0.2,58.68/2		TxRx: 2 dBi monopole Virtual cube array 25 [3]	Indoor(LOS 1.5 m, NLOS 14 m); TxH/RxH:1.5 m	AS,DS,PDP
Aalto University	62/4 (measurement), 0.2 (analysis) [49]	VNA+ up-/down-converter	Tx: VHUPA(7*7) with biconical Rx: VVUPA(7*7) with OEW	Indoors: office; TxH:~2.3 m, RxH: ~1 m	All large/ small scale parameters
	61~65 [1]		Tx:20 dBi horn antenna, rotating in azimuth with 3° step; Rx: biconical	Shopping mall, cafeteria, open square; TxH/RxH:~2 m	All large/ small scale polarized parameters, without zenith angle
	Multi-frequency: 14~14.5,27~27.9,59~63 [11]		Tx: vertically polarized biconical Rx:19 dBi dual-polarized horn antenna,rotating in azimuth with 5° step;	Street canyon (19~121 m); TxH/RxH:2.57 m	Polarized PDP, PADP,Frequency dependence.DS, ASA,PL
	27.45/0.9 [11]		Antenna same as above,except rotating with 7.2° step	Open square access; TxH:1.6 m, RxH:5 m	

(continued)

Table 3.8 (continued)

Research institute	Frequency (GHz)/Bandwidth (GHz)	Platform type	Antenna and array	Scenario	Analysis items
	Dual-frequency: 15/2,28.5/3 [11]		Antenna same as above,except rotating with 5° step	Airport check-in area; TxH:1.6 m, RxH:5 m	
	81~86 [106–108]	SFG +VNA+ up/down-converter, sync with GPS	45 dBi parabolic antenna, 20 dBi horn antenna	Outdoors: Roof to street, street canyon, TxH/RxH:5 m	CIR (CPR/XPR)
Orange,FT at Belfort [11]	Multi-frequency:3.6~3.725, 10.5~10.625,17.3~17.425	VNA+up-/down-converter	Tx/Rx: 2 dBi@3.6/10.5 GHz, 7.5 dBi@17 GHz vertically polarized(4 orientations)	O2I(12~25 m); TxH:2.5 m, RxH:1.5 m	PDP, Blockage loss
CEA-LETI [11]	83.5/6	VNA+up-/down-converter	Tx:10 dBi horn , rotating in azimuth with 45° step; Rx:20 dBi horn,rotating in azimuth with 10° step and 3*2 planar array with 40 cm spaced	Office,conference room(<8 m);TxH/RxH:1.3 m	PDP,PADP,angular distribution
Tongji University/Huawei [113]	72/2	VSG+ VSA	25 dBi horn antenna, Tx/Rx point to each other(LOS); 6 orientations in azimuth (NLOS)	Indoors	Composite and per Cluster ASA, ZSA and relations with distance, cluster numbers
Tongji University/ETRI [114]	28/0.5	Customized broadband,sliding correlation processing	Tx: 24.4 dBi horn fixed orientation, Rx:9.9 dBi horn, rotating in azimuth with 10° step, 3 elevation settings:[0, ±10°]	Large open office; TxH/RxH:1.5 m	PADP, DS,AS,per cluster DS and AS, cluster numer,ray number within a cluster
BUPT [112]	28/1	VNA+ amplifier	25 dBi horn antenna	Indoors(Tx-Rx distance <30 m); TxH:1.93 m, RxH:1.75 m	PL, PL index, PDP, DS

Southeast University [51, 110]	45/1	PSG+VNA	23.7 dBi horn antenna, OEW, dipole	Conference room, office with cubicles, living room, TxH:1.95 m,RxH:1 m	PL index, PDP, PAS
NCEPU	26/1, 39.5/1	AWG + PSA	24.3 dBi horn(@26 GHz), 27 dBi horn(@39.5 GHz)	Dinming hall, TxH/RxH: 1.3/1.3 m,Tx-Rx distance 15 m	Blockage loss
Shandong University [111]	60/2	PSG+VNA	Tx virtual hollow cubic array (with monople) /3D rotating a 20 dBi horn antenna, Rx 10/20 dBi horn antenna	Office with cubicles TxH/RxH: 1.6 m	PDP,DS,ASD, ZSD,SV small scale parameters

In the early stage, some measurement results for urban and suburban environments at 9.6/28.8/57.6 GHz [84], 55 GHz [85], 60 GHz [86], 62 GHz [87] and other frequencies were published, in which the narrowband channel characteristics including path loss, rain attenuation and oxygen absorption were concerned. At the end of the 1990s, the transmission and reflection coefficients of some typical materials including walls, floors, ceilings, windows, etc, at 57.5 GHz, 78.5 GHz and 95.9 GHz bands were measured by NICT [88]. The measurement results were compared with the reflection model of multi-layer materials. The early measurements were carried out at single frequecy band, till recent the measurements at multiple frequecy band in the same environment were carried out in order to establish the fequency dependence of various prarmeters.

The main measurements of IEEE 802.15.3c channel model were carried out by NICT [47], NICTA in Australia, IMST in Germany, France Telecom (FT), IBM, University of Massachusetts (UMass) [8]. Except that IBM used a broadband sounder with 600 MHz bandwidth [89], the other institutes used VNA based sounder to channel measurement. Measurements mainly concentrated on ten indoor scenarios, including the living room, office, desktop, library, corridor, aircraft cabin, etc, in different in LOS and NLOS conditions. The Angle information was obtained by rotating the standard gain horn antenna with 5° step in the azimuth direction. Finaly, the cluster parameters (the number, arrival rate, power attenuation index, AS) and intra cluster ray parameters (arrival rate, average time, power attenuation rate, etc.) were given by analyzing the measurement data. See Sect. 6 in [8] for details.

Prof. Rappaport's team used the separate components and modules to build a broadband sounder based on the concept of sliding correlation [43]. The chip rate of PN sequence could be changed to 200 MHz, 400 MHz to 750 MHz (the null-to-null bandwidth is equal to two times of the chip rate). Some measurements at 28 GHz, 38 GHz, 60 GHz and 73.5 GHz band were successively carried out in VT, AUT, and NYU [90–99], from which the path loss, delay, angle, and other parameters were analyzed [20]. All measurement activities were designed and targeted to a specific measurement environment, but the similar equipments and procedures were used. Directional antennas were adopted at the BS (the transmitter of a channel sounder) and UT (the receiver of a channel sounder) to send and receive signals respectively. The directional antenna was installed on a 3D rotating tripod. For every setting of BS antenna orientation or UT antenna height, the UT antenna was steppedly rotated in the whole solid angle to scan the channel. The related work are outlined in [20] and Sect. 2.1.2 of [4]. Based on the measurements, an outdoor mmWave channel model [100, 101] was established by using a time cluster-spatial lobe modeling method, in which the path loss, the power and delay parameters of time clusters and rays within a cluster, and the angular parameters represented by the number and spread of spatial lobes were included. This model did not support the parametric modeling of elevation angle i.e., it was a 2D model. Later by adopting the ray tracing technique, a 3D mmWave channel model under the framework of 3GPP (WINNER) was proposed [102]. Herein in which the average power in dB scale of the cluster or intra-cluster rays are Gaussian distributed and decay exponentially with delay as SV model. For a ray within a time cluster, a spatial lobe is randomly assigned to it to be used to generate the angle values. The angles of spatial lobes or rays followed Gaussian or Laplacian distribution. Three path loss models, i.e, directional, beam combining, and omni-directional, with two model structures of close-in reference distance and floating intercept, were proposed.

The main measurements to produce IEEE 802.11ad channel model [9] were carried out by Intel Nizhny Novgorod Lab (INNL) in Russia [48], as well as Technical University of Braunschweig (TUB) [103], TUI [104], Fraunhofer HHI [105], MEDAV, and IMST in Germany. A customized broadband sounder composed of a VSG, a VSA and two commercial 60 GHz up/down-converter modules was used by INNL. Each 18 dBi standard gain antennas that could steppedly rotate in the whole 3D space was used at the transmitter and receiver. A VNA-based sounder was used by TUB to measure the human blockage loss. A channel sounder used by Fraunhofer HHI was comprised of arbitrarily waveform generator (AWG), oscilloscope and up/down-converter. TUI, MEDAV, and IMST cooperated with each other to build a 1T2R (one transmitting port and two receiving ports) broadband (up to 3 GHz) sounding system composed by an ultra broadband receiver, a frequency synthesizer and two up/down-converter. Two types of link, access and D2D, were measured in indoor environments, such as living rooms, meeting rooms and offices with cubicles. The measurement results showed that obvious clustering exists in the channel, which matched well with the results of ray tracing simulation. Fraunhofer HHI laboratory and Intel Mobile Communications (IMC) laboratory in Germany took part in the MiWEBA project [4]. Fraunhofer HHI used self-developed baseband sounder with 250 MHz bandwidth, commercial 60 GHz up/down-converter module and approximate omni-directional (2 dBi) antennas to measure the access channel for street canyon scenario. The transmit antenna was 3.5 m high and the antenna of the mobile receiver was 1.5 m. The power and delay of several paths were measured and analyzed. IMC used the same equipment as INNL to measure the access channel on campus. A 19.8 dBi rectangular horn antenna or a 34.5 dBi lens antenna was used depending on the Tx-Rx distance less than or more than 35 m and placed at 6.2 m high. The mobile receive antenna was a 12.3 dBi circular horn antenna placed at 1.5 m high. The measurement results showed that two strongest paths were LOS and the ground reflection path, while the other paths were 15~20 dB lower. The measurement results matching well with the ray tracing calculation prompted the project team to use the Q-D modeling method to construct the channel model, which is comprised of deterministric D rays and random R rays. For more details please refer to the MiWEBA model [4]. The researchers in Fraunhofer HHI measured the propagation at 10 GHz and 60 GHz simultaneously and and compared. In the LOS case, they found the propagation loss at 60 GHz was 15.6 dB higher than that at 10 GHz, which conformed to the Friis' free space path loss equation. These two channels showed similar characteristics, but at 60 GHz there were less resolvable paths due to smaller dynamic range. In the NLOS case, some paths simultaneously occured at both frequency bands, while some paths occured at only one frequency band. In general, similar to the LOS case, there were less resolvable paths at 60 GHz.

Some measurements at 15 GHz, 28 GHz, 60 GHz, and E band (81~86 GHz) were carried out by Aalto University in Finland and outlined as follows.

1. The measurements at 60 GHz in conference room were carried out by using a VNA-based sounder. At the transmitter side, a virtual horizontal uniformly planar array (VHUPA) with 7×7 elements was formed by translating a vertically polarized biconical antenna. At the receiver side, a vertically polarized OEW was used to form a virtual vertical uniformly planar array (VVUPA) with 7×7 elements. The high resolution estimation algorithms were used to obtain channel parameters, based on which a SV-like model were proposed [49].
2. In the METIS project, multiple measurements at 61~65 GHz (4 GHz bandwidth) in the indoor shopping mall, indoor cafeteria and outdoor square were carried out by using a VNA-based sounder [1]. A 20 dBi horn antenna was horizontally rotated with a step of $3°$ at the transmitter side, while a 5 dBi biconical antenna was used at the receiver side, so that only single-directional and 2D channel parameters were obtained. In indoor cafeteria and outdoor square, more channel parameters were obtained by adopting the point-cloud field prediction technique (a laser ranging system was used to build a 3D electronic map called as a point cloud). After calibrating with the measurement results, a 3D channel model for these two scenarios were proposed.
3. In the mmMAGIC project, they completed many measurements, e.g., in the street canyon at multiple frequencies, in the open square at 27 GHz, in the airport check-in area at 15 GHz and 28 GHz.
4. With Sweep Frequency Generator (SFG) as the transmitter, and the VNA as the receiver, the wireless channel at 81~86 GHz E-band was measured [106–108]. A 24 dBi and a 45 dBi directional antennas were used in the measurement. The scenarios included street canyon, root-to-street and the maximum measurement range was up to 1100 m. It was found that there still existed the MPCs even in the LOS condition and by using high-gain directional antennas, and the first MPC was 20 dB lower than the LOS component.

The researchers in Ericsson Ltd. used a VNA-based sounder to complete a series of measurements in participating the METIS, mmMAGIC, and 5GCM projects. (1) The human blockage measurements in indoors at 57.68-59.68 GHz band with two 10 dBi (60°elevation beam width, 30°horizontal beam width) antennas were carried out and found that the blockage loss was as high as 10–20 dB. (2) The measurements in indoor medium range and a long corridor of the office environment were performed at multiple frequencies to find the path loss exponent and the diffraction components contributed the main power for indoor mmWwave transmission in the NLOS case. (3) Multi-frequency measurements at 2.44, 14.8 & 58.68 GHz were carried out in street microcell enviroments in the LOS and NLOS case. The measurements found that in the NLOS case the path loss is less than expected by knife edge diffraction, and the frequency dependency is not as obvious as the diffraction, which revealed that there were other reflection or diffuse components dominating the propagation in outdoor NLOS scenario. (4) The penetration loss through the wall at multiple frequencies were measured. (5) The rich directional scattering of mmWave channel in indoor scenario was measured at multiple frequencies using an extreme size virtual cubic array (25 × 25 × 25) antenna formed by means of using a 3D positioning robot to place a dipole antenna with 0.4 wavelengths space.

Another two measurements at the mmWave band were completed in the METIS project [1]. (1) The path loss measurements in the city blocks were carried out narrowband (single frequency) sounding system by NTT DOCOMO at 0.8 and 2.2, 4.7, 8.45, 27, and 36 GHz bands. The cesium clocks were used as the reference source to ensure the frequency consistency between the transmitter and the receiver. (2) TIT in Japan used a self-developed modular software radio device to carries out the sounding of 11 GHz channel [109]. Two antenna schemes were used to meaure the channel in room and hall scenarios. In the first scheme, the antenna elements was a 4 dBi dual-polarized dipole antenna. The transmitter used 6 elements to form a uniform linear array (with 2λ spaced) and the receiver used 2 elements. The system was used to measure a 12 × 4 dual-polarized MIMO channel. In the first scheme, the antenna elements was a 6 dBi dual-polarized dipole antenna. Transmitter and receiver used 12 elements respectively to form uniform circular array antennas (with 0.44λ spaced), which was used to measure the 24 × 24 dual-polarized MIMO channel. The polarized path loss, DS, cross-polarization power ratio (XPR), coherent bandwidth and Ricean K-factor, etc. were mainly analyzed.

In the 5GCM project, Nokia and Aalborg University in Denmark completed the path loss measurement at 10 GHz and 18 GHz [12]. In the mmMAGIC project, there were other measurement activities. (1) Belfort lab of Orange (a French telecom operator) completed the measurements for O2I scenarios at multiple frequencies. (2) CEA-LETI in France completed indoor channel measurements at 83.5 GHz. (3) Bristol University in UK completed channel measurements in a shopping mall at 60 GHz.

Some measurement campaigns carried out in China are briefly outlined as follows. The channel measurement at 45 GHz band for three indoor scenarios defined by IEEE 802.11aj were carried out by Southeast University and China Telecommunication Technology Labs (CTTL) [51, 110]. A microwave signal generator and VNA were used as the transmitter and the receiver respectively. Path loss and DS at 45 GHz were mainly analyzed. The researchers in Shandong University adopted a similar system to carry out the channel measurements at 18, 28, 38 & 60 GHz in the laboratory with cubicles. Two types of virtual antenna array formed by rotating horn antenna and translating omni-directional antenna in 3D axes, were used at the transmitter, while the horn antennas were used at the receiver. The main parameters such as delay, angle, and others were obtained from the measurement data to be used in the channel modeling [111]. The researchers in BUPT used a VNA-based sounder to carry out the measurement and analysis the radio propagation at 28 GHz in the indoor environment [112]. The measurement results showed that parameters such as path loss exponent, AOA, and others had a strong correlation with the propagation environment. Xuefeng Yin et al. cooperated with Huawei Ltd. to conduct channel sounding for the indoor enviroments at 72 GHz [113] and cooperated with ETRI of South Korea to complete the 28 GHz channel measurements in a large open office in the LOS and NLOS cases [114], in which the PADP, composite and per cluster DS & AS, cluster numer, ray number within a cluster were analyzed. Xiongwen Zhao et al. of North China Electric Power University (NCEPU) measured the blockage attenuation at 26 and 39.5 GHz. The Vogler's multiple KED (Knife Edge Diffration) model was used to model the blockage effects and found the loss at 26 GHz was smaller than that at 39.5 GHz [115].

3.4   Channel Data Processing

In channel sounding, the measurement data, or the received signals, are superposition of altered versions of the transmit signal passing through multiple paths with different delays, gains, and directions. The processing to be applied to the measurement data includes two aspects: path parameters extraction and channel statistical analysis. Path parameters extraction is to extract the parameters of each path from the received signals using a high resolution parameter estimation algorithm. These path parameters include complex amplitude, delay, arrival angle, departure angle, Doppler shift and so on. Channel statistical analysis is to further analyze the obtained path parameters and find the statistical characteristics and descriptions of the sounded channel.

3.4.1   Path Parameters Extraction

There are mainly two categories of commonly used channel parameters extraction algorithms, i.e., subspace-based methods and maximum likelihood methods. Within the first category are MUltiple Signal Classification (MUSIC) algorithm [116], Estimation of Signal Parameter via Rotational Invariance Techniques (ESPRIT) algorithm [117–118] and its variant Unitary ESPRIT [119–121], etc. In the second category, there are Expectation Maximization (EM) algorithm [122, 123], Space Alternating Generalized EM (SAGE) algorithm [124–129] and frequency-domain SAGE algorithm [130], Sparse Variational Bayesian SAGE (SVB-SAGE) algorithm [131], and Richter's Maximum Likelihood Framework For Parameter Estimation (RIMAX) algorithm [132], etc.

ESPRIT algorithm and its variant are been developed to either jointly estimate delay and azimuth angle, or jointly estimate azimuth and elevation angles [120]. MEDAV Company's MIMO channel sounder RUSK took ESPRIT as the core algorithm at early stage, and later it adopts the RIMAX algorithm. EM algorithm alternates E (expectation) step and M (maximization) step iteratively to get the optimal solution but with a relatively slow convergence speed. SAGE algorithm bases on EM algorithm, separates all parameters into several subsets and updates a parameters subset each time in the iterative process, which reduces the computational complexity and expedite the convergence speed significantly. As an extension of the classical SAGE algorithm, SVB-SAGE algorithm obtains the distributions of the MPC parameters instead of parameter point estimates by utilizing the idea of variational bayesian inference. Meanwhile the number of MPC is jointly estimated using several sparsity priors [131]. RIMAX algorithm also use the ideal of SAGE to alternatively estimate the parameters of specular component (SC) and dense multipath component (DMC). Gradient based algorithm like Gauss-Newton or Levenberg-Marquardt is used to estimate iteratively the required parameters and achieve a convergence rate even faster than SAGE algorithm [132].

3.4.1.1 EM Algorithm

Basing on the signal model, one can write the probability density function (PDF) of the measurement data in terms of signal parameters (aka the likelihood function of parameters). Maximum likelihood (ML) parameter estimation algorithms choose the parameter values that can maximize the likelihood function. Usually, the expression of the likelihood function is too complex to directly obtain the parameters maximizing it. Thus, the iterative algorithms are often resorted. EM algorithm can effectively estimate parameters by iteratively carrying out the E step and M step [122, 123]. The classical EM algorithm can be described as follows. Firstly, in the E step, the posterior distribution of the complete (unobservable) data is found with respect to the observable measurement data (incomplete data) and assumed (or updated in the last steps) parameters and used to calculate the expectation of the complete-data log- likelihood. Then the parameters maximize the expectation is searched in the M step.

For channel sounding, EM algorithm splits the optimization problem that jointly estimate multiple superimposed paths into multiple separate optimization problems for estimating a single path. Here we illustrate the EM algorithm used in a SIMO (single transmitter and M receivers) sounding system. The received signal is superposition of L paths signals, which can be expressed as

$$y(t) \triangleq [y_1(t), \dots, y_M(t)]^{\mathrm{T}} = \sum_{l=1}^{L} s(t; \boldsymbol{\vartheta}_l) + N(t)$$

$$= \sum_{l=1}^{L} c(\phi_l)\alpha_l \exp\{j2\pi v_l t\} u(t - \tau_l) + N(t) \qquad (3.6)$$

where $u(t)$ is time-domain sounding signal with transmit power P_u and period T_a. I snapshots are measured with time intervals T_f. $\boldsymbol{\vartheta}_l \triangleq [\tau_l, \phi_l, v_l, \alpha_l]$ denotes the vector containing the parameters of the l-th path, including path delay τ_l, azimuth AOA ϕ_l, Doppler shift v_l and amplitude α_l. $c(\phi) \triangleq [c_1(\phi), \dots, c_M(\phi)]^{\mathrm{T}}$ denotes the steering vector of antenna array reflecting two aspects of information: antenna radiattion pattern and geometric structure of the array. $n(t) \triangleq [n_1(t), \ldots, n_M(t)]^T$ denotes M dimensional complex white Gaussian noise.

Define complete (unobservable) data sets $x_l(t) \triangleq s(t; \vartheta_l) + \sqrt{N_0/2} n_l(t)$ and incomplete (observable) data sets $y(t) = \sum_{l=1}^{L} x_l(t)$. First of all, assuming that $x_l(t)$ is known, the log likelihood function of complete data $x_l(t)$ in terms of the parameter vector ϑ_l can be expressed as follows,

$$\Lambda(x_l; \vartheta_l) \triangleq \frac{1}{N_0} \left[2 \int_{D_0} \Re\{s^T(t'; \vartheta_l) x_l(t')\} dt' - \int_{D_0} \|s(t'; \vartheta_l)\|^2 dt' \right] \quad (3.7)$$

The parameters that maximize the ML function are simply written as,

$$\left(\widehat{\vartheta}_l\right)_{ML} (x_l) \in \arg \max_{\vartheta_l} \{\Lambda(x_l; \vartheta_l)\} \quad (3.8)$$

E Step: In fact $x_l(t)$ is unknown, thus, we need to estimate $x_l(t)$ firstly based on the incomplete data $y(t)$. One direct way is to use the conditional expectation of $x_l(t)$ with repects to $Y(t)$ and the predefined or old parameters. Specifically, we can use successive interference cancelation method to obtain $x_l(t)$ assuming the parameters about other paths are known,

$$\hat{x}_l(t; \widehat{\vartheta}') \triangleq y(t) - \sum_{\substack{l'=1 \\ l' \neq l}}^{L} s(t; \widehat{\vartheta}'_{l'}) \quad (3.9)$$

M Step: The l-th path parameters ϑ_l can then be re-estimated by computing its MLE based on the estimation of the l-th signal

$$\widehat{\vartheta}''_l = \left(\widehat{\vartheta}_l\right)_{ML} \left(\hat{x}_l(t; \widehat{\vartheta}')\right) \quad (3.10)$$

In particular, the ML estimator can be expressed as

$$\left(\tau_l, \widehat{\varphi_l}, \nu_l\right) = \arg \max_{\tau, \phi, \nu} \left\{\left|z(\tau, \phi, \nu; \hat{x}_l)\right|\right\}$$

and

$$z(\tau, \phi, \nu; x_l) = \sum_{i=1}^{I} \int_{D_i} u^*(t' - \tau) e^{-j2\pi\nu t'} c^H(\phi) \hat{x}_l(t') dt' \quad (3.11)$$

Where D_i is observation time interval of the i-th snapshots, $i = 1,2,\ldots I$. The calculation of z can be seen as correlations (match filtering) in both time domain and spatial domain. The spatial correlation is just the conventional Bartlett beamforming. The M step in EM algorithm updates all the parameters at the same time, which requires a large number of calculations. It is validated by mathematical proof that the convergence speed of EM algorithm is inversely proportional to the amount of Fisher information in complete data space, so that a faster asymptotic convergence can be achived with less Fisher information. Because of updating all the parameters together in each iteration step, EM algorithm gets a very slow convergence speed.

3.4.1.2 SAGE Algorithm

SAGE algorithm is developed based on the classical EM algorithm. However, in M step, it divides all parameters of one path into several sets and updates only one set of parameters at a time while keeping other sets of parameters fixed. The process continues till all the parameters of one path are updated. Then it run E step and M step to re-estimate the parameters of next path. The procedure of updating all parameters of all components in this way constitutes a single update cycle of the algorithm. The update cycles are repeated anew until convergence or reach a given maximum cycles. In each step, it updates a parameter subset and hence has less Fisher information, which not only reduces computation, but also has a faster convergence speed.

B. H. Fleury, professor of University of Aalborg in Denmark, firstly applied the SAGE algorithm to the problem of channel parameter extraction [126]. Here we still consider a SIMO sounder, the M step of SAGE algorithm can be written as follows:

$$\widehat{\tau}_l'' = \arg\max_{\tau} \left\{ \left| z\left(\tau, \widehat{\phi}_l', \widehat{\nu}_l'; \hat{x}_l(t; \widehat{\boldsymbol{\vartheta}}')\right) \right| \right\}$$

$$\widehat{\phi}_l'' = \arg\max_{\phi} \left\{ \left| z\left(\widehat{\tau}_l'', \phi, \widehat{\nu}_l'; \hat{x}_l(t; \widehat{\boldsymbol{\vartheta}}')\right) \right| \right\}$$

$$\widehat{\nu}_l'' = \arg\max_{\nu} \left\{ \left| z\left(\widehat{\tau}_l'', \widehat{\phi}_l'', \nu; \hat{x}_l(t; \widehat{\boldsymbol{\vartheta}}')\right) \right| \right\}$$

$$\widehat{\alpha}_l'' = \frac{1}{I\left\|c\left(\widehat{\phi}_l''\right)\right\| T_a P_u} z\left(\widehat{\tau}_l'', \widehat{\phi}_l'', \widehat{\nu}_l''; \hat{x}_l(t; \widehat{\boldsymbol{\vartheta}}')\right) \tag{3.12}$$

As for the initialization of SAGE algorithm, the initial parameters of each path can be set all zeros or obtained using the MUSIC algorithm. In the M step, since no a priori knowledge about the phase of the complex amplitudes, the maximization procedures for τ and ϕ are replaced by two non-coherent estimation programs.

$$\widehat{\tau}_l'' = \arg\max_{\tau} \left\{ \sum_{m=1}^{M} \sum_{i=1}^{I} \left| \int_{D_i} u^*(t' - \tau)\hat{x}_{l,m}(t'; \widehat{\boldsymbol{\vartheta}}') dt' \right|^2 \right\}$$

$$\widehat{\phi}_l'' = \arg\max_{\phi} \left\{ \sum_{i=1}^{I} \left| \int_{D_i} u^*\left(t' - \widehat{\tau}_l''\right) c^H(\phi) \hat{x}_l(t'; \widehat{\boldsymbol{\vartheta}}') dt' \right|^2 \right\} \qquad (3.13)$$

Whereas the calculation of Doppler shift still uses the same procedure as the above iterative process. Correspondingly, since only the $l-1$ paths are estimated, in the initialization process the E step of estimating l-th path parameters is replaced by

$$\hat{x}_l(t; \widehat{\theta}') = y(t) - \sum_{l'=1}^{l-1} s(t; \widehat{\boldsymbol{\vartheta}}_{l'}') \qquad (3.14)$$

Issues related to the measurement, including the number of sounding channels (such as SIMO and MIMO), sounding modes (such as parallel sounding and serial sounding), and sounding sequences (such as PN sequence, frequency sweep in frequency domain or OFDM signal), the polarization measurement requirements (such as one or all of the polarization pairs $\theta\theta$, $\theta\phi$, $\phi\theta$, $\phi\phi$) will result in greatly different signal models, which requires to develop corresponding SAGE algorithm. Based on the traditional SIMO SAGE algorithm, B. H. Fleury, Xuefeng Yin et al., pioneered a series of work: extended the SAGE algorithm to estimate 3D angle (angle of azimuth and zenith) parameters at the receiver [127]; extended to estimate 3D angles at both receiver and transmitter, and proposed Initialization and Search Improved SAGE (ISI-SAGE) algorithm, whose idea is to reduce the search range adaptively so as to speed up the calculation [128]; estimated the four polarized complex gain parameters. Since one complex contains two real values, the real part and the imaginary part, the algorithm can estimate upto 14 real parameters of each path [129]. Besides, researchers have considered the parameters estimation problem for near field communications (supporting spherical wavefront), where the distances between the transmitter/receiver arrays and the scatterers (the distance from scatterers to the array center) are added to the signal model, and the corresponding SAGE algorithm was derived [135, 136].

CRLB Performance Bounds

For the single path signal, Fisher information matrix can be derived from its likelihood function, Equ (3.5), and then the parameters performance can be got. SIMO sounding is still taken as an example, i.e., one transmit antenna and uniformly linear array comprised of M receive antennas. It is assumed that the sounding sequence $u(t)$ containing K chips with the chip duration of T_p and the sequence length of $T_a = KT_p$. One channel sounding is carried out in each snapshot and there is a total of I snapshots with snap spacing of T_f. Then Cramer Rao lower bounds (CRLBs) of the path parameters, AOA ϕ, Doppler shift ν, delay τ and path gain α, are given by

$$\mathrm{CRLB}(\phi) = \frac{1}{\gamma_O} \frac{3}{2(\pi d/\lambda)^2 \sin^2(\phi)(M^2 - 1)}$$

$$\mathrm{CRLB}(\nu) = \frac{1}{\gamma_O} \frac{3}{2\pi^2 T_f^2 (I^2 - 1)}$$

$$\mathrm{CRLB}(\tau) = \frac{1}{\gamma_O} \frac{1}{8\pi^2 B_u^2}$$

$$\mathrm{CRLB}\left(\frac{\alpha}{|\alpha|}\right) = \frac{1}{\gamma_O} \qquad\qquad (3.15)$$

where $B_u = (1/\pi T_p)\sqrt{(N_s/2)(1 + 1/K)}$ is the Gabor bandwidth of the transmitted signal $u(t)$. γ_O is the signal-to-noise (SNR) at the output of the correlator and $\gamma_O = MIKN_s\gamma_I$. γ_I is the SNR at the input of each antenna port and $\gamma_I = P_u|\alpha|^2/(N_0/T_s)$. d is the space between antenna elements. When d is equal to half a wavelength, $\mathrm{CRLB}(\phi)$ coincides with the expression (27) in [126]. It can be seen from the above formula that each parameter is inversely proportional to γ_O, which means the estimation error can be reduced by increasing antenna number M, snapshot number I and sequence length K. Meanwhile, increasing M can further reduce the errors in AOA estimation. ϕ is the angle between the arrival beam direction and the array axis. The $sin^2(\phi)$ in the AOA performance bounds shows that estimation error is minimum when beam impinging along the boresight of the array. $\mathrm{CRLB}(\phi)$ gradually reaches to infinity when the beam direction gradually changes from boresight to broadside of the array. Since ϕ is always between 0 and π, the estimation error should be bounded. Therefore when ϕ is close to 0 or π, the estimation ambiguity exists and $\mathrm{CRLB}(\phi)$ in the above expression is no longer applicable. Increasing snapshot number I and snapshot interval T_f will reduce the Doppler shift estimation error. However, T_f is limited by the Doppler shift according to the Nyquist sampling theorem, i.e., $T_f < 1/(2\nu)$.

Separability of Multipath Signals

The resolution of several commonly used estimation techniques for delay, angle and Doppler shift, such as correlation, beamforming and Fourier methods, is limited by the intrinsic parameters of the measurement system. These resolutions are defined to be the half-widths of main lobe of the magnitude of their corresponding correlation function. For uniform linear arrays (M antenna units), these half-widths are respectively

$$\tau_c = T_p, \phi_c = \frac{360^\circ}{\pi M}, \nu_c = \frac{1}{IT_f} \qquad\qquad (3.16)$$

Two paths are called separable or resolvable only when the difference of their parameters satisfy the following conditions

$$\Delta\tau > \tau_c \tag{3.17}$$

Shorter chip length (corresponding to higher bandwidth) has higher resolution capabilities for delay. Increasing the antenna unit number M can provide high spatial angle resolution. Increasing the measurement snapshot number I and snapshot interval T_f can improve the resolution of Doppler shift. The facts is more optimistic than the claims above. According to the simulation results of the SAGE algorithm [126], if the parameter difference between two paths satisfies $\Delta\tau \gtrsim \tau_c/5$, or $\Delta\phi > \phi_c/2$, or $\Delta\nu > \nu_c/2$, then the root mean squared error (RMSE) of parameter estimation of each path are close to CRLBs of single path, which means the two paths are resolvable.

3.4.1.3   FD-SAGE Algorithm

The SAGE algorithm mentioned is applicable for channel sounding using time domain PN excitation signal. In some channel sounding, frequency excitation signal is commonly used, such as broadband multi-carrier excitation or VNA-based single frequency stepping excitation. Correspondingly the frequency domain SAGE (FD-SAGE) algorithm is developed for parameter extraction [130]. Here we still use a SIMO (single transmitter and multiple receiver) system as an example, and assume that the entire measurement environment is stationary (i.e., no need to estimate the Doppler shift). According to the configuration, the received signal in frequency domain is given by

$$Y(f) \triangleq \sum_{l=1}^{L} c(\phi_l)\alpha_l e^{-j2\pi f\tau_l} + N(f). \tag{3.18}$$

Assuming a total of $K+1$ (K is an even number) frequency points is soundered, the signal model at every frequency point can be expressed as

$$Y(k\Delta f) \triangleq \sum_{l=1}^{L} c(\phi_l)\alpha_l e^{-j2\pi k\Delta f\tau_l} + N(k\Delta f), k = -\frac{K}{2}\dots+\frac{K}{2} \tag{3.19}$$

Using the derivation similar to the time-domain EM algorithm, the complete (aka admissible hidden) data $X_l(k\Delta f)$, i.e., the frequency domain response of the l-th path signal, can be estimated from the incomplete data $Y(f)$ by

$$\hat{X}_l\left(k\Delta f; \widehat{\vartheta}'\right) = Y(f) - \sum_{\substack{l'=1 \\ l' \neq l}}^{L} X\left(k\Delta f; \widehat{\vartheta}'_{l'}\right) \tag{3.20}$$

$\hat{X}_l\left(k\Delta f; \widehat{\vartheta}'\right)$ is stacked into a $(K+1)$-by-M matrix, in which each column represents the $K+1$ frequency responses of each antenna element, and each row represents the frequency response of M antenna elements on a certain frequency point. The maximum likelihood estimation of parameters is given by

$$\left(\widehat{\tau_l, \varphi_l, \nu_l}\right) = \arg\max_{\tau, \phi, \nu} \left\{ |z(\tau, \phi, \nu; x_l)| \right\}$$

$$z\left(\tau, \varphi; \hat{X}_l\right) = \sum_m c_m^*(\varphi) \sum_k e^{j2\pi k\Delta f \tau} \hat{X}_l(k, m) \tag{3.21}$$

The EM algorithm updates all parameters simultaneously, resulting in slow convergence speed. Whereas, the SAGE algorithm updates only a subset of parameters each time in M step,

$$\widehat{\tau}_l^{\,\prime\prime} = \arg\max_{\tau} \left\{ \left| z\left(\tau, \widehat{\phi}_l^{\,\prime}; \hat{X}_l\right) \right| \right\}$$

$$\widehat{\phi}_l^{\,\prime\prime} = \arg\max_{\phi} \left\{ \left| z\left(\widehat{\tau}_l^{\,\prime\prime}, \phi; \hat{X}_l\right) \right| \right\}$$

$$\widehat{\alpha}_l^{\,\prime\prime} = \frac{1}{(K+1)M} z\left(\widehat{\tau}_l^{\,\prime\prime}, \widehat{\phi}_l^{\,\prime\prime}; \hat{X}_l\right) \tag{3.22}$$

Assuming the full zero initialization method is used, in the M step, the path delay τ is obtained by a non-coherent estimation procedure given by

$$\widehat{\tau} = \arg\max_{\tau} \left\{ \sum_m \left| \sum_k e^{j2\pi k\Delta f \tau} \hat{X}_l(k, m) \right| . \right\} \tag{3.23}$$

The calculation of AOA ϕ is the same as the iterative process. The item in braces in (3.23) can be regarded as summing of the power delay profile of every antenna element over all antenna elements. Using Fourier transform directly will lead to a large number of sidelobe components in delay domain. The superposition of multipath signals may produce multiple false peaks, which may lead to improper delay. Usually frequency-domain windowing, such as Kaiser, Hanning or Gaussian window, can be used to reduce the sidelobe in time domain [130]. That is, $\hat{X}_l(k, m)$ is replaced by $\hat{X}_l'(k, m) = \hat{X}_l(k, m)W(k)$, where $\{W(k), k = 1, \ldots\}$ is the window function with normalized power.

Accordingly, in the initialization phase, since only the previously estimated path parameters are available, the E step is changed to

$$\hat{X}_l\left(k\Delta f; \widehat{\vartheta}'\right) = Y(t) - \sum_{\tilde{l}=1}^{l-1} X\left(k\Delta f; \widehat{\vartheta}'_{\tilde{l}}\right) \tag{3.24}$$

As the time-domain version of SAGE algorithm, increasing the number of antenna M and frequency points (bandwidth) improves the estimation accuracy of all parameters. Here only a SIMO sounding system is considered, the signal model and estimation algorithm need to be extended according to the actual configurations and requirements in specific sounding applications.

3.4.1.4 DMC and Its Estimation

The signal emitting from transmit antenna and arriving the receiving antenna is composed of line-of-sight (LoS) component, specular components (SCs), and dense multipath components (DMCs). Each SC corresponds to a discrete and strong propagation path formed by an independent scatterer reflecting electromagnetic waves. The propagation paths can be described by simple geometric relationships. However, the contents of DMCs are far more complex and illustrated in Fig. 3.8. It is usually known that the DMCs consists of Diffuse components (DC) reflected off rough surface having wavelength comparable to radio wave, diffraction, reflection from different layers of scatterers, echoes among scatters, and so on.

The contribution of DMC in the total reveived power varies with the distance between the transmitter and receiver as well as the environment. Intuitively, the proportion of DMC is larger in indoor scenarios or in the NLOS case. Recent studies have indicated that DMCs contribute 20% to 80% to the total receiving power in indoor environment [137] and Industrial Environment [138]. Usually DMC has larger angle distribution region than SC. Moreover, DMCs have longer delay spread than the SCs and the power of DMC cluster approximately decays exponentially with delay [139]. Ignorning DMCs in the channel modeling will underestimate the channel capacity. Therefore, DMC has enormous significance for channel modeling.

The earliest DMC modeling assumed that the DMCs have a "white" spectrum in azimuth angle while they have an exponentially decayed Power Delay Profile (PDP), which can be expressed as [133]

$$\psi(\tau) = E\left[|x(\tau)|^2\right] = \begin{cases} 0, & \tau < \tau_d \\ \alpha_d/2, & \tau = \tau_d \\ \alpha_d e^{-B_d(\tau - \tau_d)} & \tau > \tau_d \end{cases} \tag{3.25}$$

where B_d is the coherence bandwidth, α_d is maximum power, and τ_d is base time delay. The Correlated power spectral density of DMCs is the Fourier transform ($\tau \mapsto \Delta f$) of PDP which is given by

Fig. 3.8 Ilustration of SC and DMC

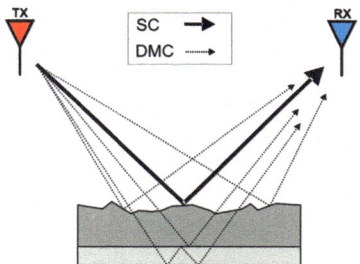

$$\psi(f) = \frac{\alpha_d}{\beta_d + j2\pi\Delta f} e^{-j2\pi\Delta f \tau_d} \tag{3.26}$$

where $\beta_d = B_d/(Mf_0)$ is the normalized correlation bandwidth, f_0 is the sampling interval in frequency domain. Equ (3.26) represents the power spectrum in continuous infinite bandwidth. In practice, it is more convenient to use the discrete power spectrum of finite bandwidth as follows,

$$\kappa = \frac{\alpha_d}{M} \left[\frac{1}{\beta_d} \quad \cdots \quad \frac{e^{-j2\pi(M-1)\tau_d'}}{\beta_d + j2\pi(M-1)/M} \right] \tag{3.27}$$

where $\tau_d' = \tau_d f_0$. The parameters describing DMCs are denoted by $\boldsymbol{\theta}_d = \begin{bmatrix} \alpha_d & \beta_d & \tau_d' \end{bmatrix}$. The covariance matrix of DMCs sampled in frequency domain is a Toeplitz matrix composed of κ, i.e. $R_d(\boldsymbol{\theta}_d) = toep(\kappa, \kappa^H)$. In [133], Gauss-Newton algorithm is used to iteratively update the DMCs parameters, while the initial estimation is found by so called global search strategies beginning with estimated PDP. A simpler least-square based method can be used. First obtain the residual CIR containing DMC by substracting SC component $H_{sc}(\tau)$ (obtained by high resolution estimation algorithm such as SAGE) from the total CIR $H_{Rx}(\tau)$. Then, averaging the residual CIR over multiple antennas to get the residual PDP

$$\psi_{res}(\tau) = \frac{1}{N_t N_r} H_{Rx}(\tau) - H_{sc}(\tau)_F^2 \tag{3.28}$$

where $\| \cdot \|_F^2$ denotes the square of Frobenius norm of a matrix. The optimal DMC parameters are those that can make parameterized PDP approach to the residual PDP,

$$\widehat{\boldsymbol{\theta}}_d = \arg\min_{\boldsymbol{\theta}_d} \sum_{m=0}^{M-1} |\psi_{res}(m\Delta\tau) - \psi_d(m\Delta\tau, \boldsymbol{\theta}_d)|^2 \tag{3.29}$$

where $\psi_d(\boldsymbol{\theta}_d)$ is the PDP of DMCs expressed in discrete time domain, which can be obtained from $R_d(\boldsymbol{\theta}_d)$,

$$\psi_d(\boldsymbol{\theta}_d) = \text{diag}\left[\mathscr{F}^{-1} R_d(\boldsymbol{\theta}_d) \mathscr{F} \right], \tag{3.30}$$

where $\mathscr{F}$ denotes Fourier transform matrix, and diag[·] denotes the diagonal elements of the matrix.

More measurements show that DMC is not white (colored) spectrum in azimuth angle. In [134], the estimation and analysis of DMC directional features were presented. Herein we explain its basic ideas as follows. In delay domain, the power of DMCs is still modeled as exponential decay function. In angular domain, DMCs are modeled as the sum of multiple DMC clusters and each DMC cluster corresponds to a SC cluster. The azimuth angle and elevation angle are described by

Fisher-Bingham distribution (or Kent distribution). Some literatures also models DMCs being von Mises-Fisher distributed [140]. Unlike (3.28), traditional beamforming algorithm is used to obtain the residual channel power angular spectrum,

$$h(\theta, \varphi, \tau) = \frac{\sum_{i=1}^{N_t N_r} h_i(\tau) B_i(\theta, \varphi) W_i}{\sqrt{\sum_{i=1}^{N_t N_r} |B_i(\theta, \varphi) W_i|^2}}. \tag{3.31}$$

Where $B_i(\theta, \varphi)$ is the 3D radiation pattern of the i-th antenna element and W_i is the window function used to reduce sidelobe. Firstly, DMCs are clustered in angle domain according to the azimuth anglular spread and elevation anglular spread of SC. $(\theta, \varphi)_{DMC,k} \in (\theta, \varphi)_{SC,k} + 2\sqrt{2}\left[-\sigma_{(\theta,\varphi)_{SC,k}}, \sigma_{(\theta,\varphi)_{SC,k}}\right]$, where $(\theta, \varphi)_{SC,k}$ and $\sigma_{(\theta,\varphi)_{SC,k}}$ are the average angles and angular spreads of the k-th SC cluster. Each DMC cluster has an independent measure of PDP, which is calculated as

$$\psi_{res,k}(\tau) = \int_{\min\varphi_{DMC,k}}^{\max\varphi_{DMC,k}} \int_{\min\theta_{DMC,k}}^{\max\theta_{DMC,k}} |h(\theta, \varphi, \tau)|^2 \sin\theta d\theta d\varphi \tag{3.32}$$

Then according to (3.29), the parameters of each DMC cluster are calculated.

3.4.1.5 Mobile Channel Estimation and Tracking

In the measurement of a dynamic channel, either the receiver or the transmitter, or both, are changing over time, so are the channel parameters. To model the dynamic channel accurately, it is required to estimate and track the number of paths and multipath parameters. There are several commonly used path tracking algorithms, e.g. state space method [140, 141] and particle filter method [142]. The state space method is used to establish state equation and observation equation for the tracked parameters. Due to the are the nonlinearity of the signal model, Kalman Filter Extended (EKF) is used to carry out the linearization processing based on the Taylor series of the observation equation. EKF is only suitable for the situations where linearization approximates the nonlinear observation equation. When the channel coefficients are subject to high fluctuations in mobile enviroments, the linear approximation will be very inaccurate, so that the changes in the channel can not be catched up in time. There are two approaches to combat the nonlinear problems. One is Kalman Enhanced Super resolution Tracking algorithm (KEST) proposed by Thomas Jost et al. , which combines Kalman filter and high resolution estimation algorithm (such as SAGE) to carry out channel parameters tracking [141]. Xuefeng Yin et al. apply the Particle Filter (PF, aka Sequential Monte Carlo) method to track dynamic channel parameters [142]. PF can avoid calculating the second-order derivatives of the received signal with respect to the path parameters when be applied to nonlinear observation model. Some researchers use PF to track the path parameters indirectly. For example, K. Saito used PF to track four parameters including two-dimensional positions of discrete scatterers, delay from transmitter to scatterer, and AOD. Therein SAGE algorithm was used to obtain the path parameters used to update the PF state [143].

3.4.1.6 Automatic Clustering Algorithm

It has been found that the channel MPCs tend to appear in clusters, i.e. in groups of multi-path components (MPCs) with similar parameters. Meanwhile, describing and subsequently processing based on clusters will significantly reduce the number of parameters appearing in the channel modeling. This is also one reason why clusters are used as the basis for the GSCM and SV modeling methods. In this part, we will describe the automatic clustering algorithm. It involves two problems, one is how to cluster all of the MPCs when the number of clusters is given, and the other is how to determine the optimal number of clusters and the optimal criteria. Clustering algorithms, such as K Means, standard or fuzzy CMeans and other algorithms [144, 145], can be used to deal with the first problem. For the second problem, the commonly used methods at present are to use a variety of indexes to determine the number of clusters, such as the Kim-Park index [146], Davis-Bouldin index, Dunn index, and Calinski-Harabasz index [147]. The following gives a simple description on automatic clustering.

1. Clustering. Assuming there is L MPCs in total, each of which has four parameters: delay τ, AOA θ, AOD φ and power P. The parameter set of each path is denoted by $X_l = [\tau_l, \theta_l, \varphi_l, P_l]$ and the parameter set of all L MPCs are denoted by $\Theta = \{X_l, l=1,\ldots,L\}$. The expected number of clusters is K and each MPC belongs to one of the K clusters. According to K-Means algorithm, clustering is executed as follows. Firstly, K MPCs are randomly selected in Θ as the center of K clusters and denoted as $c_k \in \Theta$. Then the Multipath Component Distance (MCD), $MCD(X_l, c_k)$, between each MPC in Θ to the K cluster centers is calculated. MCD denotes the distance between two paths. For i-th and j-th MPC, their MCD can be expressed by [148]

$$MCD(X_i, X_j) = \sqrt{\|MCD_{AOA,ij}\|^2 + \|MCD_{AOD,ij}\|^2 + MCD_{\tau,ij}^2} \qquad (3.33)$$

There are three MCD components, two for angle and one for delay. The calculation of angular and delay MCD component are different. Angular MCD component is related to the distance between 3D direction vector of the two paths, and the delay MCD component is the normalized delay difference,

$$\mathrm{MCD}_{AOA/AOD, ij} = \frac{1}{2} \left| \begin{pmatrix} \sin(\theta_i)\cos(\varphi_i) \\ \sin(\theta_i)\sin(\varphi_i) \\ \cos(\theta_i) \end{pmatrix} - \begin{pmatrix} \sin(\theta_j)\cos(\varphi_j) \\ \sin(\theta_j)\sin(\varphi_j) \\ \cos(\theta_j) \end{pmatrix} \right|$$

$$\mathrm{MCD}_{\tau, ij} = \xi \frac{|\tau_i - \tau_j|}{\Delta\tau_{max}} \frac{\tau_{std}}{\Delta\tau_{max}} \tag{3.34}$$

Where θ_i and φ_i are elevation angle and azimuth angle respectively, τ_{std} is the standard deviation of all path delay, $\Delta\tau_{max}$ is the maximum excess delay. ξ is a proportional factor, usually setting to 1, which is used for adjusting the proportion of delay in the calculation of distance.

Next, the index of cluster with its center being nearest to each path is taken as cluster identification of the path, so that Θ can be divided into K clusters. Then new cluster centers are calculated, which is the power weighted average in delay, AOA and AOD components of all the MPCs belonging to the same cluster, i.e.,

$$c_k^{i+1} = \frac{\sum_{l \in c_k^i} P_l [\tau \quad AOA \quad AOD]_l}{\sum_{l \in c_k^i} P_l} \tag{3.35}$$

2. Determinating the optimal number of clusters. The idea of using Kim-Park index to determine the cluster number is to minimize the sum of normalized intra-cluster distance and inter-cluster distance [146]. The average intra-cluster distance is calculated as follows

$$\nu_u(K) = \frac{1}{K} \sum_{k=1}^{K} \frac{\sum_{l \in c_k} MCD(X_l, c_k)}{|c_k|} \tag{3.36}$$

Where $|c_k|$ is cardinality of c_k, namely the number of MPCs in the k-th cluster. The average intra-cluster distance reflects the compactness of a cluster. When the number of clusters is excessive, $\nu_u(K)$ becomes very small. On the contrary, when the cluster number is too small, many MPC that are distinct or less correlated are divided into the same cluster, so $\nu_u(K)$ becomes large. The measure of inter-cluster distance is calculated as follows

$$\nu_o(K) = \frac{K}{\min_{i \neq j} MCD(c_i, c_j)} \tag{3.37}$$

When the number of clusters is too small, the minimum inter-cluster distance $min_{i \neq j} MCD(c_i, c_j)$ becomes large, so $\nu_o(K)$ becomes very small. On the contrary, when the number of clusters is excessive, $\nu_o(K)$ becomes very large. Thus, there will be a optimal point in $\nu_u(K) + \nu_o(K)$ when varing the number of clusters. That is the key of K-P index. Going through all possible cluster numbers, $K=2,\ldots,K_{max}$, the optimal number of clusters K_{opt} is determined by

$$K_{opt} = \arg\max_K \left\{ \frac{\nu_u(K) - \min\limits_K \nu_u(K)}{\max\limits_K \nu_u(K) - \min\limits_K \nu_u(K)} + \frac{\nu_o(K) - \min\limits_K \nu_o(K)}{\max\limits_K \nu_o(K) - \min\limits_K \nu_o(K)} \right\}. \quad (3.38)$$

Herein normalizd inter-cluster distance $\nu_u(K)$ and inter-cluster distance $\nu_o(K)$ are used.

3.4.2 Channel Statistical Analysis

In this subsection, we will introduce the main channel statistical characteristics that need to be acquired and the commonly used statistical analysis methods. The main contents include the basic distribution fitting methods, the basic correlation analysis, and those items calculated from delay, angular and cluster domains. For more details please refer to WINNER II document [82].

3.4.2.1 Basic Distribution Fitting Methods

The channel parameters, such as amplitude, delay and angle, are usually Random Variables (RVs) following a certain distribution function expressed in PDF or cumulative distribution function (CDF). According to the measured data, some criteria are used to determine the suitable distribution function and statistical parameters so as to describe the parameters accurately in a sense of probability. These commonly used criteria include maximum likelihood (ML), least square (LS), minimum mean squared error (MMSE) or uniformly minimum variance unbiased estimation (UMVUE). For a channel parameter, RV X, it is assumed that the PDF of X is known but with unknown parameters, which is denoted by $f(x;\theta)$. Parameters estimation is to use the known observed points for X to estimate the unknown parameter θ. Taking the normal distribution as an example, assuming there are n observed points x_i ($1 \leq i \leq n$) for a random variable X, the mean and variance of X is given by

$$\widehat{\mu} \triangleq E[X] = \frac{1}{n}\sum_{i=1}^{n} x_i, \widehat{\sigma^2} \triangleq Var[X] = \frac{1}{n}\sum_{i=1}^{n} \left(x_i - \widehat{\mu}\right)^2 \quad (3.39)$$

In statistical analysis of channel parameters, the commonly used PDF that are related with amplitude, angle, time and arrival are listed in Table 3.9, 3.10 and 3.11. The estimation methods for the parameters and generating methods of RVs are also given in the three tables.

Table 3.9 Commonly used distribution function for amplitude

Distribution type	PDF	Parameters estimation	Random Variable generation
Gaussian or Normal	$f(x;\mu,\sigma) = \dfrac{1}{\sqrt{2\pi}\sigma}\exp\left(-\dfrac{(x-\mu)^2}{2\sigma^2}\right)$	$\hat{\mu} = E[X]$ ML: $\sigma^2 = \widehat{Var[X]}$ UMVU: $\sigma^2 = \dfrac{N}{N-1}\widehat{Var[X]}$	Box–Muller method [149]: $X = \mu + \sigma\sqrt{-2\ln U}\cos{(2\pi V)}$, or $X = \mu + \sigma\sqrt{-2\ln U}\sin{(2\pi V)}$ U and V are uniform random variables: $U,V \sim \text{Uniform}[0,1)$
Rayleigh	$f(x;\sigma) = \dfrac{x}{\sigma^2}\exp\left(\dfrac{x^2}{2\sigma^2}\right)$	$\sigma^2 = \dfrac{\widehat{E[X^2]}}{2}$	$X = \sigma\sqrt{-2\ln{(U)}}$
Ricean	$f(x;\sigma,K) = \dfrac{x}{\sigma^2}\cdot$ $\exp\left(-K - \dfrac{x^2}{2\sigma^2}\right)I_0\left(\dfrac{\sqrt{2K}}{\sigma}x\right)$	MoM method [150]: $P = X^2$ $\hat{K} = \dfrac{1}{\sqrt{(E[P])^2 - \widehat{Var[P]}}} - 1$ $\dfrac{E[P]}{}$ $\sigma^2 = E[P]/2/(\hat{K}+1)$	$R = \sqrt{X^2 + Y^2}$ X,Y are Gaussian random variables: $X \sim N(2K\sigma^2\cos\theta, \sigma^2)$ $Y \sim N(2K\sigma^2\sin\theta, \sigma^2)$ θ is any real number
Nakagami	$f(x;m,\Omega) = \dfrac{2}{\Gamma(m)}\left(\dfrac{m}{\Omega}\right)^m \cdot$ $x^{2m-1}\exp\left(-\dfrac{m}{\Omega}x^2\right)$	MoM method [151]: $\hat{m} = \dfrac{E^2[X^2]}{\text{Var}[X^2]}$, $\hat{\Omega} = E[X^2]$	$X = \sqrt{(\Omega/2m)Y}$ Y is a Chi distribution random variable $Y \sim \chi(2m)$ with the DOF of $2m$; $Y = \sqrt{\displaystyle\sum_{i=1}^{2m} X_i^2}, X_i \sim N(0,1)$
Lognormal	$f(x;\mu,\sigma) = \dfrac{1}{\sqrt{2\pi}\sigma x}\exp\left(\dfrac{(\ln x - \mu)^2}{2\sigma^2}\right)$	$\hat{\mu} = E[\ln X]$ $\hat{\sigma}^2 = \text{Var}[\ln X]$	$X = \exp(X_G)$ X_G is Gaussian random variable: $X_G \sim N(\mu, \sigma^2)$
Weibull	$f(x;k,\lambda) = \dfrac{k}{\lambda}\left(\dfrac{x}{\lambda}\right)^{k-1}\exp\left(-\left(\dfrac{x}{\lambda}\right)^k\right)$	$\hat{\lambda}^k = E[X^k]$ $\hat{k}^{-1} = \dfrac{E[X^k \ln X]}{E[X^k]} - E[\ln X]$	$X = \lambda[-\ln(U)]^{1/k}$ U is uniform random variable: $U \sim \text{Uniform}[0,1)$

Note: $I_0(x)$: Zero order modified Bessel function of the first kind.

Table 3.10 Commonly used distribution function for angle, $X \in [-\pi, \pi)$

Distribution type	PDF	Parameters estimation	Random variable generation
Uniform	$f(x) = \frac{1}{2\pi}$	None	$X = 2\pi U - \pi$ U is uniform random variable: $U \sim \text{Uniform}[0,1]$
Wrapped Normal	$f(\theta; \mu, \sigma) = \frac{1}{\sigma\sqrt{2\pi}} \cdot$ $\sum_{k=-\infty}^{\infty} \exp\left[\frac{-(\theta - \mu + 2\pi k)^2}{2\sigma^2}\right]$	$Z = e^{iX}, \hat{\mu} = \arg(E[Z])$, $\widehat{\sigma^2} = \ln\dfrac{N-1}{N\lvert E[Z]\rvert^2 - 1}$	$X = \text{mod}(X_G + \pi, 2\pi) - \pi$ X_G is Gaussian random variable: $X_G \sim N(\mu, \sigma^2)$
Laplacian	$f(x; \mu, \sigma) = \frac{1}{\sqrt{2}\sigma}\exp\left(-\frac{\sqrt{2}\lvert(x-\mu)\rvert}{\sigma}\right)$	MLE [152]: $\hat{\mu} = E[X]$ $\hat{\sigma} = \sqrt{2}E[\lvert X - \hat{\mu}\rvert]$	$X = \mu - \frac{\sigma}{\sqrt{2}}\text{sgn}(U)\cdot$ $\ln(1 - \lvert U\rvert)$ $U \sim \text{Uniform}[-1, 1)$
von Mises	$f(x; \mu, \kappa) = \frac{1}{2\pi I_0(\kappa)}\exp(\kappa \cos(x - \mu))$	Reference [153] $Z = e^{iX}, \hat{\mu} = \arg(E[Z])$, $\dfrac{I_1(\hat{\kappa})}{I_0(\hat{\kappa})} = \sqrt{\dfrac{N\lvert E[Z]\rvert^2 - 1}{N - 1}}$	Reference [154]
von Mises-Fisher (p=2 is von Mises)	$f_p(\boldsymbol{x}; \boldsymbol{\mu}, \kappa, p) = \frac{\kappa^{p/2-1}}{(2\pi)^{p/2}I_{p/2-1}(\kappa)}\exp(\kappa\boldsymbol{\mu}^T\boldsymbol{x})$ $\boldsymbol{\mu}$: average direction, $\mu = 1$ κ: focus parameters, $\kappa \geq 0$	Reference [155]: $\hat{\mu} = E[X]/E[X]$, $\dfrac{I_{p/2}(\hat{\kappa})}{I_{p/2-1}(\hat{\kappa})} = E[X]$	Reference [156, 157]
Fisher–Bingham(Kent)	$f(\boldsymbol{x}; \kappa, \beta, \gamma_1, \gamma_2, \gamma_3) = \frac{1}{c(\kappa,\beta)} \cdot$ $\exp\left\{\kappa\gamma_1^T\boldsymbol{x} + \beta\left[(\gamma_2^T\boldsymbol{x})^2 - (\gamma_3^T\boldsymbol{x})^2\right]\right\}$ $c(\kappa,\beta) = 2\pi.$ $\sum_{j=0}^{\infty} \frac{\Gamma(j+\frac{1}{2})}{\Gamma(j+1)} \beta^{2j}\left(\frac{\kappa}{2}\right)^{-2j-\frac{1}{2}} I_{2j+\frac{1}{2}}(\kappa)$ $\boldsymbol{x}$ is 3d unit vector, denoted by a point on the unit sphere	Reference [158]	Reference [159]

Note: $I0(x)$: v order modified Bessel function of the first kind

Table 3.11 Distribution function related with time and arrival

Distribution type	PDF	Parameters estimation	Random variable generation
Exponential	$f(x; \lambda) = \lambda \exp$ $(-\lambda x)$	ML:$\widehat{\lambda} = \dfrac{1}{E[X]}$ UMVU:$\widehat{\lambda} = \dfrac{(N-1)}{(NE[X])}$	$X = \dfrac{-\ln(U)}{\lambda}$ U is uniform random variable: $U \sim \text{Uniform}(0,1]$
Poisson	$f(k; \lambda) = \Pr(X = k)$ $= \dfrac{\lambda^k e^{-\lambda}}{k!}$	$\widehat{\lambda} = E[X]$	Reference [160–161]

Among many candidates of distribution function, Good of Fitness (GoF) tool is used to validate the distributions and find the best matching one. The commonly used GoF tools include Kolmogorov-Smirnov (K-S) test and Anderson-Darling (A-D) test etc. [162].

3.4.2.2 Basic Correlation Analysis Method

Correlation reflects the similarity between two random variables (RVs) in one or more dimensions. The correlation calculation does not need to know the distribution function of the RVs. It is assumed that the RVs are ergodic, so the correlation can be calculated using ensemble average. Assume there are N_x and N_y collected values for x and y respectively, the correlation value between x and y can be calculated as

$$R_{xy}(k) = \frac{1}{N_2 - N_1} \sum_{n=N_1+1}^{N_2} x(n+k)y(n)^*, \quad \begin{array}{l} k = [-N_y + 1, N_x - 1] \\ N_1 = \max(-k, 0) \\ N_2 = \min(N_y, N_x - k) \end{array}. \quad (3.40)$$

This estimation is unbiased, but the estimated variance of $R_{xy}(k)$ increases with k. A biased estimation can be used instead when the variance is small

$$R_B(k) = \frac{N_2 - N_1}{\min(N_y, N_x)} R_{xy}(k) \quad (3.41)$$

The calculation of covariance is similar to that of correlation, which just removes the respective average value before calculating

$$C_{xy}(k) = \frac{1}{N_2 - N_1} \sum_{n=N_1+1}^{N_2} (x(n+k) - \mu_x)(y(n) - \mu_y)^*$$

$$\mu_x = \frac{1}{N_x} \sum_{n=1}^{N_x} x_n, \mu_y = \frac{1}{N_y} \sum_{n=1}^{N_y} y_n \tag{3.42}$$

Correlation coefficients can be calculated based on covariance

$$\rho_{xy}(k) = \frac{C_{xy}(k)}{\sqrt{C_{xx}(0)}\sqrt{C_{yy}(0)}} \tag{3.43}$$

Correlation distances (time) are symmetrical on both sides of maximum covariance, which are proportional to the area of the covariance function. It can be calculated as

$$d_R = \frac{\sum_{k=-N_y+1}^{N_x-1} C_{xy}(k)}{2\max\{C_{xy}(k)\}}. \tag{3.44}$$

When spatial correlation is calculated, if the spatial distance between all snapshots is not constant, .e.g., the measured routes are not always along a straight line or the speed is varying, then the above calculation will be problematic. The correlation value will not show a certain relationship with snapshot index k. Therefore, we must group the snapshot by distance, denoted by $k(\Delta)$, and the correlation will be

$$R(\Delta) = \frac{1}{N(\Delta)} \sum_n \sum_{k(\Delta)} x(n + k(\Delta))y(n)^*, \tag{3.45}$$

where $N(\Delta)$ is the number of valid data pairs $[x(n + k(\Delta)), y(n)]$ in the distance Δ.

3.4.2.3   Delay Domain Analysis

The items to be analyzed in this part include CIR, time-varying PDP, PDP, MED, DS, total power, path loss, SF standard deviation, and Ricean K-factor. At this stage, all calculations are based on the CIR and no angular information is needed. The CIRs can be got by correlation of the receiving samples with the PN sequence (known for the receiver) in time domain sounding, or can be got directly by Fourier transformation in frequency domain excitation including broadband OFDM signal or single carrier in VNA based system. Of course, we can also use the high resolution parameter estimation algorithm to extract angular the information, and get CIRs through accumulating in the angle domain. The former is called direct method, and the latter is called synthesis method.

1. Channel Impulse Response

CIR, denoted by $h(t,\tau,s,u,p)$, is a function of time (snapshot) t, delay τ, transmitting antenna element s, receiving antenna element u, and polarization p. If the channel $h(\tau)$ for (t,s,u,p) dimension does not satisfy SNR criterion, it will not be used in subsequent analysis. This is mainly caused by too large path loss, too small transmit power, or insufficient receiver sensitivity, etc. The criterion is that if the peak power of $h(\tau)$ exceeds the noise threshold by predefined value, such as 20 dB, it is considered to be a valid CIR. For such valid CIR, the MPCs with path gain being lower than noise threshold (sometimes 3 dB above the noise threshold) is set to 0. There are n_t snapshots collected in a stationary interval. The number of CIRs contained in each snapshot depends on the number of antenna pairs, and there should be at least one valid CIR per snapshot. It should be noted that all of the snapshot time t, delay τ, transmit antenna s and receive antenna u, are discrete. The snapshot time t depends on snapshot cycle and delay τ depends on the delay resolution of the sounding system.

2. Time-variant Power Delay Profile

Averaging the power of CIRs in the antenna pair (u,s) dimension to get time-variant PDP. Because the propagation environment and Tx-Rx distance change over time during channel sounding, the delay of peak path will undergo minor but unignorable drift. Therefore, it is necessary to align highest peaks of all CIRs. The specific operation is shown as follows

$$P(t,\tau',p) = \frac{1}{N_s N_u} \sum_{s=1}^{N_s} \sum_{u=1}^{N_u} |h(t, \tau - \tau_{u,s}(t,p), s, u, p)|^2 \qquad (3.46)$$

where $\tau_{u,s}(t,p) = argmax_\tau |h(t,\tau,p)|^2$ indicates the delay adjustment for each CIR. $\tau - \tau_{u,s}(t,p)$ is denoted by τ'. Hereafter, τ is used for delay instead of τ' for clarity.

3. Power Delay Profile

The time-variant PDPs of n_t snapshots in stationary interval are averaged to get the PDP.

$$P(\tau,p) = \frac{1}{n_t} \sum_{i=1}^{n_t} P(t_i, \tau, p) \qquad (3.47)$$

4. Maximum Excess Delay (MED).

MED is the maximum delay with non-zero power in PDP during a stationary interval, i.e.,

$$\tau_{max}(p) = \underset{\tau}{argmax} \{P(\tau,p) > 0\} \qquad (3.48)$$

5. Mean Square Delay Spread

DS partially reflects frequency selectivity of a channel. If the delay PDF is replaced by the PDP, then DS is the second-order centeral moment of the delay distribution. The calculation of DS is performed in a stationary interval with n_t snapshots. If the paths in PDP having power lower than a predefined value (such as 20 dB) below the peak path power will be skipped, i.e.,

$$P^*(\tau_i, p) = \begin{cases} P(\tau_i, p) & P(\tau_i, p) > \max\{P(\tau_i, p)\}/100 \\ 0 & other \end{cases} \tag{3.49}$$

Where the delay τ_i get discrete value because the measurement system has a certain delay resolution. The delay distribution function is estimated using PDP as follows,

$$p(\tau_i, p) = \frac{P^*(\tau_i, p)}{\sum_{i=1}^{N_x} P^*(\tau_i, p)} \tag{3.50}$$

Mean delay $\bar{\tau}$, second-order moments of delay $\overline{\tau^2}$ and mean square DS $DS(p)$ can be got from delay distribution function by

$$\sigma_{DS}(p) = \sqrt{\overline{\tau^2} - \bar{\tau}^2}$$

$$\bar{\tau}(p) = \sum_{i=1}^{N_x} \tau_i p(\tau_i, p)$$

$$\overline{\tau^2}(p) = \sum_{i=1}^{N_x} \tau_i^2 p(\tau_i, p) \tag{3.51}$$

Based on the mean square DS obtained in different positions, the DS correlation between two locations can be calculated, so can the DS correlation distance. Details can refer to the subsection of Basic Correlation Analysis Method.

6. Total path power

Total path power is the sum of PDP over all the delay τ_i and given by

$$P(p) = \sum_{i=1}^{N_x} P(\tau_i, p) \tag{3.52}$$

7. Path loss

To establish the relations between path loss and other propagation parameters (such as Tx-Rx distance, carrier frequency, etc.) is a very important task in channel modeling. The following takes analyzing the dependence of Path Loss on the distance as an example.The Tx-Rx distance d for every snapshot should be recorded during measuring. For each total path power there is a corresponding distance,

$P(p) \xrightarrow{d} P(d,p)$. Considering the other parameters of the measurement system, such as the transmitting power P_{tx}, transmitting and receiving antenna gain $\sum_i G_i$, the total cable attenuation $\sum_i A_i$, we can obtain the path loss

$$PL(d,p) = P_{tx}(p) - \sum_i G_i + \sum_i A_i - P(d,p) \qquad (3.53)$$

where all the parameters are in unit of dB. Multiple measurements at different distances can obtain more observations of PL values with respect to the distances. With these observations, linear regression analysis can be used to find slope $A(p)$ and intercept point $B(p)$ of the path loss model,

$$\begin{bmatrix} PL(d_1,p) \\ \vdots \\ PL(d_n,p) \end{bmatrix} = \begin{bmatrix} log_{10}(d_1) & 1 \\ \vdots & 1 \\ log_{10}(d_n) & 1 \end{bmatrix} \begin{bmatrix} A(p) \\ B(p) \end{bmatrix} \qquad (3.54)$$

Namely, the path loss model in terms of $A(p)$ and $B(p)$ can be given by,

$$\overline{PL}(d,p) = A(p)log_{10}(d) + B(p) \qquad (3.55)$$

In fact, path loss is also affected by other factors such as carrier frequency, antenna height and environmental,etc. Therefore, large amounts of measurements in various frequencies and antenna heights should be conducted to find their relationships with PL. Multiple regression analysis may be used. Path loss modeling always is the most important work in the channel modeling, and we will not give much details due to space limitations.

8. Shadow Fading (SF) Standard Deviation

For N measured data in the path loss analysis, the measured SF values can be got by substracting the expected path loss $\overline{PL}(d_n,p)$ from the measured values $PL(d_n,p)$, $SF(d_n,p) = PL(d_n,p) - \overline{PL}(d_n,p)$. Collecting these values, the standard deviation of SF can be obtained by

$$\sigma_{SF}(p) = \sqrt{\frac{1}{N-1} \sum_{n=1}^{N} |SF(d_n,p)|^2} \qquad (3.56)$$

Using correlation analysis method, we can also calculate the correlation coefficients of SF at two distant locations, thus the correlation distance of SF.

9. Ricean K-factor

Ricean factor K is the power ratio of the LOS component to all other components. In WINNER, IMT-A and 3GPP models, K-factor is calculated through the method of moments (MoM) [150]. But the MoM method is only suitable for narrowband channel. For broadband channel, MoM method is required to be used in each of the coherent bandwidth. Actually the K-factor can be calculated directly by its definition and given by

$$K = \frac{\sqrt{(E\{P\})^2 - Var\{P\}}}{E\{P\} - \sqrt{(E\{P\})^2 - Var\{P\}}} \tag{3.57}$$

where $E\{P\}$ is the average power, $E\{P\} = (1/n_t)\sum_{i=1}^{n_t} P(i)$. $Var\{P\}$ is the power variance, $Var\{P\} = (1/(n_t - 1))\sum_{i=1}^{n_t} (P(i) - E\{P\})^2$. The value of K can be used to estimate the conditional probability of LOS propagation. The obtained K can be further used to calculate variance and correlation distance.

Angular Domain Analysis

With high-resolution parameter estimation algorithm, the parameters of multiple MPCs can be extracted, such as delay, the complex gain of each polarization direction and angle, etc. According to the capabilities of measurement system and measuring requirements, the angles to be investigated include one or more of the AOA, AOD, Zenith angle of Arrival (ZOA, or EOA) and Zenith angle of Departure (ZOD, or EOD). In this part, the analysis on power angle spectrum (PAS) and angular spread (AS) are given.

PAS (ϕ /θ, p) is mainly used to describe the power distribution in angular domain. When focusing on one angular domain, we accumulate the channel power over other dimensions (including delay) to get corresponding PAS. The angle wrapping will be involved in the calculation of angle statistics, so that a special processing method is used here. All angle information is added a small angle offset Δ to be determined. Taking AOD for example, we have $AOD_i(\Delta) = AOD_i + \Delta$. For azimuth angle, the wrapping range is $[-\pi, \pi)$, so the wrapping operation is given by

$$AOD_i'(\Delta) = \mod(AOD_i(\Delta) + \pi, 2\pi) - \pi \tag{3.58}$$

For the elevation angle, the wrapping range is $[-\pi/2, \pi/2)$. Then the anglular power distribution function (normalized PAS) is estimated from the PAS,

$$pas\left(AOD_i'(\Delta)\right) = \frac{PAS\left(AOD_i'(\Delta)\right)}{\sum_{i=1}^{N_x} PAS\left(AOD_i'(\Delta)\right)}. \tag{3.59}$$

The first-order moment (mean) of angle distribution are calculated and removed from the individual angle to get an unbiased value, again the wrapping process is carried out.

$$AOD_i''(\Delta) = \mathrm{mod}\left(AOD_i'(\Delta) - \left(\sum_{i=1}^{N_x} AOD_i'(\Delta)pas\left(AOD_i'(\Delta)\right) \right) + \pi, 2\pi \right) - \pi$$

$$(3.60)$$

Then the second-order moment of the angle distribution are calculated,

$$\sigma(\Delta) = \sqrt{ \sum_{i=1}^{N_x} \left[AOD_i''(\Delta)\right]^2 pas\left(AOD_i'(\Delta)\right) } \tag{3.61}$$

ASD is defined as the minimum second-order moment of the angle $\sigma(\Delta)$ that can be obtained by varing Δ.

$$ASD = \min_{\Delta} \sigma(\Delta) \tag{3.62}$$

With the ASDs obtained at multiple locations, the mean and variance of ASD as well as the correlation distance can be obtained. The same processing can be carried out for other angles.

3.4.2.4   Cluster Analysis

After parameter extraction and cluster classification, some features of the clusters can be tracked in mobile channel, such as the cluster life cycle, cluster generation rate, cluster drifting in delay and angular domain. Clusters tracking is often ignored in the existing models, however, it is particularly important for the new 5G channel model. In mobile enviroments the parameters of the clusters will change slowly. The whole life cycle of a cluster can be recorded from its appearance to disappearance, so that it is easy to compute the average survival time and birth and death rate of the cluster. In a stationary interval, the change of delay and angle of the clusters from one snapshot to another snapshot is also very important, which can be used to analyze the time evolution of the channel.

3.5 Existing Channel Models

This section will review several existing channel models, and point out their main characteristics and restrictions. These channel models include those proposed by various research project or groups, such as the WINNER famliy [13, 163, 164], COST 259 [165]/ 273 [166]/ 2100 [167]/ IC1004 [168], METIS [1], MiWEBA [4], QuaDRiGa [19], mmMAGIC [11], 5GCM [12], and proposed by the standardization organizations, such as ITU-R IMT-Advanced [14] and IMT-2020 [178] channel model, 3GPP SCM [169]/3D [170]/D2D [3]/HF [6], IEEE 802.15.3c [8], IEEE 802.11 TGn [171]/ac [172]/ad [9]/ay [10].

3.5.1 3GPP SCM/3D/D2D/HF

Spatial Channel Model (SCM) is a GSCM developed for cellular MIMO system by the Third Generation Partnership Project (3GPP) in 2003 [169]. The importance of SCM lies in that it is convenient for expansion and parameterization and provides an infrastructure for subsequent channel models. The frequency range and system bandwidth supported is 1~3 GHz and 5 MHz respectively. The SCM channel model suits for three scenarios including Suburban Macro (SMa), Urban Macro (UMa), and Urban Micro (UMi). The path loss (PL) model uses modified COST 231 Hata urban model (SMa, UMA), COST 231 Walfish-Ikegami NLOS model (UMi NLOS) and COST 231 Walfish-Ikegami block model (UMi LOS) [173]. In terms of LSPs, it takes the DS, AS and SF as lognormal distributed or normal distributed and gives their auto-correlation and cross-correlation coefficient, inter-station SF correlation coefficient and so on. With respect to the SSPs, a channel is composed of six clusters, each of which contains 20 rays. The model gives the characteristics of each cluster with angles being Gaussian distributed, delays being exponential distributed, and power decaying exponentialy with delay.

Based on the parameters presented by WINNER II and WINNER+, 3GPP proposed two new channel models, 3D-MIMO and D2D. 3GPP 3D MIMO [170] is suitable for the scenario of UMi and UMa, both of which include the case of Outdoor-to-Outdoor (O2O) and Outdoor-to-Indoor (O2I). Its applicable frequency range, bandwidth and maximum mobility is 1~4 GHz, 10 MHz and 3 km/h respectively. It gives new path loss parameters (see Table 4.1 in [170]), the correlation coefficient of intra-station LSPs, and the updated small scale (SS) parameters (Table 7.3.6, 7.3.7, 7.3.8 in [170]), in which the azimuth angle follows wrapped Gaussian distribution, zenith angle follows Laplacian distribution, and the zenith angle parameters partially refer to the WINNER+ channel model. 3GPP D2D Model [3] support the scenarios of UMi (both O2O and O2I) and Indoor-to-Indoor (I2I). The main contribution is the support for dual mobility and the core of the model is calculating the Doppler shifts. Two system simulation scenarios are defined: generic scenarios (2 GHz, mobility speed of 3 km/h) and public security scenario (700 MHz, maximum speed of 60 km/h). The bandwidth is set as 10 MHz for uplink and downlink respectively in frequency division duplexing mode, and 20 MHz in time division duplexing mode. Most of the parameters and the channel generation process refer to the existing standard models.

The 3GPP-HF channel model (6-100 GHz) is the first standarderized high frequency band channel model released to the public. Besides those typical scenarios specified in the 5GCM [12], additional scenarios including backhaul, D2D/V2V, stadium and gymnasium, are also taken into consideration. The 3GPP-HF channel model is developed based on the 3GPP-3D MIMO channel model and adopts both measurement-based and RT modeling methods. Though the modeling methods adopted and the parameter values are the same as those in the 5GCM, 3GPP-HF gives more detailed implementation. These revisions include multi-frequency correlation to support carrier aggregation (cluster delays and angles are the same for all frequence bands, while other parameters are frequency dependent or frequency independent), finer rays modeling for powers, delays, and angles to support large bandwidth, and delay drifting over the array for each ray to support massive MIMO (still not support spherical wavefront). Based on the modeling method supporting spatial consistency proposed in the 5GCM, the 3GPP-HF provides the correlation distance required by spatial consistency, and generates the spatial-domain contineous RVs used to express LOS/NLOS probability, indoor/outdoor probability, type of building and related SSPs. Time evolution of the channel was modeled based on locations of clusters. The 3GPP-HF channel model adoptes the blockage modeling method proposed in the 5GCM and refines the expression in the polar coordinates, i.e., the blockages are divided into two catagories. Furthermore, the 3GPP-HF also proposes several simplified channel models, i.e., five CDL models and five TDL models for link-level simulations. Moreover, a map-based hybrid model developed by ZTE ltd. is adopted by combining the deterministric clusters obtained by ray tracing and the random clusters obtained by stochastic modeling. The model can be used in the case when the system performance is desired to be evaluated or predicted with the use of digital map in order to investigate the impacts of environmental structures and materials.

3.5.2 WINNER I/II/+

The aim of Working Package 5 (WP5, i.e., the channel model group) of WINNER I is to present a broadband MIMO channel model in 5 GHz frequency range [163]. At the beginning of the project, two channel models are considered, i.e. 3GPP SCM for outdoor scenario and 802.11n IEEE model for indoor scenario. Because SCM only supports a bandwidth of 5 MHz, WP5 extends the SCM model to the SCME channel model. But it still can't meet the requirements of simulation. In the year of 2004~2005, seven partners (Elektrobit, HUT, Nokia, KTH, ETH, TUI) completed a large number of measurement activities, based on which the WINNER I channel model was presented. This model supports frequency range of 5 GHz, maximum bandwidth of 100 MHz, and six scenarios: indoor, UMa, UMi, SMa, rural macro cell (RMa) and fixed relay. It provides the correlation coefficients for four LSPs like DS, AS, and SF, in which AS contains angular spread in AOA and AOD. At the same time, a Cluster Delay Line (CDL) model with fixed delay is proposed to simplify the simulation. It is worth noting that the model provides simple 3D parameters (zenith angle) for the A1 scenario.

The Working Group 1 (WP1) of WINNER II continued to extent the WINNER I Model. Six partners (Elektrobit, Finland Oulu University, Wireless communications Center of Oulu University, TUI, Nokia and Communications Research Center of Canada) completed a large number of measurements activities, and ultimately obtained WINNER II Channel Model [13, 176]. The model supports frequency range of 2~6 GHz, and includes 13 scenarios like indoor, outdoor, urban, rural, macro cell, micro area, fixed, mobile, hotspot, etc. All of the scenarios include NLOS and LOS cases. 3D features are supported in part of these scenarios (A1, A2/B4/C4) (see Table 4.5, 4.6 in [13]). HSR propagation is supported in D2a scenario with a maximum speed of 350 km/h. WINNER + Channel Model [164] further extends the operating frequency down to 450 MHz, in which it refers to Okumura Hata Model in 450~1500 MHz and refers to COST 231 COST-Hata Model in 1500~2000 MHz. The model also provides several path loss models for hexagonal layout in the NLOS case in addition to the Manhattan grid based model, and improves zenith dimension parameters for the five scenarios including indoor, O2I (UMi and UMA), UMi, UMA and SMa (see Table 4.3, 4.4, 4.5 and 3.13 in [164]), which making it a real practicable 3D MIMO channel model.

WINNER family channel Model considers the correlation of intra-station and inter-station LSPs, and provides the related correlation values based on the measured data. Intra-station correlation refers to the auto-correlation and cross-correlation of several LSPs of a UT. Inter-station correlation is the LSP correlation of two UTs with a certain distance apart, or the correlation of LSP at different locations in the track of a mobile UT. The model uses a function exponentially decaying with distance to characterize the inter-station correlation. The number of clusters varies with the environment, and the number of rays within a cluster is set to be 20 as SCM. In order to simplify the simulation, a fixed delay CDL model is also provided. WINNER model also present a conceptual idea to support time evolution of the channel, which can be realized by an old cluster ramping down and a new corresponding cluster ramping up when transiting from segment to segment.

WINNER model is a stochastic model, which embodys its randomness in the following three aspects. The first is the randomness of the LSPs such as SF, DS and AS. The second is the random SSPs such as power, arrival angle and departure angle. The last is the initial random phase of each MPC, after drawing initial phase randomly, different channel samples can be generated.

In general, the channel models of WINNER family describe the wireless channel in different environments including outdoor, indoor, and O2I. Its biggest feature is the use of unified mathematical model with different parameter sets. These parameterized models is built from a large number of measurement, which make them widely be accepted and used. The models support the validation of various techniques, such as multi-antenna, 3D MIMO, polarization, multi-user, multi-cell and multi-hop network.

3.5.3   ITU IMT-Advanced, IMT-2020

Based on the field measurement by member organization and the WINNER II channel model, the International Telecommunications Union (ITU) proposed a IMT Advanced (IMT-A) channel model (aka ITU-R M.2135 channel model for its document name) [14]. This channel model supports five scenarios, including indoor hotspot (InH), typical UMa, UMi (including O2I), SMa and RMa. The model adopts the ideas, methods and stochastic channel generation step from the WINNER II model, however, it is only a 2D model and its parameter values have changed a lot (see Table A1-2, A1-3 and A1-7 in [14]). Frequency band range of 2~6 GHz is supported for all scenarios, except frequency down to 450 MHz is supported for RMa. In addition, another extended Time-Spatial Propagation (TSP) model is suggested. The TSP model is a geometry-based double directional channel model and is targeted for up to 100 MHz RF bandwidth. The path loss, PDP, AOD are modelled by several closed-form functions in terms of the following key parameters such as city structures (street width, average building height), BS height, bandwidth and the distance between the BS and the UT. The model lack the support of spatial consistency of large- and small-scale parameters.

Currently ITU is promoting the standarization of IMT-2020 (5G). In order to ensure the process of 5G technical standard and rationality and fairness of the follow-up technical assessment, a draft group (DG) for the IMT-2020 channel model was established at the 25th meeting of the ITU's Radiocommunication Sector (ITU-R) Working Party 5D (WP5D). The DG participants agreed to take the 3GPP related channel models as the reference models in IMT-2020 Evolution Report. The reference models support frequency band of 0.5~100 GHz and consist of a Primary Module, an optional Extension Module and an optional Map-based Hybrid Channel Module [178]. The Primary module inherits from 3GPP models and is used for evaluation of Radio Interface Techniques (RITs) for the following test environments, including eMBB(InH, UMi, RMa), URLLC (UMa) and mMTC (UMa). The proposed Primary Model contains Model A recommended by China and Model B recommended by Intel, Nokia, NTT DOCOMO, Sumsang, Ericsson and Telstra. The former inherits the IMT-A model and 3GPP-3D models and supplements the elevation parameters for InH and RMa in the sub-6 GHz, while it inherits the 3GPP-HF models in the above 6 GHz. The latter uses unified parameterization for the full frequency band instead of separate expressions for different bands. The optional map-based hybrid channel model proposed by ZTE and also adopted by the 3GPP uses both the ray-tracing-based deterministic and measurement-based statistical modelling methodologies to better reflect the electromagnetic propagation features in terms of SC and DC in the actual propagation environments. This hybrid channel model provides the frequency consistency between high and low frequency domains, spatial consistency and dual mobility. The optional Extension Module inherits from the IMT-A TSP model and uses the elevation angle calculation methods in Primary Module which makes it being a 3D model. The model is only applicable for the frequency band of 0.4~6 GHz and bandwidth of 0.5~100 MHz.

3.5.4   COST 259/273/2100/IC1004

The telecommunication related actions supported byEuropean Cooperation in Science and Technology (COST) organization have proposed several important channel models for wireless communication systems. The first proposed channel model is COST 207 for the second generation mobile communication system GSM, which is based on a simple TDL structure. The Following COST 231 channel model [173] concentrates on path loss modeling and provide a COST Walfish-Ikegami model and a COST Hata model which are still widely used by the WINNER channel models and 3GPP channel models.

COST 259 Action focuses on the spatial characteristics of the wireless channel and proposes a directional channel model [165]. The frequency range of the model is 450 MHz~6 GHz and the bandwidth is less than 10 MHz. COST 259 classify the radio propagation environment into three levels. The first level is the cellular type, including the macrocell, microcell and pico cells. The second level is the wireless environment. Each cellular type contains several communications environment, such as macro cell consists of four propagation environments like typical urban, unfavorable urban, rural and hilly environments. The third level is propagation scenario for practical multipath environment simulation, mainly related to the distribution of scatterers.

COST 259 assumes that the scatterers are fixed and only the UTs are movable. All scatterers in the model are classified into two types, i.e., local clusters and far-end clusters. Local clusters are located uniformly (Gaussian or others distriubted) around the UT. The location of local scatterers will change with the UT location. The far clusters are located far away from the BS and UT and can be seen by them, such as high buildings, mountains, etc. The far-end scatterers will usually be fixed in a particular area. COST259 presents the concept of visibility region (VR). Only when the UT enters the VR of a scatterer, the scatterer can contribute to the channel, which makes the adjacent users see the partially identical propagation environment. Stochastic CIRs can be generated from the delay-angular distribution obtained from the measurement, or generated by GSCM based on the locations of the scatterers. The document of COST 259 give the transformation of two methods.

COST 259 is only a single directional channel model. COST 273 Working Group continued to extend this model, and presented a double directional MIMO channel model [166]. COST 273 introduces three categories of cluster, namely the local clusters, single-bounce clusters and twin-clusters. The latter two are the differentiations of far-end scatterers in COST 259. The model provides a method to calculate the size and the distance of a cluster departing from the BS and UT based on its delay, delay and angular spread. The supported frequency range is 1~5 GHz, and the bandwidth is less than 20 MHz. The parametric modeling for different environments is difficult for COST model arising from the hardness to extract characteristics of scatterers from measurements, which limits the application of the COST 273 model. Although COST 273 defines 22 typical environments, only three of which are parameterized.

COST 2100 channel model [167, 174] continues to adopt the concept of VR proposed by COST 259 and three clusters proposed by COST 273, and further extends. Its main contribution are the suppots of polarization characteristics, DMC, the time evolution and multilink simulink. A SC cluster is modeled as a spheroid with a certain axis length and orientation according to the delay and angular information of the cluster, while a DMC cluster is modeled as a biger spheroid concentric with its corresponding SC. Although the modeling idea is advanced, the reference code provided for downloading [175] is still very simple, many advanced ideas have not been implemented, which limits its application.

Differing from the WINNER/IMT-Advanced model, the COST channel model emphasizes that the scatterers exist objectively in the environment, rather than only belong to one cluster. COST Action IC1004 focuses on the cooperative radio communications for green smart environments and ends in 2016. The major objective of its Working Group 1 (WG1) is to develop an integrated radio channel model (including mmWave band, D2D/V2V and massive MIMO) and submit the models to standardization organizations. In the application of massive MIMO, assigning different VRs for different antenna groups at both BS and UT may model the antenna non-stationary of the channel. Because the locations of the scatterers are known, the COST model can be easily extended to support the smooth time evolution and spherical wavefront [167, 174]. The existing COST model is also designed for the scenario with one end fixed. It also needs to be further extended for the D2D/V2V scenario and the extension is relatively easy. Based on the advanced modeling ideas and methods, COST model is likely to be a reasonable 5G channel model through appropriate extending, so it is worthwhile to concern.

3.5.5 IEEE 802.11 TGn/TGac

IEEE 802.11 TGn channel model [171] is a spatial (antenna) correlation-based stochastic channel model (CSCM). It also adopts the basic ideas of Saleh-Valenzuela (SV) stochastic channel model. The channel model supports frequency band of 2.4 GHz and 5 GHz with a maximum bandwidth of 40 MHz and at most 4 antennas. Specifically, it chooses three models A, B, C with small delay spread among the five channel models presented by HiperLan/2 [177] and addes three additional models for typical small office, large indoor and outdoor open space, to form a total of six models A~F. The DSs of six models are 0ns (narrow-band model), 15 ns and 30 ns, 50 ns, 100 ns and 150 ns respectively. The six models are composed of 1, 2, 2, 3, 4, 6 clusters respectively. The paths (taps) within a cluster have different delays but have the same average departure and arrival angle and AS. The AS and DS of a cluster show strong correlation with correlation coefficient of 0.7. For each taps, the PAS of a cluster can be assumed as uniformly distributed, truncated Gaussian distributed, or truncated Laplacian distributed. Based on PASs, the correlation matrix of the transmit antenna elements and that of the receiver antenna can be calculated and denoted by R_T and R_R respectively. The correlation matrix of the channel is expressed as the Kronecker product of R_T and R_R. The model supports UT at a maximum speed of 1.2 km/h, which will lead to a Doppler shift of 6 Hz at 5.25 GHz and 3 Hz at 2.4 GHz. Doppler power spectrum is a Bell-shape instead of common U-shape in classic Jakes model. Noise filtering method instead of Sum of Sinusoids (SoS) is adopted to generate the channel samples.

In order to support higher bandwidth, transmission rate and serve the multi-user scenarios (MU-MIMO), IEEE 802.11 TGac working group proposed a TGac channel model based on the TGn model [172], which can support up to 1.28 GHz bandwidth by interpolating the TGn channel taps. For multi-user case, the angular parameters of each UT are independently generated, which does not coincide with the actual situation, so the performance of MU-MIMO may be overestimated. In addition, according to the actual situation, the smaller moving speed of 0.089 km/h is proposed, which corresponds to a coherent time of 800 ms or mean square Doppler spread of 0.414 Hz.

3.5.6 QuaDRiGa, mmMAGIC

QuaDRiGa [19] is a channel model library proposed by Fraunhofer HHI Laboratory and two partners, which greatly extends the WINNER model by introducing some new features, including the support for time evolution (adjusted delay, AOA, AOD, polarization, SF and K-factor according to the UT location in the same channel segmet), scenario transition (using the idea of WINNER in different channel segments), and variable speed of UTs (including acceleration and deceleration, implemented by resampling at equidistant locations followed by interpolation rather than sampling at equal time instants). It establishes a unified framework for LOS and NLOS simulation which is convenient for simulation for a moving UT undergoing both the LOS and NLOS cases. It can calculate the polarization components for the LOS and NLOS case by using a ray-geometric approach. By adopting two diagonal filters in addition to the two horizontal/perpendicular filters used in WINNER II, the model get a more smooth correlated LSPs. The latest version of this model supports spherical wavefront modeling. Besides the six scenarios defined in the WINNER II model, namely A1, B1, B4, C1, C2, and C4, it also adds new UMa measurement scenario in Berlin and 4 scenarios of satellite ground mobile link. At present, QuaDRiGa does not support precise modeling of massive MIMO.

The mmMAGIC project, mainly constituted by several European research institutions including Fraunhofer HHI, Ericsson, and Aalto University, aims to develop new wireless access technologies for the 5G communications in the 5-100 GHz frequency band. The WG2 focuses on channel modeling and channel measurement. The mmMAIGC concentrates on the scenarios like street canyon, open square, indoor office, shopping mall, airport hall, subway station, O2I, and gymnasium. The mmMAGIC model is developed based on the 3GPP-3D model and adopts a combination modeling methods composed of measurement, RT and point cloud field prediction. The model is implemented based on the QuaDRiGa model library, adopts the KED blockage modeling method and more finer modeling for the power, delay, and angle of the rays within a cluster. Currently it only synthesizes the parameters from several existing channel models such as WINNER、3GPP-3D、METIS, and 5GCM and provides some initial model parameters.

3.5.7   IEEE 802.15.3c/ IEEE 802.11ad/aj/ay

IEEE 802.15.3c is the first 60 GHz channel model in the world proposed by IEEE 802.15 (WPAN) working group [8]. It extends the traditional SV channel model to support arriving angles. However, it only provides the azimuth information, so it is a 2D channel model. The main measurement to produce the model was completed by several institutes including Japanese NICT and German IMST, etc. It contains ten channel models CM1~CM10 for six scenarios including living room, office, library, conference room, desktop and corridor. The model does not specify the applicable frequency range and frequency band. According to the configuration of the measurement equipment, we can infer its applicable frequency range is about 59~64 GHz, and the maximum bandwidth is no more than 3 GHz.

802.11ad IEEE channel model [9] is the channel model proposed for WLAN system with ultra high data rate operating at unlicensed 60 GHz. It extends the SV model to support the azimuth and zenith angle at both Tx and Rx, so it is a double-directional 3D channel model. The model supports three kinds of indoor scenarios (conference room, office, living room) and two link types of access and D2D. The modeling methods adopted include RT, measurement, empirical distribution and theoretical models. RT is used to determine the delay and average azimuth / zenith angle of the clusters. The main clusters include LOS, one-order and two-order reflection components. Empirical distribution is used to describe the amplitude and inter-cluster angle distribution of the reflection paths. Theoretical model is used to describe polarization characteristics. The main parameters of rays within a cluster are obtained by measurements. Apart from the classical SV model, in IEEE

802.11ad model the rays within a cluster are divided into pre-cursor and post-cursor rays. Parameter fitting is done in the two part separately. The azimuth and zenith angles of the rays within a cluster are independently normal distributed. Different from the other channel models such as WINNER in which path loss, SF and small-scale path gain together determine the strength of the clusters and rays, IEEE 802.11ad channel model doesn't distinguish path loss and small scale fading, but independently generates a path gain for each ray. In general, the model can provide accurate characteristics of the channel in space and time domain, and support beamforming, polarization, and consider the blockage loss caused by human body.

In 2012, IEEE 802.11 Working Group established the IEEE 802.11aj Task Group, which targets the next generation WLAN standards for mmWave band of 45 GHz in China. A series of channel measurements were carried out by Key Lab of mmWave at Southeast University. A path loss model for three indoor access scenarios was proposed and the delay spread was analyzed. In March 2015, IEEE 802.11ay Working Group was established to develop a standard for the next generation 60 GHz transmission system, which intended to extend the application scope of IEEE 802.11ad with the support for Backhaul and Fronthaul as well as mobility and the minimum band width of 4 GHz [179]. The channel model developed has several features: extending the indoor SISO channel models of IEEE 802.11ad to the MIMO channel models, using Quasi-Deterministic (Q-D) methodology to build channel models for new scenarios, including open area outdoor hotspot access, outdoor street canyon hotspot access, large hotel lobby, ultra short range, and D2D communications. Note that except the D rays and R rays as like the MiWEBA model, a third type of rays (F-rays) that appear for a short period of time, e.g., a reflection from the moving cars and other objects, may be introduced and described with the same way as the R-rays for the special non-stationary environments.

3.5.8   MiWEBA

Millimetre-Wave Evolution for Backhaul and Access (MiWEBA) [4] is a research project committes to the promotion and application of mmWave communications. Its main participants include Fraunhofer HHI, Intel Mobile communications (IMC), and several other universities and research institutes in Japan, France, and Italy. The WG5 focused on communications and antenna and released its first version of channel model in the mmWave band in June 2014 [4]. It models the channel of mmWave band of 57~66 GHz for three indoor and outdoor scenarios, including university campus, street canyon and hotel lobby, each of which supports three link types, i.e., access, Backhaul/Fronthual and D2D. Fraunhofer HHI and IMC have completed a large number of measurements, which reveal that in the outdoor environment the diffraction components at 60 GHz is very small and can be neglected. A Q-D modeling approach is used to represent mmWave channel being composed of LOS and a few reflection paths (called Quasi-Deterministic rays) as well as several stochastic clusters (called R rays). The D rays and its parameters like path delay, power, angle and polarization can be determined through ray tracking according to the propagation environment. Whereas the parameters of lower power R rays reflected from far walls or random objects (cars, lampposts etc.) or second-order reflected can be obtained through measurement and analysis. This approach follows the IEEE 802.11ad model, but there exist some differences. The reflection coefficient of a Q-D ray is calculated using the Fresnel equation in addition to considering the roughness of reflection surface. Moreover, there are only post-cursor rays within a cluster.

3.5.9   METIS

In November 2012, as a pioneer in 5G Research and Development (R&D), EU launched Mobile and wireless communications Enablers for the Twenty-twenty Information Society (METIS) project with 29 participating entities including Huawei Company of China. One of the important tasks of METIS is to propose a set of channel models suitable for 5G R&D. METIS proposes a flexible and scalable channel modeling architecture to meet the requirements of accuracy and computational complexity. The whole channel model consists of a map-based deterministic channel model, a stochastic channel model or a mixed model of both. The map-based model, which is based on ray tracing using a simplified 3D geometric description of the propagation environment, supports the frequency range of 2~100 GHz. The significant propagation mechanisms of diffraction, reflection, diffuse reflection and blockage are taken into considerations. The model claims to meet all the requirements of the 5G channel model, including massive MIMO (pencil beamforming, spherical wavefront, and antenna non-stationary characteristics), D2D double mobility, dynamic modeling, etc. The METIS used the KED method to model the blockage. The map-based model is verified by comparison with the measurement results in some specific scenarios. At the project beginning stage, METIS participants have completed extensive measurements of 2~60 GHz in the scenarios of dense UMa, UMi, indoors, shopping malls, D2D, and vehicle link. Based on these measurements and adopting the existing channel models like WINNER II/+, 3GPP 3D/D2D and IMT-Advanced, METIS present a GSCM based stochastic channel model and channel parameter list for various propagation scenarios. Moreover, it describes 60 GHz mmWave channel under the framework of GSCM model and gives a stochastic channel model for the scenarios of shopping center, coffee house and square at 50~70 GHz, in which minority of the parameters are obtained by measurement and most of the parameters are calculated by point cloud field prediction method [180]. METIS also improves the modeling methods such as correlated large-scale paremeters generation and angle calculation of rays within a cluster. The GSCM (GGSCM) proposed in the first version [181] which supports spatial consistency is adopted by several subsequent channel models.

Mixed model provides a flexible and scalable channel modeling framework, which balances between simulation complexity and accuracy. For example, SF is generated based on map-based model while small scale fading is based on a stochastic model.

3.5.10 5GCM

The 5GCM is a 5G mmWave channel model alliance initiated by the U.S. National Institute of Standards and technology (NIST) and includes many companies and universities such as NIST, NYU, AT&T, Qualcomm, CMCC, Huawei, and BUPT. The typical scenarios in the 5GCM are UMi (urban street canyon and open square), O2O (outdoor-to-outdoor)/O2I, UMa O2O/O2I, and InH (open or closed indoor office and shopping mall). The 5GCM is developed based on the 3GPP-3D channel model and adopt the muti-frequency channel measurement and RT modeling method. Path loss, LSPs, penetration losses, and blockage models in the LOS and NLOS cases for several scenarios have been obtained. The path loss is modeled by the CI, CIF (CI with frequency), and ABG model. Dual-slope CIF model and ABG model are also provided for InH scenario in the NLOS case. For the O2I penetration loss, two models were provided, i.e. low loss model and high loss model for the external wall with different glass. Without doubt the path loss and penetration loss will depend on the frequency band. However, the LSPs do not show a certain frequency dependency by analyzing the measurement results. A weak frequency dependence of LSPs is only discoveried by the RT technique. The correlated distances and correlation coefficients among multiple LSPs are inherited from the 3GPP-3D model parameters. Three methods are proposed to support spatial consistency, i.e., (1) spatial-time-frequency consistency RVs used to generating the LOS/NLOS state, path gain, delay, and angles of rays, etc, are obtained by interpolating the RVs on the regular grids at the UT location; (2) Dynamic evolution method similar to the one presented in QuaDriGa; (3) the GGSCM modeling method proposed by METIS. For the blockage modeling, the KED expressions in the Cartesian coordinates and polar coordinates are provided. The parameters of two typical blockages, i.e., human body and vehicle, are suggested. Since the measurement campaigns have not been finished, the current model is not the final version.

3.5.11 Comparison of Existing Models

With the progress of 5G research and development, more channel model are built by the methods combing the deterministic modeling based on Ray Tracing and stochastic modeling based on channel measurement. The existing channel models still cannot meet the requirements of 5G mobile communications. Though it is claimed that the METIS map-based model can support all requirements, it needs to be further verified and it is too complex to be used in real-time simulation. Compares and summarizes on the existing wireless channel models meeting the various requirements of the future 5G channel models were shown in Table 3.12. It is based on Table 4.1 in [1], as well as some other literatures and our knowledge.

The frequency bands at which the channel are measured and built for those channel models mainly include sub-6 GHz, 10~11 GHz, 14~15 GHz, 18 GHz, 26 GHz, 28~29 GHz, 38~40 GHz, 45 GHz, E band (71~76 GHz, 81~86GHz) and 60 GHz band. IEEE 802.11ad, MiWEBA and METIS have only measured the channel at 60 GHz band. The complete measurements and modeling for the full mmWave bands are ongoing. The frequency dependence of LSPs and SSPs are being obtained by channel measurements and ray tracing simulations. Higher bandwidth and carrier aggregation techniques are required for the future wireless communication. Both the GSCM and map-based model in the 3GPP-HF and its descendent 3GPP TR 38.901 [182] provide efficient methods to support big bandwidth and frequency consistency. With respect to the high mobility, WINNER model can support HSR environment with maximum speed of 350 km/h. But as the HSR speed continues to increase (e.g., 500 Km/h) and the scenarios that a train traveles become more diverse (mountains, hills, etc.), it is required to conduct enough channel measurements and channel modeling for these new HSR scenarios. Many models don't support dual mobility (D2D/V2V), spatial consistency and time evolution (dynamic simulation). For spatial consistency, it is firstly to gurantee the correlation continuity of LSPs, without which we will get incorrect results [183]. To guarantee the continuity, WINNER calculates a correlation table in advance at each location and height (with resolution from a few meters to tens of meters). But with the diversification of dual locations and antenna heights in D2D, the size of the correlation table will be extremely large which will lead to the dramatic increase in computational complexity. The spatial consistency also ask for the spatial continuity of SSPs in dynamic simulations. It is also needed to describe the channel for close links, e.g, if two UTs are close to each other, they should see the similar scattering environment, and thus have similar AOA and AOD. If neglecting this, the performance of multi-antenna technology will be overestimated. We have seen the achievement in this aspects made by those channel models like 3GPP-HF, mmMAGIC, 5GCM and IMT-2020. With respect to large antenna arrays, almost all the models support pencil beamforming, yet the models in the framework of GSCM don't support spherical wavefront and antenna non-stationarity, however both of which are supported by the ray tracing (map-based) models and COST model based on the location of scatterers.

Table 3.12 Comparison of the existing models

Characteristics		Frequency band [GHz]	Bandwidth [MHz]	Maximum speed [km/h]	Massive MIMO Beamforming (≤1°)	Spherical wavefront	Antenna non-stationarity	3-D	D2D	Backhaul	Dynamic simulation	Spatial consistency	Frequency consistency / dependency
3GPP SCM		1~3	5	–									
3GPP 3D		1~4	100	3	Limited			Yes					
3GPP D2D		1~4	100	60					Limited			Limited	
3GPP-HF		6~100	10% fc	500 (Not yet)	Yes	Not yet	Not yet	Yes	Yes	Not yet	Yes	Yes	Yes
WINNER II/+		0.45~6	100	350	Limited			Yes		Yes	Very Limited	Limited	
IMT-2020		0.5~100	100 (<6 GHz) 10% fc	500 (Not yet)	Yes	Not yet	Not yet	Yes	Yes	Not yet	Yes	Yes	Yes
QuaDRiGa/ mmMagic			100~2 GHz (Not yet)	350	Yes	Yes		Yes	Rx Side		Yes	Not yet	Not yet
IEEE 802.11n/ac		2.4/5	1.28 GHz	Very Low				Yes					
IEEE 802.11ad/ ay		57~68	2.16 GHz, 2.64 GHz (ay)	3~5	Yes			Yes	Yes	Yes	Limited		
MiWEBA		57~66	–	Slow	Yes			Yes	Yes	Yes	Limited	Yes	
COST		0.45~100	–	–	Yes	Yes	Yes	Yes	Yes	Yes	Yes	Yes	
5GCM		0.5~100	100 (<6 GHz) ~2 GHz (>6 GHz)	350	Yes		Partly	Yes			Yes	Yes	Yes (one side fixed)
METIS	GSCM	0.45~6,61~65	100 (<6 GHz) 4 GHz (63 GHz)	200	Yes			Yes	Limited	<6 GHz	Yes		
	Map-based	≤100	10% fc	250	Yes	Yes	Yes	Yes	Yes	Yes	Yes	Yes	Yes

Note: *fc* carrier frequency

3.6 Stochastic Channel Generation

Stochastic channel samples are required in system simulations. Taking 3GPP 3D channel model as an example, this section gives the general process of stochastic channel generation, as shown in Fig. 3.9, which is similar to that of WINNER II and METIS (GSCM). The following channel generation process is suitable for downlink simulation, however, it can be modified for the case of uplink simulation by simply exchanging DOA and AOA.The whole generation process is composed of four parts: definition of geometric parameters of simulation scenarios, generation of LSPs, generation of SSPs and generation of channel coefficients.

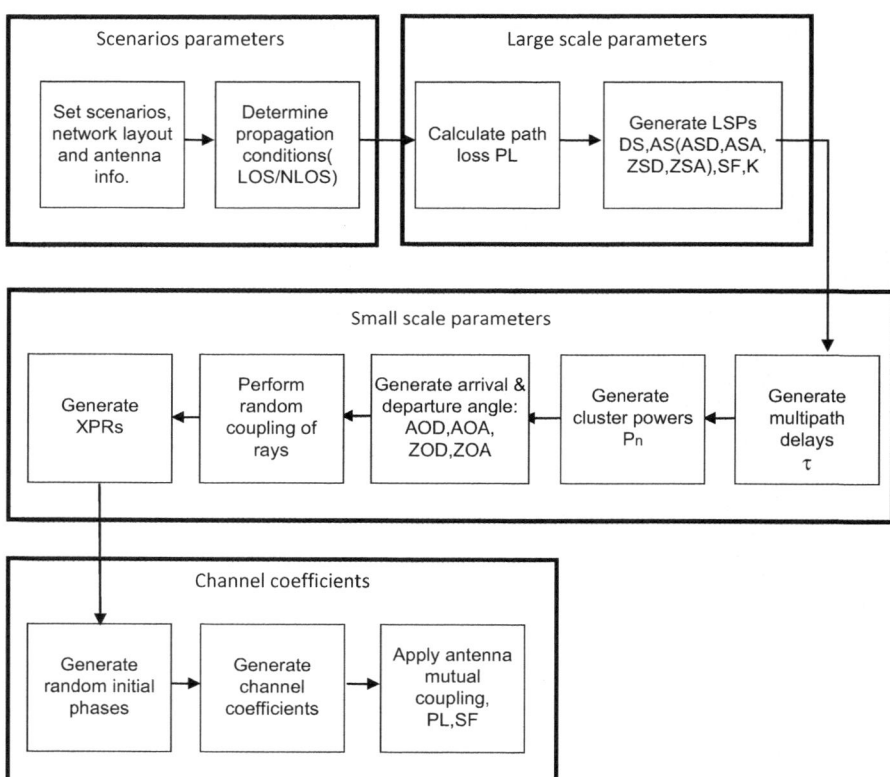

Fig. 3.9 Channel coefficients generation procedure of 3GPP TR36.873 [170]

3.6.1 Definition of Simulation Scenarios

This part is mainly related to the configuration of simulation environment, network layout and antenna parameters. The specific steps are as follows.

1. Determining system operating center frequency f_c. The effective frequency can be set to 450 MHz~6 GHz.
2. Choosing simulation scenarios, 3D-Uma or 3D-UMi.
3. Determining number of BSs and UTs in the network layout.
4. Determining the antenna 3D polarimetric radiation patterns $F_{tx}(\theta,\phi)$ and $F_{rx}(\theta,\phi)$ of BSs and UTs in the Global Coordinate System (GCS) and array geometries.
5. Generating 3D geometric locations of BSs. In system level simulation, the entire network is usually laid out in hexagonal mesh, as shown in Fig. 3.10. The BSs are placed on mesh points. The inter-site distance (ISD) is typically set to 200 m (3D-UMi) and 500 m (3D-Uma) [14]. The height of BS, h_{BS}, is usually set to 10 m (3D-UMi) and 25 m (3D-Uma).
6. Generating 3D geometric locations of UT. The horizontal two-dimensional coordinates of UTs are distributed uniformly in the whole network, and the heights depend on whether the scenario is indoor or outdoor. In 3GPP TR36.873 [170], it is assumed that 80% of the users are located indoors. The heights of outdoor users are fixed on 1.5 m above the ground while the heights of indoor users are highly dependent on the floors that they stay in the buildings. For each

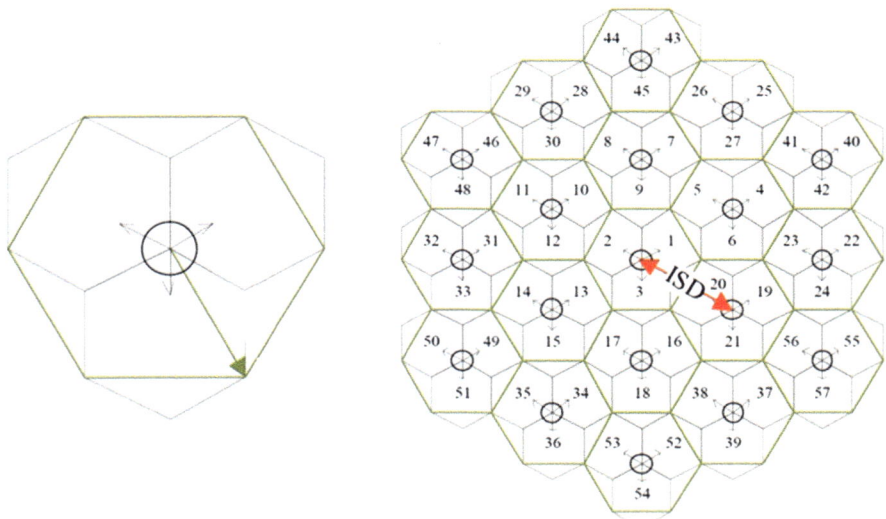

Fig. 3.10 Cellular mesh layout [14]

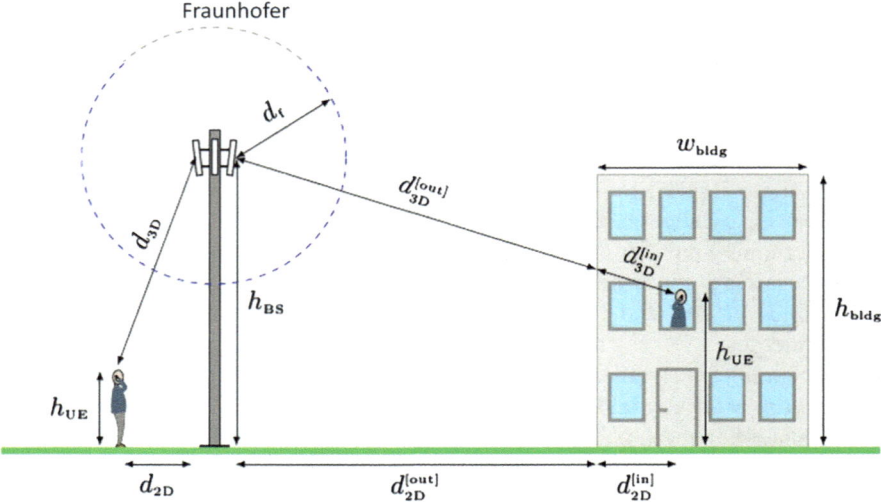

Fig. 3.11 2D and 3D distances of outdoor (left) and indoor (right) users

building, the number of floors, N, are uniformly distributed in 4~8 with floor height of 3 m. The floors that the UTs may stay, n_{floor}, are uniformly distributed between 1~N so that the heights of indoor users can be written as $h_{UT} = 3(n_{floor}-1)+1.5$. The distances between indoor users and the BS can be expressed as (Fig. 3.11)

$$d_{3D} = d_{3D-out} + d_{3D-in} = \sqrt{(d_{2D-out} + d_{2D-in})^2 + (h_{BS} - h_{UT})^2}. \qquad (3.63)$$

It is necessary to ensure that the UT is within the Fraunhofer Region (FR) of the BS antenna, that is, d_{3D} should be larger than the Fraunhofer distance. In the past, it was guaranteed by ensuring that the horizontal distances between UT and BS are more than 35 m (UMa) or 10 m (UMi). If the locations of BS and UT are given, AoD ($\theta_{LOS,ZOD}$, $\phi_{LOS,AOD}$) and AOA ($\theta_{LOS, ZOA}$, $\phi_{LOS, AOA}$) of the LOS path can be determined. In D2D scenario, it is necessary to generate two locations for two UTs in each link.

7. Generating array orientations of the BSs with respect to the GCS. The BS array orientations is defined by three angles, i.e., bearing angle α_{BS}, downtilt angle β_{BS} and slant angle γ_{BS}. The bearing angles can be determined in the phase of network deployment. The mechanical downtilt angles is usually set to 12 degrees, while the slant angle is usually set to 0 degrees. These three angles constitute the 3D rotation vector ($\alpha_{BS}, \beta_{BS}, \gamma_{BS}$) of BS Local Coordinate System (LCS) relative to the GCS.

8. Generating array orientations of the UTs with respect to the GCS. The array orientations of each UT is determined by three angles: bearing angle α_{UT},

Table 3.13 LOS transmission probability (3GPP TR36.873 [170])

Transmission scenarios	LOS transmission probability
3D UMi	Outdoor users: $P_{LOS} = \begin{cases} 1, & d_{2D} \leq 18\text{ m} \\ \dfrac{18}{d_{2D}} + \dfrac{d_{2D} - 18}{d_{2D}}\exp\left(-\dfrac{d_{2D}}{36}\right), & d_{2D} > 18\text{ m} \end{cases}$ Indoor users: replacing d_{2D} with d_{2D_out}
3D UMa	Outdoor users: $P_{LOS} = \begin{cases} 1 + C(d_{2D}, h_{UT}), & d_{2D} \leq 18\text{ m} \\ \left(\dfrac{18}{d_{2D}} + \dfrac{d_{2D} - 18}{d_{2D}}\exp\left(-\dfrac{d_{2D}}{63}\right)\right)(1 + C(d_{2D}, h_{UT})), & d_{2D} > 18\text{ m} \end{cases}$ $C(d_{2D}, h_{UT}) = \begin{cases} 0, & h_{UT} < 13\text{ m} \\ \left(\dfrac{h_{UT} - 13}{10}\right)^{1.5} g(d_{2D}), & 13\text{ m} \leq h_{UT} \leq 23\text{ m} \end{cases}$ $g(d_{2D}) = \begin{cases} \dfrac{1.25}{10^6}(d_{2D})^2\exp\left(-\dfrac{d_{2D}}{150}\right), & d_{2D} > 18\text{ m} \\ 0, & d_{2D} \leq 18\text{ m} \end{cases}$ Indoor users: replacing d_{2D} with d_{2D_out}

downtilt angle β_{UT} and slant angle γ_{UT}, which can be randomly generated or based on the practical situation. These three angles constitute the 3D rotation vector $(\alpha_{UT}, \beta_{UT}, \gamma_{UT})$ of UT LCS relative to the GCS.

9. Determining moving velocity vector of UT in the GCS, including velocity v and motion direction (θ_v, ϕ_v). The motion direction is set to be random in horizontal plane. The typical velocity can be set to 3 km/h, or be set according to practical situation. In D2D scenario, the velocity vector of both ends are generated with this method.

10. Determining the LOS/NLOS conditions of a link. For indoor and outdoor users, the propagation conditions (LOS/NLOS) are determined according to Table 3.13.

Going through the above steps, a complete GSCM simulation environment can be built. All the location information and array configuration of BSs and UTs are defined in this simulation environment, in which all transmission links are determined as well.

3.6.2 Generation of Large Scale Parameters

This subsction contains two aspects. One is to generate the path loss of all links according to the definition of simulation environment. The other is to generate LSPs of all links, such as SF, Ricean K-factor, DS, AS, etc.

Table 3.14 3D-UMi path loss model (3GPP TR36.873 [170])

	Path loss model f_c[GHz], d [m]	SFσ_{SF} [dB]	Applicable distance [m]
LOS	$PL = 22.0\log_{10}(d_{3D}) + 28.0 + 20\log_{10}(f_c)$	3	$10 < d_{2D} < d'_{BP}$
	$\begin{aligned} PL = &40\log_{10}(d_{3D}) + 28.0 + 20\log_{10}(f_c) \\ &-9\log_{10}\left((d'_{BP})^2 + (h_{BS} - h_{UT})^2\right) \end{aligned}$	3	$d'_{BP} < d_{2D} < 5000$
NLOS	$PL = \max(PL_{NLOS}, PL_{LOS})$ $PL_{NLOS} = 36.7\log_{10}(d_{3D}) + 22.7 + 26\log_{10}(f_c)$ $-0.3(h_{UT} - 1.5)$	4	$10 < d_{2D} < 2000$
O2I	$PL = PL_b + PL_{tw} + PL_{in}$ $\begin{cases} PL_b = PL_{3D-UMi}(d_{3D-out} + d_{3D-in}) \\ PL_{tw} = 20 \\ PL_{in} = 0.5d_{2D-in} \end{cases}$	7	$10 < d_{2D} < 1000$ $0 < d_{2D-in} < 25$

3.6.2.1 Path Loss

The large scale fading models differ with the propagation scenarios and conditions. Table 3.14 shows the path loss model in 3D-UMi scenario. Unlike IMT-A path loss model, 3D-UMi scenario only takes the traditional cellular scenarios into account and does not involve the Manhattan grid deployment. In addition, the SF in 3D-UMi scenario is lognormal distributed and its variance is also given in Table 3.14.

When the UT is located outdoors, the link can be either LOS or NLOS. In NLOS conditions, the PL_{LOS} is the path loss assuming the propagation condition is LOS at the same location apart from BS. d'_{BP} is the 2D breakpoint distance of PL model.

$$d'_{BP} = \frac{4(h_{BS} - h_E)(h_{UT} - h_E)}{\lambda} \tag{3.64}$$

where h_E is the effective height (in meters) of surrounding environment that the link exists, which is set to 1 in 3D-UMi scenario. h_{BS} and h_{UT} are the antenna height of BS and UT, respectively. Generally, h_{BS} is lower than the average height of surrounding buildings, and h_{UT} is in range of 1.5~22.5 m. When the user is located indoors, PL_b is the basic path loss, which is equal to PL_{3D-UMi}, namely the path loss assuming the UT is outdoors. PL_{tw} is the penetration loss and set to 20 dB in cellular cell. PL_{in} is the indoor path loss, which is related to d_{2D-in}, i.e. the perpendicular distance between the wall and the UT. $d_{2D} = d_{2D-in} + d_{2D-out}$, with d_{2D-in} uniformly distributed in 0~25 m. The Antenna height can be expressed as $h_{UT} = 3(n_{floor} - 1) + 1.5$, where n_{floor} is the floor on which the UT locates, with the value usually set to 1~8. $n_{floor} = 1$ corresponds to the ground floor.

In 3D-UMa scenarios, the path loss model is shown in Table 3.15. Its SF is also a logarithmic model, and specific fading parameters are also given in the table.

In NLOS transmission, W is the width of the street (in meters), which is generally set to 5~50 m. h is the average height of buildings (in meters), which is generally set

Table 3.15 3D-Uma path loss model (3GPP TR36.873 [170])

	Path loss PL [dB], f_c [GHz], d [m]	$SF\sigma_{SF}$ [dB]	Applicable distance [m]
LOS	$PL = 22.0\log_{10}(d_{3D}) + 28.0 + 20\log_{10}(f_c)$.	4	$10 < d_{2D} < d'_{BP}$
	$PL = 40\log_{10}(d_{3D}) + 28.0 + 20\log_{10}(f_c)$ $\quad -9\log_{10}\left((d'_{BP})^2 + (h_{BS} - h_{UT})^2\right)$	4	$d'_{BP} < d_{2D} < 5000$
NLOS	$PL = \max(PL_{NLOS}, PL_{LOS})$ $PL_{NLOS} = 161.04 - 7.1\log_{10}(W) + 7.5\log_{10}(h)$ $\quad -\left(24.37 - 3.7(h/h_{BS})^2\right)\log_{10}(h_{BS})$ $\quad +(43.42 - 3.1\log_{10}(h_{BS}))(\log_{10}(d_{3D}) - 3)$ $\quad +20\log_{10}(f_c) - \left(3.2(\log_{10}(17.625))^2 - 4.97\right)$ $\quad\quad -0.6(h_{UT} - 1.5)$		$10 < d_{2D} < 5000$
O2I	$PL = PL_b + PL_{tw} + PL_{in}$ $\left\{\begin{array}{l} PL_b = PL_{3D-UMa}(d_{3D-out} + d_{3D-in}) \\ PL_{tw} = 20 \\ PL_{in} = 0.5 d_{2D-in} \end{array}\right.$	7	$10 < d_{2D} < 1000$ $0 < d_{2D-in} < 25$

to 5~50 m. The path loss model in LOS condition is similar to that of 3D-UMi scenario except the value of h_E. When the link is LOS transmission, the probability of h_E equaling 1 m is, $p(h_E = 1m) = 1/(1 + C(d_{2D}, h_{UT}))$, where the definition of $C(d_{2D}, h_{UT})$ has been described in Table 3.13. Otherwise h_E will get values uniformly from the set of $\{12, 15, \dots, (h_{UT}-1.5)\}$. In addition, when the UT is located inside the building and the link is O2I transmission, the path loss model and the definition of its parameters are the same as the O2I path loss model in 3D-UMi scenario.

3.6.2.2   Other Large Scale Parameters

Other LSPs, including SF, Ricean K-factor, DS, and AS (ASA, ASD, ZSA, ZSD) remain to be generated. Different LSPs of the same UT show certain correlation between them, which are known as intra-station correlations. In multi-link simulation, different links (generally only the UTs connected to the same BS are considered) of each LSP show also certain correlation between them, which are known as inter-station correlations. Assume there are K links corresponding to K user located at (x_k, y_k) and each link has M LSPs. There are a total of $N=M*K$ large scale parameters. Usually these parameters are lognormal random variables, and they are correlated with each other. The correlation matrix is denoted as an N-by-N matrix C. N LSPs random variables can be generated by multiplying $\sqrt{C}$ with an N-by-1 column vector formed by N independent standard normal RVs. Obviously, the computation of LSPs with tens or hundreds of links in system level simulation is extremely huge. Therefore, WINNER suggests to calculate the intra-station correlations and the inter-station correlations separately.

1. For each LSP parameter, the inter-station correlation is calculated first, which can be completed through the LSP map. The specific procedure of this method is as follows. Firstly, generating a two-dimensional mesh based on UT locations. The region of the mesh is outwardly extention of a rectangle covering all UTs by double correlation distances. For each mesh grid, M standard normally distributed RVs (corresponding to M LSPs) are assigned and filtered by their corresponfing 2D FIR filter separately. Impulse response of the filter for the m-th LSP can be expressed as:

$$h_m(d) = \exp\left(-\frac{d}{d_{m,cor}}\right) \tag{3.65}$$

where d is distance. $d_{m,cor}$ is the correlation distance of the m-th LSPs, which is dependent on simulation environments and links condition. The specific parameters are given in Table 3.16 and 3.17. Finally, M filtered LSPs, $\xi_m(x_k, y_k)$, are obtained at every user locations (grids). The above procedure can only ensure that the inter-station correlation decreases negatively exponentially with the distance in horizontal and perpendicular directions. QuaDriGa proposes a more reasonable method, considering the correlation between the two diagonal directions. It uses two sets of filters. One set is applied in horizontal and perpendicular directions, just as WINNER, and the other is applied in the two diagonal directions. Assuming the grids spacing is d_{px}, the filter coefficients at distance kd_{px} for two filter sets are

$$a_m(k) = \frac{1}{\sqrt{d_{m,cor}}} \exp\left(-\frac{kd_{px}}{d_{m,cor}}\right)$$

$$b_m(k) = \frac{1}{\sqrt{d_{m,cor}}} \exp\left(-\frac{k\sqrt{2}d_{px}}{d_{m,cor}}\right). \tag{3.66}$$

This method can discretize the whole two-dimensional plane into finer mesh grids (less than 1 m), and the LSPs of UTs can be obtained by interpolating the LSPs at adjacent mesh grids [19]. Whereas in WINNER, the smallest location resolution of the UT coordinates and mesh grids is one meter.

The above method is only effective for 2D plane. When the UTs are distributed in 3D space, especially when there exists two-way movement in D2D/V2V scenarios, there will be totally 6D coordinate locations for both ends. Therefore, the filtering method will be very difficult to use. METIS model introduces a new method of Sum of Sinusoids (SoS) to describe the inter-station LSP correlations [184]. Specifically, taking SF as an example, it can be expressed as

$$\text{SF} = \sqrt{\frac{2\sigma_{SF}^2}{M}} \sum_{m=1}^{M} \sin\left(\bar{D} \cdot \bar{\beta}_m + \theta_m\right) \tag{3.67}$$

Table 3.16 Large-scale parameters 3GPP-3D channel model (3GPP TR36.873 [170])

Scenario		3D-UMi			3D-UMa		
		LOS	NLOS	O-to-I	LOS	NLOS	O-to-I
DS $\log_{10}([s])$	μ_{DS}	−7.19	−6.89	−6.62	−7.03	−6.44	−6.62
	ε_{DS}	0.40	0.54	0.32	0.66	0.39	0.32
AOD spread (ASD) $\log_{10}([°])$	μ_{ASD}	1.20	1.41	1.25	1.15	1.41	1.25
	ε_{ASD}	0.43	0.17	0.42	0.28	0.28	0.42
AOA spread (ASA) $\log_{10}([°])$	μ_{ASA}	1.75	1.84	1.76	1.81	1.87	1.76
	ε_{ASA}	0.19	0.15	0.16	0.20	0.11	0.16
ZOA spread (ZSA) $\log_{10}([°])$	μ_{ZSA}	0.60	0.88	1.01	0.95	1.26	1.01
	ε_{ZSA}	0.16	0.16	0.43	0.16	0.16	0.43
SF [dB]	σ_{SF}	3	4	7	4	6	7
K−factor (KF) [dB]	μ_K	9	N/A	N/A	9	N/A	N/A
	σ_K	5	N/A	N/A	3.5	N/A	N/A
Cross−correlation	ASD vs DS	0.5	0	0.4	0.4	0.4	0.4
	ASA vs DS	0.8	0.4	0.4	0.8	0.6	0.4
	ASA vs SF	−0.4	−0.4	0	−0.5	0	0
	ASD vs SF	−0.5	0	0.2	−0.5	−0.6	0.2
	DS vs SF	−0.4	−0.7	−0.5	−0.4	−0.4	−0.5
	ASD vs ASA	0.4	0	0	0	0.4	0
	ASD vs K	−0.2	N/A	N/A	0	N/A	N/A
	ASA vs K	−0.3	N/A	N/A	−0.2	N/A	N/A
	DS vs K	−0.7	N/A	N/A	−0.4	N/A	N/A
	SF vs K	0.5	N/A	N/A	0	N/A	N/A
	ZSD vs SF	0	0	0	0	0	0
	ZSA vs SF	0	0	0	−0.8	−0.4	0
	ZSD vs K	0	N/A	N/A	0	N/A	N/A
	ZSA vs K	0	N/A	N/A	0	N/A	N/A
	ZSD vs DS	0	−0.5	−0.6	−0.2	−0.5	−0.6
	ZSA vs DS	0.2	0	−0.2	0	0	−0.2
	ZSD vs ASD	0.5	0.5	−0.2	0.5	0.5	−0.2
	ZSA vs ASD	0.3	0.5	0	0	−0.1	0
	ZSD vs ASA	0	0	0	−0.3	0	0
	ZSA vs ASA	0	0.2	0.5	0.4	0	0.5
	ZSD vs ZSA	0	0	0.5	0	0	0.5
Correlation distance (in horizontal plane) [m]	DS	7	10	10	30	40	10
	ASD	8	10	11	18	50	11
	ASA	8	9	17	15	50	17
	SF	10	13	7	37	50	7
	K	15	N/A	N/A	12	N/A	N/A
	ZSA	12	10	25	15	50	25
	ZSD	12	10	25	15	50	25

Table 3.17 ZSD and ZOD offset of 3D-UMa and 3D-UMi (3GPP TR36.873 [170])

Scenario		3D-UMa		3D-UMi	
		LOS/ O-to-I LOS	NLOS/ O-to I NLOS	LOS/ O-to-I LOS	NLOS/ O-to I NLOS
ZOD spread (ZSD) $log_{10}([°])$	μ_{ZSD}	max$[-0.5, -2.1$ $(d_{2D}/1000)$ -0.01 $(h_{UT} -$ $1.5)+0.75]$	max$[-0.5, -2.1$ $(d_{2D}/1000) -0.01$ $(h_{UT}-1.5)+0.9]$	max$[-0.5,$ $-2.1(d_{2D}/$ $1000)+0.01|$ $h_{UT}-h_{BS}|$ $+0.75]$	max$[-0.5, -2.1$ $(d_{2D}/1000)$ $+0.01$max$(h_{UT}$ $-h_{BS},0) +0.9]$
	ε_{ZSD}	0.40	0.49	0.4	0.6
ZOD offset	$\mu_{offset,}$ $_{ZOD}$	0	$-10^\wedge$ $\{-0.62log_{10}($max $(10, d_{2D}))$ $+1.93-0.07$ $(h_{UT}-1.5)\}$	0	$-10^\wedge$ $\{-0.55log_{10}($max $(10, d_{2D}))+1.6\}$

where M is the number of sinusoids. σ_{SF} is the target standard deviation. $\bar{D}$ is the 6D coordination vectors of a transmitter and a receiver. θ_k is the random phase uniformly distributed in $[0,2\pi)$. $\bar{\beta}_k$ is the k-th wave vector with arbitrarily direction and norm of $\bar{\beta} \approx 7\pi\sqrt{2}/2d_{cor}$. Figure 3.12 shows an example that only contains a 3D correlation diagram at one end. Other LSPs can also be obtained by this method.

2. Next, for k-th UT, linear transformation can be used to get M LSPs with intra-station correlations.

$$\tilde{s}(x_k, y_k) = \sqrt{C_{MxM}(0)}\xi(x_k, y_k) \tag{3.68}$$

The specific parameters of the elements in correlation matrix $C_{MxM}(0)$ are also given in Table 3.16. All seven LSPs can be represented as normally distributed RVs in logarithm domain. But they are somewhat different. DS and four ASs use $log_{10}(\cdot)$ processing method, i.e., $log_{10}(X) \sim \mathcal{N}\left(\mu_{lgX}, \sigma_{lgX}^2\right)$, which is equivalent to $X \sim 10^{\mathcal{N}\left(\mu_{lgX}, \sigma_{lgX}^2\right)}$. SF and K use $10 \, log_{10}(\cdot)$ processing method, i.e. $10 \, log_{10}(X) \sim \mathcal{N}(\mu, \sigma^2)$, which is equivalent to $X \sim 10^{\mathcal{N}\left(\mu_{lgX}, \sigma_{lgX}^2\right)/10}$. The mean and variance of each LSP are also given in Table 3.16. In the above processing, we start from the standard normally distributed RVs and get the random variables $\tilde{s}(x_k, y_k)$ that can reflect the inter-station and intra-station correlations. Since the diagonal element of correlation matrix is one, $\tilde{s}(x_k, y_k)$ is still a standard normally distributed RV. The final LSPs can be expressed as $X = 10^{\mu+\sigma\tilde{s}(x_k, y_k)}$ (applicable to DS and four ASs) and $X = 10^{(\mu+\sigma\tilde{s}(x_k, y_k))/10}$ (applicable to SF and K).

The AS generated randomly is limited. ASA and ASD should not exceed 104°, i.e., ASA = min(ASA,104°), ASD = min(ASD,104°). Zenith angular Spread of Arrival (ZSA, or ESA) and Zenith angular Spread of Departure (ZSD, or ESD) should not exceed 52°, i.e., ZSA = min(ZSA, 52°), ZSD = min(ZSD, 52°).

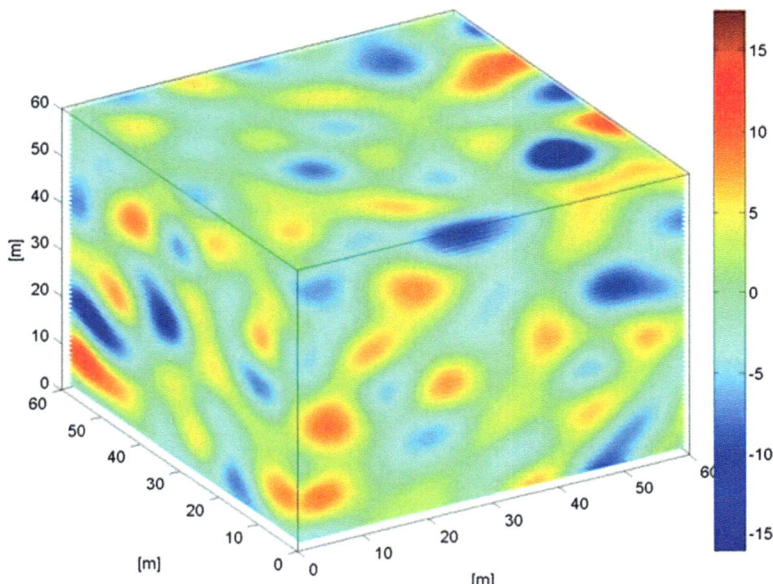

Fig. 3.12 An example of 3D SF [1]

3.6.3 Generation of Small Scale Parameters

This subsection will discuss the generation the SSPs, including cluster parameters (delay, power, angle, XPR) and intra-cluster parameters (delay, power and angle of rays in clusters).

1. Generating the relative delay of multipath τ.

In 3GPP 3D channel model, the cluster delays follow exponential distribution, which can be generated by the methods specified in Table 3.11, i.e.,

$$\tau_n' = -r_\tau \sigma_\tau \ln (X_n). \tag{3.69}$$

Where τ_n' is the absolute delay of the n-th cluster. r_τ is the ratio of standard deviation of delay distribution σ_{delays} over DS σ_τ, i.e. $r_\tau = \sigma_{delays}/\sigma_\tau$. X_n is a RV uniformly distributed in [0, 1]. A total of N cluster delays are generated. The value of N depends on the environment, which is given in Table 3.18. The delays are being normalized and sorted in ascending orde, i.e., $\tau_1=0$, $\tau_{n-1} \leq \tau_n$, n = 2, ..., N

$$\tau_n = \text{sort}(\tau_n' - \min(\tau_n')). \tag{3.70}$$

In the case of LOS conditions, additional scaling of delays is required to compensate for the impact of LOS power on the delay spread,

Table 3.18 Small-scale parameters of 3GPP-3D channel model (3GPP TR36.873 [170])

	3D-UMi			3D-UMa		
Scenario	LOS	NLOS	O-to-I	LOS	NLOS	O-to-I
Delay distribution	Exp	Exp	Exp	Exp	Exp	Exp
AOD and AOA distribution	Wrapped Gaussian			Wrapped Gaussian		
ZOD and ZOA distribution	Laplacian			Laplacian		
Delay scaling factor r_τ	3.2	3	2.2	2.5	2.3	2.2
XPR[dB] μ	9	8.0	9	8	7	9
σ	3	3	5	4	3	5
Number of clusters	12	19	12	12	20	12
Number of rays per cluster	20	20	20	20	20	20
Cluster ASD	3	10	5	5	2	5
Cluster ASA	17	22	8	11	15	8
Cluster ZSA	7	7	3	7	7	3
Per cluster shadowing std ζ [dB]	3	3	4	3	3	4

$$\tau_n^{LOS} = \frac{\tau_n}{C_{DS}}. \tag{3.71}$$

The scaling factor is related to the Rice factor K, that is,

$$C_{DS} = 0.7705 - 0.0433KF + 0.0002KF^2 + 0.000017KF^3, \tag{3.72}$$

with KF being the Ricean K-factor in dB scale generated in the phase of LSPs generation.

2. Generating cluster power P.

Cluster powers are calculated assuming a single slope exponential power delay profile, which is a basic assumption for all GSCMs. The average cluster powers decay exponentially with the increase of delay, given by

$$P'_n = \exp\left(-\tau_n \frac{r_\tau - 1}{r_\tau \sigma_\tau}\right) \cdot 10^{\frac{-CSF_n}{10}}, \tag{3.73}$$

where $CSF \sim \mathcal{N}\left(0, \sigma_{CSF}^2\right)$ is the per cluster shadow fading (CSF) in dB scale, whose statistical parameters are also specified in Table 3.18. Note that the delays specified in Equ. (3.70) are used to generate cluster powers, while not the scaled delays, i.e. Equ. (3.71), even in the case of LOS conditions. The sum power of all clusters is normalized to one, i.e.

$$P_n = \frac{P'_n}{\sum_{n=1}^{N} P'_n}. \tag{3.74}$$

In the case of LOS conditions, additional specular component is added to the first cluster. The cluster powers are given by

$$P_n^{LOS} = \frac{1}{KF+1}P_n + \delta(n-1)P_{1,LOS}$$
$$= \frac{1}{KF+1}P_n + \delta(n-1)\frac{KF}{KF+1}, \tag{3.75}$$

where $P_{1,LOS}$ is the power of LOS component and KF is Ricean K-factor in linear scale. $\delta(n)$ denotes Dirac function. Generally, except the two strongest clusters, the cluster power is uniformly allocated to every ray within the cluster. Assuming a cluster contains M_n rays, the power of each ray within a cluster n is given by P_n/M_n. Some weaker clusters, such as the cluster with power being 25 dB lower than the maximum cluster power, can be removed.

3. Generating AOA and AOD

The Power Angular Spectrum (PAS) determines the spatial correlation properties of the channel. It reflects the power distribution in different directions and is regarded as an important parameters of a MIMO channel model. At both the transmitter and the receiver, 3GPP 3D channel model assumes that the composite PAS in azimuth of all clusters is modeled as Wrapped Gaussian distribution, while the composite PAS in zenith of all clusters is modeled as Laplacian distribution, as shown in Fig. 3.13. It can be seen that the peak of Laplacian distribution is more abrupt than that of Wrapped Gaussian distribution, which means that the energy is more concentrated in zenith, while the energy is more dispersed in azimuth.

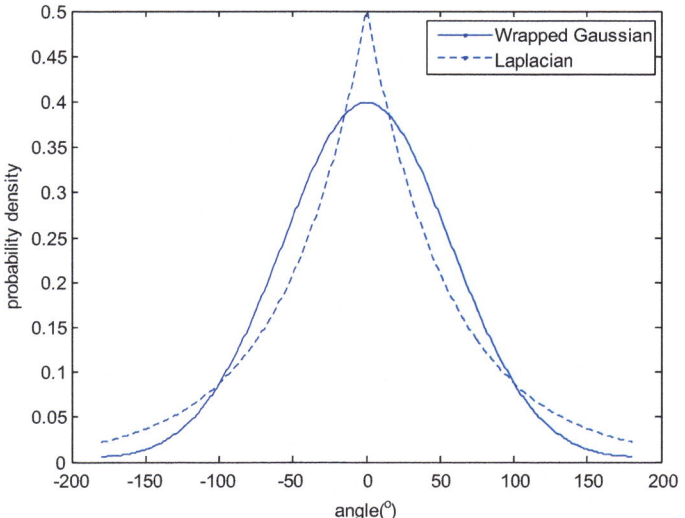

Fig. 3.13 Laplacian distribution and wrapped Gaussian distribution

Table 3.19 Scaling factor for AOA and AOD generation

N	4	5	8	10	11	12	14	15	16	19	20
C_{AS}	0.779	0.860	1.018	1.090	1.123	1.146	1.190	1.211	1.226	1.273	1.289

(a) Generating AOA and AOD

This part takes the generation of AOAs as an example. The generation of AODs is similar to that of AOAs. The AOAs are determined by applying the inverse Gaussian function (Ref. to Table 3.10) with cluster average power P_n and RMS azimuth angle spread of arrival *ASA*, i.e.,

$$\widehat{\varphi}_{n,AOA} = \frac{ASA}{0.7C_{AS}} \sqrt{-\ln\left(\frac{P_n}{\max(P_n)}\right)}. \tag{3.76}$$

The constant C_{AS} is a scaling factor related to the number of clusters N and given by Table 3.19.

In the case of LOS conditions, additional scaling of angles is required to compensate for the impact of LOS power on the angular spread, so that constant C_{AS} is substituted by Ricean K-factor dependent scaling constant C_{AS}^{LOS},

$$C_{AS}^{LOS} = C_{AS}\left(1.1035 - 0.028KF - 0.002KF^2 + 0.0001KF^3\right) \tag{3.77}$$

with *KF* being the Ricean K-factor in dB. The AOAs of clusters can be generated for the case of LOS and NLOS respectively

$$\varphi_{n,AOA} = \begin{cases} X_n\widehat{\varphi}_{n,AOA} + Y_n + \varphi_{LOS,AOA} & ,NLOS \\ X_n\widehat{\varphi}_{n,AOA} + Y_n - X_1\widehat{\varphi}_{1,AOA} - Y_1 + \varphi_{LOS,AOA} & ,LOS \end{cases} \tag{3.78}$$

where random variable X_n is set to -1 or $+1$ with equal probability, and Y_n is a Gaussian RV with zero mean and variance of $(ASA/7)^2$, $Y_n \sim \mathcal{N}\left(0, (ASA/7)^2\right)$. $\varphi_{LOS, AOA}$ is the AOA of LOS direction determined in the stage of network layout. In the case of LOS, the additional items are to enforce the first cluster to the LOS direction. Finally, the AOA of ray *n,m* is calculated by adding offset angles to the cluster AOA,

$$\varphi_{n,m,AOA} = \varphi_{n,AOA} + CASA \cdot \alpha_m. \tag{3.79}$$

Where *CASA* is the RMS ASA of each cluster and given in Table 3.18. α_m is the offset angles of *m*-th ray within a cluster and given in Table 3.20. With this method, only rough angular resolution can be obtained, which does not meet the high angle resolution requirements of massive MIMO or pencil beamforming. METIS suggested a more precise offset angles generation method, i.e. sampling the Gaussian function directly. For more details please refer to [185].

Ray path m	Ray path offset angle α_m [°]
1,2	±0.0447
3,4	±0.1413
5,6	±0.2492
7,8	±0.3715
9,10	±0.5129
11,12	±0.6797
13,14	±0.8844
15,16	±1.1481
17,18	±1.5195
19,20	±2.1551

Table 3.20 Ray offset angles within a cluster, given for 1° RMS AS

(b) Generating zenith angle of arrival (ZOA)

The ZOAs are determined by applying the inverse Laplacian function (Ref. to Table 3.10) with cluster average power P_n and RMS zenith angle spread of arrival ZSA

$$\widehat{\theta}_{n,ZoA} = -\frac{ZSA}{C_{ES}} \ln \left(\frac{P_n}{\max(P_n)} \right) \tag{3.80}$$

The constant C_{ES} is a scaling factor related to the number of clusters N and given by

$$C_{ES} = \begin{cases} 1.104, & N = 12 \\ 1.184, & N = 19 \\ 1.178, & N = 20 \end{cases} \tag{3.81}$$

In the case of LOS conditions, constant C_{ES} is substituted by Ricean K-factor dependent scaling constant C_{ES}^{LOS},

$$C_{ES}^{LOS} = C_{ES}(1.3086 + 0.0339K - 0.0077K^2 + 0.0002K^3), \tag{3.82}$$

The ZOAs of clusters can be generated for the case of LOS and NLOS respectively

$$\theta_{n,ZOA} = \begin{cases} X_n\widehat{\theta}_{n,ZOA} + Y_n + \theta_{ZOA} & ,NLOS \\ X_n\widehat{\theta}_{n,ZOA} + Y_n - X_1\widehat{\theta}_{n,ZOA} - Y_1 + \theta_{ZOA} & ,LOS \end{cases} \tag{3.83}$$

where random variable X_n is set to -1 or $+1$ with equal probability and $Y_n \sim \mathcal{N}\left(0, (ZSA/7)^2\right)$. θ_{ZOA} is determined dependent on the location of UT. If the UT locates indoors, $\theta_{ZOA} = 90°$, otherwise $\theta_{ZOA} = \theta_{LOS,ZOA}$, namely the ZOA of the

LOS direction determined at the stage of network layout. Finally, the ZOA of ray n, m is calculated by adding offset angles to the cluster ZOA,

$$\theta_{n,m,ZOA} = \theta_{n,ZOA} + CZSA\alpha_m \qquad (3.84)$$

where $CZSA$ is the RMS ZSA of each cluster and also given in Table 3.18. The value of α_m is illustrated in Table 3.20. It should be noted that, the value of $\theta_{n,m,ZOA}$ is wrapped in $[0, 360°]$ in the calculation above, while usually $\theta_{n,m,ZOA} \in [0,180]$. If $\theta_{n,m,ZOA}$ falls within $[180°, 360°]$, additional processing is applied to set $\theta_{n,m,ZOA} = 360° - \theta_{n,m,ZOA}$.

(c) Generating zenith angle of departure (ZOD)

The process of generating the cluster ZOD is similar to the process of generating ZOA, in which only an extra offset is introduced, i.e.,

$$\theta_{n,ZOD} = \begin{cases} X_n\widehat{\theta}_{n,ZOD} + Y_n + \mu_{offset,ZOD} + \theta_{LOS,ZOD}, & NLOS \\ X_n\widehat{\theta}_{n,ZOD} + Y_n - X_1\widehat{\theta}_{n,ZOD} - Y_1 + \theta_{LOS,ZOD}, & LOS \end{cases} \qquad (3.85)$$

where X_n is set to -1 or $+1$ with equal probability, $Y_n \sim \mathcal{N}\left(0, (ZSD/7)^2\right)$. ZSD is the RMS zenith angle spread of departure. For its calculation please refer to the section of LSPs generation. $\mu_{offset,ZOD}$ is a modified factor in the case of NLOS conditions, which is related to Tx-Rx distance and antenna height, and given in Table 3.17. Finally, the ZOD of ray n,m is calculated by

$$\theta_{n,m,ZOD} = \theta_{n,ZOD} + \frac{3}{8}10^{\mu_{ZSD}}\alpha_m, \qquad (3.86)$$

where μ_{ZSD} is the mean of ZSD with lognormal distribution, which is also given in Table 3.17. It is also dependent on the Tx-Rx distance and antenna height.

(d) Coupling of rays within a cluster for both azimuth and elevation

Couple randomly $\varphi_{n,m,AOA}$ to $\varphi_{n,m,AOD}$ within a cluster n (or within the sub-clusters in the case of two strongest clusters). Couple randomly $\theta_{n,m,ZOA}$ to $\theta_{n,m,ZOD}$ using the same procedure. Then couple randomly $\varphi_{n,m,AOD}$ to $\theta_{n,m,ZOD}$. As a result, four angles of each ray within a cluster n are determined ($\varphi_{n,m,AOD}$, $\theta_{n,m,ZOD}$) and ($\varphi_{n,m,AOA}, \theta_{n,m,ZOA}$).

4. Generating XPRs

Generating XPR for each ray within each cluster. XPR is lognormal distributed. Draw XPR values as

$$\kappa_{n,m} = 10^{X/10}, \qquad (3.87)$$

where $X \sim N(\mu, \sigma^2)$ is Gaussian distributed with mean μ and variance σ^2, which is given in Table 3.18 for different simulation environment.

Till now, the LSPs and SSPs are generated for each ray m within each cluster n. Next we will discuss the generation of channel coefficients.

3.6.4   Generation of Channel Coefficient

Firstly, draw random initial phase$\{\ \Phi_{n,m}^{\theta\theta}, \Phi_{n,m}^{\theta}, \Phi_{n,m}^{\varphi\theta}, \Phi_{n,m}\ \}$for four polarization pairs $(\theta\theta, \theta\varphi, \varphi\theta, \varphi\varphi)$ of ray n,m. The initial phases are uniformly disturbed in $[-\pi,\pi)$. In the case of LOS conditions, draw a random initial phase Φ_{LOS} for both $\theta\theta$ and $\varphi\varphi$ polarization pairs.

Next, by using the array radiation pattern at both ends for every transmission link, the channel coefficients are to be generated for each cluster n and each receiver and transmitter antenna element pairs u,s. Since WINNER and 3GPP channel model seperate the twenty rays of the two strongest clusters into 3 sub-clusters, the processing are carried out with two ways.

1. For the $N-2$ weakest clusters, the channel coefficients are given by:

$$
H_{u,s,n}^{NLOS}(t;\tau) = \sqrt{\frac{P_n}{M}} \sum_{m=1}^{M} \begin{bmatrix} F_{rx,u,\theta}(\theta_{n,m,ZOA}, \, \phi_{n,m,AOA}) \\ F_{rx,u,\varphi}(\theta_{n,m,ZOA}, \, \phi_{n,m,AOA}) \end{bmatrix}^T
$$

$$
\cdot \begin{bmatrix} e^{j\Phi_{n,m}^{\theta\theta}} & \sqrt{\kappa_{n,m}^{-1}}e^{j\Phi_{n,m}^{\theta\varphi}} \\ \sqrt{\kappa_{n,m}^{-1}}e^{j\Phi_{n,m}^{\varphi\theta}} & e^{j\Phi_{n,m}^{\varphi\varphi}} \end{bmatrix} \cdot \begin{bmatrix} F_{tx,s,\theta}(\theta_{n,m,ZOD}, \phi_{n,m,AOD}) \\ F_{tx,s,\phi}(\theta_{n,m,ZOD}, \phi_{n,m,AOD}) \end{bmatrix}
$$

$$
\cdot e^{\left(j2\pi\lambda_0^{-1}\left(\overline{\Omega}_{rx,n,m}^T\left(\bar{d}_{rx,u}+\bar{v}_{rx}t\right)+\overline{\Omega}_{tx,n,m}^T\left(\bar{d}_{tx,s}+\bar{v}_{tx}t\right)\right)\right)} \delta(\tau - \tau_{n,m}) \qquad (3.88)
$$

Where $F_{tx,s,\theta}$ and $F_{tx,s,\varphi}$ are the vertical and horizontal polarimetric radiation pattern of s respectively. $F_{rx,u,\theta}$ and $F_{rx,u,}$ are the vertical and horizontal polarimetric radiation pattern of u respectively. $\overline{\Omega}_{tx,n,m}$ and $\overline{\Omega}_{rx,n,m}$ are the unit direction vectors of the ray n,m at transmitter and receiver respectively. $\bar{d}_{tx,s}$ and $\bar{d}_{rx,u}$ are the location vectors of s and u respectively. $\kappa_{n,m}$ is the cross polarisation power ratio in linear scale, and λ_0 is the wavelength of the carrier. $\tau_{n,m}$ is the delay of ray n,m. $\bar{v}_{tx}$ and $\bar{v}_{rx}$ are the velocity vectors of the transmitter and receiver relative to the-first bounce and last-bounce scatter respectively, so that this model is applicable for D2D/V2V scenarios.

2. For the two strongest clusters, twenty rays of a cluster are spread in delay into three sub-clusters, and the relative delays of each sub-cluster is as follows,

Table 3.21 Sub-clusters power and delay information (3GPP TR36.873 [170])

Sub path	Ray path	Average power ratio	Delay offset
1	1,2,3,4,5,6,7,8,19,20	10/20	0 ns
2	9,10,11,12,17,18	6/20	5 ns
3	13,14,15,16	4/20	10 ns

$$\begin{aligned}\tau_{n,1} &= \tau_n + 0 \text{ ns} \\ \tau_{n,2} &= \tau_n + 5 \text{ ns} \\ \tau_{n,3} &= \tau_n + 10 \text{ ns}\end{aligned} \tag{3.89}$$

Where τ_n is the relative delay of the cluster to be seperated. Table 3.21 shows the power allocation for the rays with different index in the three sub-clusters.

The strongest clusters separating into three sub-clusters may increase the delay resolution of the channel model, and correspondly support larger bandwidth up to 100 MHz compared with 5 MHz bandwidth of SCM model.

In the case of LOS, an LOS component needs to be added and the power of each ray is scaled down in terms of Ricean K-factor. The channel coefficients are given by

$$H_{u,s,n}^{LOS}(t) = \sqrt{\frac{1}{KF+1}} H_{u,s,n}^{NLOS}(t) + \delta(n-1)\sqrt{\frac{KF}{KF+1}}$$

$$\cdot \begin{bmatrix} F_{rx,u,\theta}(\theta_{LOS,ZOA},\varphi_{LOS,AOA}) \\ F_{rx,u,\varphi}(\theta_{LOS,ZOA},\varphi_{LOS,AOA}) \end{bmatrix}^T \begin{bmatrix} e^{j\Phi_{LOS}} & 0 \\ 0 & -e^{j\Phi_{LOS}} \end{bmatrix} \begin{bmatrix} F_{tx,s,\theta}(\theta_{LOS,ZOD},\varphi_{LOS,AOD}) \\ F_{tx,s,\varphi}(\theta_{LOS,ZOD},\varphi_{LOS,AOD}) \end{bmatrix}$$

$$\cdot e^{\left(j2\pi\lambda_0^{-1}\left(\bar{d}_{rx,LoS}^T(\bar{d}_{rx,u}+\bar{v}_{rx}t)+\bar{d}_{tx,LoS}^T(\bar{d}_{tx,s}+\bar{v}_{tx}t)\right)\right)} \tag{3.90}$$

In the applying of massive MIMO technology, the coupling between the antenna elements will be nonneglibible, so it should be considered in modeling. According to the method in [31], the channel coefficients are given by

$$\tilde{H}_{u,s,n}(t) = 2r_{11}R_l^{1/2}(Z_l+Z_r)^{-1}H_{u,s,n}^{LOS/NLOS}(t)R_t^{-1/2}. \tag{3.91}$$

where R_t is the real part of the impedance matrix of transmitting antennas Z_t, $R_t = \mathrm{Re}\{Z_t\}$. R_l is the real part of load impedance Z_l, $R_l = \mathrm{Re}\{Z_l\}$. Z_r is the impedance matrix of receiving antennas. r_{11} is t real part of the self-impedance of single antenna z_{11}, $r_{11} = \mathrm{Re}\{z_{11}\}$. Finally, apply path loss and shadow fading on the generated channel coefficients to get

$$H_{u,s,n}(t) = \sqrt{PL \cdot SF}\tilde{H}_{u,s,n}(t) \tag{3.92}$$

The delay values of obtained clusters (rays) are continuous, which are not sure to coincide with the periodically sampling instants of digital communications system.

Therefore, for system simulation in time domain, it is necessary to carry out interpolation to generate the channel coefficients, which can be used to convolve with discrete transmitted signal to obtain discrete received signal.

3.6.4.1 Channel Simulation Examples

This subsection gives an example of channel simulation following the above procedure. The simulation code use a framework provided by WINNER+ and support four scenarios specified in 3GPP 3D MIMO channel model. Here, we only show the simulation results for two scenarios, 3D-UMi and 3D-UMi-O2I. Single BS and four UTs are used in the simulation to form four links. Two of the UTs locate in the same building but on different floors (1st Floor and 7th Floor), and the link conditions are NLOS. The other two UTs locate outdoors, with one link being LOS transmission and another link being NLOS transmission. The BS adopts a uniformly circular array composed of 8 antennas (UCA-8). All the 4 UTs are equipped with 2-element linear array antenna (ULA-2). The two-dimensional layout of the simulation network is shown in Fig. 3.14. Relevant simulation parameters are listed in Table 3.22. We make some visual observations from one channel realization produced by following the procedure above. It should be noted that these observations are only valid for this channel realization. Different simulation conditions and runs will get different observations.

First, we show the distribution of the rays' parameters, including the delays, path powers (gain), and AODs and ZODs at BS.

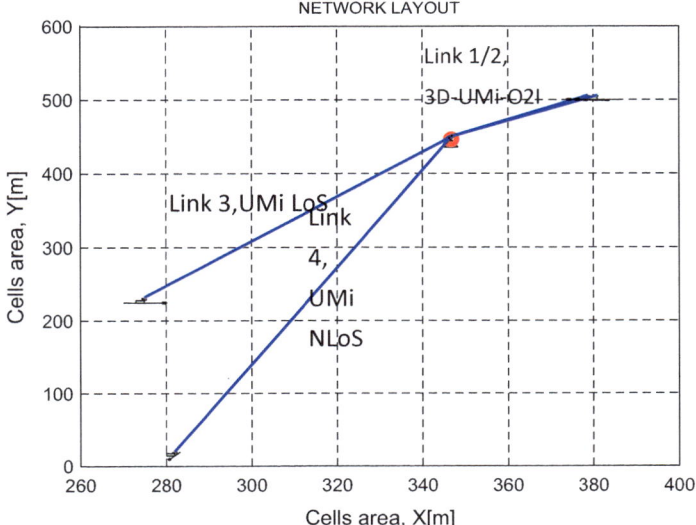

Fig. 3.14 2D layout of simulation network

Table 3.22 Simulation parameters

Link No.	1	2	3	4
Scenario	3D-UMi-O2I		3D-Umi(LoS)	3D-Umi(NLoS)
Carrier frequency (GHz)	3.5			
BS antenna type	UCA-8(Diameter $\lambda/2$)			
BS antenna height(m)	10			
UT antenna type	ULA-2(Interval $\lambda/2$)			
UT antenna height(m)	1.5	19.5	1.5	1.5
UT floor	1	7	-	-
UT indoor distance	22.5	6.1	-	-
UT moving direction (°)	90	90	195.5	83.6
UT moving speed (m/s)	0.1	0.1	10	10
BS-UT 2D distance (m)	66.2	68.1	217.7	431.9
LOS AoDs (°)	61.1	60.0	-108.4	-98.6
LOS AoAs (°)	-118.9	-120.0	71.6	81.3
LOS ZoDs (°)	97.3	82.1	92.1	91.1
LOS ZoAs (°)	90.0	90.0	87.9	88.9
Straightline propagation delay (us)	0.22	0.24	0.73	1.44
Path loss (dB)	-131.8	-118.7	-92.3	-130.6
KF(dB)	N/A	N/A	8.52	N/A

Figure 3.15 shows the spatial and temporal distribution of these rays. A ray is marked with tag 'o', with marker size reflecting its power intensity and color (red, blue, black, pink for link 1~4 respectively) indicating the link it belongs. There is the large power difference among rays. In order to show the rays clearly, here we carry out a simple process.In the LOS case, the rays powers are normalized by scaling between 2 and 30, where in the NLOS case, the power range is between 2 and 10. The subfigure (a) presents a three-dimensional distribution of the rays. The subfigure (b) gives the two-dimensional view of ZoDs-AoDs. The subfigure (c) shows the two-dimensional view of the ZoDs-delays. The subfigure (d) shows the two-dimensional view of AoDs-delays. Several points can be got obviously from these figures. (1) Link 3 is LOS transmission, and the LOS components can be identified clearly from the figure. (2) Link 1 and 2 have shorter communications distance, so that they have minimum propagation delay. However their DSs are not small. Link 1 has the largest DS. Link 4 has farthest communications distance, so its propagation delay is largest while its DS is smallest. (3) All links have obvious angular spread in the zenith dimension. The span of zenith angles of the four links are approximately 70, 10, 30 and 50 degrees respectively. The zenith AS are independent with the propagation conditions, such as the link 3 (LOS) has a midium zenith AS. (4) Link 1 and 2 have almost the same average azimuth angle and azimuth AS because the UTs are in the same building. (5) All links have almost the same azimuth AS which is greater than the zenith AS.

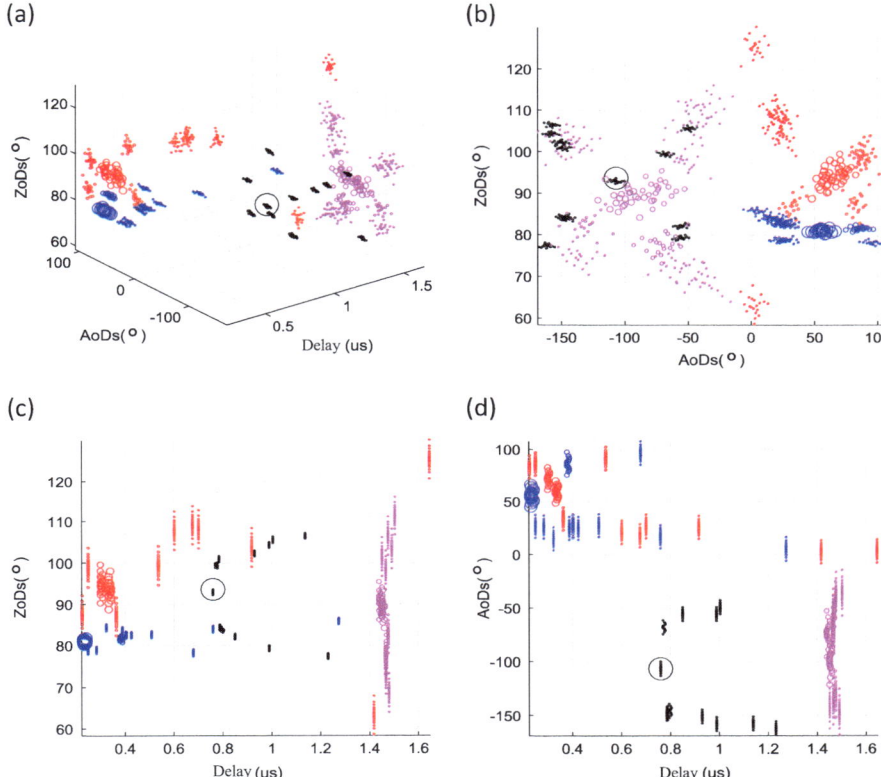

Fig. 3.15 Rays' parameters distribution for one simulation: angles at BS, delays and powers. (a) 3D view; (b) ZoDs-AoDs view; (c) ZoDs-delay view; (d) AoDs-delay view.

Figure 3.16 shows the channel impulse response of four different links observed from the selected antenna pair, i.e. the 2nd antenna of BS array and the 1st antenna of every UT. The path loss and shadow fading are also integrated into the channel impulse response. In simulation, it is assumed that propagation conditions, delays and angles of the multipath rays are kept constant. Only the effect of Doppler shift caused by the movement of UTs is investigated. The figure can reveal: (1) All channels have obvious time-varying characteristics. Because the UTs in Link 1 and 2 are at a low speed, the two channels change slowly in the observation time span of 0~70 s. The channels of Link 3 and 4 change violently even in a short observation time span of 0~0.7 s. (2) Link 3 is in the LOS condition. LOS path is dominant in CIR. (3) Link 4 has more stronger NLOS multipath components and smaller DS, but the rays scatter in angle domain, which make the channel appear to be more time-selective.

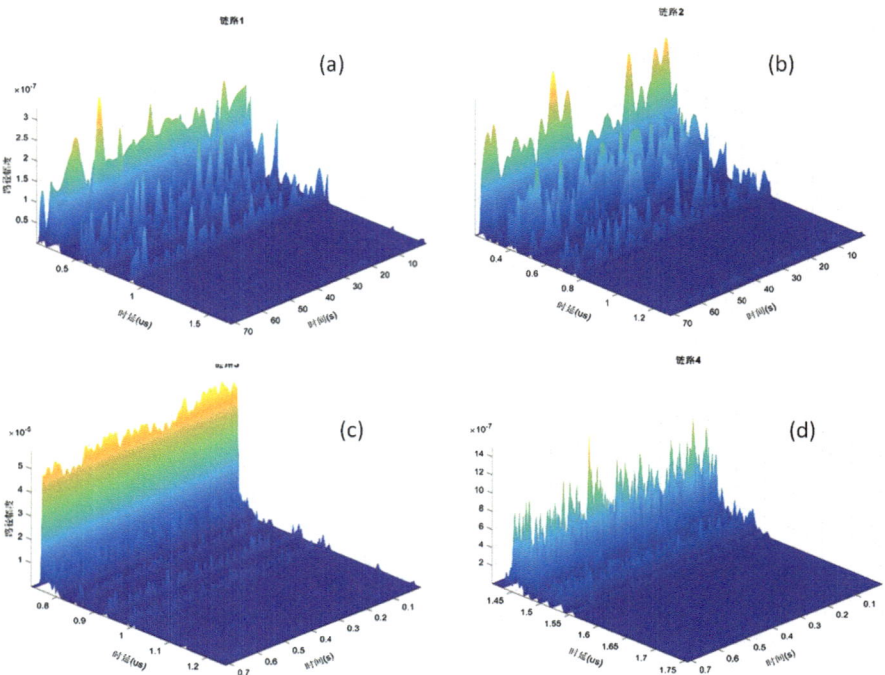

Fig. 3.16 Downlink time-varing channel from 2nd antenna at BS to 1st antenna at UT

3.7   Chapter Summary

Channel model is crucial for research and development for 5G and future mobile communications. The 5G channel model needs to face up with a variety of challenges and requirments, including massive MIMO, mmWave, high mobility and diversified application scenarios. This chapter begins with the requirement of 5G channel model, and reviews five modeling methods: measurement-based GSCM, regular shape-based GSCM, CSCM, extended SV and ray tracing. Especially, ray tracing has caught more and more attention. With the combination of ray tracing and other modeling methods, several new channel models are proposed. The formation and verification of all the channel models cannot be separated from channel sounding, which can provide real measurement data for channel modeling and verify the presented channel model. This chapter introduces the measurement methods and several channel sounders, as well as three types of 5G channel measurement activities (massive MIMO, mmWave, and high mobility). High resolution parameter estimation algorithms are used to extract path parameters from the measured data. These post-processed data are further handled by statistical analysis to obtain probability distribution functions (PDFs) and the corresponding numerical values of statistical parameters. More than ten existing channel models are introduced and compared according to the requirements of the 5G channel model. At the lat section, we take 3GPP-3D channel model as an example to gives the general process of stochastic channel generation, and finally give a simple instantiation of channel simulation.

References

1. METIS D1.4 V1.0, "METIS Channel Models," ICT-317669, METIS project, Feb. 2015. https://www.metis2020.com/
2. "5G Vision and Requirements", IMT-2020 (5G) Promotion Group. May 2014.
3. 3GPP TR36.843, "Study on LTE Device to Device Proximity Services, Radio Aspects," 3rd Generation Partnership Project, V12.0.1, March 2014. http://www.3gpp.org
4. MiWEBA D5.1, "Channel Modeling and Characterization," FP7-ICT-608637, V1.0,June 2014. http://www.miweba.eu
5. METIS D5.1, "Intermediate description of the spectrum needs and usage principles," V1.0, ICT-317669, METIS project, August 2013. https://www.metis2020.com
6. 3GPP TR 38.900, "Study on channel model for frequency spectrum above 6 GHz", v14.0.0 2016
7. W. Roh, Ji-yun seol, Jeongho park, et al. Millimeter-wave beamforming as an enabling technology for 5G cellular communications: theoretical feasibility and prototype results. *IEEE Communications Magazine*, vol. 52, no. 2, pp.106-113, 2014.
8. Su-Khiong Yong, "IEEE P802.15 Wireless Personal Area Networks - TG3c Channel Modeling Sub-committee Final Report," IEEE 15-07-0584-01-003c,March 2007
9. A. Maltsev, V. Erceg, E. Perahia, C. Hansen, R. Maslennikov, A. Lomayev, A. Sevastyanov, A. Khoryaev, G. Morozov, M. Jacob, S. Priebe, T. Kürner, S. Kato, H. Sawada, K. Sato and H. Harada, "Channel Models for 60 GHz WLAN Systems," IEEE 802.11ad 09/0334r8, 2010.
10. A. Maltsev, A. Pudeyev, Y. Gagiev, et.al., "Channel Models for IEEE 802.11ay", IEEE 802.11-15/1150r9, 2016
11. H2020-ICT-671650-mmMAGIC/D2.1. "Measurement Campaigns and Initial Channel Models for Preferred Suitable Frequency Ranges", 2016. https://bscw.5g-mmmagic.eu/pub/bscw.cgi/d94832/ mmMAGIC_D2-1.pdf
12. 5GCM White Paper, "5G Channel Model for bands up to 100 GHz", v2.0, 2016. http://www.5gworkshops.com/
13. WINNER II D1.1.2, "Channel models," IST-4-027756 V1.2, Sept. 2007. Online] Available: http://www. istwinner.org/deliverables.html
14. ITU-R M.2135-1, "Guidelines for evaluation of radio interface technologies for IMT-Advanced," International Telecommunication Union (ITU), Geneva, Switzerland, Technical Report, December 2009.
15. K. Zheng, S. Ou, and X. Yin, "Massive MIMO channel models:A survey," *International Journal of Antennas and Propagation*, vol. 2014, 2014
16. C. A. Balanis, *Antenna Thory:Analysis and Design*, John Wiley & Sons, Hoboken, NJ, USA, 2012.
17. S. Wu, C.-X. Wang, H. Aggoune, M. M. Alwakeel, and Y. He, "A non-stationary 3D wideband twin-cluster model for 5G massive MIMO channels," *IEEE J. Sel. Areas Commun.*, vol. 32, no. 6, pp. 1207-1218, June 2014.
18. X. Gao, F. Tufvesson and O. Edfors, "Massive MIMO channels:Measurements and models. " in *Proc. of 2013 Asilomar Conference on Signals, Systems and Computers*, 2013

19. S. Jaeckel, L. Raschkowski, K. Börner and L. Thiele, "QuaDRiGa: A 3.D Multicell Channel Model with Time Evolution for Enabling Virtual Field Trials," *IEEE Transactions on Antennas Propagation*, 2014. http://quadriga-channel-model.de

20. T. S. Rappaport, G. R. Maccartney, M. K. Samimi, et al. Wideband Millimeter-Wave Propagation Measurements and Channel Models for Future Wireless Communication System Design.*IEEE Transactions on Communications*, vol.63, no. 9, pp.3029-3056,2015.

21. G. Calcev, D. Chizhik, B. Goeransson, S. Howard, H. Huang, A. Kogiantis, A. F. Molisch, A. L. Moustakas, D. Reed and H. Xu, "A Wideband Spatial Channel Model for System-Wide Simulations," *IEEE Trans. Vehicular Techn.*, March 2007.

22. Matthias Patzold, *Mobile Radio Channels*, 2nd Edition, John Wiley & Sons, Chichester, UK, 2012

23. Y. Yuan, C.-X. Wang, Y. He, M. M. Alwakeel, and H. Aggoune, "Novel 3D wideband non-stationary geometry-based stochastic models for non-isotropic MIMO vehicle-to-vehicle channels," *IEEE Transactions on Wireless Communications*, vol 13, no. 1, pp. 298-309, 2014.

24. A. Ghazal, C.-X. Wang, B. Ai, D. Yuan, and H. Haas, "A non-stationary wideband MIMO channel model for high-mobility intelligent transportation systems," *IEEE Trans. Intell. Transp. Syst.*, vol. 16, no. 2, pp. 885-897, Apr. 2015.

25. A. M. Sayeed, "Deconstructing multiantenna fading channels," *IEEE Trans. S ignal Process.*, vol. 50, no. 10, pp. 2563–2579, Oct. 2002.

26. W. Weichselberger, M. Herdin, H. Ozcelik, and E. Bonek, "A stochastic MIMO channel model with joint correlation of both link ends," *IEEE Transactions on Wireless Communications*, vol. 5, no. 1, pp. 90-100, Jan. 2006.

27. J. Hoydis, S. ten Brink, and M. Debbah, "Massive MIMO in the UL/DL of cellular networks: how many antennas do we need?"*IEEE Journal on Selected Areas in Communications*, vol. 31, no. 2, pp. 160–171, 2013.

28. C. Masouros, M. Sellathurai, and T. Ratnarajah, "Large-scale MIMO transmitters in fied physical spaces:the effct of transmit correlation and mutual coupling," *IEEE Transactions on Communications*, vol. 61, no. 7, pp. 2794–2804, 2013

29. B. Clerckx, C. Craeye, D. Vanhoenacker-Janvier, and C. Oestges, "Impact of antenna coupling on 2 × 2 MIMO communications," *IEEE Transactions on Vehicular Technology*, vol. 56, no. 3, pp. 1009–1018, 2007.

30. C. Masouros, J. Chen, K. Tong, M. Sellathurai, and T. Ratnarajah, "Towards massive-MIMO transmitters:on the effcts of deploying increasing antennas in fied physical space," in *Proceedings of the Future Network and Mobile Summit*, pp. 1–10, 2013.

31. Y. Fei, Y. Fan, B. K. Lau, and J. S. Thompson, "Optimal single-port matching impedance for capacity maximization in compact MIMO arrays," *IEEE Trans. Antennas Propagat.*, vol. 56, no. 11, pp. 3566–3575, Nov. 2008.

32. R. Srinivasan, J. Zhuang, et. al., "IEEE 802.16m Evaluation Methodology Document (EMD) ," IEEE 802.16m-08/004r2,July 2008.

33. A. Saleh and R. Valenzuela, "A Statistical Model for Indoor Multipath Propagation,"*IEEE J. Select. Areas Commun.*, Vol. SAC-5, No. 2, pp. 128-137, Feb. 1987.

34. Wu Zhizhong, Mobile Communications Radio Waves Propagation, Beijing: People'S Posts And Telecommunications Publishing House, 2002. (in Chinese)

35. Nan Wang, Modern Uniform Geometrical Theory of diffraction, Xi'an: Xi'an Electronic Sience &Technology University Press, 2010. (in Chinese)

36. J. Pascual-García, M.-T. Martinez-Ingles, J. M. Molina Garcia-Pardo, J. V. Rodríguez, and L. JuanLlácer, "Using tuned diffuse scattering parameters in ray tracing channel modeling," in Proc. 9th European Conf. Antennas and Propagation (EuCAP 2015), Lisbon, Portugal, pp. 1–4.

37. Pekka Kyösti and Tommi Jämsä, "Complexity Comparison of MIMO Channel Modelling Methods," *ISWCS'07*, Trondheim, Norway, October 2007

38. J. Chen, X. Yin, L. Tian and M. D. Kim,"Millimeter-Wave Channel Modeling Based on A Unified Propagation Graph Theory." IEEE Communications Letters 21(2): 246-249, 2017.

39. Elektrobit Ltd., "Propsound - multi-dimensional radio channel sounder," System specifications document, Concept and specifications, Technical report, 2004.
40. Channelsounder.de, MEDAV GmbH, [Online]. Available:http://www.channelsounder.de/
41. V. Kolmonen, J. Kivinen, L. Vuokko, and P. Vainikainen, "5.3.GHz MIMO radio channel sounder," *IEEE Trans. Instrum. Meas.*, vol. 55, no. 4, pp. 1263–1269, Aug. 2006.
42. K. Kitao, K. Saito, Y. Okano, T. Imai and J. Hagiwara, "Basic study on spatio-temporal dynamic channel properties based on channel sounder measurements,"*Asia Pacific Microwave Conference (APMC)*, pp. 1064-1067, 2009.
43. W. Newhall, T. Rappaport, and D. Sweeney, "A spread spectrum sliding correlator system for propagation measurements," *RF Design*, pp. 40–54, Apr. 1996.
44. S. Salous, R. Lewenz, I. Hawkins, N. Razavi-Ghods, and M. Abdallah,"Parallel receiver channel sounder for spatial and MIMO characterization of the mobile radio channel," in *Proc. Inst. Elect. Eng. Commun.*, vol. 152, no. 6, pp. 912–918, Dec. 2005.
45. Y. Konishi, M. Kim, M. Ghoraishi, J. Takada, S. Suyama, and H. Suzuki, "Channel sounding technique using MIMO software radio architecture," in *Proc. 5th EuCAP*, Rome, Italy, Apr. 2011, pp. 2546–2550.
46. K. Minseok, J. Takada and Y. Konishi, "Novel Scalable MIMO Channel Sounding Technique and Measurement Accuracy Evaluation With Transceiver Impairments,"*IEEE Transactions on Instrumentation and Measurement*, 61(12): 3185-3197, 2012.
47. H. Sawada, Y. Shoji, H. Ogawa, "NICT propagation data," IEEE 802.15-06/0012-01-003c, Jan. 2006
48. A. Maltsev, R. Maslennikov, A. Sevastyanov, A. Khoryaev, and A. Lomayev, "Experimental investigations of 60 GHz wireless systems in office environment," *IEEE J. Sel. Areas Commun.*, vol. 27, no. 8, pp.1488-1499, Oct. 2009.
49. C. Gustafson, K. Haneda, S. Wyne and F. Tufvesson, "On mm-wave multi-path clustering and channelmodeling," *IEEE Trans. Antennas Propag.*, vol. 62, no. 3, pp. 1445 -1455, 2014.
50. J. I. Tamir, T. S. Rappaport, Y. C. Eldar, and A. Aziz, "Analog compressed sensing for rf propagation channel sounding," in *2012 I.E. International Conference on Acoustics, Speech and Signal Processing (ICASSP)*, pp. 5317-5320, March 2012.
51. Zhu Jin, Wang Haiming and Hong Wei, "Large-Scale Fading Characteristics of Indoor Channel at 45-GHz Band," *IEEE Antennas and Wireless Propagation Letters*, 14:735-738, 2015
52. X. Gao, O. Edfors, F. Rusek, and F. Tufvesson, "Linear precoding performance in measured very-large MIMO channels," in *Proceedings of the 74th IEEE Vehicular Technology Conference (VTC '11)*, pp.1–5, Budapest, Hungary, Sept. 2011.
53. S. Payami and F. Tufvesson, "Channel measurements and analysis for very large array systems at 2.6 GHz," in *Proceedings of the 6th European Conference on Antennas and Propagation (EuCAP '12)*, pp. 433–437, Prague, Czech Republic, March 2012.
54. X. Gao, F. Tufvesson, O. Edfors, and F. Rusek, "Measured propagation characteristics for very-large MIMO at 2.6 GHz," in *Proceedings of the 46th IEEE Asilomar Conference on Signals, Systems and Computers (ASILOMAR '12)*, pp. 295–299, Pacific Grove, Calif, USA, November 2012.
55. X. Gao, O. Edfors, F. Rusek and F. Tufvesson, "Massive MIMO performance evaluation based on measured propagation data,"*IEEE Transactions onWireless Communications*, PP (99): 1-1, 2015.
56. J. Flordelis, X. Gao, G. Dahman, F. Rusek, O. Edfors and F. Tufvesson, "Spatial Separation of Closely-Spaced Users in Measured Massive Multi-User MIMO Channels, in *Proc. IEEE International Conference on Communications (ICC)*, London, June 2015.
57. J. Hoydis, C. Hoek, T. Wild and S. Ten Brink, "Channel measurements for large antenna arrays." *2012 International Symposium on Wireless Communication Systems (ISWCS)* , 2012
58. A. O. Martinez, E. De Carvalho and J. O. Nielsen, "Towards very large aperture massive MIMO:A measurement based study". *Globecom Workshops (GC Wkshps)*, 2014

59. L. Liu, C. Tao, D. Matolak, Y. Lu, B. Ai, H. Chen, "Stationarity Investigation of a LOS Massive MIMO Channel in Stadium Scenarios," in *Proc. of IEEE 82th Vehicular Technology Conference (VTC Fall)*, 2015

60. D. Fei, R. He, B. Ai, B. Zhang, K. Guan, and Z. Zhong,"Massive MIMO Channel Measurements and Analysis at 3.33 GHz," *ChinaCom*, 2015

61. IEEE P802.11p-2010:Part 11:Wireless LAN Medium Access Control (MAC) and Physical Layer (PHY) Specifications:Amendment 6:Wireless Access in Vehicular Environments, Jul. 15, 2010, DOI:10.1109/IEEESTD.2010. 5514475.

62. A. Roivainen, P. Jayasinghe, J. Meinilau, V. Hovinen and M. Latva-Aho, "Vehicle-to-vehicle radio channel characterization in urban environment at 2.3 GHz and 5.25 GHz,"In *Proc. of IEEE 25th Annual International Symposium on Personal, Indoor, and Mobile Radio Communication (PIMRC)*, 2014

63. J. Karedal, F. Tufvesson, N. Czink, A. Paier, C. Dumard, T. Zemen, C. F. Mecklenbrauker and A. F. Molisch, "A geometry-based stochastic MIMO model for vehicle-to-vehicle communications,"*IEEE Transactions on Wireless Communications*, 8(7): 3646-3657, 2009.

64. T. Abbas, J. Karedal, F. Tufvesson, A. Paier, L. Bernado and A. F. Molisch, "Directional Analysis of Vehicle-to-Vehicle Propagation Channels. in*Proc. of IEEE 73rd Vehicular Technology Conference (VTC Spring)*,2011

65. L. Bernado, T. Zemen, F. Tufvesson, A. F. Molisch and C. F. Mecklenbrauker, "Delay and Doppler Spreads of Nonstationary Vehicular Channels for Safety-Relevant Scenarios,"*IEEE Transactions on Vehicular Technology*, 63(1): 82-93, 2014.

66. L. Bernado, T. Zemen, F. Tufvesson, A. F. Molisch and C. F. Mecklenbrauker, "Time- and Frequency-Varying K-Factor of Non-Stationary Vehicular Channels for Safety-Relevant Scenarios," *IEEE Transactions on Intelligent Transportation Systems, 16(2): 1007-1017, 2015.

67. O. Renaudin, V. M. Kolmonen, P. Vainikainen and C. Oestges, "Wideband Measurement-Based Modeling of Inter-Vehicle Channels in the 5-GHz Band,"*IEEE Transactions on Vehicular Technology, 62(8): 3531-3540, 2013.

68. He Ruisi, O. Renaudin, V. M. Kolmonen, K. Haneda, Zhong Zhangdui, Ai Bo and C. Oestges, "Characterization of Quasi-Stationarity Regions for Vehicle-to-Vehicle Radio Channels,"*IEEE Transactions on Antennas and Propagation*, 63(5): 2237-2251, 2015.

69. M. Walter, U. C. Fiebig and A. Zajic, "Experimental Verification of the Non-Stationary Statistical Model for V2V Scatter Channels," in *Proc. of IEEE 80th Vehicular Technology Conference (VTC Fall)*, 2014

70. M. Boban, J. Barros and O. K. Tonguz, "Geometry-Based Vehicle-to-Vehicle Channel Modeling for Large-Scale Simulation,"*IEEE Transactions on Vehicular Technology*, 63(9): 4146-4164, 2014.

71. He Ruisi, A. F. Molisch, F. Tufvesson, Zhong Zhangdui, Ai Bo and Zhang Tingting, "Vehicle-to-Vehicle Propagation Models With Large Vehicle Obstructions,"*IEEE Transactions on Intelligent Transportation Systems*, 15(5): 2237-2248, 2014.

72. K. Amiri,Y. Sun, P. Murphy,C.Hunter, J.R.Cavallaro, andA. Sabharwal, "WARP, a unified wireless network testbed for education and research," in *Proc. IEEE MSE*, 2007, pp. 53–54.

73. B. Ai, X. Cheng, T. Kürner, Z. D. Zhong, K. Guan, R. S. He, L. Xiong, D. W. Matolak, D. G. Michelson, C. Briso-Rodriguez, "Challenges Toward Wireless Communications for High-Speed Railway." *IEEE Transactions on Intelligent Transportation Systems* , vol. 15, no. 5: 2143-2158.2014

74. R.S. He, Z. Zhong, Bo Ai, G. Wang, J. Ding, A.F. Molisch , "Measurements and Analysis of Propagation Channels in High-Speed Railway Viaducts,"*IEEE Transactions on Wireless Communications*, vol.12, no.2, pp.794,805, February 2013

75. Tao Zhou; Cheng Tao; Liu Liu; Zhenhui Tan, "Ricean K-Factor Measurements and Analysis for Wideband Radio Channels in High-Speed Railway U-Shape Cutting Scenarios,"*Vehicular Technology Conference (VTC Spring), 2014 I.E. 79th*, pp.1,5, 18-21 May 2014

76. Guan Ke, Zhong Zhangdui, Ai Bo and T. Kurner, "Propagation Measurements and Analysis for Train Stations of High-Speed Railway at 930 MHz,"*IEEE Transactions on Vehicular Technology*, 63(8): 3499-3516, 2014.
77. Guan Ke, Zhong Zhangdui, Ai Bo and T. Kurner, "Propagation Measurements and Modeling of Crossing Bridges on High-Speed Railway at 930 MHz,"*IEEE Transactions on Vehicular Technology*, 63(2): 502-517, 2014.
78. Guan Ke, Zhong Zhangdui, J. I. Alonso and C. Briso-Rodriguez, "Measurement of Distributed Antenna Systems at 2.4 GHz in a Realistic Subway Tunnel Environment,"*IEEE Transactions on Vehicular Technology*, 61(2): 834-837, 2012.
79. R. He, Z. Zhong, B. Ai, K. Guan, B. Chen, J. I. AIonso, and C. Briso,"Propagation channel measurements and analysis at 2.4 GHz in subway tunnels," *IET Microwaves, Antennas & Propagation*, vol. 7, no. 11, pp. 934–941, 2013
80. K. Guan, B. Ai, Z. Zhong, C.F. Lopez, L. Zhang, C. Briso-Rodriguez, A. Hrovat, B. Zhang, R. He, T. Tang, "Measurements and Analysis of Large-Scale Fading Characteristics in Curved Subway Tunnels at 920 MHz, 2400 MHz, and 5705 MHz,"*Intelligent Transportation Systems, IEEE Transactions on* , vol.PP, no.99, pp.1,13, 2015
81. J. Li, Y. Zhao, J. Zhang, R. Jiang, C. Tao, Z. Tan, "Radio channel measurements and analysis at 2.4/5GHz in subway tunnels,"*Communications, China*, vol.12, no.1, pp.36,45, Jan. 2015
82. WINNER II D1.1.2 "WINNER II Channel Models Part II Radio Channel Measurement and Analysis Results," IST-4-027756 V1.0, Sept. 2007. http://www. istwinner.org/ deliverables. html
83. Qian Wang, Chunxiu Xu, Min Zhao and Deshui Yu, "Results and analysis for a novel channel measurement applied in LTE-R at 2.6 GHz,"in *Proc. of Wireless Communications and Networking Conference (WCNC)*, 2014 IEEE,2014
84. E. J. Violette, R. H. Espeland, R. O. DeBolt and F. K. Schwering, "Millimeterwave propagation at street level in an urban environment,"*IEEE Transactions on Geoscience and Remote Sensing, vol. 26*, pp. 368-380, 1988.
85. H. J. Thomas, R. S. Cole and G. L. Siqueira, "An experimental study of the propagation of 55 GHz millimeter waves in an urban mobile radio environment,"*EEE Transactions on Vehicular Technology, vol. 43*, pp. 140- 146, 1994.
86. L. M. Correia, J. J. Reis and P. O. Frances, "Analysis of the average power to distance decay rate at the 60 GHz band," in *IEEE in Vehicular Technology Conference*, 1997.
87. A. M. Hammoudeh, M. G. Sanchez and E. Grindrod, "Experimental analysis of propagation at 62 GHz in suburban mobile radio microcells," *IEEE Transactions on Vehicular Technology, vol. 48*, pp. 576-588, 1999.
88. K. Sato, H. Kozima, H. Masuzawa, T. Manabe, T. Ihara, Y. Kasashima, K. Yamaki, "Measurements of reflection characteristics and refractive indices of interior construction materials in millimeter-wave bands,"in *Proc. of IEEE 45th Vehicular Technology Conference(VTC Spring 1995)*, vol.1, pp.449,453, 25-28 Jul 1995
89. Thomas Zwick, Troy J. Beukema, and Haewoon Nam, "Wideband Channel Sounder With Measurements and Model for the 60 GHz Indoor Radio Channel,"*IEEE Trans. On Vehicular Technology*, vol. 54, no. 4, pp.1266-1277, JULY 2005.
90. X. Hao, T. S. Rappaport, R. J. Boyle and J. H. Schaffner, "38-GHz wide-band point-to-multipoint measurements under different weather conditions,"*IEEE Communications Letters*, vol. 4, pp. 7-8, 2000.
91. H. Xu, V. Kukshya, and T. S. Rappaport, "Spatial and Temporal Characteristics of 60 GHz Indoor Channels," *IEEE J. Sel. Areas Commun.*, vol. 20, no. 3, pp. 620–630, Apr. 2002.
92. T. S. Rappaport, E. Ben-Dor, J. N. Murdock and Q. Yijun, "38 GHz and 60 GHz angle-dependent propagation for cellular & peer-to-peer wireless communications," in *IEEE International Conference on Communications (ICC)*, 2012.
93. E. Ben-Dor, T. S. Rappaport, Q. Yijun and S. J. Lauffenburger, "MillimeterWave 60 GHz Outdoor and Vehicle AOA Propagation Measurements Using a Broadband Channel Sounder," in *IEEE Global Telecommunications Conference (GLOBECOM)*, 2011.

94. T. S. Rappaport, Q. Yijun, J. I. Tamir, J. N. Murdock and E. Ben-Dor, "Cellular broadband millimeter wave propagation and angle of arrival for adaptive beam steering systems," in *IEEE Radio and Wireless Symposium (RWS)*, 2012.
95. T. S. Rappaport, F. Gutierrez, E. Ben-Dor, J. N. Murdock, Q. Yijun and J. I. Tamir, "Broadband Millimeter-Wave Propagation Measurements and Models Using Adaptive-Beam Antennas for Outdoor Urban Cellular Communications,"*IEEE Transactions on Antennas and Propagation,* vol. 61, pp. 1850-1859, 2013.
96. J. N. Murdock, E. Ben-Dor, Q. Yijun, J. I. Tamir and T. S. Rappaport, "A 38 GHz cellular outage study for an urban outdoor campus environment," in *IEEE Wireless Communications and Networking Conference (WCNC)*, 2012.
97. Y. Azar, G. N. Wong, K. Wang, R. Mayzus, J. K. Schulz, Z. Hang, F. Gutierrez, D.Hwang, T. S. Rappaport, "28 GHz propagation measurements for outdoor cellularcommunications using steerable beam antennas in New York city," in *IEEEInternational Conference on Communications (ICC)*, 2013.
98. T. S. Rappaport, S. Shu, "Multi-beam antenna combining for 28 GHz Cellular linkImprovement in urban environments," in *IEEE Global Telecommunication Conference (Globecom).*, Atlanta, 2013.
99. M.R. Akdeniz, Y. Liu, M.K. Samimi, S. Sun, S. Rangan, T.S. Rappaport, E. Erkip, "Millimeter Wave Channel Modeling and Cellular Capacity Evaluation," *IEEE J. Sel. Areas on Comm.*, Aug. 2014
100. M. K. Samimi and T. S. Rappaport, Ultra-wideband statistical channel model for non line of sight millimeter-wave urban channels. *Global Communications Conference* (GLOBECOM), 2014 IEEE,2014
101. T. S. Rappaport, R. W.Heath, Jr., R. C.Daniels, and J. N.Murdock, MillimeterWave WirelessCommunications.Pearson/Prentice Hall, 2015.
102. T. A. Thomas, Nguyen Huan Cong, G. R. Maccartney and T. S. Rappaport, 3D mmWave Channel Model Proposal. *Vehicular Technology Conference (VTC Fall), 2014 I.E. 80th*,2014
103. M. W. M. Peter, W. K. M. Raceala-Motoc, R. Felbecker, M. Jacob, S. Priebe and T.Kürner, "Analyzing human body shadowing at 60 GHz:Systematic wideband MIMOmeasurements and modeling approaches," in *Proc. of IEEE 6th European Conference on Antennas and Propagation*, 2012.
104. A.P. Garcia, W. Kotterman, R.S. Thomä, U. Trautwein , D. Brückner,W. Wirnitzer, J. Kunisch, "60 GHz in-cabin real-time channel sounding,"*3rd International Workshop on Broadband MIMO Channel Measurement and Modeling - IWonCMM 2009*, Xi'an, China, August 25, 2009.
105. M. Peter, W. Keusgen, A. Kortke, M. Schirrmacher, "Measurement and Analysis of the 60 GHz In-Vehicular Broadband Radio Channel,"*Proc. of IEEE Vehicular Technology Conference*, 2007 *(VTC-2007)*, pp. 834 - 838
106. M. Kyro, S. Ranvier, V. Kolmonen, K. Haneda and P. Vainikainen, "Long range wideband channel measurements at 81-86 GHz frequency range," in *European Conference Antennas and Propagation (EuCAP)*, 2010.
107. M. Kyro, V. Kolmonen and P. Vainikainen, "Experimental Propagation Channel Characterization of mm-Wave Radio Links in Urban Scenarios,"*IEEE Antennas and Wireless Propagation Letters,* vol. 11, pp. 865-868, 2012.
108. A. V. Raisanen, J. Ala-Laurinaho, K. Haneda, J. Jarvelainen, A. Karttunen, M. Kyro, V. Semkin, A. Lamminen and J. Saily, "Studies on E-band antennas and propagation," in *Antennas and Propagation Conference (LAPC).*, Loughborough, 2013.
109. Kim Minseok, Y. Konishi,Yu-yuan Chang, et al., Large Scale Parameters and Double-Directional Characterization of Indoor Wideband Radio Multipath Channels at 11 GHz. *IEEE Trans. Antennas Propagation*, vol. 62, no. 1, pp.430-441, 2014.
110. WANG H M, HONG W, et al. "Channel Measurement for IEEE 802.11aj (45 GHz)," IEEE 802.11-12/1361r3, 2013.

111. X. Wu, C.X. Wang, J. Sun, J. Huang, R. Feng,Y. Yang, X. Ge, 60 GHz Millimeter-Wave Indoor Channel Measurements and Modeling for 5G Systems, *IEEE Trans. Antennas Propag*, 2017, vol. 65, no. 4, pp. 1912-1924, 2017.

112. Mingyang Lei, Jianhua Zhang, Tian Lei, Detao Du, "28-GHz Indoor Channel Measurements and Analysis of Propagation Characteristics,"*IEEE 25th International Symposium on Personal, Indoor and Mobile Radio Communications (PIMRC 2014)*,2014

113. Zhang Nan, Yin Xuefeng, S. X. Lu, Du Mingde and Cai Xuesong, "Measurement-based angular characterization for 72 GHz propagation channels in indoor environments," *Globecom Workshops (GC Workshps)*, 2014

114. YIN Xuefeng, LING Cen and KIM Myung-Don, "Experimental Multipath-Cluster Characteristics of 28-GHz Propagation Channel". *IEEE Access*, 3, 2015. 3138-3150

115. X. Zhao, Q. Wang, S. Li, S. Geng, M. Wang, S. Sun, and Z. Wen, "Attenuation by human bodies at 26 and 39.5 GHz millimeter wave bands," IEEE Antennas Wireless Propag. Lett., vol. 19, pp. 1229 – 1232, Nov. 2016.

116. R. O. Schmidt, "Multiple emitter location and signal parameter estimation," *IEEE Trans. Antennas Propagat.*, vol. AP-34, pp. 276–280, Mar.1986.

117. R. Roy and T. Kailath, "ESPRIT—Estimation of signal parameters via rotational invariance techniques," *IEEE Trans. Acoust., Speech, Signal Processing*, vol. 37, pp. 984–995, July 1989.

118. A. van der Veen, M. Vanderveen, and A. Paulraj, "Joint angle and delay estimation using shift-invariance properties," *IEEE Signal Processing Lett.*, vol. 4, pp. 142–145, May 1997.

119. M. Haardt and J. Nossek, "Unitary ESPRIT: How to obtain increased estimation accuracy with a reduced computational burden," *IEEE Trans. Signal Processing*, vol. SP-43, pp. 1232–1242, May 1995.

120. M. Zoltowski, M. Haardt, and C. Mathews, "Closed-form 2-D angle estimation with rectangular arrays in element space or beamspace via unitary ESPRIT," *IEEE Trans. Signal Processing,* vol. SP-44, pp. 316–328, Feb. 1996.

121. J. Fuhl, J.-P. Rossi, and E. Bonek, "High-resolution 3.D direction-of-arrival determination for urban mobile radio," *IEEE Trans. Antennas Propagat.*, vol. AP-45, pp. 672–682, Apr. 1997.

122. G. McLachlan and T. Krishnan, *The EM Algorithm and Extensions. Probability and Statistics*. New York:Wiley, 1996

123. A. P. Dempster, N. M. Laird, and D. B. Rubin, "Maximum likelihood from incomplete data via the EM algorithm," *J. Royal Statist. Soc.*, Ser. B, vol. 39, no. 1, pp. 1–38, 1977.

124. J. A. Fessler and A. O. Hero, "Space-alternating generalized expectation- maximization algorithm," *IEEE Transactions on Signal Processing*, Vol 42, no 10, pp 2664-2677, 1994.

125. B. H. Fleury, D. Dahlhaus, R. Heddergott, and M. Tschudin, "Wideband angle of arrival estimation using the SAGE algorithm," in *Proc. of the IEEE Fourth Int. Symp. on Spread Spectrum Techniques and Applications (ISSSTA '96)*, Mainz, Germany, pp. 79-85, Sept. 1996.

126. B. H. Fleury, M. Tschudin, R. Heddergott, D. Dahlhaus, and K. I. Pedersen, "Channel parameter estimation in mobile radio environments using the SAGE algorithm," *IEEE J. Sel. Areas Commun.*, vol. 17, no. 3, pp. 434-450, Mar. 1999.

127. M. Tschudin, R. Heddergott and P. Truffer, "Validation of a high resolution measurement technique for estimating the parameters of impinging waves in indoor environments," in *Proc. of The Ninth IEEE International Symposium on Personal, Indoor and Mobile Radio Communications*(PIMRC 1998),1998

128. B. H. Fleury, Yin Xuefeng, K. G. Rohbrandt, P. Jourdan and A. Stucki, "Performance of a high-resolution scheme for joint estimation of delay and bidirection dispersion in the radio channel,"*IEEE 55th Vehicular Technology Conference(VTC Spring 2002)*, 2002

129. Yin Xuefeng, B. H. Fleury, P. Jourdan and A. Stucki, "Polarization estimation of individual propagation paths using the SAGE algorithm,"*14th IEEE Proceedings on Personal, Indoor and Mobile Radio Communications(PIMRC 2003)*, 2003

130. C. C. Chong, D. I. Laurenson, C. M. Tan, S. Mclaughlin, M. A. Beach and A. R. Nix, Joint detection-estimation of directional channel parameters using the 2-D frequency domain SAGE algorithm with serial interference cancellation,"*IEEE International Conference onCommunications(ICC 2002)*, 2002

131. D. Shutin, B.H. Fleury, "Sparse Variational Bayesian SAGE Algorithm With Application to the Estimation of Multipath Wireless Channels," *IEEE Transactions on Signal Processing*, vol 59, no 8, pp. 3609 – 3623, Aug. 2011

132. A. Richter, "*Estimation of Radio Channel Parameters:Models and Algorithms*," Ph. D. dissertation, 2005,nel parameters:Models and algorithms" Ph.D. dissertation, Technischen Universität Ilmenau, Ilmenua, Germany, May 2005, ISBN:978-3.938843.02-4 [Online]. Available:www.db-thueringen.de.

133. Andreas Richter, Reiner S. Thoma, "Parametric Modeling and Estimation of Distributed Diffuse Scattering Components of Radio Channels", COST273, TD(03)198

134. F. Quitin, C. Oestges, F. Horlin, and P. De Doncker, "A Polarized Clustered Channel Model for Indoor Multiantenna Systems at 3.6 GHz,"*IEEE Transactions on Vehicular Technology*, vol. 59, no. 8, pp. 3685-3693, Oct. 2010.

135. Zhang Yan, Hu Xinwei, Zhou Shidong ,Wang Jing, Near Field Channel Parameter Estimation Method Based on the SAGE, Journal of System Simulation, vol. 23, 9 September 2011. (in Chinese)

136. J. Chen, S. Wang and X. Yin, "A Spherical-Wavefront-Based Scatterer Localization Algorithm Using Large-Scale Antenna Arrays." *IEEE Communications Letters* 20(9): 1796-1799, 2016.

137. F. Quitin, C. Oestges, F. Horlin, and P. De Doncker, "Diffuse multipath component characterization for indoor mimo channels," in *Antennas and Propagation (EuCAP), 2010 Proceedingsof the Fourth European Conference on*, Apr. 2010, pp. 1–5.

138. Tanghe E., Gaillot D. P., Lienard M. and et al. Experimental Analysis of Dense Multipath Components in an Industrial Environment. *IEEE Transactions on Antennas and Propagation*, 62(7): 3797–3805, 2014.

139. J. Poutanen, F. Tufvesson, K. Haneda, L. Liu, C. Oestges and P. Vainikainen, "Adding Dense Multipath Components to Geometry-Based MIMO Channel Models," [Online] Available: http://lup.lub.lu.se/record/2218357/file/2218371.pdf

140. J. Salmi, A. Richter and V. Koivunen, "Detection and Tracking of MIMO Propagation Path Parameters Using State-Space Approach," *IEEE Transactions on Signal Processing*, 57(4): 1538–1550, 2009.

141. T. Jost, Wei Wang; U. Fiebig, F. Perez-Fontan, "Detection and Tracking of Mobile Propagation Channel Paths," *IEEE Transactions on Antennas and Propagation*, vol 60, no 10, pp. 4875 – 4883, Oct. 2012

142. Xuefeng Yin, G. Steinbock, G.E. Kirkelund, T. Pedersen, P. Blattnig, A. Jaquier, B.H. Fleury,"Tracking of Time-Variant Radio Propagation Paths Using Particle Filtering,"*IEEE ICC '08*, pp. 920-924, May 2008.

143. K. Saito, K. Kitao, T. Imai and Y. Okumura, Dynamic MIMO channel modeling in urban environment using particle filtering. *2013 7th European Conference on Antennas and Propagation (EuCAP)*, 2013

144. J.C. Bezdek, *Pattern recognition with fuzzy objective function algorithms*, New York, 1981

145. N. Czink, P. Cera, J. Salo, E. Bonek, J.-P. Nuutinen, J. Ylitalo, "A framework for automatic clustering of parametric MIMO channel data including path powers," in *Proc. IEEE 64th Vehicular Technol. Conf.*, Sep. 25–28, 2006, pp. 1–5.

146. D.-J. Kim, Y.-W. Park, and D.-C. D.-J. Park, "A novel validity index for determination of the optimal number of clusters," *IEICE Trans. Inf. Syst.*, vol. E84-D, no. 2, pp. 281–285, 2001.

147. U. Maulik and S. Bandyopadhyay, "Performance evaluation of some clus-tering algorithms and validity indices," *IEEE Trans. Pattern Anal. Mach. Intell.*, vol. 24, no. 12, pp. 1650–1654, Dec. 2002.

148. N. Czink, P. Cera, J. Salo, E. Bonek, J.-P. Nuutinen, J. Ylitalo, "Automatic clustering of MIMO channel parameters using the multi-path component distance measure," WPMC'05, Aalborg, Denmark, Septemter 2005

149. G. E. P. Box and Mervin E. Muller,"A Note on the Generation of Random Normal Deviates,"*The Annals of Mathematical Statistics* (1958), Vol. 29, No. 2 pp. 610–611

150. Greenstein et al, "Moment method estimation of the Ricean K-factor," *IEEE Comm. Lett.* Vol 3, Issue 6, p.175, 1999.

151. R. Kolar, R. Jirik, J. Jan (2004) "Estimator Comparison of the Nakagami-m Parameter and Its Application in Echocardiography", Radioengineering, 13 (1), 8–12

152. R. M. Norton,"The Double Exponential Distribution:Using Calculus to Find a Maximum Likelihood Estimator,"*The American Statistician (American Statistical Association)* 38 (2): 135–136, May 1984. doi:10.2307/2683252.

153. Borradaile, Graham. *Statistics of Earth Science Data*. Springer. ISBN 978-3.540-43603.4. Dec 2009.

154. D. J. Best, N. I. Fisher, "Efficient Simulation of the von Mises Distribution,"*Applied Statistics*, vol. 28, No. 2, pp 152-157, 1979

155. Sra, S. "A short note on parameter approximation for von Mises-Fisher distributions And a fast implementation of $I_s(x)$," *Computational Statistics* 27:177–190. 2011.doi:10.1007/s00180-011-0232-x.

156. Andrew T.A Wood, "Simulation of the von mises fisher distribution," *Communications in Statistics - Simulation and Computation*, 23(1):157-164, 1994.

157. S. Jung, "Generating von Mises Fisher distribution on the unit sphere (S^2) ,"Oct. 2009. [Online] Available at http://www.stat.pitt.edu/sungkyu/software/randvonMisesFisher3.pdf

158. J. T.Kent, "The Fisher–Bingham distribution on the sphere,"*J. Royal. Stat. Soc.*, Sr. B, 44:71–80, 1982

159. John T. Kent, Asaad M. Ganeiber and Kanti V. Mardia, "A new method to simulate the Bingham and relateddistributions in directional data analysis with applications,"Arxiv. [Online] Available at http://arxiv.org/abs/1310.8110

160. L. Devroye, *Non-Uniform Random Variate Generation*,Springer-Verlag.1986

161. Donald E. Knuth, *Art of Computer Programming, Volume 2:Seminumerical Algorithms (3rd Edition)*, Addison-Wesley Professional, November 14, 1997

162. F. Babich and G. Lombardi, "Statistical analysis and characterization of the indoor propagation channel," *IEEE Trans. Commun.*, vol. 48, no.3, pp. 455–464, Mar. 2000.

163. WINNER1 WP5:"Final Report on Link Level and System Level Channel Models" Deliverable D5.4, 18.11.2005

164. WINNER+ D5.3, "Final channel models," V1.0, CELTIC CP5-026 WINNER+ project.http://projects.celticinitiative.org/winner+/deliverables_winnerplus.html, 2010.

165. M. Steinbauer, A. F. Molisch, *Spatial channel models, in Wireless Flexible Personalized Communications*, L. Correia (ed.), John Wiley & Sons, Chichester, 2001.

166. L. Correia, Ed., *Mobile Broadband Multimedia Networks*. Academic Press, 2006.

167. L. Liu, J. Poutanen, F. Quitin, K. Haneda, F. Tufvesson, P. D. Doncker, P. Vainikainen, and C. Oestges, "The COST2100 MIMO channel model," *IEEE Wireless Commun.* , vol. 19, no. 6, pp. 92–99, Dec. 2012.

168. C. Oestges,C. Brennan, F. Fuschini, M. L. Jakobsen,S. Salous, C. Schneider and F. Tufvesson, "Radio Channel Measurement and Modelling Techniques", Chapter 9 of *Cooperative Radio Communications for Green Smart Environments*, River Publishers 2016

169. 3GPP TR 25.996,"Technical Specification Group Radio Access Network; Spatial channel model for Multiple Input Multiple Output (MIMO) simulations," V11.0.0, 2012-09. http://www.3gpp.org

170. 3GPP TR36.873, "Study on 3d channel model for lte," 3rd Generation Partnership Project, v12.1.0. 2015. http://www.3gpp.org

171. V. Erceg, et al. "IEEE P802.11 Wireless LANs -TGn Channel Models," IEEE 802.11-03/940r4, May 2004.

172. G. Breit, et al. "IEEE P802.11 Wireless LANs - TGac Channel Model Addendum," IEEE 802.11-09/0308r12, March 2010.
173. Cost Final Report, http://www.lx.it.pt/cost231/
174. R. Verdone and A. Zanella, "Pervasive mobile and ambient wireless communications," COST Action 2100, Springer, 2012.
175. L. Liu, "Implementation of the COST 2100 model," v2.2.5 [Online]. Available:http://code. google.com/p/cost2100model/
176. L. Hentilä, P. Kyösti, M. Käske, M. Narandzic, and M. Alatossava. (2007, December) MATLAB implementation of the WINNER Phase II Channel Model ver1.1 [Online]. Available:https://www.ist-winner.org/phase_2_model.html
177. J. Medbo and P. Schramm, "Channel models for HIPERLAN/2," ETSI/BRAN document no. 3ERI085B.
178. ITU-R Document 5D/TEMP/332(Rev.1): Preliminary draft new Report ITU-R M.[IMT-2020.EVAL] Channel model (part), 27th meeting of WP 5D, Niagara Falls, Canada, June 2017
179. Jian Luo, et. al., Channel Sounding for 802.11ay, IEEE 802.11-15/0631r0, May 2015
180. J. Järveläinen and K. Haneda, "Sixty gigahertz indoor radio wave propagation prediction method based on full scattering model," *Radio Science,* vol. 49, no. 4, pp. 293-305, 2014.
181. METIS D1.2, Initial channel models based on measurements, v1.0, 2014. https://www.metis2020.com/
182. 3GPP TR 38.901. Study on channel model for frequencies from 0.5 to 100 GHz (Release 14). 3rd Generation Partnership Project (3GPP), V14.0.0, Mar. 2017.
183. P. Agrawal and N. Patwari, "Correlated link shadow fading in multihop wireless networks," *IEEE Trans. Wireless Commun.*, vol. 8, no. 8, pp. 4024–4036, Aug. 2009.
184. T. Jämsä and P. Kyöti, "Device-to-device extension to geometry-based stochastic channel models," *EuCAP 2015*, Lisbon, Portugal, 2015.
185. W. Fan, T. Jämsä, J. O. Nielsen and G. F. Pedersen, "On Angular Sampling Methods for 3.D Spatial Channel Models," *IEEE Antennas and Wireless Propagation Letters*, Vol. 14, 2015.

Chapter 4
Software Simulation

Starting from the requirements and technical indicators of 5G network system, this chapter analyzes the technical challenges in designing and realizing 5G software simulation system, and introduces the link level simulation and system simulation techniques with the focus on test evaluation methods, key technologies and applications, which provides technical reference for the design and realization of 5G network system software testing.

4.1 Overview on Software Simulation

The modern wireless communications system is rather complicated in terms of application scenarios, network structure, functions, characteristics, etc. Take 4G LTE network as an example. A network that can work in reality needs eNodeB, Mobility Management Entity (MME), Packet data network GateWay (PGW), Serving GateWay (SGW), Home Subscriber Server (HSS) and many other network elements as well as related interfaces and protocol stacks. When compatibility and interoperability with other systems are considered, its network structure and networking methods would be more complicated. In addition, the wireless communications system runs in a real environment which is complicated and volatile. Its difficulty and complexity further increase with propagation features and user features like SF, multipath effect, high mobility, strong interference, etc. Due to plenty of realistic factors above, in reality it's difficult to build a complete and accurate model to analyze the design of wireless communications system. Usually, it's more feasible and effective to set up a software simulation platform via computer and mathematical modeling, on which software simulation programs can make performance evaluation for the wireless communications modules or system.

5G candidate technologies are more abundant and its application scenarios are more complicated. Accordingly, the software evaluation for the performance of 5G

© Springer International Publishing AG 2018
Y. Yang et al., *5G Wireless Systems*, Wireless Networks,
DOI 10.1007/978-3-319-61869-2_4

technical schemes is facing the unprecedented challenges. Therefore, the academia and industry have been committed to improving the software simulation platform's computational performance, flexibility and accuracy, and have achieved abundant research fruits and practical results, providing important reference for building the 5G simulation platform.

GPU is one of the important methods for parallel acceleration. For example, Ben et al. [1] propose a distributed computation architecture based on CPU-GPU to solve the problems including the high price, the heavy inter-processor synchronization overheads, and the huge communications overheads for inter-server distributed coordination in the multi-core CPU based parallel computing simulation platform, which aims at meeting the simulation requirements of the large scale mobile network. This new architecture designs the hardware and software jointly, decomposes the computational tasks through master-slave node model, and makes layered design for the whole software based on modular design of application module, protocol stack, message service, connectivity management, mobility management and system interface function. The master node deployed on CPU takes charge of event scheduling, data abstraction design, scenario management and other functions. It has solved the problem of low parallel synchronization through events synchronization, data decoupling by functions, memory hierarchical management and other mechanisms. The slave nodes deployed on GPU execute simulation tasks, which fully utilize GPU's computing power and reduce the communications data effectively. Measured results prove that this architecture can improve the operating efficiency significantly.

Open source software based simulation development and design make full use of the unique open and sharing characteristics of open source community, and accelerates the realization of simulation. For example, Capozzi et al. [2] propose a LTE-Sim based open source tool, which supports the HetNet scenario with macro eNodeB and femtocell coexisting. This open source based new model can meet the simulation requirements of specific scenarios quickly and flexibly. It is one of trends for future simulation platform development.

Due to the fast development of 5G, the simulation objects, scenarios and technologies are developing as well, and the design and evaluation methods also need to develop synchronously. For example, Goga et al. [3] have analyzed the characteristics of cloud computing, proposed design architecture and simulation method for cloud computing application, and concluded the key optimization, modeling and design problems in the follow-up research.

Next, a complete simulation system is introduced briefly, as shown in Fig. 4.1.

- Requirements analysis. Traditional simulation process starts with original requirements. Generally, a networking scenario is needed to be given, including network type, network element type, network interface function, network configuration parameter, services characteristics, performance indicators, etc. These original requirements give the simulation objects and corresponding behaviors. Through analyzing original requirements, the input requirements for simulation system modeling and network performance evaluation criteria can be obtained.

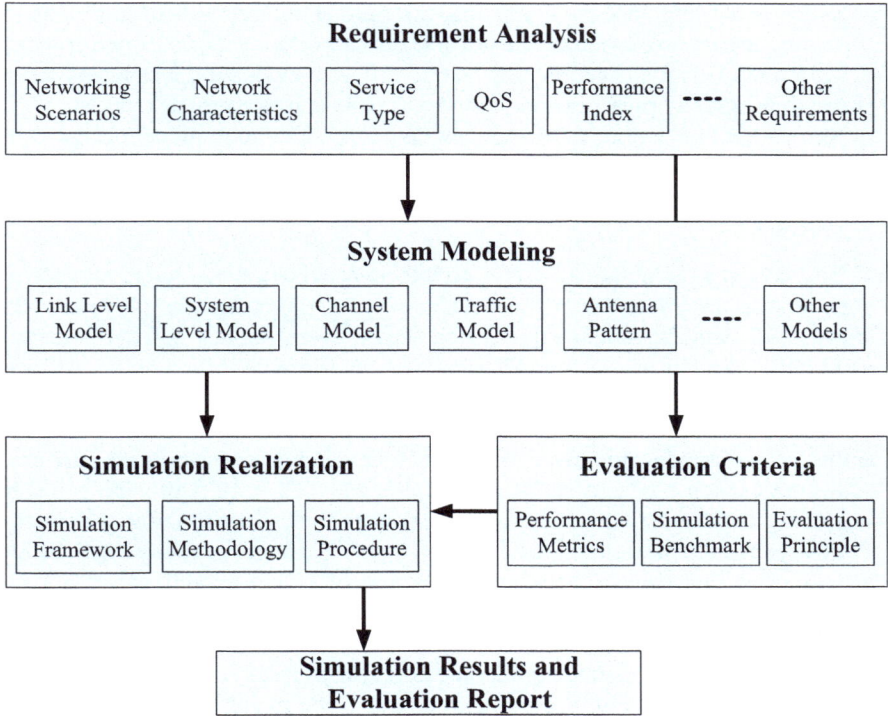

Fig. 4.1 Composition and implementation process of simulation system

- System modeling. System modeling includes abstracting and modeling the network and system components, and then outputting the simulation model. These simulation models describe the input/output relations of the network and system components. The models not only need to maintain the basic characteristics of the modeled objects, but also cannot be too complicated. Thus there is a tradeoff between accuracy, complexity and calculation, which is where the difficulty and emphasis of simulation design lie in. Generally, complex models design adopts the way that combines top-down with bottom-up. The former allows the designer to seize the design process line from requirements to system, subsystem and modules, which helps to reduce the unnecessary duplicated work and improves work efficiency to the greatest extent. The latter can guarantee the accuracy of basic simulation objects and help to break down and implement in parallel the simulation work.
- Evaluation criteria. Through the analysis of original requirements, the input information of simulation evaluation can be obtained, which includes performance indicators, simulation benchmark and evaluation criteria. Performance indicators are used to evaluate the measurable indicators of the system to be simulated. Simulation benchmark provides benchmark data for evaluation,

which is the reference for comparing different simulation schemes. Evaluation criteria are used to evaluate the advantages and disadvantages of performance. Different evaluation criteria may lead to different evaluating conclusion. Common evaluation criteria include fairness criteria and throughput criteria.

- Simulation realization. After the completion of simulation model design is the simulation realization stage. Simulation realization includes three elements [4], namely simulation system architecture, simulation methodology and simulation process.

The design of simulation system architecture defines the simulation system's logic architecture, development architecture, running architecture, physical architecture and data architecture [5]. Logic architecture defines software function partitioning, design of categories and objects, interface definition and other basic logic functions of simulation system design. Architecture development mainly considers the scalability, reusability, comprehensibility, testability and other basic ability during the development. Architecture development also contains the practical organization mode of software modules. For 5G, it must provide sufficient scalability to adapt to the change of system architecture. Running architecture mainly considers the simulation system's performance, usability, and safety, including the concurrency and synchronization of system modules, and process, thread and other running concepts. 5G simulation system significantly increases the computation complexity, which is a big challenge for the design of development architecture. Physical architecture focuses on how to install or deploy software on the physical machine. For 5G simulation system, the model selection of physical computation platform and parallelizing ability of software are the key factors that influence the computational performance. Data architecture mainly concerns about the data requirements, including data storage, duplication, transmission, synchronization, etc.

Simulation methodology is used to guide the definitions of link level and system level simulation model and relevant parameters.

Simulation processes have defined the processes and steps of simulation. A typical system simulation process includes: generating network element and network topology, generating users and allocating users to the serving cells randomly, generating channel model, generating service packets according to service model, resource scheduling, service packets processing, interference computation, users' Signal to Interference plus Noise Ratio (SINR) computation and outputting simulation results.

- Simulation results and report. Simulation system can output simulation data after it runs as the required process; and based on simulation data's analysis results, the simulation report can be got eventually. Actually, simulation data needs further analysis to determine whether the simulation data is reasonable, whether the output indicators meet the expectation and whether the design and realization of simulation system need improvement, etc. Generally, the whole simulation system needs calibration before the authentic report is obtained, that is, observing whether the output result is same with recognized benchmark system at the given input parameters and models so as to check whether the simulation system realization is right. Due to the introduction of massive MIMO and other new technologies, 5G simulation system needs to implement plenty of calibration work according to new simulation parameters and models to guarantee the correctness of the simulation output result.

From the above analysis, we can know that the software simulation process of the whole wireless system is very complex. Thus the workload of building simulation system software is huge. To improve the efficiency, simulation designers can develop the simulation system with the help of proven commercial software package. Common commercial software tool packages include MATLAB, OPNET [6], NS2/NS3 [7], QualNet [8], etc. All of these packages provide all kinds of basic parts that software framework of communications system needs, including modeler, model library, simulation kernel and postprocessor, but the way of realizing these parts and focus range of provided model library might be something different. The design and development of 5G software simulation testing system inherit and develop from4G and earlier technologies. Early simulation platforms are of great reference value for building 5G simulation software. Meanwhile, since 5G system will introduce more new functions and new technologies, designers and developers need to deeply analyze the characteristics and realization plans of candidate technologies so that they can design and realize 5G software simulation system efficiently.

4.2 5G Software Simulation Requirements

4.2.1 Simulation Requirements Analysis

As an entirely new network, 5G is the wireless communications system that faces the information society in 2020. Currently, 5G has entered the key stage of key technologies breakthrough and standards formulation. Thus corresponding key technologies simulation requirements will also help the 5G simulation to be promoted to a new height on system architecture and simulation efficiency. Evaluating the performance indicators of 5G network comprehensively, fast and accurately is the important requirement for designing and realizing 5G software simulation systems.

Comprehensiveness is the functional requirement of the software simulation system, which mainly refers to the comprehensive support for 5G performance indicators and candidate technologies. 5G performance indicators include Key Performance Indications (KPIs) such as peak data rate, guaranteed minimum user data rate, connection density, service traffic volume density, wireless delay, end-to-end delay [9], etc. It also includes secondary indicators that support these KPIs such as SINR, paging success rate, access success rate, handover success rate, etc. Software simulation system must provide corresponding modeling, statistic and evaluation for these new performance indicators. 5G candidate technologies can be classified into two types, namely air interface technology and network technology. Air interface technology includes EE-SE co-design, massive antenna, full duplex, novel multiple access, new waveform, new modulation and coding, software defined air interface, sparse mining, high frequency band communication, spectrum sharing and flexible application, etc. Network technologies include C-RAN, SDN/NFV, Self-Organizing Network (SON), Ultra Dense Network (UDN), multi-network convergence (multi-RAT) and D2D [9], etc. These new technologies' impact on existing network design and realization can be divided into three types in terms of architecture, namely architecture level, network element level and module level. New technologies in architecture level, such as SDN and SON, have the greatest impact on the design of software simulation system. Because it needs to model the new network architecture, design new network element types, interfaces between new network elements and new protocol stacks. All these changes are basic and we still need to make large-scale changes on the entire network design, which make the design more difficult and bring software simulation more workload. New technologies in network element level, such as massive MIMO technology, involve coordinated design between multiple modules and the influence is mainly limited in the network elements, so there is no need to make big changes on network architecture and interface between network elements. In comparison, the influence scope of new technologies in module level is much smaller, which is limited in one function module within the network element and other related modules can support it with small modifications, for instance, new modulation coding technologies. The main work of software simulation system includes comprehensively analyzing every 5G candidate technologies, proposing corresponding resource models, and designing and implementing schemes, which serve as significant guarantees for the completeness of 5G software simulation system functions.

Rapidity is the time efficiency requirement for software simulation system. This requirement is embodied in two aspects. One is that it can substantially reduce simulation time of 5G network traffic and performance evaluation. The other is that it can quickly adapt to the change and adjustment of network architecture. Part of 5G technologies' [10] have huge simulation objects, for example, the base station antenna scale of massive MIMO technology can be more than 128, and the number of cells under UDN reaches hundreds or even more. Some computational overheads of the simulation characteristics are very large. For example, massive MIMO precoding calculation involves a lot of more than 128×30 dimensions matrix operations. Large simulation objects scale and computational complexity require the computational performance of software simulation system based on existing system to have great improvement. Only in this way can the requirements of simulation tasks and timely evaluation be met. Sharp growth of requirements for computational performance requires the systematic brand-new design and implementation of the hardware platform, software platform and the simulation program. For the software simulation system, the key problem is how to complete the software system concurrent design and coding implementation on the new and powerful hardware platform with powerful computing power, which require the simulation program to realize the greatest concurrent on key computation paths, such as the concurrent processing of granularity that reaches the subcarrier level. Rapidity is also reflected in the fast response to the change of 5G network architecture with a relatively small cost. Different from previous wireless communications systems, 5G has many network level candidate technologies with great changes in network architecture and rich supporting scenarios, so the simulation system needs to be flexible enough. Its architecture design, module design, and interface design should have characteristics like decoupling, modularity, interface scalability and ease of integration, etc., which put forward higher requirements on network resource model design, software function modular design, and the interface design between network elements and modules. In general, the complete cycle of development and validation of a simulation task should be as short as possible, so that it can meet the rapid verification and selection requirements of 5G candidate technologies.

Accuracy is performance requirement for software simulation systems. One of main factors influencing the simulation accuracy is the simplified method introduced to network modeling, simulation service abstraction, design and realization. Those simplified methods are usually the results of the compromise between the simulation authenticity and the simulation calculation cost, i.e., reducing the simulation calculation cost at the cost of simulation authenticity. Take the widely used EESM/MIESM link mapping method as an example. This method makes equivalent mapping for link model, in which the simulation process of the link function is equivalently mapped to several performance system level indicators, so as to simplify the computational process of link realization in the system simulation process; yet meanwhile system errors are also introduced. In order to reflect the impact of link process on system simulation more accurately, hardware and software methods can be combined to truly reflect the link process. The link to be simulated is deployed in hardware (such as FPGA, channel simulator) for real-time calculation, and the simulation accuracy can be improved through the hardware simulation calculation. Other simulation processes can do similar processing. When computational overheads allows, the simulation accuracy can be improved at the cost of increasing computational resources.

On the basis of summarization of 5G network software simulation system's requirements, objectives and implementation ideas, the overall technical roadmap can be obtained in detail, as shown in Fig. 4.2. All the key supporting techniques involved will be discussed in detail in subsequent chapters.

Based on the above simulation requirements analysis, we will take the requirement for rapidity as an example, to briefly introduce some common solutions. All the present communications networks have complex system architectures. It usually takes very long time to do the simulation calculation to accomplish a specific task. The time needed for communications system simulation is at least a few days, and can be as long as a month or even longer. Aiming at this problem, in addition to the advanced simulation technology shown in Fig. 4.2, as the old saying goes, "Good tools are prerequisite to the successful execution of a job". We should also include the following common ways to improve the simulation efficiency from the perspective of hardware configuration.

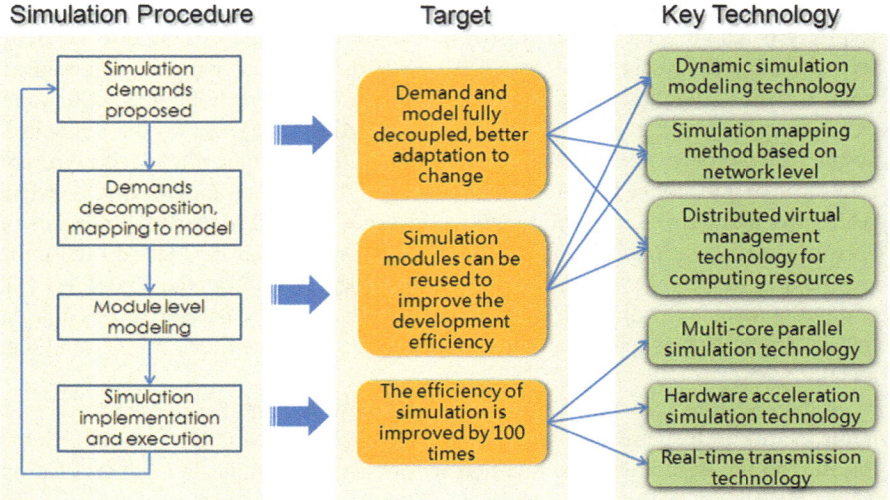

Fig. 4.2 The overall technical roadmap of 5G software simulation

1. High configuration desktop. The most simple and direct method to improve the simulation efficiency is to buy high configuration computer, such as choosing multi-core high-speed processors, large memory and high speed hard disk, which is the most common practice to improve the simulation efficiency.
2. Multicore servers. Improvement approach typically refers to the purchase of multicore servers to improve the simulation efficiency. To achieve the best effect with the server simulation, it also needs to design parallel program, and make each CPU run programs respectively.
3. High performance supercomputer. Supercomputers have higher system rate, there are special parallel tools and job scheduling system, which can be very convenient for users. Renting a supercomputer is not only convenient and fast but also cost controllable. It can be charged through the actual number of CPU cores used and the time in the simulation program.

4.2.2 Technological Impact Analysis

The main factors influencing the system simulation software design and implementation can be divided into three categories: architecture, function and performance. Architecture reflects the design of components and interfaces of the system simulation software as well as the definition of the interfaces. Function embodies simulation objects and behavior scope of the system simulation software. Performance shows the computational performance that can be reached by system simulation software and its platform. As shown in Table 4.1, the effects of main new 5G candidate technologies on the simulation software design and implementation can also be analyzed from the above three aspects.

Table 4.1 Analysis on the impact of 5G key technologies on system simulation software

Key technologies	Overview	Impact analysis		Computational performance
		Software architecture	Simulation function	
Massive MIMO	In the base station end hundreds of antennas (128, 256 or more) are installed to realize the data transmission of massive MIMO. Specifically, it includes precoding design, pilot frequency design, feedback design, receiver algorithm, large-scale active array design, and channel measurement and modeling, etc.	No impact	Functional enhancements of channel model and the physical layer are added, including precoding algorithm, channel estimation and signal detection, etc.	Antenna number increases greatly, interference calculation, precoding, receiver, scheduling and other simulation modules show sharply increased calculation strength
EE-SE joint-design	Energy efficiency and spectrum efficiency are jointly designed, providing comprehensive instruction and design framework for the choice of transmission technology, radio frequency link and antenna configuration, wireless resource management and network deployment, etc.	No impact	New algorithm function	Depending on the computational complexity
Full-duplex communication	improve spectrum efficiency through Co-time Co-frequency Full Duplex (CCFD)	No impact	The physical layer module supports full-duplex mode; resource scheduling module is enhanced	Calculation intensity is enhanced in interference calculation, scheduling, receiver modules.
New multiple access technologies	New multiple access technologies superpose the signals from different users at the same time frequency domain, in order to provide higher spectrum efficiency and improve the system access and the capacity; the receiver uses the advanced algorithm for demodulation	No impact	The physical layer module supports the new multiple access function; receiver, resource scheduling and other modules are enhanced	Computational complexity increases at the receiver module

(continued)

Table 4.1 (continued)

Key technologies	Overview	Impact analysis		
		Software architecture	Simulation function	Computational performance
New coding and modulation	Including LDPC codes of FQAM, Raptor type, Polar code, Gray coding of non-uniform distribution APSK, diversity technology of joint coding modulation and other new coding modulation technologies	No impact	Physical layer module supports new coding and modulation; the link adaptive module is enhanced	No impact
High frequency communication	Communications are realized through short-wavelength millimeter wave band, which is able to provide more than ten times bandwidth than 4G	No impact	Channel model, beamforming, and some other functions are added	Calculate intensity increases with the increase of system bandwidth
C-RAN	A green wireless access network architecture based on centralized processing, collaborative radio and real-time cloud computing architecture	Virtualization technology requires the various modules of the simulation system to decouple fully	New network types such as virtual cell are supported; topology management, wireless resource management, resource, and resource scheduling modules are enhanced	Calculation intensity depends on the complexity of the centralized management algorithm
UDN	The deployment density of low power station in hot spots in the cellular network is increased to improve system capacity and network coverage and to reduce the delay and energy consumption	It is necessary to support the ultra dense nodes deployment, virtual cell and other characteristics	Topology management, interference calculation, and handover modules should be enhanced	Calculation intensity increases with the increase of number of base stations
SDN	In the SDN architecture, control plane and data plane are separated; control function is centralized; the bottom network infrastructure is abstracted; and interfaces for the application and network services are provided	Control plane and data plane separated: centralization and virtualization management for control modules:, unified interface is opened	New type network element is supported; protocol stack module, resource management module, resource scheduling module all need to be enhanced	Calculation intensity is decided by the network scale and the complexity of centralized management algorithm

4.3   5G Software Link Level Simulation

4.3.1   Link Technology Overview

Major breakthroughs in basic theories and key technologies of the wireless transmission have been leading the evolution and innovation of the wireless communications system and its standardization. A series of novel multiple access technologies and transmission technologies (such as TDMA/FDMA/CDMA/NOMA, OFDM/MIMO, etc.) have brought about the significant improvement of the information transmission rate and opened up a new era of the wireless technology revolution [11]. Facing the massive growth of the future wireless data and the rich variety of services and experience requirements, 5G transmission technologies will explore a series of new types of multiple access and transmission mechanisms, to greatly improve the spectrum efficiency and the energy efficiency of the wireless systems [9].

The academia and industry at home and abroad have put forward a number of candidate technologies for 5G wireless transmissions. Massive MIMO, high-efficiency modulation and coding, non-orthogonal multiple access, full duplex and other new transmission technologies have caused great concern in 5G [12]. Furthermore, energy efficiency becomes one of the important performance indicators of the 5G system [13]. Compared with the 4G technologies, the signal processing mechanism adopted by the 5G transmission system will be even more complicated, and performance evaluation indicators would be more multidimensional, which will challenge the 5G air interface technology standardization, the program system design and the simulation verification.

As shown in Fig. 4.3, massive MIMO has become one of the important 5G key transmission technologies [14]. Massive MIMO can bring the huge array gain and the interference suppression gain through large-scale antenna arrays, thus greatly improving the system spectrum efficiency and the edge user spectrum efficiency. Current researches on the massive MIMO transmission mainly focus on system

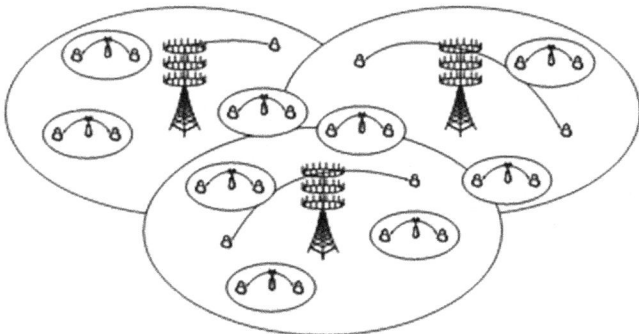

Fig. 4.3 An example of massive MIMO in heterogeneous wireless networks scenario

capacity, performance analysis, precoding, pilot pollution and other issues. The evaluated performance results based on the link level simulation for the massive MIMO technology will present important information for the massive MIMO system design.

Compared with the traditional multiple access technology, the non-orthogonal access technology based on the advanced waveform design can achieve much higher spectral efficiency. Non-orthogonal accesses in the frequency domain, the time domain as well as the code domain, such as Interleave Division Multiple Access (IDMA), Low Density Spreading Multiple Access (LDSMA), spatial coupling, and Power Domain Multiple Access (PDMA), have attracted wide research attentions [15]. The performance of multiple access technologies, as the key components of the wireless system physical layer transmission, needs comprehensive evaluation and analysis with the help of link level simulation.

In addition, the future 5G network deployment will have distinctive heterogeneous characteristics [16], which include heterogeneous connections through various air interface technologies, the heterogeneous network coordination and interference management, etc. In the massive MIMO and dense coverage scenario, the highly-efficient coordinated interference signal processing methods such as interference alignment will have great expected performance in improving the system's coverage, the spectrum efficiency and the user service experience, etc. The complex interference modeling and interference processing mechanism will also play important roles in improving the precision of the simulation results, which will be covered in both the 5G link level and system level simulation.

Overall, with the breakthrough in technologies such as massive MIMO, the mobile communications industry has been in the stage of the 5G era. In terms of 5G transmission technologies, on the one hand, we need to carry out the theoretical research and algorithm design. On the other hand, we also need to put forward the corresponding technical evaluation mechanism to provide the detailed performance evaluation results through constructing the efficient and accurate link level simulation platform.

4.3.2 Link Simulation Realization

Simulation Key Factors

The link level simulation is mostly used to evaluate the physical layer transmission performance of the wireless communications system. It usually includes the downlink simulation and the uplink simulation under certain configurations for the simulated wireless channel and the adjacent cell interference environment [11]. Through modular simulation design, the link level simulation can realize the performance comparison of different transmission schemes with a variety of transmitter structures and receiver algorithms. Therefore, it is able to provide the important reference for choosing the appropriate physical layer design or implementation schemes.

Performance metrics are important in the link level simulation, which mainly include the Bit Error Rate (BER), the BLock Error Rate (BLER) or Frame Error Rate (FER) and spectrum efficiency (bps/Hz), etc. As the increasing concern on the power consumption, the performance evaluation of energy efficiency is also introduced in the 5G system. The above metrics are used to evaluate the reliability (BER, BLER, FER) and effectiveness (spectrum efficiency, energy efficiency) of the system transmission. It mainly provides the performance results when the above performance metrics change with the SNR or the SINR. The provided simulation results will be used to determine whether the designed physical layer transmission structure and the receiving algorithm meet the system performance requirements. It also provides the design directions for the algorithm improvement.

Compared with 4G, the 5G transmission will pose higher requirements for the computing abilities and the simulation speed of the link level simulation. In addition, when transmitting wireless signals at high frequency bands, the physical environment of 5G transmissions will be more complicated. In order to improve the validity and authenticity of the link level simulation, the researchers will model the wireless channel and the interference signals more accurately based on the actual measure data. This will no doubt increase the complexity and the processing time of the wireless link simulation. For the 5G link level simulation, the 5G software needs to adopt new simulation methods, such as the high speed parallelization, to cope with the challenge posed by the technical complexity of 5G new technologies. It will be able to perform a series of rapid evaluation and verification for candidate transmission technologies.

Concerning all the above mentioned problems brought by massive MIMO, dense network nodes, and the more complicated wireless transmission environment, the link level simulation needs to consider a variety of advanced simulation methods and implementation mechanisms. Meanwhile, since there are a large number of candidate technologies for the 5G transmission, it is necessary to make a detailed analysis on technical features of candidate technologies to determine the corresponding technologies' application scenarios as well as the evaluation process. The factors that need most consideration in the link level simulation of the main candidate technologies include the following aspects.

- **Simulations for the massive MIMO channel**
  The channel model plays vital important roles in the link level simulations. In terms of the 5G link level simulation, it is necessary to carefully consider various characteristics of the massive MIMO channel such as the spatial correlation, the coupling, the near field effect, etc. [17]. The empirical channel model can also be constructed through analysis, comparison and fitting of the measured channel data in combination with the theoretical analysis.
- **Simulations for the neighboring interference**
  In 5G scenarios with the ultra-dense nodes, the co-channel interference will limit the network capacity [18].For the link level simulation, it needs to accurately reflect the influence of the interference links from neighboring cells. On the one hand, the real interference signals can be used in which the real-time

signals are generated through multiple interference links. As an alternative method, the interference signals can also be theoretically modeled through theoretical derivation in order to simplify the simulation complexity.

- **Simulations for the novel multiple access technology**

  In order to support the new physical layer schemes such as the non-orthogonal multiple access, it's necessary to build the corresponding link layer simulation to evaluate the actual performance of the corresponding technology. It needs a comprehensive evaluation on the impact on the spectral efficiency and the impact on the robustness of the non-ideal factors such as frequency offset and channel estimation errors. Meanwhile, the performance mapping from the link level to the system level corresponding to this physical layer must be obtained in order to accelerate the execution speed of the system level simulation.

- **Simulations with high-performance multi-core parallelization**

  As mentioned above, the 5G system will be configured with massive numbers of antennas. Besides, it will also consider more complex channel models, interference signals, the non-orthogonal signal processing and other more complex transmission environments. Therefore it requires much higher capabilities to realize the link simulation. The multi-thread parallel simulation, the high-performance multi-core server, the software and hardware co-simulation and other advanced simulation methods will be conducive to the fast and accurate link level simulation evaluation.

Simulation Process Overview

The main role of the physical layer transmission is to ensure the reliability and effectiveness of the information wireless transmission. The design of each function module of the physical layer is usually completed after a lot of discussions on the performance analysis, the algorithm evaluation and the standardization [19]. Take the downlink system of wireless transmission as an example. The signal processing modules at the transmitter mainly include channel coding, constellation mapping, multiple antenna precoding, multiple access, reference signal generation, framing, etc. The signal processing modules at the receiver mainly include cell search and synchronization, de-framing, channel estimation, multiple antenna detection, demodulation, channel decoding, etc. [20–21]. In the link level simulation, programming is required to realize each of the above mentioned function modules. In addition, it also needs to implement the modules for the wireless fading channel and the interference links. During the establishment of the link level simulation platform, in order to guarantee the correctness of the simulation, it's also necessary to go through multiple levels of tests, in turn, including functional module testing, subsystem testing, and integrated system testing.

The wireless link level simulation system usually includes two sets of simulation systems for downlink and uplink. In order to improve the efficiency and transportability, it usually needs to complete the software architecture design in advance.

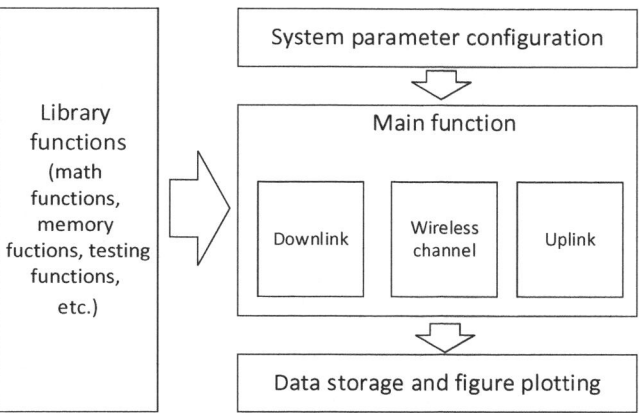

Fig. 4.4 Example of the link level simulation platform architecture

Figure 4.4 shows an example of the architecture design of the link level simulation. In it, the library functions provide to the whole simulation system the public library functions, respectively supporting functions like math calculation, memory allocation, testing comparison, etc. In addition, the section of the system parameter configuration covers the system configuration parameters required by simulations. It may include the system-level parameters such as the frame structure, the system bandwidth, cell information, resource allocation methods, etc. It can also include the transmitter configuration parameters such as control channel parameters and data channel parameters. In addition, the receiver configuration parameters can also be included such as configurations for the receiving algorithms. The part of the main function completes the whole link process in accordance with the definitions and transmission mechanisms defined as the standard protocols. Real-time simulation results will be saved, and the simulation evaluations curves are drawn when necessary. It is important for the architecture design to be with capabilities of flexibility, detectability and transportability.

The realization of each function module in uplink and downlink and the interface definition between modules are the most basic parts of the link level simulation. It will realize the indispensable functions for the physical layer transmission. They includes function modules for channel coding and decoding, modulation and demodulation, multiple antenna precoding and detection, multiple access, reference signal generation and channel estimation, cell search and synchronization, and measurement and feedback [21–22].

With the introduction of massive MIMO, non-orthogonal multiple access and other candidate technologies to 5G transmission system, the implementation way of multiple antenna processing module and multiple access module would undergo big changes compared with that in 4G transmission system in the link level simulation. Figure 4.5 gives a general process of the transmitter for the single user transmission. As shown in this figure, the resulting implementation for massive MIMO and

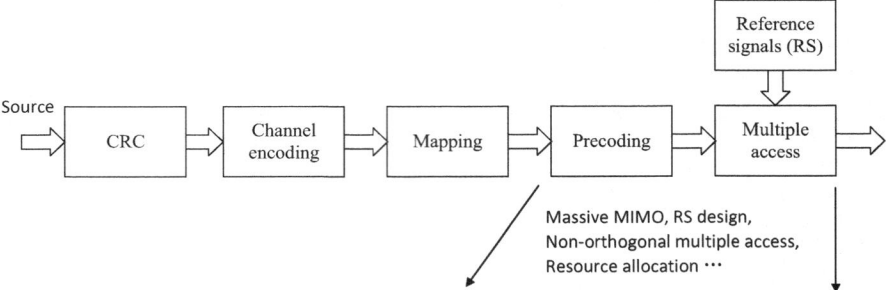

Fig. 4.5 Example of a general process of transmitter link simulation

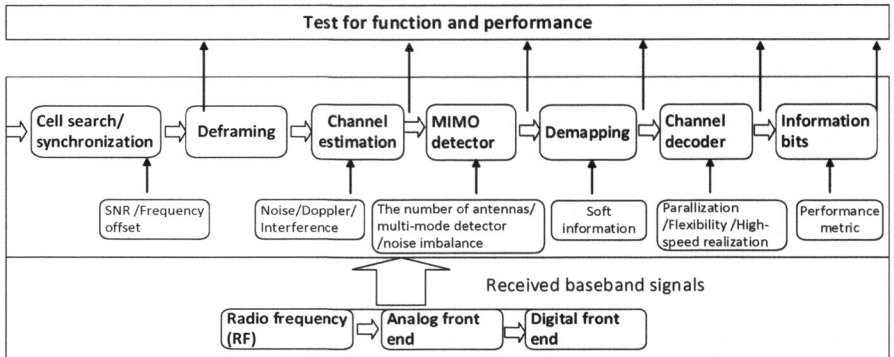

Fig. 4.6 Example of a general process of the receiver link simulation

non-orthogonal multiple access will affect the simulation modules of the reference signals, the multi-user resource allocation, etc.

In the link level simulation, the receiver often has higher algorithm complexity and therefore requires higher computing capability. The receiver usually involves RF, analog front end, digital front-end, baseband processing and other components. In the link level simulation, it usually pays more attention to baseband simulation implementation and evaluation. The non-ideality brought by the RF link, analog front-end and digital front-end is modeled and introduced to the baseband simulation. Figure 4.6 shows the main processing of a receiver. For simplicity, it simplifies the example of data extraction modules such as de-frame, resource demapping, etc. For the receiver, each module has multiple choices for receiving algorithms. And different algorithms concern different aspects of the system design in terms of complexity and performance. Theoretical analyses are usually able to provide some conclusions on the algorithm performance. But due to some ideal assumptions, the performance results usually have great gap with the actual situations. At this point, the link level simulation will play an important role in the performance evaluation of the receiver algorithm.

In addition to the general simulation process of the transmitter and the receiver, the physical layer transmission also needs to distinguish between the data channel and the control channel. They have great differences in key performance indicators. For example, the requirements for the transmission reliability of the control channel are usually much higher than those of the data channel. Therefore, the link level simulation also needs simulation verification for a wide variety of physical channels respectively. Besides, the performance and complexity of the receiver algorithm will also directly affect the choice of the transmitter for each function module. After the completion of the transmitter standards establishment, how to design a cost-effective receiver is also important for chip and equipment manufacturers. Therefore, detailed and reliable link level simulation results are of great significance for evaluating the physical transmission technology, transmission system design, standardization and product design. The link level simulation results of each algorithm module and the overall link under different scenarios and parameters configurations will provide important reference basis for the scheme design and system performance optimization.

4.3.3   Introduction to Simulation Cases

Case 1: Massive MIMO Performance Simulation

This simulation evaluates the link level performance of the massive MIMO system and the 8×8 MIMO system. For simplicity, the ITU multipath channel model is adopted in the simulation (ITU Vehicular A).The channel state information is assumed to be known. Turbo 1/3 code rate and Quadrature Phase Shift Keying (QPSK) modulation are used. In addition, Adaptive Modulation and Coding (AMC) and HARQ mechanism are not initiated. In the simulation, BER, FER and throughput are performance evaluation criteria. In the massive MIMO simulation, the number of the base station antenna is set to 128, and the number of users took $K = 20, 30, 40, 50$ respectively. When the number of the user changes, the simulation results of different users can be obtained through evaluation. The total users' throughput under Zero Forcing (ZF) precoding transmitting scheme can be obtained through simulation. In the 8×8 MIMO system simulations, the base station is equipped with 8 antennas, the terminal is configured with single antenna, and the number of users is 8. Table 4.2 shows the main parameters configuration in the simulation.

Figures 4.7, 4.8 and 4.9 respectively show the performance of BER, FER, as well as throughput. Since the massive number of antennas can bring a higher degree of freedom, the performance of massive MIMO is far superior to the performance of the existing small 8×8 MIMO system.

Table 4.3 shows the corresponding required Eb/N0 and the corresponding throughput of different users under FER = 5%.For example, when the number of single antenna users in a massive MIMO system is 20 and FER = 5%, the required

Table 4.2 Simulation parameters configuration

Parameter	8×8 MIMO system	Massive MIMO system
Carrier frequency	2.0GHz	2.0GHz
system bandwidth	5 MHz	5 MHz
sub-carrier number	512	512
CP length	32	32
Sampling rate Fs	7.68 MHz	7.68 MHz
The number of base station antennas	8	128
The number of users	8 single-antenna users	20 ~ 50 single-antenna users
Base station precoder	ZF precoding	ZF precoding
Channel model	ITU Vehicular A	ITU Vehicular A
Coding type	Turbo 1/3	Turbo 1/3
Modulation type	QPSK	QPSK
CSI	Ideal	Ideal

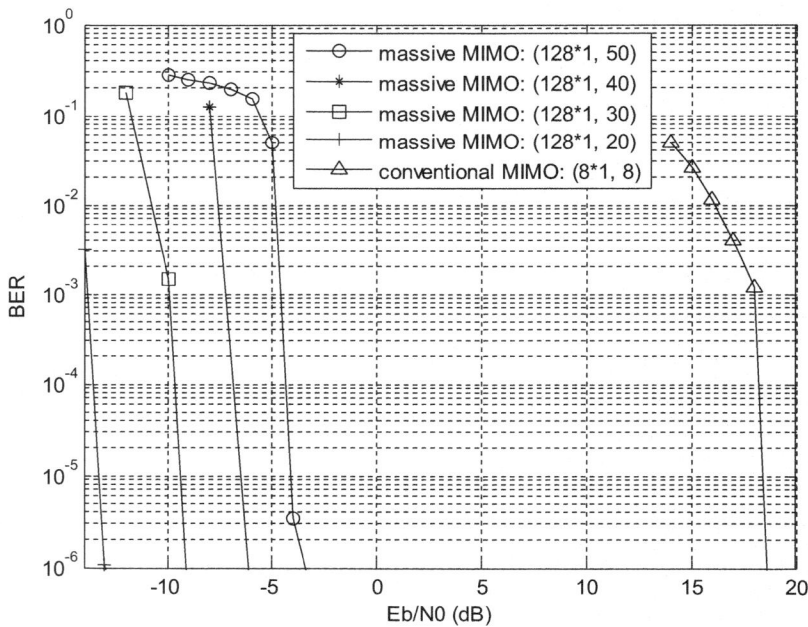

Fig. 4.7 BER performance under different users

Eb/N0 is -13.5 dB, and the corresponding throughput is 280 Mbit. For the 8x8 MIMO system, when FER $=$ 5%, the required Eb/N0 is 16.5 dB and the corresponding throughput is 125Mbit. As it shows, when FER is 5%, the SNR of massive MIMO system can be reduced to 30 dB and its throughput improves by $280/125 = 2.24$ times. As a result, the massive MIMO system can greatly reduce the power consumption of the base station and therefore improves energy efficiency and spectrum efficiency.

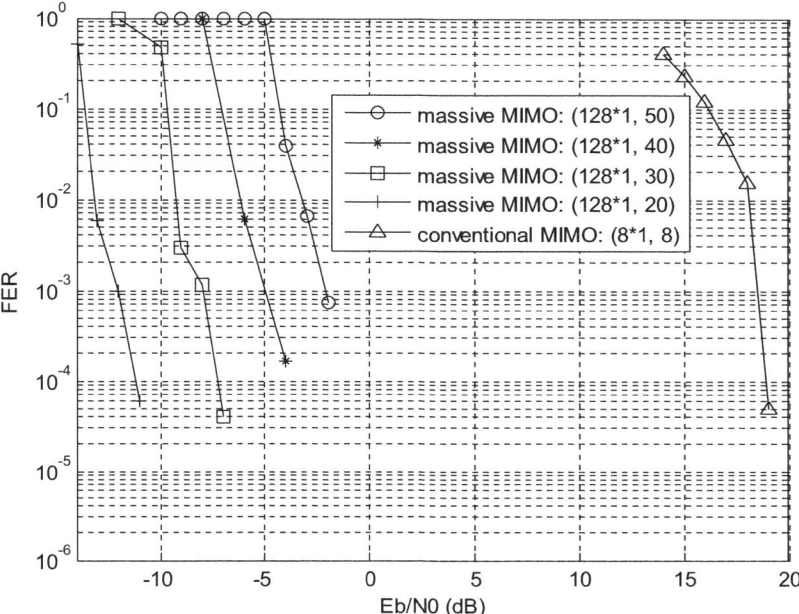

Fig. 4.8 FER performance under different users

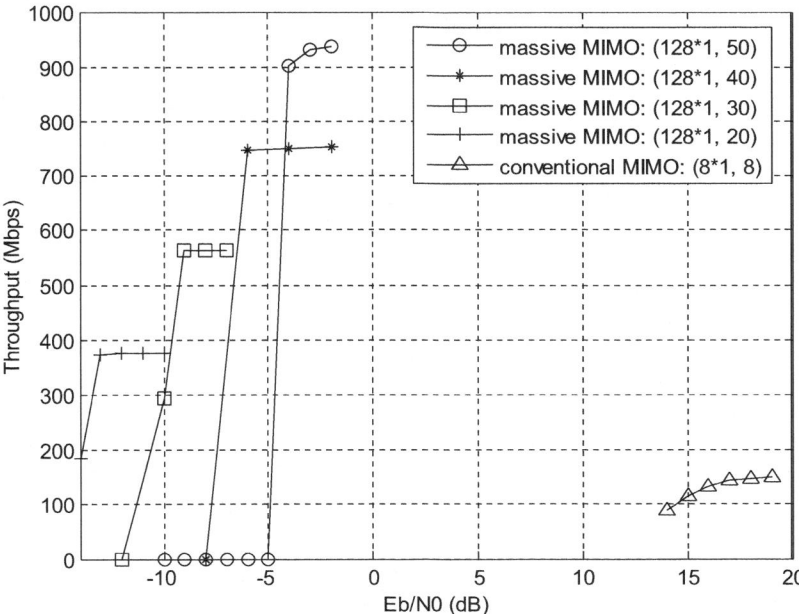

Fig. 4.9 Throughput performance under different users

Table 4.3 The required Eb/N0 and corresponding throughput under different users when FER = 5%

Single antenna UE number	20	30	40	50	8×8 MIMO system
Eb/N0 (dB)	−13.5	−9.6	−6.8	−4.2	16.5
Throughput (Mbit)	280	410	440	700	125

Table 4.4 Parallel acceleration results of link simulation (numerical unit: second)

Computational scheme	$K = 20$	$K = 30$	$K = 40$	$K = 50$
Serial (single core CPU)	296,642	356,757	455,895	699,380
Parallel (40-core CPU server)	9215	10,697	11,992	18,137
Acceleration ratio (40-core parallel vs serial)	32	33	38	38

In terms of the simulation time, the general serial simulation method often requires a long time since the link level evaluation needs to simulate the performance of multiple SNR points and each SNR point needs to simulate fading channel fully with ergodicity. Reasonable parallel simulation can greatly reduce the simulation time. For the above simulation, the following is a brief experiment for comparison.

The traditional serial simulation method obtains the data of one point when running once. The program is running on a single CPU. One point on a massive MIMO link needs 4 days' running time and the complete massive MIMO link simulation needs as many as 100 sampling points.

To accelerate the process, the parallel optimization is carried on. The link simulation consists of two layers of cycles, namely the SNR cycle and the frame cycle. Data in the SNR cycle and the frame cycle are independent. Thus the simulation programs on the SNR and frame cycles can be parallel optimized. And the theoretical parallel speedup result of the new parallel simulation method can be faster in orders of the number of SNRs times by the number of the frames. Parallelization of the SNR and frame cycles, of course, is the relatively simple and effective way to improve the simulation timeliness. With additional internal parallel operations, the simulation time can be further shortened, which will not be discussed here in detail. The results after the parallel acceleration optimization are shown in Table 4.4.

Case 2: Heterogeneous Network Energy Efficiency Simulation

Energy efficiency is one of the key performance indicators in 5G systems. And the wireless heterogeneous network is an effective means to alleviate the contradiction between the data growth and the energy consumption [23]. In respect of mobile intelligent terminals, different access methods and network planning schemes will have different influence on the energy efficiency of mobile intelligent terminals. K. Wei et al. [24] have proposed an energy efficiency indicator with comprehensive consideration of the terminal and the network. The indicators have considered the impact of the new base station layout on the energy consumption increase at the network side and the energy consumption changes at the terminal side at the same time. So it can provide a quantitative reference criterion for operators to decide whether laying out a new base station is a good choice in terms of energy efficiency. Under this indicator, it further analyzes the changes of energy consumption at the terminal side as well as the energy efficiency of the entire network after the micro base station is laid out in the macro cell. And it has obtained the relation between the terminal energy consumption and the low power station configuration parameters. Finally, it also derives the optimal cell radius to obtain the maximum terminal energy saving and the highest energy efficiency.

In order to verify the rationality and validity of the proposed energy efficiency indicators, in [24], the energy consumption and energy efficiency indicators have been simulated after the deployment of different configuration schemes of pico base stations and micro base stations. In the simulation process, it assumes that the coverage of the low power consumption base station is always within the coverage of a macro base station. Table 4.5 provided the detailed simulation parameters.

Figure 4.10 shows the relations between the terminal energy saving with different micro cell radius as well as the distance (d) between the macro base station and the micro base station. It can be seen that the results of the approximate derivation and the simulation results are very similar. The figure also shows that with the increase of the distance between two base stations, the terminal will save more

Table 4.5 Simulation parameters

Macro-cell parameters	
Carrier frequency	2.0 GHz
Bandwidth	20 MHz
Pathloss model	COST 231-Walfish-Ikegami Model
Cell radius	5 Km
Minimum SNR requirement	10 dB
Noise	−160 dBm/Hz
Low power consumption cell parameters	
Carrier frequency	2.0 GHz
Bandwidth	20 MHz
Pathloss model	COST 231-Walfish-Ikegami Model
Minimum SNR requirement	10 dB
Noise	−150 dBm/Hz
Micro-cell user parameters	
Active user density	10^{-5} user/m^2
Working state power consumption	1.2 W
Resting state power consumption	0.6 W
Packet size	100 bit
Micromicro-cell user parameters	
Background user density	10^{-5} user/m^2
Hot user ratio	15%

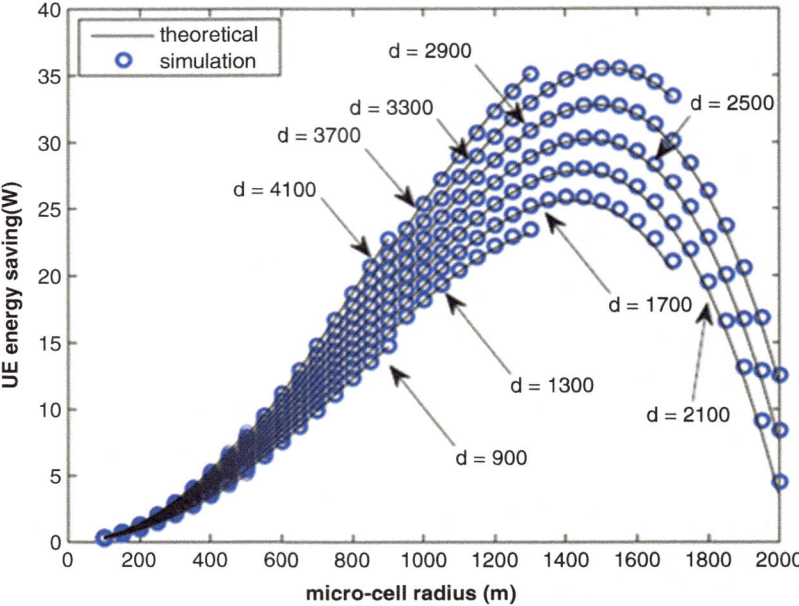

Fig. 4.10 Relation between terminal energy saving and the micro base station location as well as the coverage radius

energy under the same micro base station configuration parameters. Under the simulation parameters, the highest energy saving of the terminal is 35 W, and the energy consumption of all terminals in the entire network is 860 W. Setting up micro base stations can bring about 4% of the energy saving for general users. For users inside the micro cell, the energy saving can be as high as 18%. If active users need to download more resources, the micro base station can bring more terminal energy savings.

Figure 4.11 shows energy efficiency value of the network under different micro base station locations and micro base station radius on the basis of the proposed energy efficiency indicators. If the energy efficiency value is greater than 0, it is believed that setting the micro base station is beneficial from the perspective of energy efficiency. It can be seen from Fig. 4.8 that when the distance between the micro base station and the macro base station is less than 2100 m, since the spectrum efficiency of the micro base station is higher when served by macro station. Even within the optimal microcell radius, the network energy efficiency value is still less than 0. In this case, the setting of the micro base station is inappropriate in terms of energy efficiency. The above analysis shows that the proposed energy efficiency indicator provides a numerical reference criterion for operators to judge whether a new base station is appropriate in terms energy efficiency.

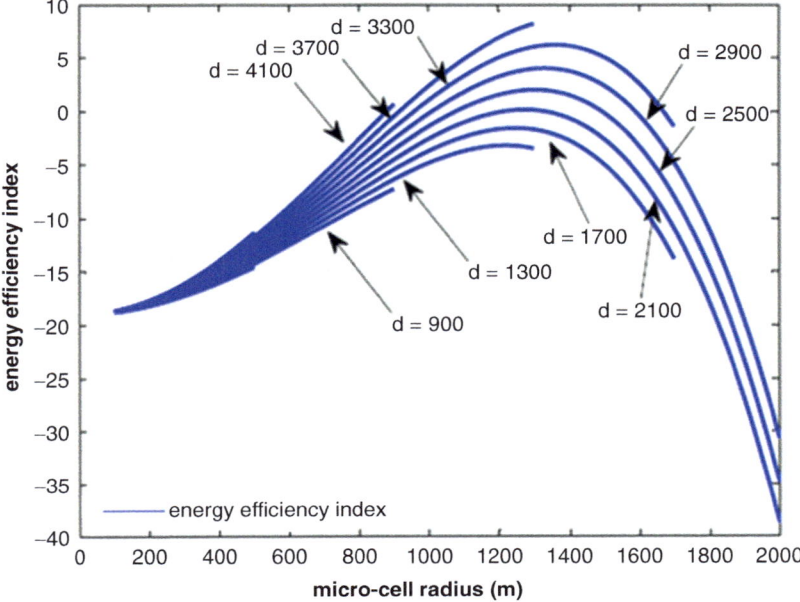

Fig. 4.11 Relation of different micro base station positions and their radius as well as energy efficiency indicators

Case 3: Tail-Biting Convolution Code Decoder Simulation

The future 5G system not only can greatly improve the traditional cellular network performance, but also plays an important role in the technology applications in vertical industries. This trend will become more obvious with the emergence of the IoT. For example, various communications systems related with IoT will have much greater demand for short packets. Thus, how to perform channel encoding and decoding with high efficiency and low complexity has become an important research direction. In the short packet transmission, short codes are usually needed for channel coding such as tail-biting convolution codes. Tail-biting convolution codes usually use Circular Viterbi Algorithm (CVA) for decoding [25]. However, there is an important problem in the CVA-based iterative sub-optimal decoding algorithm, i.e., the circular trap problem. That is, the surviving path obtained with the current iteration is the same as the surviving path obtained before. And as the iterative sequence continues, there is no new tail-biting path that is closer to the receiving sequence while the decoder cannot terminate the iteration. The works in [26] and [27] have studied the CVA iteration process and the circular trap phenomenon existing in CVA, and put forward an effective circular trap detection scheme. Detection of the circular trap can be used to help control the CVA decoding process, so as to get the fast-convergent iterative decoding algorithm.

Table 4.6 BLER performance in the Additive White Gaussian Noise (AWGN) channel for the suboptimal decoding algorithms (WAVA and S-CTD-II) and the maximum likelihood decoding algorithms (CTD-ML decoder and two-phase ML decoder) for the tail-biting convolution code with the generator polynomial of {133,171,165} and (120,40)

	SNR			
BLER	1.0 dB	2.0 dB	3.0 dB	4.0 dB
WAVA	8.535×10^{-2}	1.372×10^{-2}	1.353×10^{-3}	7.280×10^{-5}
S-CTD-II	8.598×10^{-2}	1.377×10^{-2}	1.352×10^{-3}	7.280×10^{-5}
Maximum likelihood decoder	8.466×10^{-2}	1.363×10^{-2}	1.352×10^{-3}	7.280×10^{-5}

In order to verify the decoding performance and the decoding efficiency, in the simulation below, the simulation analysis has been made to the performance of the Circular Trap Detection based ML decoder (CTD-ML decoder)algorithm and a suboptimal Simplified CTD (SCTD) algorithm. Here SCTD corresponds to the S-CTD-II algorithm in [27]. Meanwhile, it also compares with suboptimal decoder WAVA [25] and the hybrid maximum likelihood decoding algorithm based on Viterbi and the heuristic search [28], etc. The simulation has adopted the convolution codes with two kinds of lengths respectively. One of the tail-biting convolution codes adopts the convolution encoder in the LTE control channel and the broadcasting channel. The generator polynomial is {133, 171, 165} [29] and the information sequence length is 40. This codeword is denoted by (120, 40).The other code of the tail-biting convolution codes uses the convolution encoder with the generator polynomial of {345, 237} [25]. The information sequence length L is 32. This codeword is denoted by (64, 32).In the simulation, the coded bits perform the QPSK modulation and then enter the AWGN channel. For convenience, the hybrid maximum likelihood decoding algorithm based on Viterbi and the heuristic search is denoted by the two-phase ML algorithm.

Table 4.6 gives the decoding performance of the different decoding algorithm son the tail biting convolution code (120,40), from which we can find that suboptimal decoding algorithms, such as WAVA [25] and SCTD decoders, have close BLER performance to that of optimal decoding algorithms. Maximum likelihood decoders refer to the CTD-ML decoder and the two-phase ML decoder. Since these two decoders are both maximum likelihood decoders on tail-biting trellis diagrams, their BLER performances are exactly the same.

Figures 4.12 and 4.13 respectively give the relations of the decoding complexity performance of the CTD-ML decoder and the two-phase ML decoder and the demand for storage space in the decoding process with SNR. From Fig. 4.12 we can see that for the tail-biting convolution codes of any length, the CTD-ML decoder has higher decoding efficiency than the two-phase ML decoder. Meanwhile, Fig. 4.13 shows the CTD-ML decoder has lower requirements for storage space. Therefore, the CTD- ML decoder on tail-biting trellis diagrams has significant advantages in both computational complexity and storage space reduction.

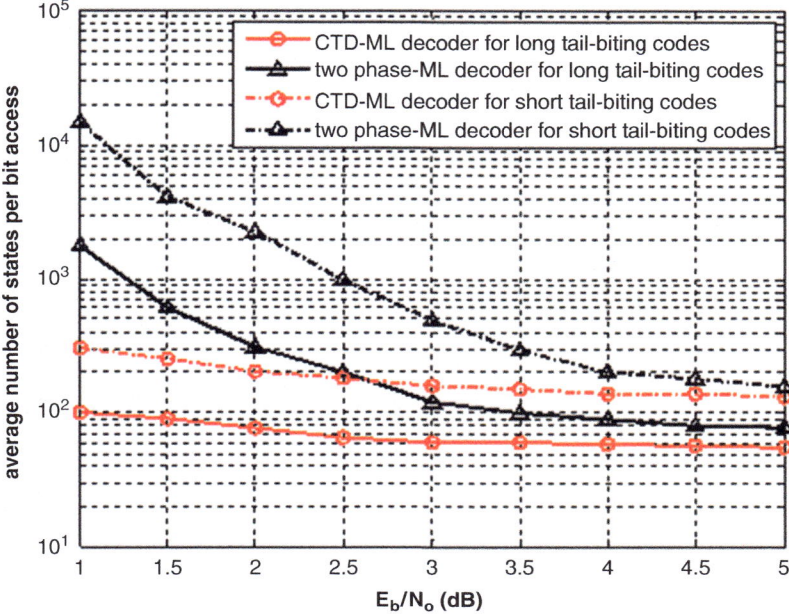

Fig. 4.12 Decoding complexity comparison of the CTD-ML decoder and the two- phase ML decoder for the long tail-biting convolutional code with (120, 40) and the short tail-biting convolutional code with (64, 32)

Case 4: Compressive Sensing Simulation

The next generation communications network will produce a large amount of data, and thus the highly efficient processing of data samples will be one of important requirements. Compressive Sensing (CS) is an important way of data processing, which can recover the raw data from extremely few sample values. The algorithm complexity of the compression process and the recovery process in CS is extremely uneven. That is to say, the computational complexity of compression sampling is much lower at the transmitter than that of the receiver. In wireless sensor network applications, for example, the sensor nodes of sensor networks can compress and sample data in the form of CS with only low computational cost. Most importantly, the compressive sensing technology is able to balance the energy consumption of sensor nodes in the network to solve the "bottleneck effect/hot spot effect". So the CS technology has a great application prospect in the sensor network data aggregation application. The CS technology [30] has aroused great interest among researchers. Zhao et al. [31] put forward a new type of Treelet-based Compressive Data Aggregation (T-CDA), and design the corresponding data collection methods. It is proved that T-CDA is better than traditional CS data aggregation algorithms in terms of energy efficiency based on the simulation of the real sensor network data.

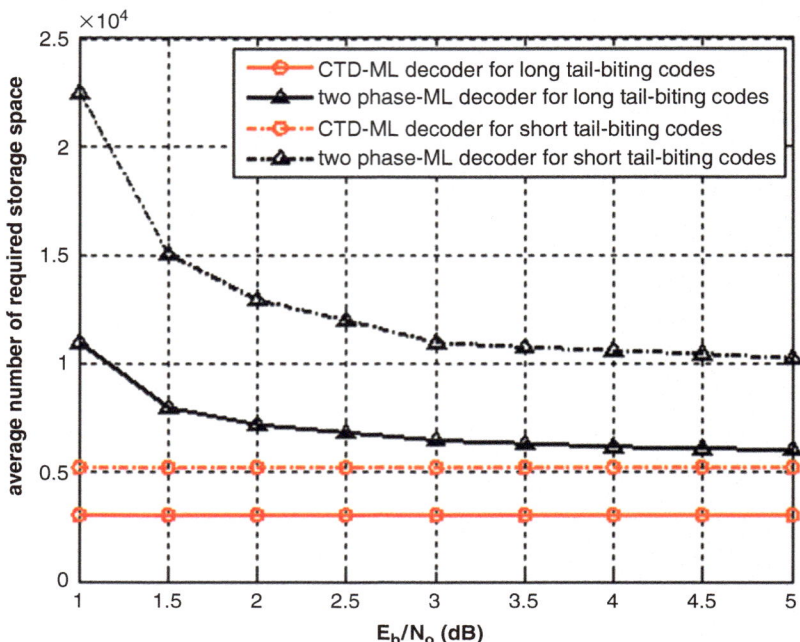

Fig. 4.13 Storage space requirements comparison of the CTD-ML decoder and the two-phase ML decoder for the long tail-biting convolutional code with (120, 40) and the short tail-biting convolutional code with (64, 32)

The simulation uses the data collected from a static sensor network in Ecole Polytechnique Federale de Lausanne University [32].This sensor network records the environmental data of 97 test points on the campus every five minutes from July 2006 to May 2007, such as temperature, humidity, illumination, etc. In [33], this section uses the highly space-time correlated temperature data as the simulation data. In the simulation, the overall network transmission overheads under different recovery precisions are used as evaluation indicators for energy consumption. Simulation has compared the performance of the following methods.

- Discrete Cosine Transform-CDA (DCT-CDA) represents the traditional method which uses DCT to make sparse compressed sensing data aggregation. CS performance depends on the sparse degree of the original signa. The existing researches usually use DCT [34] to make the actual signal sparse. However, the actual signals from the nodes are distributed in the three-dimensional world. It is difficult to prioritize them and make the data change trend smoothly. Therefore, DCT can only process smooth signals without fully exploring the actual sensor signals sparse.
- Principal Component Analysis-CDA (PCA-CDA) represents the data aggregation method with the treelet algorithm in T-CDA being replaced by PCA. PCA is a typical global data analysis method, and is introduced into the CS data

collection in [35]. PCA can get a sparse transformation matrix of the original data, which can be used to extract the main ingredients in raw data. But it is a global operation process, which overwhelms the local properties in the data. In addition, due to the global properties of the PCA algorithm, noise is also regarded as useful data. As a result, this method is extremely unstable when the observation data is small because any change of noise will affect the global result.

- T-CDA represents the proposed treelet-based compressive data aggregation method. Treelet transformation is an algorithm for data clustering and regression analysis, which can explore the local correlation characteristics in data. In sensor networks, to obtain the accurate observation data is a job with large energy consumption. It is difficult to support such global analysis algorithms like PCA. And the tree transformation is able to overcome or relieve the situation. In particular, treelet transformation can be used to utilize the correlation structure between nodes from the sensor data or training data collected in advance and then to obtain an orthogonal transformation matrix, which can fully reduce the training data sparsity. Since sensing data also shows the correlation on time dimension, the orthogonal transformation matrix can be used to reduce the sparsity of the sensing data collected within several timing sequences after the training data.

The simulation results of Figs. 4.14 and 4.15 provide the performance of the T-CDA algorithm under different system parameters. Figure 4.14 shows the impact of the selection of training sequence length (L) on the system performance. In the

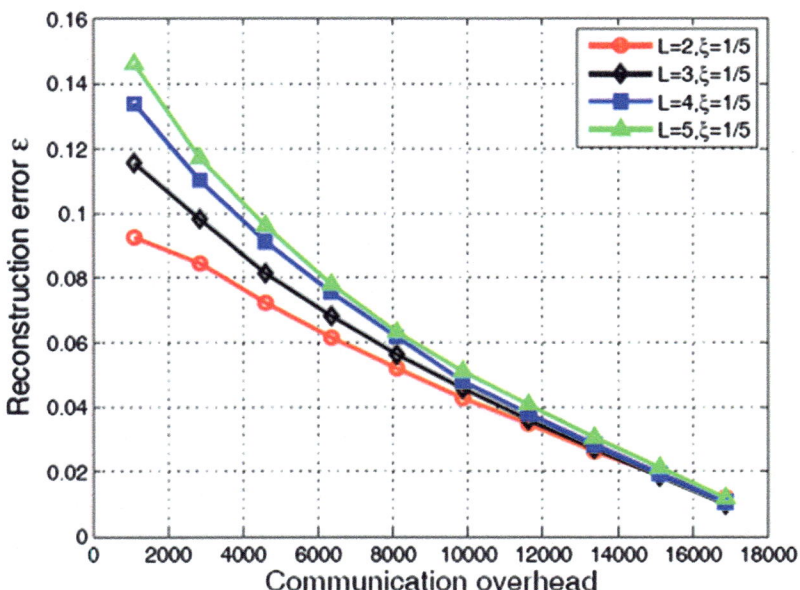

Fig. 4.14 The impact of the length selection of the training sequence on system performance

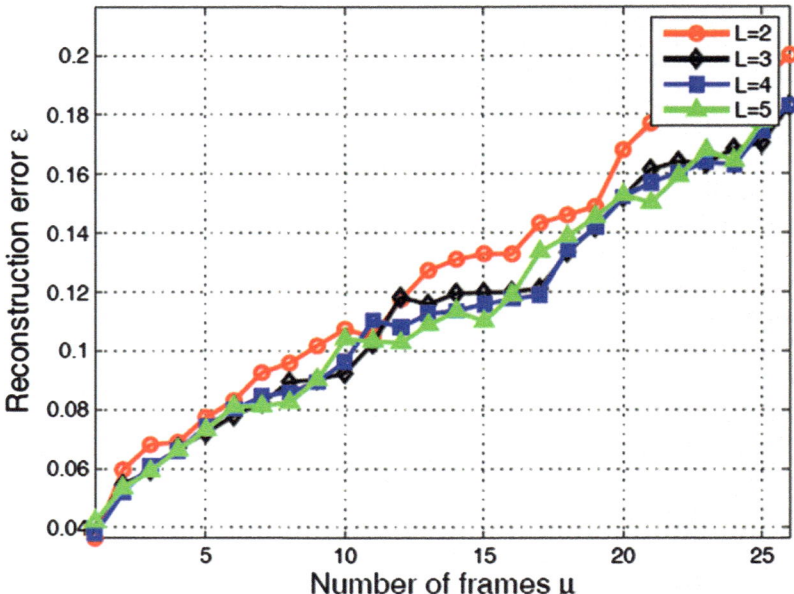

Fig. 4.15 The impact of the rounds number of continuous CS data collection on system performance

figure factor $\xi = L/\mu$, μ represents the number of rounds for carrying out T-CDA after a round of training data. As the rough trend shows in Fig. 4.14, the smaller L, the better system performance. This is because ξ is fixed, which means longer training sequence corresponds to more rounds of data collection. This suggests that the major impact of the data recovery error comes from the changes of sensor data over time, rather than the insufficient information in the training sequence. Figure 4.15 verifies the above conclusion. From Fig. 4.15, it can be seen that with the increase of the number μ of data collection rounds, the accuracy of data recovery deteriorates. Figures 4.14 and 4.15 result show that $L = 2$ is the optimal value for the system parameter selection.

The performance comparison of DCT-CDA, PCA-CDA and T-CDA is shown in Fig. 4.16. By using the actual environment monitoring data, it can be seen that the performance of DCT- CDA is the worst, and the performance of T-CDA is better than that of PCA-CDA and DCT-CDA. As a result, we can reach the following conclusions. Firstly, the method based on the training sequence is better than the traditional method. Then, the performance of T-CDA is better than PCA transformation in this application, which means the local correlation of the mining data is superior to the global correlation.

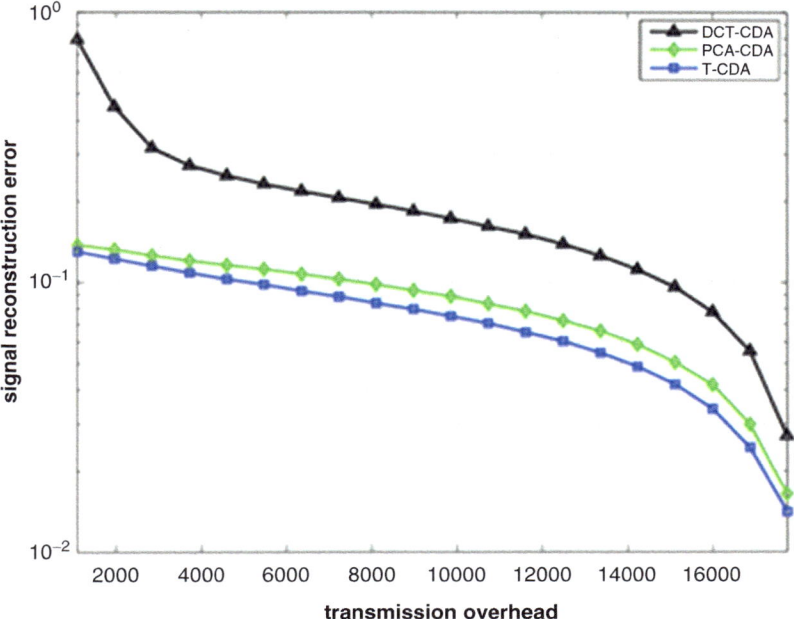

Fig. 4.16 System performance comparison under different data acquisition methods

Case 5: User-Oriented Link Adaption in D2D Network Coding Multicast

With the rapid development of wireless communications technology, smart phones, tablets and other mobile terminals have become indispensable parts in people's lives. Mobile user data requirements increase rapidly, but the available bandwidth resources of mobile networks are limited. In order to solve the contradiction between the service growth and limited resources, and to satisfy the proposed energy-saving green requirements of 5G, D2D has become a key candidate technology in 5G. As a new kind of short range direct data transmission service based on cellular mobile networks, D2D communications mainly falls into two scenarios: the point-to-point D2D communications scenario between a pair of terminals and the multicast communications scenario of D2D clusters composed of a group of terminals with the same service requirements.

One big difference between D2D clusters' multicast communications and traditional broadcast communications is the requirements for the reliability in the packet transmission. So it needs to be guaranteed by technologies like the HARQ. However, in multicast communications, one user's failure of reception will cause the repeat transmission, leading to the relatively low system efficiency. On the other hand, in the traditional broadcast/ multicast link adaptation method, the Modulation and Coding Scheme (MCS) is selected based on the worst link channel quality between all the receivers and transmitters, leading to the situation that other receivers with the good channel quality do not make full use of the channel. Through the research on link adaptive problems in the D2D cluster network coding multicast scenario, Zhou et al. [36] find that unlike traditional multicast/ broadcast, the network coding multicast has its special properties. That is, in the network coding, different information merged via the heterogeneous domain is sent to different receivers, which brings new possibilities for the adaptive modulation coding selection for the network coding multicast.

Zhou et al. [36] put forward the user-oriented link adaptive method, in which two multicast channels whose link quality corresponds to the maximum value of the modulation type and the minimum value of coding type are chosen. And then making bit mapping to the terminal with poor link quality at the transmitter to reduce the equivalent rate. So after the terminals with different purposes make network coding and decoding to the received multicast data packets, the information obtained get different equivalent modulation and coding efficiency to adapt to the quality of the two different channels. While it should be guaranteed that the users with poor channel quality can accurately decode the multicast data packets, the users with good channel quality should be able to obtain more useful information in a more efficient modulation and coding type, and eventually the overall spectral efficiency of multicast communications can be improved.

In order to verify the algorithm's BER and spectrum efficiency, link simulations are carried out to verify the different channel quality combination of two channels within D2D clusters [36]. Turbo channel coding in 3GPP LTE standards [37] is used, and the information byte length is set to 512 bits. The combinations of bit rates and modulation mode are shown in Table 4.7.

Figure 4.17 has simulated the link performance of the terminals' BLER and the overall throughputs of D2D cluster multicast communications. It can be seen that in the scheme adopted by [36], BLER of the users with poor link quality is basically in coincidence with the traditional multicast MCS. It shows that the new scheme can guarantee the users with poor link quality to get the equivalent coding efficiency in consistency with the traditional way so that they can decode the multicast packets accurately. Since the users with good link quality won more efficient modulation and coding types, their overall throughput of D2D clusters improves significantly by as high as 42%.

Figure 4.18 shows the simulation of the link performance of the overall spectrum efficiency of D2D cluster multicast communications, with the different combinations of modulation and coding types supported by two channel link qualities in

Table 4.7 Simulation parameters

MCS table of the proposed against conventional scheme		
Case	MCS of conv link adaptation	MCS of the proposed scheme
1	1/2 BPSK	1/2 QPSK
2	1/2 QPSK	1/3 16QAM
3	2/3 QPSK	1/2 16QAM
4	3/4 QPSK	2/3 16QAM

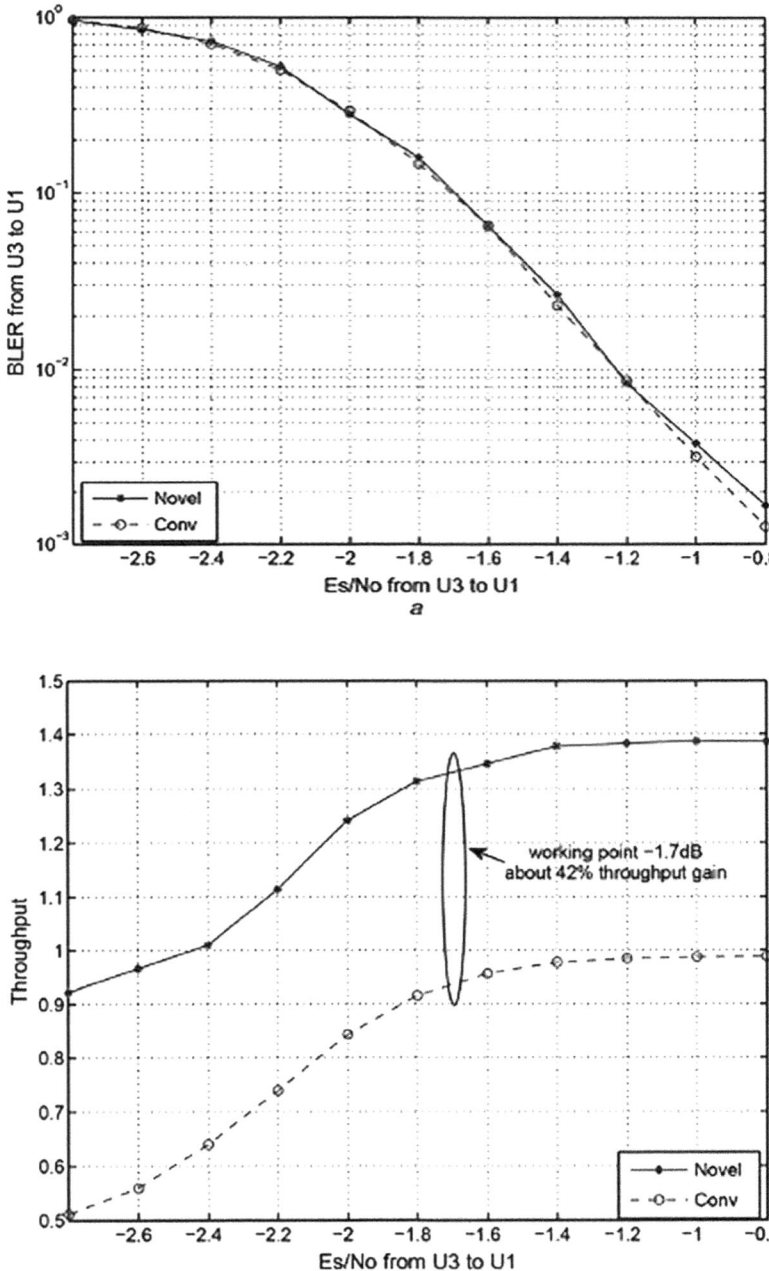

Fig. 4.17 Link performance of 1/2 BPSK and QPSK multicast link quality combination (**a**) BLER (**b**) Throughput

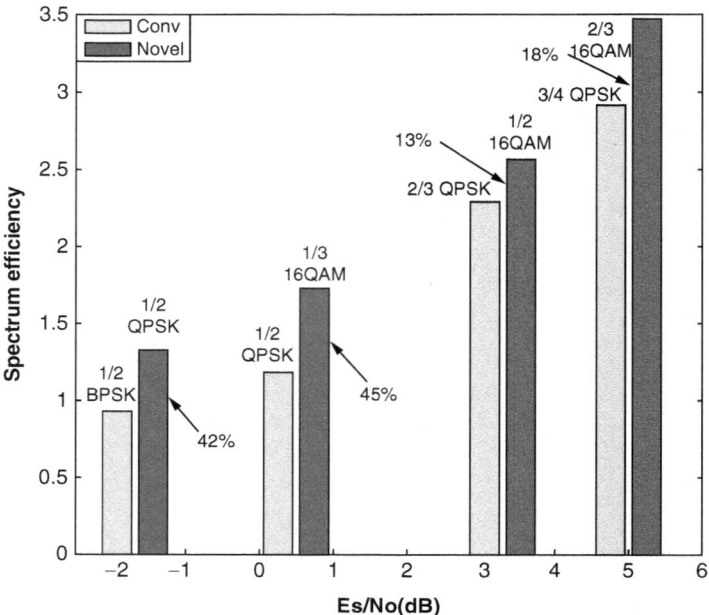

Fig. 4.18 Spectrum efficiency under different MCS combinations

D2Dmulticast communications. With the increase of SNR, new and traditional schemes adopt higher bit rates and higher modulation methods. Under the same SNR, the new scheme can support smaller bit rate and higher modulation method. And the final equivalent modulation coding efficiency is higher than the traditional scheme. So the overall throughput of D2D clusters has improved significantly. The simulation results show that the overall throughput has 13–45% performance gain under different modulation encoding combinations.

4.4   5G Software System Level Simulation

4.4.1   Test Evaluation Methods

The first problem of software simulation test is how to evaluate 5G new technologies. 5G networks have greatly different features from 4G networks. Its evaluation methods should be studied from the following aspects.

A good evaluation system requires basic features of completeness, simplicity, better usability, etc. Compared with 4G networks, 5G networks have changed not only in "quantity" such as capacity, data rate, and delay, but also in "intrinsic quality", including the basic features of network such as virtualization and definability. For the comprehensive, accurate and efficient evaluation of 5G network technologies, the study should at least include the following aspects.

- Design of new performance indicators

    Connection density, traffic volume density, minimum guaranteed rate are new KPIs introduced by 5G networks. With the deepening of the 5G technology research, it can be foreseen that there will be new evaluation indicators. The design of these directly measurable indicators, on the one hand, needs to combine the characteristics of the new services, and on the other hand, needs to fully learn from the experience of the previous KPI in network applications.

- Redesign of traditional indicators in 5G network

    Due to the development of 5G network architecture and services, the connotations of the traditional indicators are changing. For example, the virtual resources model is introduced after the introduction of the virtual technologies. How to map traditional indicators to the virtual resources and evaluate network performance accurately is a new topic in 5G networks.

    Compared with 4G networks, there are big breakthroughs and innovations in some 5G new technologies in terms of the design of the network resources model, and new concepts and designs are introduced, which are not compatible with the design in the past. Now end-to-end global design is needed for these technological models and evaluation methods. For example, the following technologies:

- Network function virtualization

    Wireless access network virtualization mainly uses the design idea of SDN [38] and NFV [39], which is similar to the virtualization of core network and cable network. There can be three layers, namely infrastructure layer, control layer and application layer based on the network abstraction of the wireless access network. This brand new architecture needs to be mapped to the corresponding network resources model, so that the functions and performance of these three layers can be comprehensively evaluated.

Infrastructure layer is supported by various kinds of network equipment nodes, such as macro eNodeB and micro eNodeB. These equipments provide virtual management functions through unified interface. Control layer is composed of a series of distributed management node. As logical nodes, the management nodes can be deployed flexibly in different types of equipment. The management nodes are responsible for the management work of the creation, updates, virtual resources scheduling, allocation and recycling of the virtual network. The application layer is made up of many different services and applications. These services, scheduled and allocated by the management nodes, realize specific functions. Control layer shields the details of the bottom resources to provide top services with unified virtual resources abstraction through virtual resources' unified description. Virtual network construction is based on the services. Different virtual resources are selected to provide services according to different services and application types, in order to better adapt to Quality of Service (QoS) requirements of the different services.

For wireless access network side, the main functions of virtual management layer include: creation and updating of service-centered virtual cell, management of virtual resources (including backhaul network resources and air interface resources), connection mobility control, and management and use of context information. This network architecture introduces the mechanism of the control and data separation. For example, based on the idea of separation of control plane and user plane, Ishii et al. present a new network architecture on the basis of the concept of virtual cell [40].

Two important problems of wireless network resources virtualization are node mapping that provides different and virtual resource allocation. Node mapping is to determine what nodes are used for a certain service. In the traditional network, a service cell is selected for the user according to the signal strength indicators such as Reference Signal Receiving Power (RSRP) / Reference Signal Receiving Quality (RSRQ). All the users' services are provided by this service cell. Under ultra-dense network scenario, a large number of irregularly deployed Small Cell will face the problems of complex interference and load distribution asymmetry. The traditional way based on signal strength indicators for cell selection will no longer be applicable. For future 5G network virtual cell, the cell can be no longer associated with the nodes, but instead take user for the center [41]. Reference signals used for data transmission will decouple with the node ID. Management and control are separated from data nodes. User-centered virtual cell can be formed at any location in the network, realizing global virtualization. For management nodes in the virtual management, virtual resources management is one of its important functions. How to avoid the interference between each node and make effective utilization of time domain, frequency domain and space domain resources are the key to the wireless resource management.

Through the above introduction to the wireless network resource virtualization technology, we can see the network resources model has undergone great changes. The virtual cell and virtual resources need to be modeled. Virtual cell needs to decouple with physical nodes, and provide services with users as the center. So, different from traditional cells, the performance of virtual cells needs to be evaluated from the user's dimension.

4.4.2 Key Simulation Technologies

This chapter describes problems and the corresponding design ideas in the platform architecture design of the 5G system simulation software. Various kinds of new technologies' simulation design, implementation and evaluation in 5G networks should be analyzed and carried out separately according to the specific services, which will not be covered in this chapter. The key technologies of 5G system simulation design are shown in Table 4.8.

Table 4.8 Key technologies of 5G system simulation design

Key technologies	Overview
Dynamic simulation modeling technology	Simulation model library and simulation parameter library are obtained according to simulation scenarios, services models, service quality requirements, carried technologies and other kinds of simulation requirement analysis. Based on the simulation model components and simulation process of the model and the parameter dynamic configuration, highly multiplexing of simulation software module and improvement in the simulation development efficiency is realized.
Management technology of virtualized computational resources	Hardware resources, software resources, and communications bus are virtualized into computational resources, storage resources and transmission resources for unified management. Meanwhile, users' simulation requirements are converted and broken down into multiple simulation tasks that can be run in parallel. Through mapping the simulation tasks dynamically into different computational resources to implement the resources dynamic configuration and virtualization management, so as to improve efficiency and scalability of parallel computing
Multi-core parallel simulation technology	The software simulation platform design based on multi-core parallel computing includes hardware, operating systems, software design, model and algorithm design, and other aspects of the design requirements

Hardware acceleration simulation technology	Hardware module is used to replace the software algorithms to take full advantage of the hardware's inherent fast features, including:
	  FPGA based high performance hardware acceleration key technology
	  Interface and middleware design with combined hardware acceleration and soft simulation platform
	  Reconfigurable FPGA hardware acceleration card design

Real-time transmission technology	Mature real-time transmission technologies at present include:
	  Real-time signal transmission based on TCP/IP technology
	  Real-time signal transmission based on Data socket
	  Real-time signal transmission based on shared variable engine
	  Signal transmission with direct optical fiber

Dynamic Simulation Modeling Technology

5G technologies have brought more complex network scenarios and services types, and also created all kinds of new technologies. It makes the simulation model and process more complicated and changeable, and leads to the exponential rise in the number of the simulation scenarios. It is embodied in the following aspects.

- Networking scenario is complicated and changeable

The 5G network architecture becomes more flexible after adopting SDN, NFV, and SON. More combinations of different network function entities come into being. And network scenarios are also more complicated and changeable. In order to adapt to these changes, flexible network oriented simulation modeling technology is needed.

- Network model is evolving

The design concept of "no more cell" has weakened the independent concepts of traditional link level and system level simulations. The design concept of virtualization requires network resource model to completely decouple in three dimensions, namely control plane, data plane, and physical resources. The channel modeling of massive MIMO and High Frequencies Band (HFB) communications derive new channel models. All these key technologies have been promoting the new network model design, and the system simulation also needs to develop synchronously with the abstraction and design of the network models.

The efficiency of the traditional simulation design pattern which is realized by encoding for a particular scene is very low, which is far from being able to meet the increasing requirements. The high multiplexing modeling technique becomes a must, and thus the dynamic simulation modeling technology is proposed.

The core idea of dynamic simulation modeling technology is network layering and modeling, and modular design of different layers of simulation object models [42]. Meanwhile, simulation model components and simulation parameters are obtained based on the simulation scenarios and services model mapping, and then through dynamic configuration, the specific simulation process is combined. As the simulation object model has realized the modular design, the main simulation design implementation can be fully multiplexed, which on the one hand, has improved the simulation design and development efficiency, and on the other hand, enhanced the scalability of the simulation platform.

Dynamic simulation modeling techniques include two key technologies.

(a) Generation function library and parameter library. The basic function modules of the simulation platform are outputted. Modeling is made according to the requirements of the simulation. Public library and characteristic library are abstracted and separated. The well-matched and practical functions are realized through intelligent interfaces, and the function scalability is met at the same time. The purpose for separating design of function library from parameter library is to ensure that the simulation model can adapt to different simulation scenarios and requirements, and be fully decoupled. The positions of the function library and parameter library in the system are shown in Fig. 4.19.

(b) Dynamic analysis and configuration mechanism. In the whole running process of the simulation, this technology provides analysis and configuration mechanism. It includes the decomposition of simulation requirements, mapping into different functional library and parameter library, and generating simulation process according to the specific simulation requirement configurations.

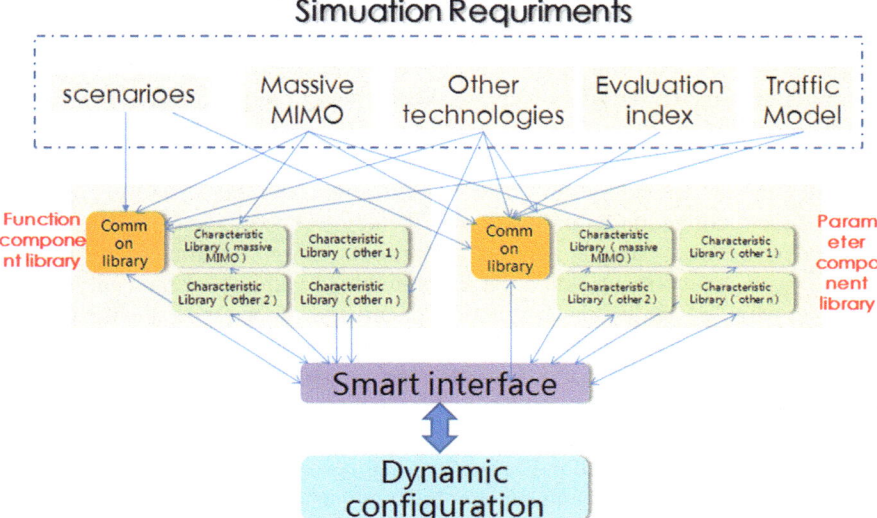

Fig. 4.19 Function library and parameter library

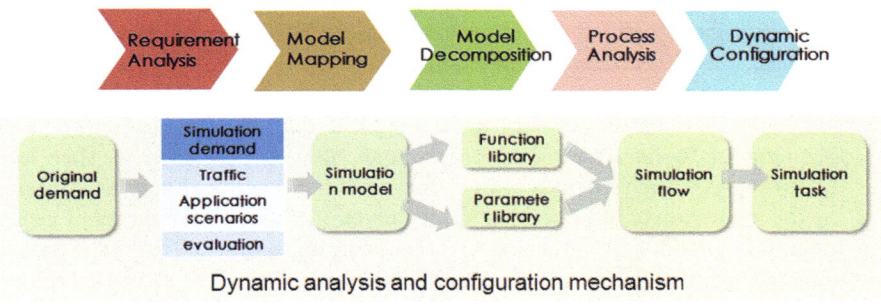

Fig. 4.20 Process of dynamic analysis and configuration mechanism

As shown in Fig. 4.20, dynamic simulation modeling process is composed of the following steps.

• According to simulation requirements, the corresponding simulation model is decomposed. According to the OSI layered architecture, the simulation model is decomposed into five layers, namely application layer, transmission layer, network layer, MAC layer and physical layer, each of which has a corresponding simulation model. In addition to the basic requirements of simulation model design according to the OSI layered architecture, each new technology of 5G has put forward more new requirements to the design of simulation model. For example, SDN requires that control plane and user plane are separated, NFV requires network functions to be decoupled from dedicated hardware devices, and the concept of virtual cell is put forward in the UDN, etc. These new technologies have put forward model design, decoupling, abstraction, resource virtualization and centralized management, and other design requirements for simulation model design, which are the key design requirements for the entire simulation platform.

- Simulation parameter library is generated according to the model and the requirements, including the system specifications and the scenario parameters, as well as configuration parameters of the key technologies. Since 5G network models are more complex, and network modes are more abundant, the network scenarios and configuration parameters involved increase dramatically. Therefore, when designing the simulation parameters library, we should take the simulation model as the center. Parameterized templates of networking scenario, network function and system specification are established based on simulation model, and with the reasonable combination of these parameter templates the complexity of the parameter library is reduced.

- The corresponding function libraries are mapped by the model. These functions libraries consist of different functional components. Each component has realized the corresponding model function. Communication interactions between components are realized through specific function interfaces. Each component can be allocated to the computational resources and can be managed as an independent computational unit. Function libraries can realize decoupling and scalability through the flexible interface design. For example, massive MIMO technology will use different precoding techniques, and each precoding technique corresponds to a precoding matrix design algorithm. The input variables and output variables of different precoding techniques have the same form, with the input of channel matrix and the output of precoding matrix. Therefore, different precoding functions can provide a unified functional interface. Realizing the separation of interface and implementation through the concept of virtual function is good for the extension new precoding algorithm and the realization of independent allocation of computational resources. For example, global precoding matrix algorithm involves the big dimension matrix inversion and multiplication calculation and it is the key path of the simulation platform calculation. On the basis of precoding algorithm decoupling design, this function can be flexibly deployed on GPU or FPGA for acceleration.

- According to the simulation requirements, the mapped function library and parameter library are organically organized to become a complete simulation process. The design of the simulation process has ensured the consistency of every step of the simulation on the whole. Simulation processes of different types are realized through the rational allocation of scenario parameters, features, and service model parameters. In UDN scenario, for example, the user handover process has different implementation scheme, in which different handover simulation process can be completed by configuring different handover algorithm parameters.

- By dynamically configuring the parameter library, function library, and the simulation process, we can get the specific simulation tasks, which directly face the users and need to provide friendly configuration management interface.

It can be seen from the above process that the key to dynamic simulation modeling lies in the design of model, library components and parameters. When the designs of stratification, encapsulation, and interface decoupling are used to solve the coupling between the conceptual model and implementation model, the goal of the minimal impact of the technological change on implementation can be achieved.

Virtual Computational Resources Management Technology

The computational efficiency in single CPU simulation environment mainly depends on the frequency of the CPU. However, the current CPU frequency is close to its limit and cannot be greatly increased, so the main means of enhancing the simulation calculation efficiency is the parallel calculation, which means to use multi-core technology to execute the computational tasks in parallel. Since the computational resources can be distributed on different physical equipment, how to distribute management resources reasonably has become the core issue.

To solve this problem, the virtual computational resources management technology is proposed, which can be broken down into three parts.

- The simulation requirements are mapped into computational tasks that can be deployed independently.

    Simulation requirements are a systematic working process, which needs to map the requirements into computational tasks and complete two parts of work: process decomposition and resources allocation.

    Process decomposition: The simulation requirements are decomposed into several serial tasks. Then these serial tasks are further decomposed into independently deployed parallel subtasks.

    Resources allocation: According to the characteristics of the parallel computing tasks decomposed by process, corresponding virtual resources are configured. Computational resources and storage resources are distributed to the locally deployed parallel computing tasks. For computational tasks deployed in slave nodes we also need to make sure to allocate enough communications resources to avoid any delay caused by data transmission.

- Various kinds of hardware resources can be virtualized into three kinds of virtual resources.

    - Computational resources: Undertaking detailed numerical calculation work. Related hardware devices are CPU, GPU, FPGA.
    - Storage resources: Undertaking data cache work, completing the calculation process by cooperating with computational resources. Related hardware includes memory, hard disk, etc.
    - Communications resources: Undertaking the transmission of control data and service data between the computational nodes. Related hardware includes high-speed bus, shared memory, etc.

- Binding the virtual resource dynamically to computational tasks.

Simulation subtasks can only run after bound with the virtual resources. Dynamic binding can make full use of resources. The virtual resources needed by simulation sub-tasks need to be assessed in advance. Different simulation subtasks have different computational complexities and different requirements for virtual resources. The running performance of the simulation subtasks can be obtained through the simulation code static analysis and statistical analysis methods. The resource requirements of each simulation subtasks can be identified according to the simulation goals. For example, in massive MIMO characteristics, precoding module and interference module have high computational intensity, which require sufficient CPU or GPU processor resources, while 3D channel model requires more storage resources.

With the development of the bottom software and hardware platforms, the available parallel virtualization technologies are relatively abundant [43]. For example, MATLAB offers parallel tools such as parfor, SPMD and MDCE, and parallel programming mechanisms which are applicable to multiple development languages, such as MPI and OpenMP. For specific simulation implementation, not only the bottom parallel technology is needed, but also the parallel design is needed for simulation applications. It's hard to give a universal method to the parallel design of simulation applications. We need to design special parallel algorithm according to specific service characteristics, so it is the key path to realize parallel virtualization for the simulation system.

Multi-core Parallel Simulation Technology

Simulation platform design based on multi-core parallel computation covers hardware, operating systems, parallel technology, simulation software, models and algorithm design, etc.

From the hardware level, parallel server scheme or high performance host system can be chosen to support higher computing power. Parallel server generally uses the following architecture types: Parallel Vector Processor (PVP), Symmetric MultiProcessor (SMP), Massively Parallel Processor (MPP), Distributed Shared Memory multiprocessor (DSM), and the Cluster of Workstation/PC (COW). Parallel server has strong parallel processing capability, but the cost is high. The high performance host system based on high-speed Internet network is able to realize high performance computing with relatively small cost. The system can consist of several multi-core servers and high-speed infiniband switch.

Operating system allocates process, storage and other hardware resources for parallel tasks, realizing the inter-process communications.

Parallel technologies generally include messaging, shared storage, and data parallel.

- Messaging: The asynchronous parallel operation mode and the distributed data storage mode are adopted with good scalability. Typical examples include MPI, PVM, etc.

- Shared storage: The asynchronous parallel operation mode and shared data storage mode are adopted with poor scalability. Typical examples include programming models like OpenMP.
- Data parallel: The loose synchronous parallel operation model and shared data storage mode are adopted with medium scalability. A typical example is HPF. HPF provides annotation-like instructions to extend the variable types so that it can control the data layout of the array in detail.

The parallelization of simulation software is the key work of multi-core parallel design of simulation platform, which needs to consider the following design requirements.

- Simulation software is decomposed in parallel from the aspects of function, algorithm and operands. Serial simulation tasks are decomposed into fully decoupled and independently-processed subtasks. Since different functions, algorithms, and operands have different computational intensity and characteristics, in parallel design we need to analyze and process them according to actual situation.
- The reasonable division design of simulation function modules can reduce the communications data between parallel subtasks. It tries to ensure that the equivalent calculation amount of each parallel subtask. And it can reduce the waiting time needed for synchronous processing.

As the evolution of the multi-core CPU parallel processing, the heterogeneous scheme of CPU + GPU has made it possible to improve the speed of the simulation calculation [1]. This is because that the CPU, as a general-purpose processor that has complex control logic, is good at complex logic operations. However, GPU is graphics processor, which often has hundreds of stream processor cores. Its design goal is to realize large throughput of data parallel computation with a large number of threads. Its single precision floating point's computing power can reach more than 10 times of CPU, which is suitable for processing parallel computing of large scale data.

Therefore, adopting the heterogeneous parallel architecture of CPU + GPU, in which the two are coordinated and multi-core parallel CPU makes complex logic calculation while GPU does data parallel tasks, can perform the maximum computer parallel processing capabilities. The world's fastest computer Tianhe-II currently adopts CPU + GPU heterogeneous polymorphic architecture.

Programming of GPU generally adopts CUDA architecture, under which the computational tasks are mapped to a large number of threads that can be executed in parallel. Its organization form is the Grid-Block-Thread 3 levels. And it has the hardware dynamic scheduling and execution.

Figure 4.21 shows a typical heterogeneous multi-core architecture. It can be seen that multi-core CPU uses OpenMP while GPU uses Compute Unified Device Architecture (CUDA) for processing. And task division is specified by program and operating system level. The two parts are interconnected with PCIee bus.

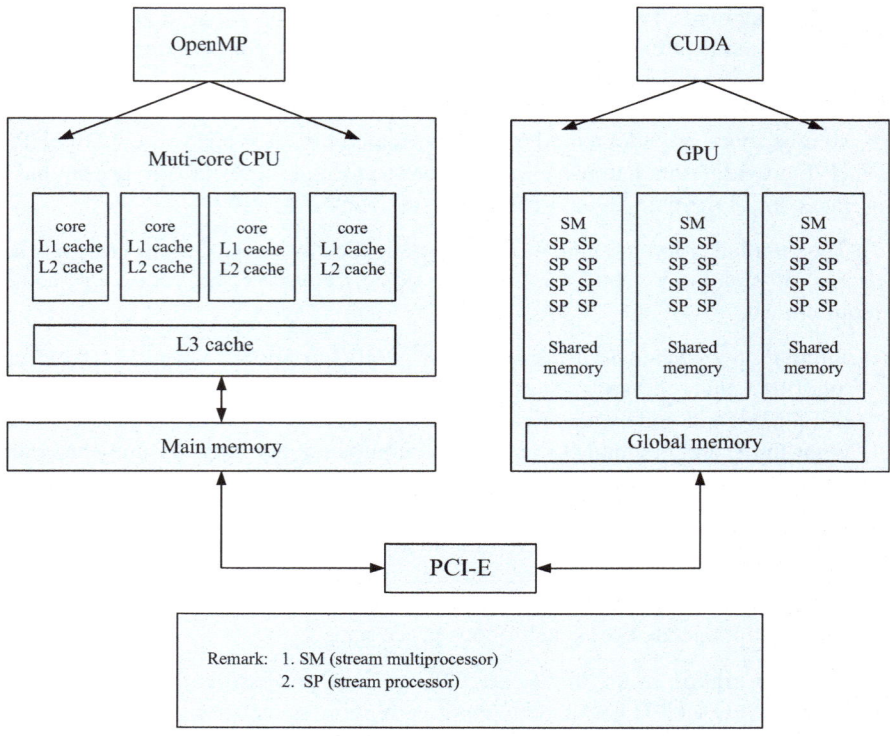

Fig. 4.21 Heterogeneous multi-core architecture

Multicore CPU-GPU heterogeneous platform has multiple layers of parallel execution ability in task level and data level. In load distribution, the computing power should be fully utilized. The simulation model structure shown in Fig. 4.22 can be considered.

The model's mapping process has the following three levels.

(a) Mapping from simulation model instance to logical process

The map divides the simulation examples into several groups, with each group assigned a logical process. Through this division, the load of each logic process should be balanced, and the communications overheads between each logical process should be as small as possible.

(b) Mapping from logical process to thread

Thread is the execution unit of the software hierarchy while the GPU thread organization is performed by cores. The computational load of each logical process is firstly divided into the CPU thread. Then the CPU thread takes the responsibility to assign the tasks that are suitable to be executed on GPU to GPU. Since GPU allows performing multiple cores at the same time, the cores under different logical process mappings have distinct task parallelism. The process can adopt the way of 1 to 1 or 1 to more.

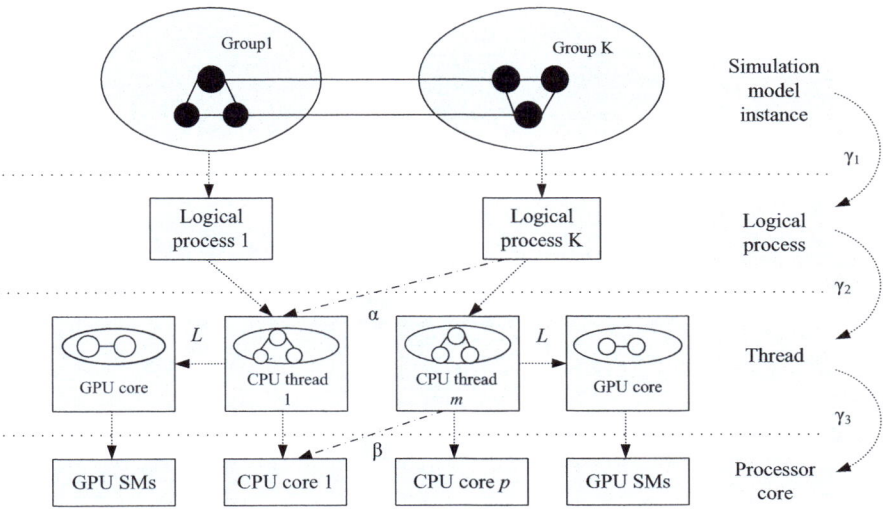

Fig. 4.22 Simulation model structure

(c) Mapping from thread to processor cores

This mapping is about how software execution unit is scheduled to hardware for execution. In order not to disrupt the operating system's optimization scheduling strategy, the operating system can take full charge of the scheduling from CPU thread to CPU execution core. Multiprocessor scheduling from GPU core to GPU stream is charged by CUDA bottom drive.

It can be seen from the above analysis that the final acceleration capability of multi-core parallel simulation technology is influenced by hardware, parallel processing mechanism, system transmission bandwidth, parallel granularity of the simulation application and many other factors. Evaluating parallel computing power of the whole simulation system is a system level problem. Due to the rapid development of the 5G technology, the simulation requirements change quickly, and the simulation application programs will also be constantly updated. How to quickly and accurately assess the overall simulation system's computing bottlenecks and provide the corresponding parallel solutions become the key issues in the multi-core parallel simulation technology. Wang et al. [44] present a parallel computing performance evaluation method based on equivalent statistical model. They propose the equivalent statistical model with double transform domain, in which the target program is continuously transformed from serial calculation domain to the equivalent serial computing domain and equivalent parallel computing domain. Then a few comprehensive indicators are derived based on Principal Component Analysis (PCA) and statistical methods so as to construct the equivalent calculation model. Thus, the optimal parallel scheme is obtained by using this model to achieve the

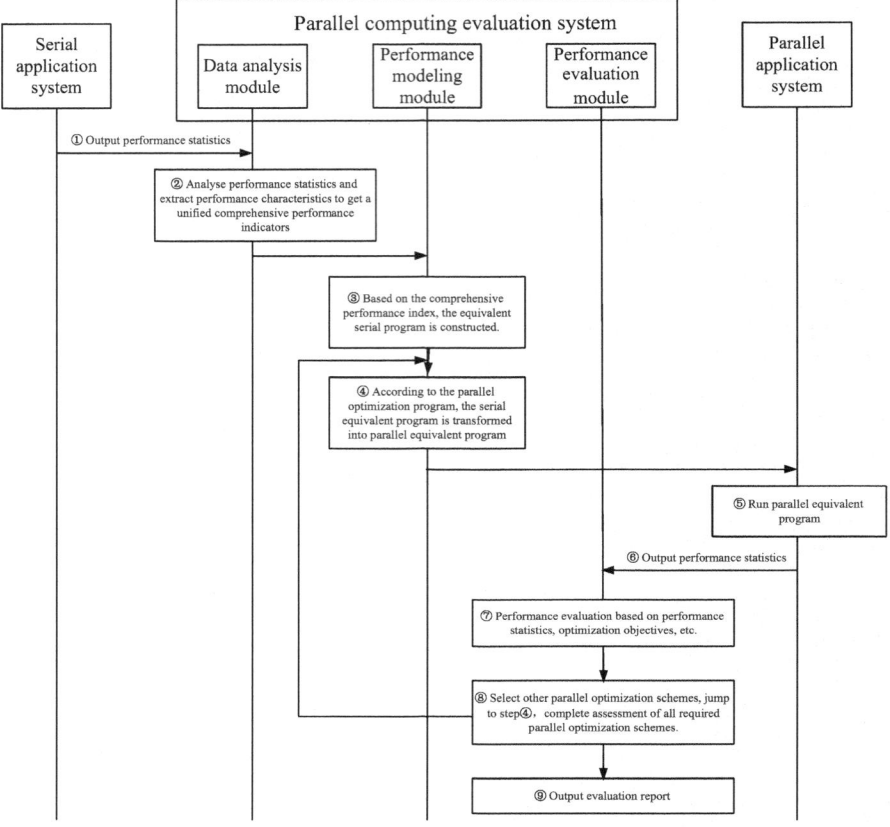

Fig. 4.23   Performance evaluation process for parallel computing

performance optimization goals. After measurement, with this method the prediction errors for simulation resource and time consumption are less than 10%. The basic working process of this method is shown in Fig. 4.23.

Hardware Acceleration Simulation Technology

Hardware acceleration simulation technology uses hardware modules rather than the software modules to make full use of the inherent fast hardware features. The hardware acceleration simulation technology has different focus from the hardware-in-loop simulating technology, with the latter focusing on the real-time implementation of function features and the former aiming at solving system bottlenecks and accelerating the overall calculation speed. Generally, hardware uses high performance FPGA board which has strong computing power and logical processing ability. FPGA board has stronger floating point computing power than

CPU server and stronger task management, resource scheduling and other logic handling abilities than GPU server. With the rapid development of FPGA development environment and compiler in recent years, the development difficulty for FPGA board is reduced greatly. The main research points of hardware acceleration simulation technology can be divided into the following three aspects.

1. Studies on key technologies of FPGA based high performance hardware acceleration

   With the development of integrated circuit technology, the computing power and storage resources of FPGA chip grow rapidly. Meanwhile, the FPGA has become the ideal algorithm accelerating implementation platform due to its programmability, powerful parallel ability, rich hardware resources, flexible algorithm adaptability and lower power consumption. FPGA can easily realize the distributed algorithm structure, which is very useful for acceleration scientific computing. For example, most matrix decomposition algorithms in scientific computation require a lot of Multiply and ACcumulate (MAC) operation. FPGA platform can effectively realize the MAC operating through distributed arithmetic structure. Today's FPGA products have entered the era of 20 nm \14 nm. Major companies are working on developing logic products with more logical units and higher performance.

   Research contents include high speed parallel processing, hardware and software simulation task partition and mapping, high precision signal processing, etc.

2. Interface and middleware design with the combination of hardware acceleration and software simulation platform

   For some calculations that are hard to complete by FPGA, we can transfer it to C language. At the end of the design, the intermediate results from the hardware are imported into C for visual analysis. Through drawing graphical waveform and eye diagram, etc., we can check whether each part of the system design is correct. Introducing the simulation based on C/C++ to the design of FPGA has solved the FPGA debugging difficulty, as well as the frequent inconsistency with the system simulation logic.

   Research contents include virtual adaptation mechanism, middleware design, and reusable calculation design.

3. The design of reconfigurable FPGA hardware accelerator board

   Reusable programming is an important characteristic of the hardware acceleration based on FPGA platform. The FPGA based system can be altered or reconfigured in the development and application phase arbitrarily, which will increases the overall gain.

   Research contents include: high speed PCIe interface design, data interaction between high-speed USB 3.0 interface and the host.

   The implementation process of hardware acceleration simulation is as follows.

   • The first is the key technology research. Starting from the simulation design requirements, we should fully combine the characteristics of the hardware parallel processing and software system architecture to reasonably divide and design hardware acceleration module, so as to significantly reduce the time overheads of simulation core module calculation.

- Then the FPGA based hardware accelerator card system is designed, involving the C simulation code based adapter, the design of middleware on interface layer (including board support package) and the design of reconfigurable calculation.
- When the host receives the signals indicating that the configuration data is loaded, the initial data is sent into memory via PCIe interface. When the initial data is sent, launch signal is sent to accelerator through PCIe interface. When the accelerator completes the processing, accelerated processing completion signal is sent to the host through PCIe interface. When the host receives the acceleration processing completion signals, the final results are read from memory.

As shown in Fig. 4.24, on system simulation platform, some links use hardware implementation. The actual test results of the delay of this link's physical bus transmission are within 10 ms. Physical bus transmission delay is defined as the delay of the whole process when data is distributed from system

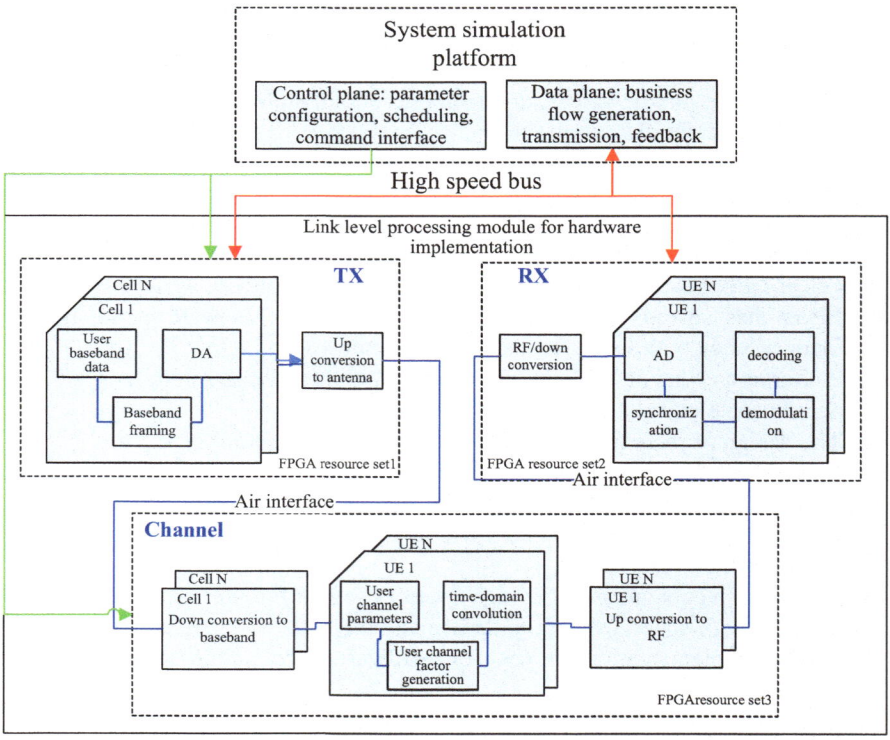

Fig. 4.24 System implementation diagram of software and hardware co-simulation

simulation platform, sent to the air interface through the link transmission module, and then through the link receiving module, received, demodulated, and decoded, and then sent back to the system simulation platform.

This kind of software and hardware co-simulation method can fully use the hardware high-speed processing capabilities, and enable some link's system simulation performance to be close to real-time. Combined with relatively perfect system function of the system simulation platform, it can better simulate the system application scenarios with high indicator requirements for system transmission delay. The implementation scheme of patent CN101924751B [45], for example, is the entire network solution scheme across the core network and access network, ranging from the physical layer of base station to every layer of core network protocol stack. The handover process is at millisecond-level scale, which is difficult to verify on the old system simulation platforms. Adopting the method of software and hardware co-simulation platform can well support system function simulation test at millisecond-level small scale.

Real-time Transmission Technology

Network architecture can extend hardware processing ability of the simulation platform through the way of extending server node. But in parallel computing solution scheme of multi servers networking, the data exchange delay between the servers becomes the key bottleneck to affect the simulation calculation speed [1], which requires the corresponding solutions.

After the high-strength computational tasks are parallelized, the computational time of each independent parallel subtask becomes shorter, which is usually within hundreds of microseconds. Calculation results of each subtask in each calculation cycle need to be collected to the center module for centralized processing. After completing the previous cycle, the calculation of the next cycle is carried out. The collection process needs the data interaction between the servers. If data interacts via IP switches,, the data exchange delay between the servers is commonly in millisecond due to the bus mechanism as well as the TCP/IP protocol stack processing overheads, which has exceeded the calculation time of subtasks. This will greatly reduce the overall efficiency, and directly influence the final parallel effect, so it is necessary to provide low latency server data sharing interactive solution. In order to reduce the data interaction delay between servers brought by server cluster and make sure that data between the multiple simulation nodes can share with each other in real-time, reflective memory technology can be used to realize memory sharing mechanism between multiple servers or multiple servers transmission networking based on high-speed InfiniBand switches. Reflective memory network is a special type of memory shared system, which aims at sharing common data sets between multiple independent computers. Reflective memory networks can store the independent backups of the whole shared memory in each subsystem. Each subsystem shall have full and unrestricted access, and can also modify local data sets with extremely high local memory writing speed. The transmission delay between reflective memory cards is up to 400 nanoseconds.

4.4.3 Introduction to Simulation Cases

Case 1: Massive MIMO System Level Simulation

This section uses massive MIMO 5G key techniques as an example, and suggests that how to apply the key technologies described before to complete the design and implementation of massive MIMO technology in simulation system, so as to reduce the simulation calculation complexity and accelerate the simulation calculation speed.

1. Simulation scenario description

   - Simulation parameter description
     Massive MIMO uses the MU-MIMO model to simulate LTE downlink system performance. The channel matrix of MU-MIMO is formed with 128 base station transmitting antennas, 1 UE receiving antenna, 15 users being scheduled in a single cell at the same time. Detailed simulation parameters can be seen in Table 4.9.
     The statistics indicators of simulation output mainly include the average cell spectrum efficiency, 5% cell edge spectrum efficiency, UE downlink SINR and some other indicators.
   - Simulation environment description
     Hardware: GPU server XR-4802GK4

     - CPU configuration: two Intel Xeon Ivy Bridg E5 (3.0 G, single 10 cores, 20 threads)
     - GPU configuration: 8 pieces of TESLA K20
     - CPU memory: 256GB
     - Bus: PCIe 3.0 ×16

     Software:

     - WINDOWS SERVER 2008R2
     - MATLAB R2014a

2. Simulation design analysis

   (a) Functional procedure
       Simulation process of MU-MIMO is shown in Fig. 4.25. The main features of MU-MIMO are embodied in the following function modules.

       - Transmitter module: It includes CSI feedback, pilot pollution suppression, antenna resource allocation, user scheduling, and transmitter precoding.
       - Transmitting: Wireless channel modeling, including 3D–Uma, 3D–Umi, 3D–UMa-H channel.
       - Receiver processing: It includes interference calculation, SINR calculation, and receiving detection algorithm.

Table 4.9 Simulation parameter settings

Parameter	Value
Base station antenna numbers	128
Base station antenna polarization modes	±45° double polarization
Terminal antenna numbers	1
Terminal polarization modes	Perpendicular horizontal dual polarization
Access user numbers of each sector	30
Networking type	Same frequency in three sectors, 19 stations, 57 cells, 30 users in each cell
Channel model	3D–UMa
Interval between horizontal antennas	0.5 wavelength
Interval between perpendicular antennas	0.5 wavelength
Moving speed	3 km/h
Precoding algorithm sent by multiple users	Zero Forcing
Terminal receiving algorithm	LMMSE
Enabling of ideal uplink channel estimation	Yes
Enabling of ideal downlink channel estimation	Yes
Constraint type of transmitting power	Consistency of total transmitting power
Pilot pollution modeling	No pilot pollution
Scheduling method	Proportional fairness
Transmitting power	Total transmitting power of the base station is 40w
Simulation bandwidth	10 M
Terminal antenna gain	0dBi
Single array gain	8dBi
Antenna gain	Determined by the mapping type from the antenna elementto the port
Thermal noise level	-174 dBm/Hz
eNB noise figure	5 dB
UE noise figure	7 dB

(b) Computation analysis

According to the simulation parameter settings and MU-MIMO features, calculations are mainly distributed in the channel calculation, precoding calculation, and receiving SINR computation. Set subcarrier number to be Nc, OFDM symbols M, the number of base station antennas Nt, users (single antenna) number Nr, receiving antenna number Nr, and the cell number in the system C.

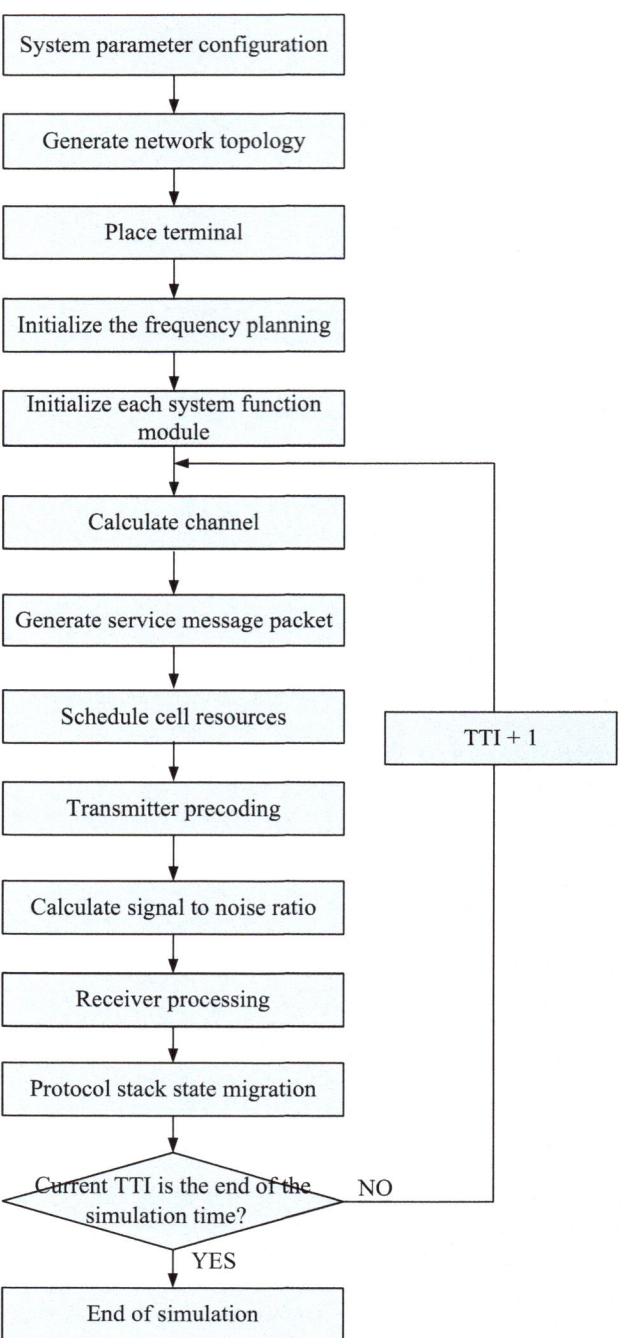

Fig. 4.25 MU-MIMO simulation process

- 3D channel

  It includes the path loss calculation of large scale, shadow and small scale. Because the number of transmitting and receiving antennas increases significantly compared with 4G, and the antenna and the channel parameter models are more complex, the computation has a huge increase compared with the commonly used TU channel in 4G. Let's analyze only the times of FFT transformation calculation from time domain channel to frequency domain channel. If 3D channel FFT transformation calculation in a cell is about $M \times Nt \times Nr$, then the calculation amount ratio is $(128 \times 15)/(2 \times 1) = 960$ compared with the scenario that antenna scale of 4G is 2×1 in the case that downlink antenna scale is 128×15. The calculation amount ratio increases linearly with the product of transmitting antenna number and receiving antenna number.

- Transmitter precoding

  According to the simulation parameter settings, the transmitter precoding scheme is forced zero algorithm, and precoding matrix calculation formula is $\mathbf{W}^{ZF} = H(H^H H)^{-1}$, $H \in \mathbf{C}^{Ntx Nr}$. Precoding computational complexity mainly lies in the matrix multiplication and inverse. Under the condition of forced zero algorithm, the main calculation includes two parts. The first part is $C \times Nc$ times $Nr \times Nr$ dimensional matrix inversion. The second part is multiplication of $C \times Nc$ times $Nt \times Nr$ dimensional matrix and $Nr \times Nr$ dimension matrix. The algorithm complexity of different kinds of matrix calculation is usually $O(n \wedge 3)$. Compared with antenna scale with 2×1 scenario in 4G, the calculation ratio of matrix inversion is $15^3/1^3 = 3375$. The calculation ratio of matrix multiplication is $(128 \times 15 \times 15)/(2 \times 1 \times 1) = 14,400$, which increases along with the three times power of antenna number.

- SINR calculation

  For the calculation of this part, the users' received signal power, the interference power in the cell, interference power between cells are calculated based on the channel matrix and precoding matrix, and then the user's SINR is obtained. According to the following MIMO signal model, the size of calculation can be roughly obtained.

$$y_{jm}^{dl} = \sqrt{\rho_{dl}} \sum_{l=1}^{L} \sqrt{\lambda_l} \mathbf{h}_{ljm}^{H} \sum_{k=1}^{K} \mathbf{w}_{lk} x_{lk}^{dl} + n_{jm}^{dl}$$

$$= \sqrt{\rho_{dl} \lambda_j} \mathbf{h}_{jjm}^{H} \mathbf{w}_{jm} x_{jm}^{dl} + \underbrace{\sqrt{\rho_{dl} \lambda_j} \mathbf{h}_{jjm}^{H} \sum_{k=1,\ k \neq m}^{K} \mathbf{w}_{jk} x_{jk}^{dl}}_{intra-cell\ interference} + \underbrace{\sqrt{\rho_{dl}} \sum_{l=1,\ l \neq j}^{L} \sqrt{\lambda_l} \mathbf{h}_{ljm}^{H} \sum_{k=1}^{K} \mathbf{w}_{lk} x_{lk}^{dl}}_{inter-cell\ interference}$$

$$+ n_{jm}^{dl}$$

Calculating the signal power of an UE in a cell requires two times Nt dimensional vector multiplication; the number of multiplication is $2Nt + 1$;

the number of multiplication getting the interference power (including interference in the cell and between cells) needs is $C \times (2Nt + 1)$; the total number of multiplication is $(C + 1) \times (2Nt + 1)$. Compared with antenna scale 2×1 scenario in 4G, the calculation amount ratio is $(2 \times 128 + 1)/(2 \times 2 + 1) = 51.4$.

It can be known from the above analysis that there are big differences of MIMO's calculation amount in different modules, which mainly lie in channel calculation, transmitter precoding and receiving SINR calculation modules. Among the modules, transmitter precoding has the largest computational amount of 3 power of antenna number, followed by channel calculation. According to the law of Amdahl, these two modules are also the key optimization goals of computational acceleration.

(c) Optimization scheme

According to the characteristics of the different modules, acceleration optimization scheme can be made combining the previous key technologies.

- Channel computational

  Given simulation parameters, wireless link channel coefficients have nothing to do with behaviors like system scheduling, so the channel calculation can be done in advance, and the results can be stored in hard disk. When simulation system is initialized, the channel matrix stored in hard disk can be read directly, and then be stored in memory. There is no need to calculate the channel matrix in the process of the simulation. Using the pre-computed results directly saves the channel calculation time. If memory is big enough, the actual time overheads only depend on the memory reading time, and the channel computational time can be neglected.

- Transmitter precoding

  Transmitter precoding mainly involves large matrix multiplication and inverse calculation, during which multicore computing capability of CPU and GPU can be fully utilized to compute in parallel in the subcarrier level. First CPU parallel computing capability is used to do parallel processing in the subcarrier granularity. Different sub-carrier precoding calculation work is allocated to different GPU boards for parallel computing Matrix inversion and multiplication calculation are completed by each GPU. As a result of the combined parallel computing of CPU and GPU, and parallel computing on the granularity of subcarrier, it can parallel to $C \times Nc$ paths in maximum. In this test example, it can parallel $57 \times 300 = 17,100$ roads. For enough GPU cores, considering the influence of the transmission bandwidth, it can meet the requirements of acceleration optimization requirements of the precoding in transmitter. In this test example, since only one GPU server is adopted and CPU and GPU hardware computational resources are limited, the actual acceleration effect is limited by the hardware resources. In multi servers networking and super computer environment, the acceleration effect can be further improved.

- SINR calculation

  This part of the calculation is mainly the vector multiplication. Its calculation amount is much smaller than channel calculation and precoding module of the transmitter. So, it is able to obtain better results with the help of CPU acceleration.

3. Simulation measurement results and analysis

   The simulation measurement results are shown in Tables 4.10 and 4.11. According to the calculation characteristics of different modules, the final acceleration effects are also different when using different acceleration program.

   SINR calculation module and message processing module adopt the CPU parallel computing scheme. Precoding module adopts CPU + GPU co-acceleration scheme. The acceleration ratio of the interference module is less than that of message processing module from the perspective of acceleration ratio. It's because interfere module needs to transfer large amounts of data between the parallel computing tasks, including signal power, interference power, SINR, channel allocation, scheduling information and other data, most of which are vast data of subcarrier granularity. The time overheads on data transmission is greater than that on message processing module, so its acceleration ratio is smaller than that of the message processing module. The further optimization of SINR calculation module includes increasing the number of parallel CPU cores, compressing transmitted data, increasing the transmission bandwidth (high-speed optical fiber transmission and memory reflection technology, etc.) and other schemes. Precoding module adopts CPU + GPU co-acceleration scheme, whose acceleration ratio can reach 127 times. Due to the limitation of hardware resources, the acceleration effect of this part is far from upper limit. The above acceleration effect is measured on a single GPU server. Because the parallel granularity of various software modules are much more than the number of the server processor units, the parallel acceleration can also still be greatly improved after improving the capacity of the hardware

Table 4.10 Computational time overheads of each MU-MIMO module in serial scheme and parallel scheme

Computational scheme	Total length of one TTI (second)	SINR calculation (second)	Message processing (second)	Precoding (second)	Other modules (with no parallel optimization) (second)
Serial	9338.72	772.64	124.88	8411.95	29.25
Parallel	153.92	50.8	4.12	66.03	/

Table 4.11 Acceleration ratio under different modules of MU-MIMO

Measurement results of acceleration ratio	Total acceleration ratio (serial/parallel)	SINR calculation	Message processing	Precoding
Acceleration ratio	60.67	15.2	30.31	127.4

computational platform. For example, the acceleration ratio of precoding module, SINR module and other modules can be further enhanced in super computer environment.

Case 2: Radio Resource Optimization of Ultra Dense Network

With the increasing number of base stations near the small base station in ultra-dense network scenario, the interference will also be more serious. In particular, the site number to be coordinated will also increase in resources allocation, making resource allocation more difficult. The site deployment in hot spot areas shows a trend of high density and no programming. With more and more deployed sites, how the system performance changes in hot spot area is rather important for both users and network operators. Liu et al. [46] simplify the resource allocation under the ultra-dense deployment scenario through network clustering. Assuming that intra-cluster channel is orthogonalized, it puts forward a semi-distributed resource allocation algorithm based on clustering, which reduces the resource allocation problem under the dense deployed scenarios. A semi-distributed resource allocation algorithm based on clustering mainly includes the following three stages.

The above optimization problem is a nonlinear mixed integer planning problem. The resource allocation in ultra-dense deployment scenarios is simplified based on the assumption of network clustering and intra-cluster channel orthogonalization. Then a semi-distributed resource allocation algorithm based on clustering is proposed to reduce the resource allocation problem under the dense deployment scenarios. A semi-distributed resource allocation algorithm process based on clustering is shown in Fig. 4.26, which mainly includes the following three stages:

In the first stage, the network clustering algorithm based on Breadth-First Search algorithm (BFS) is proposed, where the performance loss brought by the interference is taken as the similarity parameters of the site, network clustering is carried out based on BFS and the cluster heads are selected;

In the second stage, it introduces the concept of reference cluster to approximately estimate the received interference of each cluster for clustered network. In the information interaction between cluster heads, it selects the cluster with the strongest interference as a reference for a given cluster, and only takes into consideration the interference of the reference cluster in the next phase of resource allocation. Introducing the concept of reference cluster can reduce the information interaction of channel state information during the channel power control and reduce the complexity of the resource allocation problem by obtaining reasonable suboptimal solution. In addition, the clustering algorithm in first stage assigns the cells with strong interference to the same cluster. Combining the assumption of orthogonal channel within the cluster, it can guarantee that the reference cluster with the strongest interference is a reasonable approximation for all interference sources.

In the third stage, solving the sub-problems of resource allocation for cluster and reference cluster, and two-step iterative heuristic resource allocation algorithm is

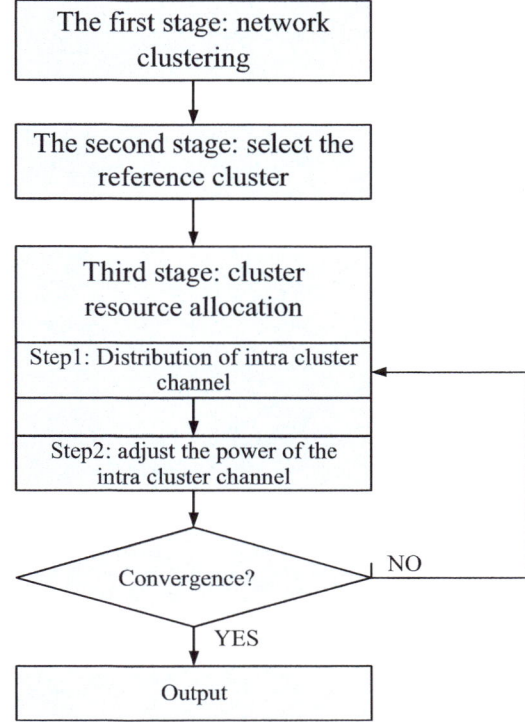

proposed. The algorithm uses heuristic Algorithm 3 to obtain channel allocation, and then uses the power allocation algorithm based on binary search to adjust power. It is iterated repeatedly until the system convergence.

In the simulation scenario, the system contains 16 small Base Stations (sBS) in the 400 m × 400 m indoor hot spot area. The radius of sBS is 100 m; each small base station has 16 available sub-channels; each of these small stations has 4 users UE randomly distributed in the coverage area of sBS. Channel gain includes large scale fading caused by path fading and shadow effect. Path fading model uses the ITU PED-B path loss model with $PL = 37 + 32\log_{10}(d)$ [47], where d is the distance between sBS and user UE. To ensure that the solutions of the problems do exist, each user uses the smaller QoS constraints. Except special instructions, other related simulation parameters are shown in Table 4.12.

Figure 4.27 has compared the system performance in scenarios with different cluster sizes considering all the interference (Interference Case 1) and estimating interference with reference cluster (Interference Case 2). The result of Fig. 4.27 is the site average simulation results for many times. In each simulation, the clusters with different sizes are realized through selecting different λ and r for 16 sBS within the area. The approximate simulation results of reference cluster interference are obtained by estimating the interference in accordance with the reference cluster concept in this chapter; and then subchannel allocation and power control are made.

Table 4.12 System simulation parameters of resource allocation based on clustering

Parameter	Value
Carrier frequency	2.0 GHz
Sub-channel bandwidth	180 KHz
Available channel number	16
Minimum user speed	10kdps
Noise spectral density	−174 dBm/Hz
Indoor shadow fading standard deviation	4 dB
Base station maximum transmitting power	15 dBm

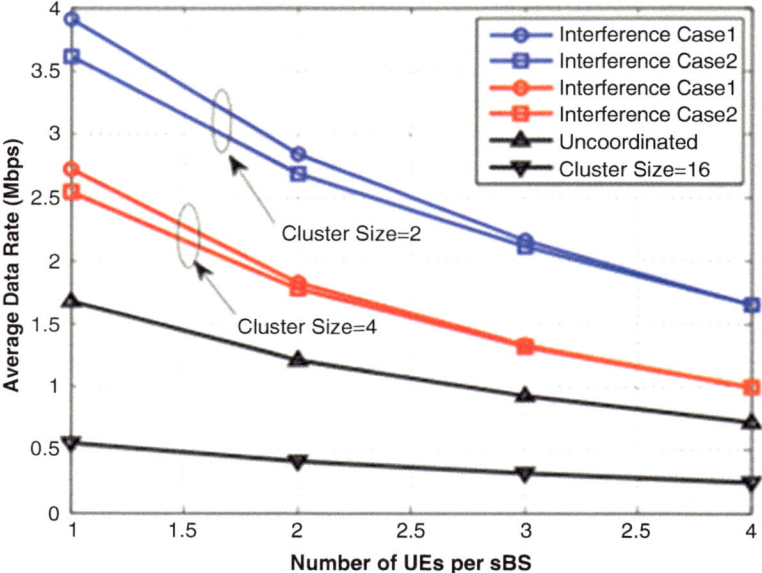

Fig. 4.27 System performance comparison between interference estimation through reference cluster and including all the interference

After multiple iteration, the solution of power and channel allocation is obtained. And then the actual average rates of each user in the computational system are obtained through computing all the interference. Simulation results including all the interference have considered all the interference of the site to solve the power allocation in iteration. The average rate of each user within the computational system is also the actual average rate of each user including all the interference. Simulation results show that reference cluster approximation and the way including all interference have a very similar system performance. But the way using reference clusters to estimate the interference is better in approximation than including all interference with the increase of the cluster size,. This is because the site number within the cell is fixed, and the cluster number decreases within the cell when the cluster size increases. In accordance with the concept of reference cluster, the cluster including the strongest interference has a better interference approximation for a cluster. In particular, when the cluster size is 16, the way of including all the interference and the way of estimating interference through reference clusters have the same result. It's because all the base stations are in the same cluster, and the way of estimating interference through reference cluster is equivalent to that including all the interference, thus the results are the same.

In the simulation shown in Fig. 4.28, the 16 sBS sites within the area are in two areas with different sizes respectively: 100 m × 100 m and 60 m × 60 m. The user number changes within in sBS from 1 to 5. When the number of in-cluster users is more than the number of subchannels within the cluster, simple Round Robin Scheduling is adopted. Fig. 4.28 has compared the performance of the proposed algorithms in this chapter, the Semi-Definite Programming Based algorithm (SDP Based) and each sBS uncoordinated resource allocation algorithm. In this figure, user average rates are all obtained by calculating all the system interference in the channel allocation and power allocation schemes obtained by using each algorithm. The interference estimation of reference cluster is only used in solving power allocation.

Figure 4.28 shows that the average rate of each user decreases with the increase of users within sBS site. This is because that the resource can be allocated in Round Robin Scheduling decreases with more users. Further, it can be seen from the figure that when the algorithm of allocating resources independently in each cell is adopted, the system performance is the worst with severe interference since the

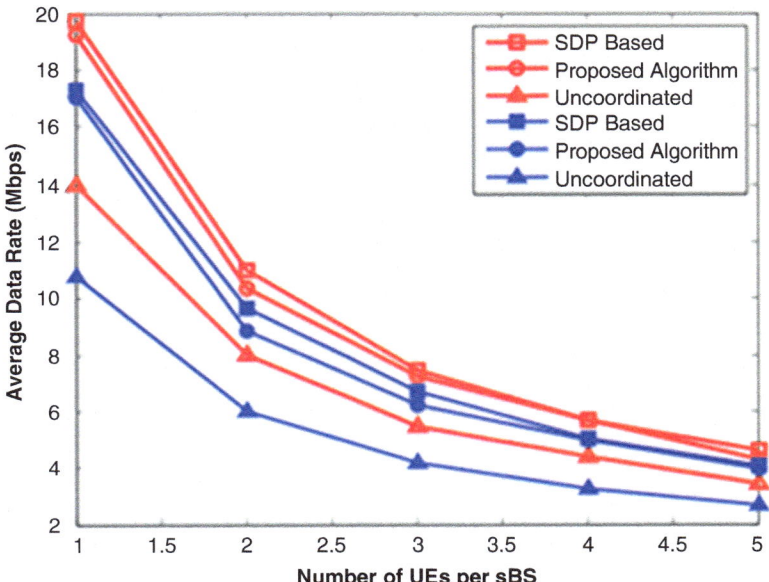

Fig. 4.28 The user performance variation when adopting different ultra-dense network resource allocation algorithms in different site densities

interference from other sites is not considered in resource allocation. Besides, with the increases in the site density, compared with 100 m × 100 m (the lines with hollow markers) and 60 m × 60 m (the lines with solid markers) areas, the system performance is even worse. The algorithm proposed in [48] can get better system performance, but with high complexity. The algorithm proposed in this chapter has the similar performance with that proposed in [48]. Through the above theoretical analysis, we can know that our algorithm has lower complexity. Figure 4.28 in fact verifies that the proposed algorithm can guarantee approximate optimal system performance with lower complexity.

Case 3: Statistical Modeling Simulation of the Uplink Interference [49]

In the Orthogonal Frequency Division Multiple Access (OFDMA) -based frequency multiplexing network, inter-cell interference has become one of the key factors that restrict the improvement of system performance. Research shows that the uplink power control for the user terminals in the cell, on the one hand, can reduce the uplink interference between cells and improve the system uplink frequency spectrum efficiency; on the other hand, can reduce the energy consumption of user terminals and bring power saving effect for terminals. However, in the past network deployment, it is often difficult to choose appropriate control parameters and only conservative settings can be made according to limited experience. In fact, in order to make parametric statistical modeling analysis on the interference characteristics between cells, we can analyze the impact of the system parameter setting on the interference between cells and the system performance according to the constructed model. Therefore, in the uplink open loop power control scenario, it is necessary to build an uplink interference parametric statistical model so as to provide theoretical guidance for the design and optimization of power control scheme.

After the parametric statistical modeling of PDF for the uplink interference between cells is obtained through theoretical analysis, firstly the statistics is carried out through a large number of samples produced by Monte Carlo simulation. Then the samples are compared with the results of the theoretical derivation to verify whether the theoretical derivation is correct. Secondly, the simulation results can be used to display the impact of system parameters on the statistical distribution of the interference between the uplink cells.

System Model

Considering the open-loop uplink power control in the terminals, the system model of the uplink interference between cells is shown in Fig. 4.29. The model is established through the location relationship between the nodes in polar coordinate system. eNB_1 is located in the origin point of coordinates system, which is (0, 0).The cell coverage area is a circle with the radius R. UEs are distributed randomly

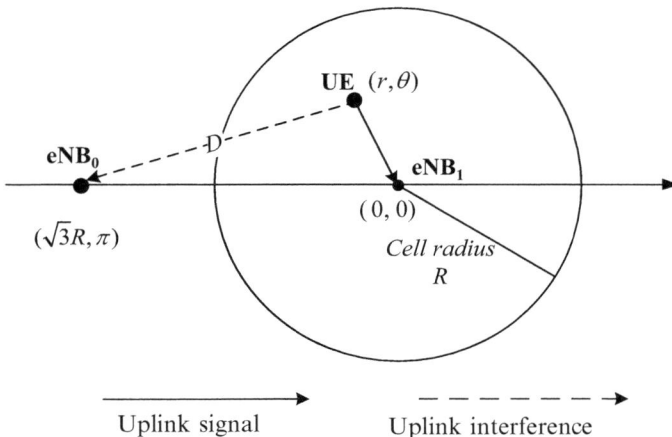

Fig. 4.29 Open loop power control uplink interference scenario

uniformly in the eNB$_1$ service region, with the coordinates of the location as (r, θ). eNB$_0$ is the interfered adjacent cell base station; the distance between eNB$_0$ and eNB$_1$ is $\sqrt{3}R$; the position in coordinates is $\left(\sqrt{3}R, \pi\right)$. The distance between UE and the interfered base station eNB$_0$ is D.

- Base station location is fixed with users distributed randomly and uniformly,

$$f(r, \theta) = \frac{r}{\pi\left(R^2 - r_{\min}^2\right)}.$$

- The transmitting power of UE is $P_t = \min(P_{\max}, P_0 + \alpha PL)$, where P_0 denotes the nominal power, and α denotes the open loop path loss compensation factor.
- PL represents the UE uplink signal link propagating loss, including path loss and shadow shading: $PL = A + B\log_{10}r + X_{\text{UE-eNB}_1}$, where A and B are path loss model parameters; r is the distance between UE and eNB1. $X_{\text{UE}} - \text{eNB}_1$ is the shadow shading that UE uplink experienced.
- Interference link propagation loss is $PL_{\text{UE-eNB}_0} = A + B\log_{10}D(r, \theta) + X_{\text{UE-eNB}_0}$, where the distance of interference link is denoted as

$$D(r, \theta) = \sqrt{r^2 + 3R^2 - 2\sqrt{3}Rr\cos(\pi - \theta)} = \sqrt{r^2 + 3R^2 + 2\sqrt{3}Rr\cos\theta}.$$

- Shadow shading: $X_{\text{UE-eNB}_1} = aX_{\text{UE}} + bX_{\text{eNB}_1}$; $X_{\text{UE-eNB}_0} = aX_{\text{UE}} + bX_{\text{eNB}_0}$, where $a^2 = b^2 = 1/2$.
- Assume that the terminal's maximal transmitter power threshold is set high, and then the uplink interference in Log-Domain is:

$$I = P_0 + (\alpha - 1)A + \alpha B\log_{10}r - B\log_{10}D(r,\theta) + \frac{(\alpha - 1)}{\sqrt{2}}X_{\mathrm{UE}} + \frac{\alpha}{\sqrt{2}}X_{\mathrm{eNB}_1}$$

$$- \frac{1}{\sqrt{2}}X_{\mathrm{eNB}_0}$$

Interference Statistic Model and Simulation Verification

After derivation, the uplink interference link can be simplified as $I = P_0 + (\alpha - 1)A + Y(r,\theta) + X$, and thus the PDF of uplink interference can be obtained through the expression below,

$$f_I(I) = \int_{-\infty}^{+\infty} f_X(I - (\alpha - 1)A - P_0 - y)f_Y(y)dy,$$

where the distribution function of X related to shadow fading and the distribution function of intermediate variable Y related to UE's position. They can be given respectively as follows.

$$f_X(x) = \frac{1}{\sqrt{2\pi(\alpha^2 - \alpha + 1)}\sigma}\exp\left[-\frac{x^2}{2(\alpha^2 - \alpha + 1)\sigma^2}\right],$$

$$f_Y(y) = \frac{dF_Y(y)}{dy}$$

$$= \begin{cases} \dfrac{2}{\pi(R^2 - r_{\min}^2)}\displaystyle\int_0^\pi \dfrac{r^2(y,\theta)\partial r(y,\theta)}{\partial y}d\theta, & Y_{\min}(\alpha, r_{\min}) \leq y \leq Y_0(\alpha) \\[2em] \dfrac{[R^2 - r^2(y,\theta_0(y))]}{\pi(R^2 - r_{\min}^2)}\dfrac{d\theta_0(y)}{dy} + \dfrac{\displaystyle\int_{\theta_0(y)}^\pi \dfrac{\partial[r^2(y,\theta)]}{\partial y}d\theta}{\pi(R^2 - r_{\min}^2)}, & Y_0(\alpha) < y \leq Y_{\max}(\alpha, R) \end{cases}.$$

In order to verify the above theoretical derivation, the theoretical interference distribution and Monte Carlo simulation of intermediate variable Y and interference I are shown in Figs. 4.30 and 4.31 respectively. The system parameters involved in theoretical calculation and simulation are shown in Table 4.13. Clearly, there is high consistency between simulation results and theoretical results according to the results in Figs. 4.30 and 4.31, which prove the correctness of the theoretical derivation.

Further, based on the derived theoretical model, the impact of the power control parameters (path loss compensation factor α and nominal power P_0) setting on the uplink interference distribution is explored. The corresponding results are shown in Figs. 4.32 and 4.33.

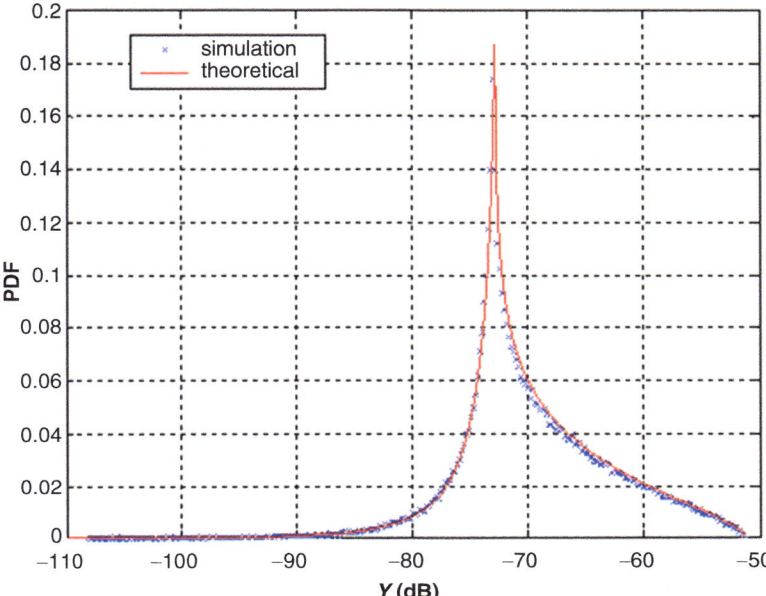

Fig. 4.30 PDF of intermediate variable Y

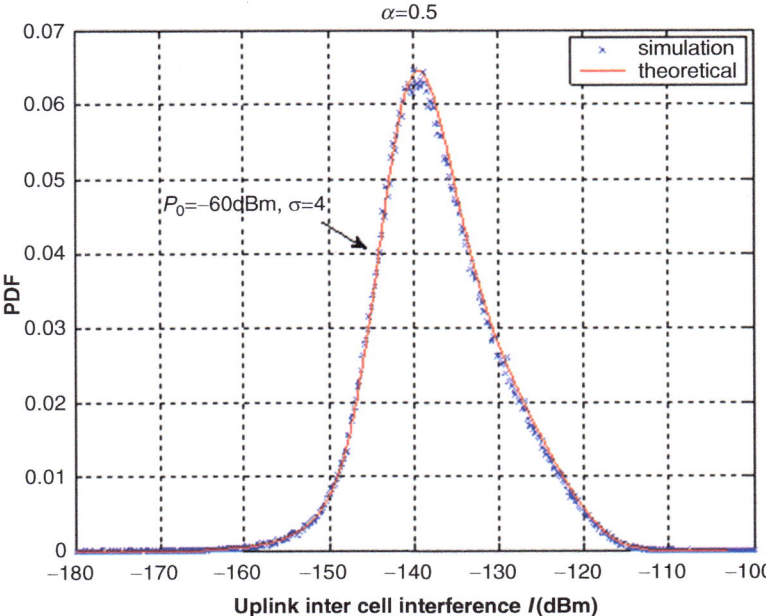

Fig. 4.31 PDF of uplink interference I

Table 4.13 Simulation parameter setting

System parameter	Value
Cell diameter	$R = 1000$ m
Nominal power	$P_0 = -60$ dBm
Minimal distance between UE and eNB	$r_{min} = 3$ m
Antenna model of UE and eNB	Omnidirectional antenna
Antenna gain of UE and eNB	0 dBi
Path loss model	$A = 15.3, B = 37.6$
UE distribution	Uniform distribution
Standard deviation of shadow shading	$\sigma = 4$ dB
Path loss compensation factor	$\alpha = 0.5$
Monte Carlo simulation times	500,000 times

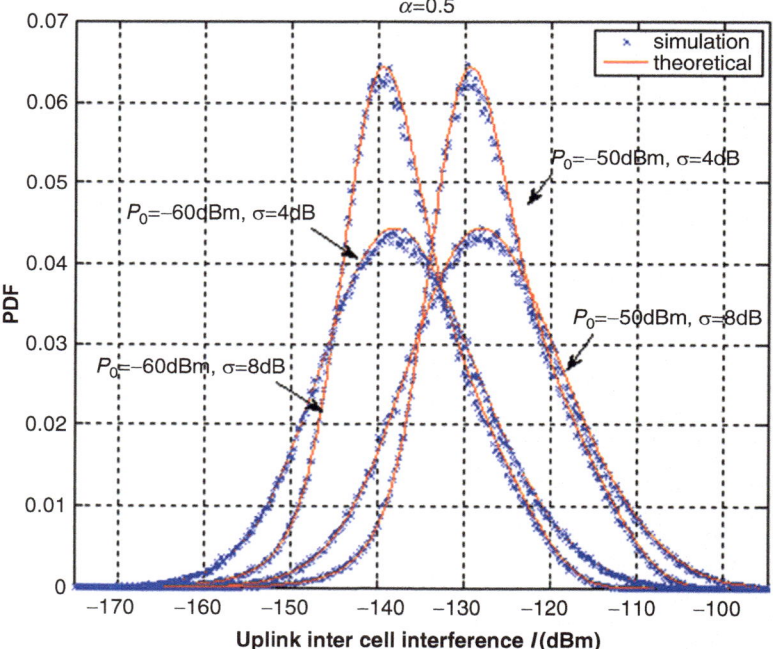

Fig. 4.32 The impact of nominal power P_0 on interference distribution

Through the comparison in Fig. 4.32, we can see that the increase in $P0$ will directly lead to the increase in the uplink interference power. The uplink interference PDF curve will translate to the region with larger interference. The translation scale is the increase scale of P_0. The shape of the distribution curve will not change with the change of P_0. For different shadow fading scenarios, the impact of the change of P_0 is consistent with the interference distribution function.

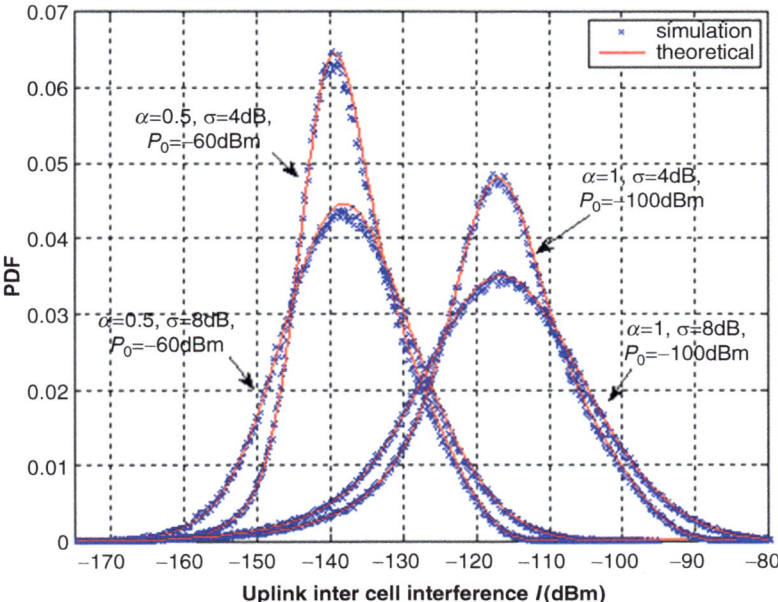

Fig. 4.33 The impact of path loss compensation factor α on interference distribution

Intuitively, for the same nominal power setting, the larger the path loss compensation factor is, the greater the user transmitting power is, then the greater the corresponding uplink interference for the adjacent cell is. It is not hard to find from Fig. 4.33 that in the scenario with shadow fading standard deviation $\sigma = 4$ dB, although P_0 is lower when $\alpha = 1$ than that when $\alpha = 0.5$ by 40 dB, the uplink interference when $\alpha = 1$ is still far greater than the uplink interference when $\alpha = 0.5$. On the other hand, from the perspective of interference distribution, when α increases, the distribution is more dispersed, and the variance distribution is bigger. In contrast, when α is smaller, the uplink interference distribution is more concentrated. For the scenario with shadow fading standard deviation $\sigma = 8$ dB, the changes of α also impact the distribution of the uplink interference. From the analysis conclusion in Fig. 4.32, the change of P_0 does not affect the shape of the uplink interference PDF, but only affects the translation of distribution curve on the horizontal axis (interference). Therefore, the increase in path loss compensation factors will not only introduce more uplink interference, but also lead to greater volatility in the uplink interference.

Case 4: Local Mobile Cloud Assisted Computation Offloading

Computation offloading technology, as an important application of mobile cloud computing, has drawn a lot of attention in recent years. Computation offloading technology offloads the user's local computational tasks to the cloud with rich

resources and extends the computing power of the mobile terminal with limited
resources, so as to use the new computation-intensive applications. Hu et al. [50]
research the computation offloading in Heterogeneous Cloud Radio Access
Network(H-CRAN) scenario, and put forward User-Centric Local Mobile Cloud
(UC-LMC) based on the D2D communications. The basic idea of this framework is:
suppose that the users reach an agreement with the network, and based on some
kind of mechanism, the users can exchange for the traffic/service priority through
providing free computational resources so as to help network provide services to
other users. In the framework of H-CRAN, base station function modules in
baseband resource pool collect the appropriate computational resources, and build
and maintain its mobile local cloud for particular requesting users. Suppose that a
mobile application is divided into a series of subtasks. UC-LMC executes each
subtask in serial order; at the same time, it will send back the calculation results on
each auxiliary user to the requesting user equipment in the short distance communications link. For battery-powered mobile equipment, energy consumption is an
important factor needed to be considered in the process of offloading. The article
studies the subtask allocation problem of minimum energy consumption. Given that
different services require different network abilities for front haul network, the front
haul link loading limitation is added to the subtask allocation algorithm.

In order to simulate and evaluate the subtask allocation algorithm put forward in
[50], this section simulates with the simulation tools of Matlab. Simulation uses
small cell dense deployment model, and the simulation parameters are listed in
Table 4.14. In the simulation, 25 small cells are set up with each small cell serving
for 5 user equipment. Assuming that each small cell is square of 10 m × 10 m. RRH
is located in the center of the small cell. And the mobile user equipment is
uniformly distributed in the small cell.

As shown in Fig. 4.34, this research simulates the influence of different data
input rates on the average energy consumption. The figure compares four kinds of

Table 4.14 Simulation parameters

Small cell number	5*5
Number of mobile user equipment in each small cell	5
Simulation times	1000
Path loss model (dB)	PL $= \begin{cases} 30 * \log10\,(distance) + k * PL\_\text{wall} + 37 \\ k\text{—number of walls} \\ PL_\text{wall} \text{— penetrationloss} \end{cases}$
PL_ wall (dB)	0,5,10,15,20,25,30
λ	0.05
n	3
D (kbits)	$1 \sim 30$
ω	0.2, 0.5, 0.8,1
N	10

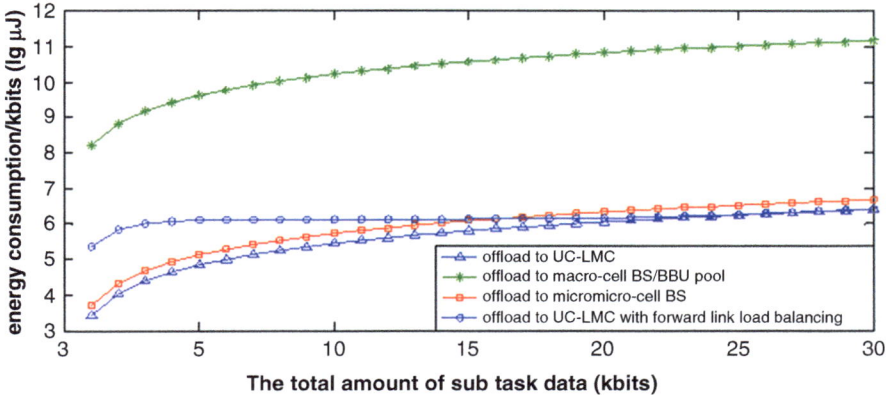

Fig. 4.34 Comparison of 4 types of computation offloading in average energy consumption

computation offloading strategies: (i) offloading to macro station (or through macro station to the network side server); (ii) offloading to a small cell station (or via a small cell station/RRH to be connected to network side server); (iii) offloading to local mobile cloud without considering the effect of the fronthaul network; (iv) offloading to local mobile cloud and considering load balancing of the prequel network in the meanwhile. It can be seen from the simulation results that the energy consumption of offloading the computational tasks to the macro station is the highest. It is because that the requesting user equipment is usually far from the macro station, its transmission energy attenuation is bigger on the air interface, and thus the energy efficiency is minimal. For the same reason, the energy consumption of offloading the computational tasks to the cell RRH is in the second place. In the strategy proposed in [50], computational tasks is offloaded to a local mobile cloud composed of the nearby auxiliary user equipment, which consumes the least energy. When considering load balancing in fronthaul network, since its main purpose is to scatter data, the performance of this strategy when the data rate is small is no better than the situation that offloads the computational task to the small base station or local mobile cloud. When the amount of data is large, channel differences between auxiliary user equipment cannot play a key role, so the fronthaul network load balancing algorithm is also able to reach a good performance.

The simulation result given by Fig. 4.35 is the impact of the set value of different fronthaul load balancing factors on mobile equipment's subtask alloca-tion. Four different values are selected: $\omega = 0.2, 0.5, 0.8, 1$. When putting the auxiliary user equipment in descending order according to the channel quality, we can see when a fronthaul load balancing factor is bigger, the allocated computa-tional data of auxiliary user equipment is mainly decided by the channel quality. The channel with better channel quality would be allocated to more data. When $\omega = 1$, i.e., not considering fronthaul network capacity, it will try to allocate the computational tasks to the auxiliary user equipment with good channel quality until it reaches the upper limit of its computational power, which will lead to large

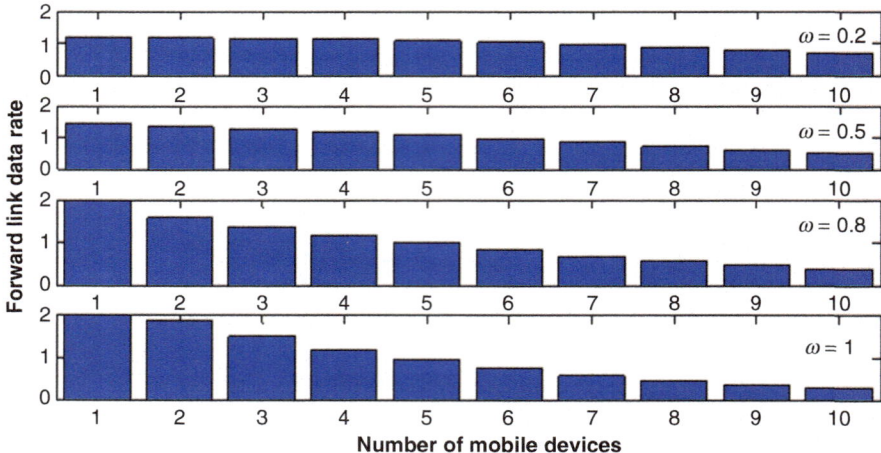

Fig. 4.35 The subtask allocation considering fronthaul network load balance

Table 4.15 The variance of energy consumption and fronthaul network load.	ω	0.2	0.5	0.8	1
	$E(\mu J/kB)$	43,400	29,429	21,537	20,547
	V		1	10	25	34

load variance, as shown in Table 4.15. When the fronthaul load balancing factor is small, the impact of channel quality will be weakened and the fronthaul capacity restriction factors will be highlighted. The user with relatively good channel quality, may be allocated with relatively less computational data than the auxiliary user equipment with poorer channel quality because the fronthaul link capacity of its correlated Remote Radio Head (RRH) is limited, but less data is assigned to the auxiliary user equipment with abundant fronthaul link capacity of its correlated RRH.

Case 5: Cross Operator Flexible Spectrum Management Simulation

ITU-R WP5D spectrum requirements report points out that the world frequency spectrum resource shortfall will reach thousands of MHZ in 2020. To meet the requirements for high traffic and high rate brought by the rapid development of intelligent terminals, it's necessary not only to develop LTE available spectrum (such as high frequency band and unauthorized frequency band), but also to continuously explore the efficient use of spectrum. The network service requirements of different operators are unevenly distributed in terms of time, space and frequency. The QoS requirements are diversified with great differences, and authorized spectrum of different operators has different center frequency, bandwidth, propagation characteristics, interference level and other characters. Cross-operator spectrum sharing can match the spectrum resources and services distribution well, which traditional spectrum planning models can't do. Besides, it enables the operators to use and manage the frequency spectrum more flexibly and dynamically. Yu et al. [51] propose a cross-operator interference coordination scheme based on peer-to-peer spectrum sharing in ultra-dense network. Each operator has the same priority for the resource in the shared spectrum pool. The shared spectrum pool is divided dynamically and flexibly through network statistical information. And asymmetric power allocation is used to coordinate between the Inter-Operator Interference (IOI). Finally, the network performance gain brought by cross-operator spectrum sharing is analyzed through numerical simulation. The simulation results show that, when the small cell is in dense deployment, the network spectrum efficiency of cross-operator spectrum sharing is promoted significantly compared with the scenario where there is no inter-operator sharing.

The simulation has set two scenarios. One is small cell dense deployment scenario. The other one is the small cell sparse deployment scenario. The density of small cell deployment is represented by the activation rate of small cell so as to verify the applicable scenarios of cross-operator spectrum sharing. Besides, several schemes are compared in the simulation performance.

- "Proposed" represents the scheme proposed in this paper. Spectrum is shared between operators. With the spectrum requirements obtained by network statistical information, the usable spectrum resources are dynamically assigned to the operators, and the inference between operators is coordinated through asymmetric power.
- "Baseline1" represents the first comparison scheme, that operators use their own spectrum rather than the spectrum sharing.
- "Baseline2" represents the second comparison scheme. Spectrum sharing is adopted between the operators, but the resources allocated to each operator are fixed. And the asymmetric power is used to coordinate the interference between the operators.
- "High power" represents that the spectrum sharing is adopted between the operators, but all the operators use the high power transmission. And the high power is the base station power in simulation parameters.
- "Middle power" represents that the spectrum sharing is adopted between the operators, but all the operators use the medium power for transmission. And the middle power in the simulation is lower than the high power by 3 dBm.
- "Low power" represents that the spectrum sharing is adopted between the operators, but all the operators use the low power for transmission. And the low power in the simulation is lower than the middle power by 3 dBm.

Specific simulation parameters are shown in Table 4.16.

Figure 4.36 is the simulation results of small cell in dense deployment scenarios; the activation rate of small cell is 1, i.e., all the rooms in the small cell are active. The line with triangle markers represents cross-operator spectrum sharing scheme using power control in this proposal. The line with dot markers represents the

Table 4.16 Simulation parameters

Parameter	Value
Deployment model	Dual-stripe model, single layer
Room size	10 m * 10 m
Activation rate	Dense 1; sparse 0.5
Operator number	2
The minimum distance between users and base stations	0.3 m
User distribution	One BS serve for one UE
Base station power	23 dBm
System bandwidth	80 MHz
Shared resource pool	8 CCs
Each carrier bandwidth	10 MHz
Received noise indicator	9 dB
Path loss model	TR 36.872
Seepage loss	Outer wall 23 dB; interior wall 5 dB
Shadow fading standard deviation	10 dB

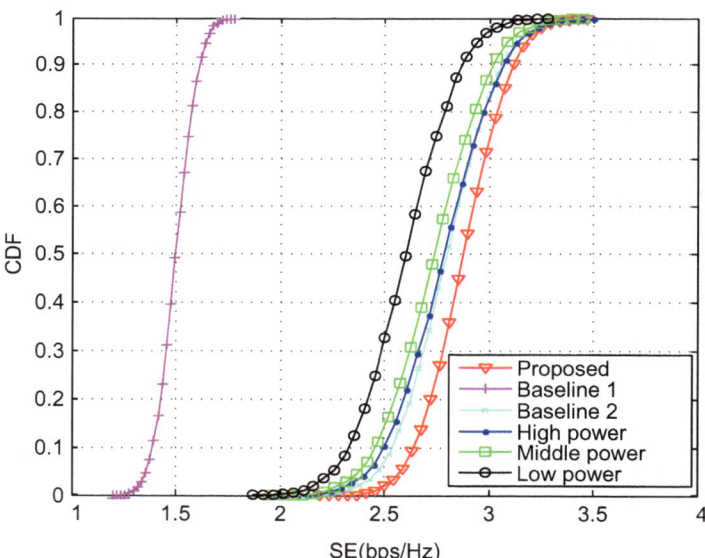

Fig. 4.36 The performance in dense deployment scenarios (small cell activation rate is 1)

spectrum sharing scheme without power control in this proposal. The line with short line markers represents the scheme without spectrum sharing between operators. It can be seen from the figure that, in a small cell dense deployment scenarios, the scheme with spectrum sharing between operators have a significant performance gain compared with the scheme without sharing. Meanwhile, with the use of power

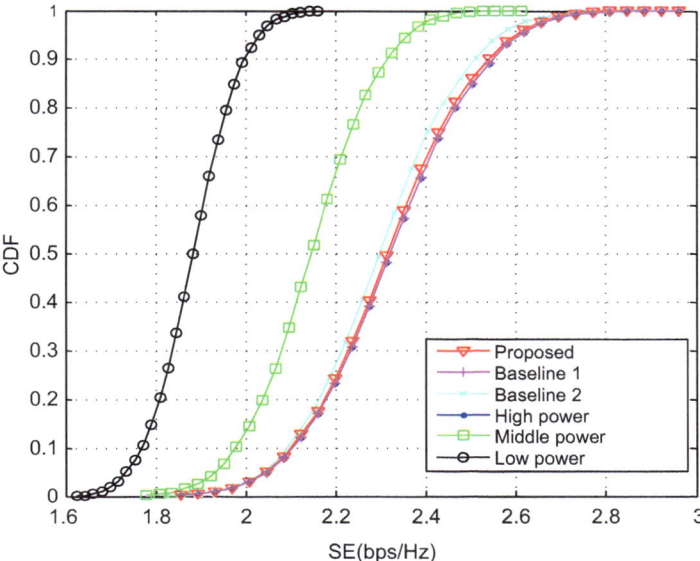

Fig. 4.37 Performance in sparse deployment scenario (small cell activation rate is 0.5)

control, interference between the operators is reduced, which makes the performance of the scheme with power control improved compared with the scheme with no power control (using the same power, such as high power, middle power and low power). The simulation shows that in small cell dense deployment scenario, the inter-operator spectrum sharing mechanism can bring certain performance gain to network performance. And the inter-operator interference can be eliminated by the rough information of operator interaction.

Figure 4.37 is the simulation results of small cell sparse deployment scenario. The activation rate of small cell is 0.5, i.e., only half of each operator's small cells in the rooms are active. In the small cell sparse deployment scenario, due to the little spectrum requirement of each station in the simulation, it can still meet all the spectrum access requirements of small cell even without cross-operator sharing scheme. As a result, the performance curves of the three schemes are basically overlapping. Meanwhile, the line with hollow circle markers and the line with hollow rectangle markers adopt lower power, which directly reduce the network spectrum efficiency. The simulation results show that in the sparse deployment scenario, there is no need for inter-operator spectrum sharing. Of course, due to the relatively low services traffic in the simulation, the intra-operator shortage of spectrum resources may also appear if the services traffic of each small cell is big. Thus the inter-operator spectrum sharing is needed.

4.5   Software Visualization of 5G Network Simulation

4.5.1   Architecture Summary

Here we introduce a kind of new simulation platform for 5G based on the universal software and hardware platform. The platform adopts the distributed master-slave parallel processing architecture, including the master node of the simulation platform, the slave node of the simulation platform, the client end, and the communications interface. As shown in Fig. 4.38, the functions of each part are as follows.

- The master computational node. It is the management center of the simulation platform, which manages the simulation task requirement from the clients, decomposes the simulation tasks, schedules the simulation task to multiple core computational resource of the master node or slave node for parallel processing, collects the simulation results of each slave node in real time, and then presents the summary results to the clients. Master nodes are also the computational nodes at the same time, which undertakes the specific simulation task. The master node also undertakes master-slave node synchronization management to ensure that the master-slave node simulation task executed synchronously. The multi-core parallel processing capabilities of master node and slave nodes have greatly increased the simulation calculation performance. The master node and slave nodes support both multi-core servers and hardware board/FPGA.
- The slave node. The slave nodes are managed by master node, take on computational tasks of the master node, and report the simulation results. A master

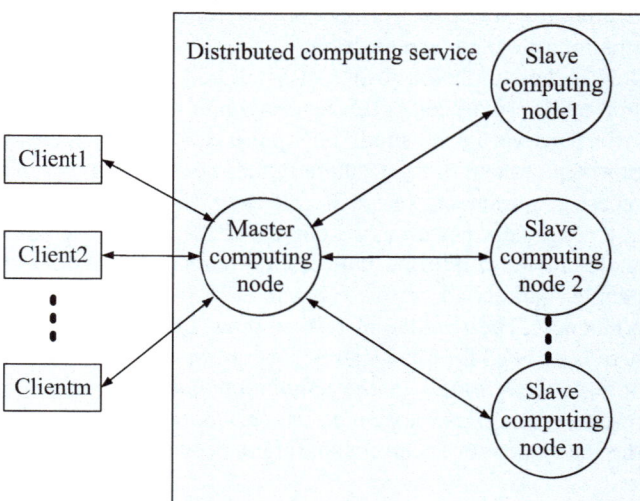

Fig. 4.38   The logical architecture of the new simulation architecture

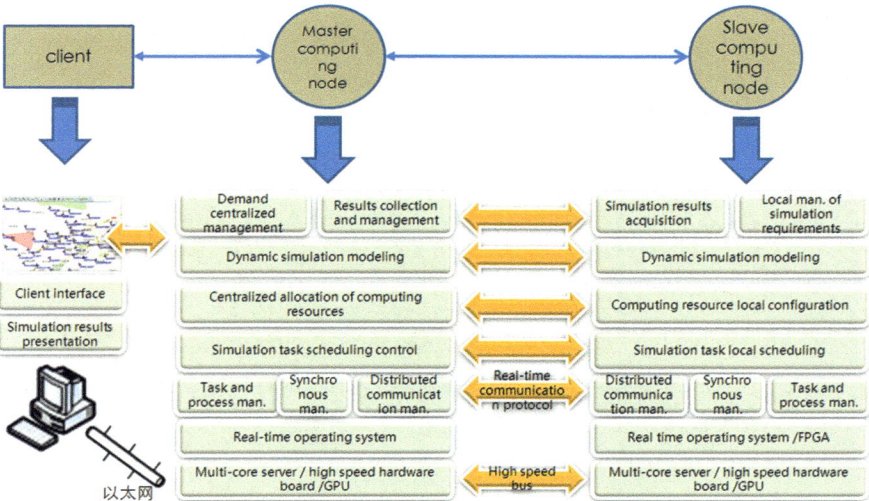

Fig. 4.39   A function deployment of the simulation platform

node can manage multiple slave nodes, so increasing the number of slave nodes can improve the computing capability of the simulation system.

- The client end. It provides human-machine interface, supports the starting up and issuing of simulation tasks and the real-time display of the simulation results on the interface.
- The communications interface. Considering the high-speed and real-time simulation services requirements, it requires the use of high speed bus and the communications protocol with higher real-time property.

Figure 4.39 gives a deployment configuration of master-slave node and the client end. Each software module of the simulation can be flexibly deployed on the master-slave nodes according to the requirements and doesn't have to be in a particular form.

The software functions of simulation system are organized according to Fig. 4.40.

4.5.2   Interface Demonstration

Demonstration of the Simulation Results of 3D MIMO Scenario

Simulation platform supports the simulation calculation of 5G key technology in massive MIMO technology and the dynamic display of the simulation. As shown below, the map shows massive MIMO area coverage effect after the user opens and executes massive MIMO project. QoS effect is shown through the comparison

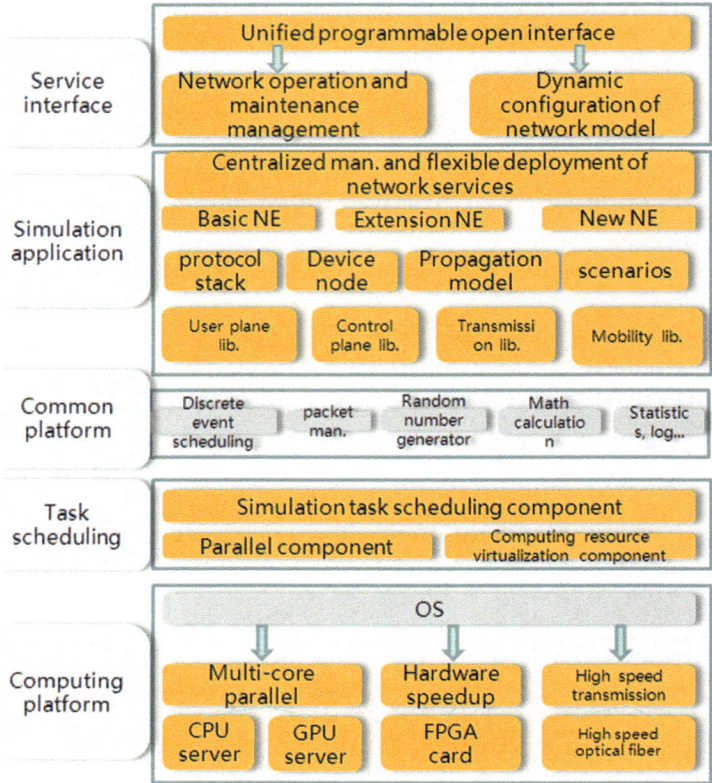

Fig. 4.40 Simulation system function hierarchy

between video display effect of massive MIMO cell and common cell, as shown in Fig. 4.41. The video on the left side is the video display effect of the users in the massive MIMO cell, where antenna configuration number of the base station is 32. The right side is the video display effect of the common cell, where the base station antenna configuration number is 8. The trend graph of the cell throughput rate in lower left shows the real-time changes in the cell throughput indicator of massive MIMO cell and common cell. It can be seen from the throughput and video that massive MIMO performance is superior to the common cells. By switching the number of massive MIMO antennas to 64, the cell beamforming ability is improved, which further enhances the performance as shown in Fig. 4.42.

Demonstration of the Simulation Results of UDN Scenario

The simulation platform supports the dynamic demonstration of the simulation calculation and simulation effects of 5G key technology and UDN technology. As

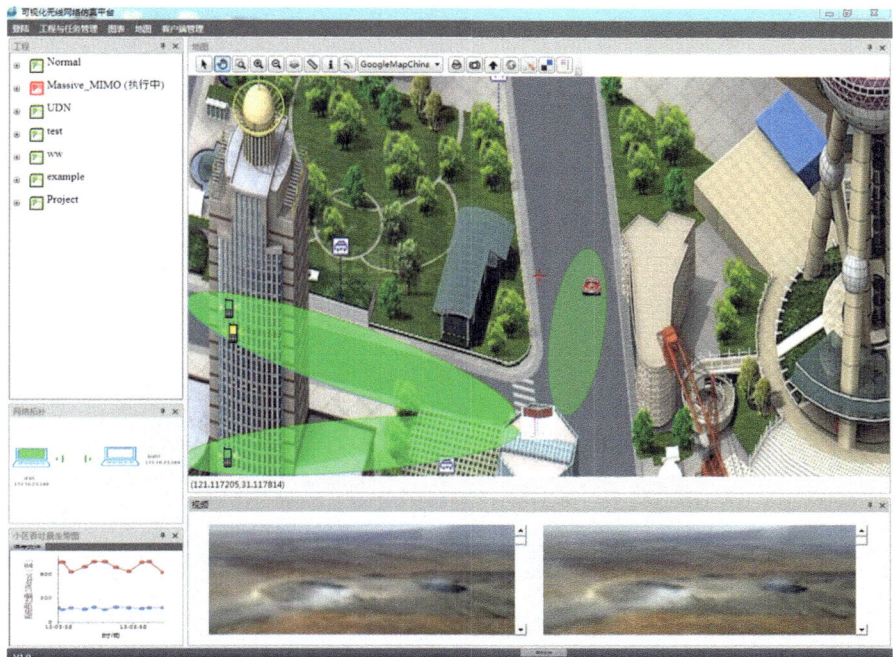

Fig. 4.41 The coverage effect of the massive MIMO cell with 32 base station transmitting antennas

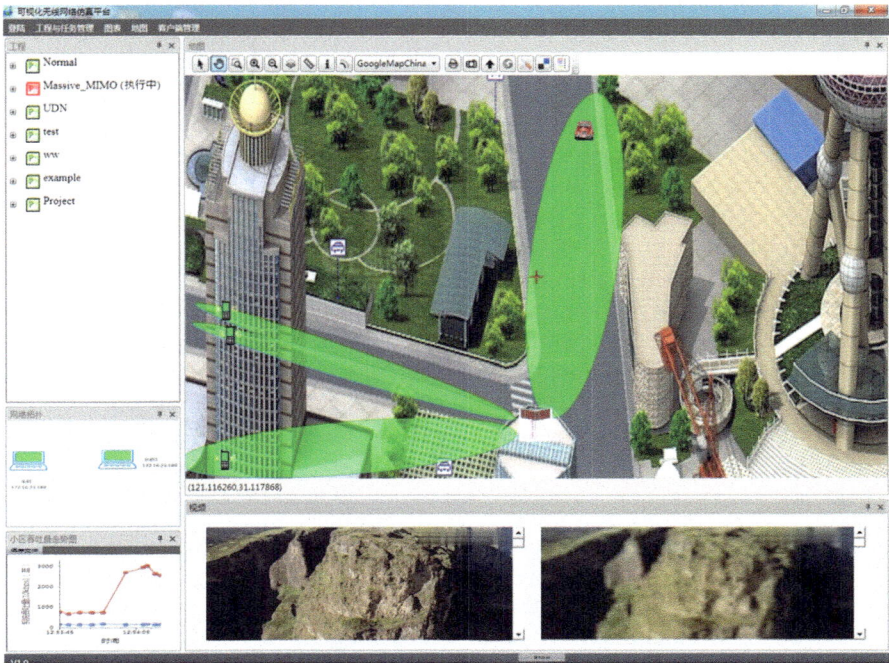

Fig. 4.42 The coverage effect of the massive MIMO cell with 64 base station transmitting antennas

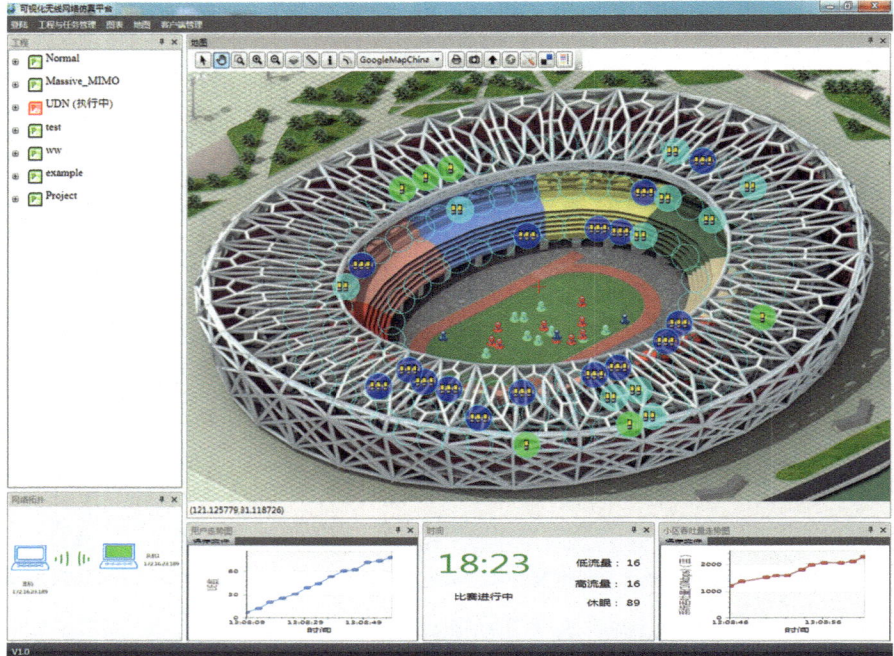

Fig. 4.43 UDN dynamic demonstration

shown in the figure below, the map shows the coverage effect of the dense network
with 121 cells after the user opens and executes UDN project. As shown in
Fig. 4.43, each circle represents a cell. The closed circle with no filled color
represents the cell is closed and services is not available. The cell with color
represent that it is open and can provide services. The deeper the color is, the
greater services volume can be provided. The change in the user number in the
whole simulation and the cell throughput trend graph are given below the main
interface. The simulation task has simulated the whole process of a football match
held in the Bird Nest Stadium, which has experienced UDN cell from the closed
state to the open state from before to during the game, and the changing process of
improving the throughput through the interference coordination algorithm, as well
as the process from the UDN cell's open state gradually to the closed state after the
match. It intuitively presented the simulation effect of UDN network in this
scenario.

Through analyzing the characteristics of the 5G network and the key technologies of system simulation software, we can see the future development trend of
software simulation test.

1. Rapidity. It can be predicted that within a period of time the parallel calculation
   based on multi-core and many-core will become the major technology means to
   improve the computing efficiency. Cloud computing and supercomputers will

gradually become the main stream calculation way of the software system simulation.

2. Flexibility. Influenced by factors like the flexible network architecture, network resource virtualization management, and parallel computational requirements, each module of the system simulation software needs full decoupling, flexible expansion. The requirements for the flexibility of software design become higher.

3. Comprehensiveness. Out of the pursuit for higher QoS, simulation evaluation indicators have expanded from the cell performance indicators to the user level QoS indicators. It requires that the simulation should be more comprehensive, be able to establish the services model closer to the real scenarios, design more refined and comprehensive statistical methods and indicators, and achieve the algorithm with more comprehensive performance.

References

1. B. R. Bilel, N. Navid, M. S. M. Bouksiaa. Hybrid CPU-GPU Distributed Framework for Large Scale Mobile Networks Simulation. IEEE/ACM International Symposium on Distributed Simulation and Real Time Applications. IEEE, 2012:44–53.
2. F. Capozzi, G. Piro, L. A. Grieco and et al. On accurate simulations of LTE femtocellsusing an open source simulator. EURASIP Journal on Wireless Communications and Networking, 2012 (1):328
3. K. Goga, O. Terzo, P. Ruiu and F. Xhafa. Simulation, Modeling and Performance Evaluation Tools for Cloud Applications. IEEE 2014 Eighth International Conference on Complex, Intelligent and Software Intensive Systems, 2014:226–232.
4. IEEE 802.16m-08/004r5: IEEE 802.16m Evaluation Methodology Document.
5. P. Kruchten. The 4+1 View Model of architecture. IEEE Software, 1995, 12(6):45–50.
6. Y. Zaki, T. Weerawardane, C. Görg and et al. Long Term Evolution (LTE) model development within OPNET simulation environment. OPNET workshop, 2011:1–5.
7. N. Baldo, M. Miozzo, M. Requena-Esteso and et al. An open source product-oriented LTE network simulator based on ns-3. Proceedings of the 14th ACM international conference on Modeling, analysis and simulation of wireless and mobile systems. ACM, 2011:293–298.
8. M. Suresh, C. Kamalanathan and S. Valarmathy. Avoiding congestion in IEEE 802.16 and IEEE 802.11b using Qualnet. 2013 Fifth International Conference on Advanced Computing (ICoAC), 2013:488–493.
9. White paper. 5G Vision and Requirements. IMT-2020 (5G) Promotion Group, China, May, 2014.
10. T. Xie, Q. Han, H. Xu, Z. Qi and W. Shen. A Low-Complexity Linear Precoding Scheme Based on SOR Method for Massive MIMO Systems, IEEE Vehicular Technology Conference (VTC Spring), 2015:1–5.
11. TR 36.201. Evolved Universal Terrestrial Radio Accesss (E-UTRA); LTE Physical layer; General Description, version 12.2.0, Mar. 2015.
12. C.-X. Wang, F. Haider, X. Gao and et al. Cellular architecture and key technologies for 5G wireless communication networks. IEEE Communications Magazine, 2014, 52(2):122–130.
13. Y. Yang, W. Zhang, K. Wei and X. Yang. Power Reduction for Mobile Devices by Low-power Base Stations Planning. IET Communications, 2014, 8(18):3372–3380.
14. S. Jin, X. Liang, K. Wong, X. Gao and Q. Zhu. Ergodic Rate Analysis for Multipair Massive MIMO Two-Way Relay Networks. IEEE Transactions on Wireless Communications, 2015, 14 (3):1480–1491.

15. D. Truhachev. Universal multiple access via spatially coupling data transmission. in Proc. of IEEE International Symposium on Information Theory Proceedings (ISIT), 2013:1884–1888.
16. X. Chu, D. Lopez-Perez, Y. Yang and F. Gunnarsson. Heterogeneous Cellular Networks: Theory, Simulation and Deployment. ISBN:9781107023093, Cambridge University Press, 2013.
17. S. Wu, C.-X. Wang, H. Haas, H. Aggoune, M. Alwakeel and B. Ai. A Non-stationary Wideband Channel Model for Massive MIMO Communication Systems. IEEE Transactions on Wireless Communications, 2015, 14(3):1434–1446.
18. R. Zhang, X. Cheng, Q. Yao, C-X. Wang, Y. Yang, and B. Jiao. Interference Graph Based Resource Sharing Schemes for Vehicular Networks. IEEE Transactions on Vehicular Technology, 2013, 62(8):4028–4039.
19. S. Sesia, I. Toufik and M. Baker. LTE-The UMTS Long Term Evolution; From Theory to Practice. ISBN:9780470697160, John Wiley & Sons, 2011.
20. TR 36.211. Evolved Universal Terrestrial Radio Accesss (E-UTRA); Physical channels and modulation, version 12.5.0, Mar. 2015.
21. TR 36.212. Evolved Universal Terrestrial Radio Accesss (E-UTRA); Multiplexing and channel coding, version12.4.0,Mar. 2015.
22. J. C. Ikuno, M. Wrulich and M. Rupp. System level simulation of LTE networks. IEEE Vehicular Technology Conference (VTC 2010-Spring), 2010, 11(18):1–5.
23. R. Q. Hu, Y. Qian, S. Kota, G. Giambene. Hetnets - a new paradigm for increasing cellular capacity and coverage. IEEE Wireless Communications, 2011, 18(3):8–9.
24. K. Wei, W. Zhang and Y. Yang. Optimal Microcell Deployment for Effective Mobile Device Power Saving in Heterogeneous Networks. Proceedings of IEEE International Conference on Communications (ICC 2014), 2014:4048–4053.
25. R. Shao, S. Lin, and M. Fossorier. Two Decoding Algorithms for Tailbiting Codes. IEEE Transactions on Communications, 2003, 51(10):1658–1665.
26. X. Wang, H. Qian, W. Xiang, J. Xu and H. Huang. An Efficient ML Decoder for Tail-biting Codes Based on Circular Trap Detection. IEEE Transactions on Communications, 2013, 61 (4):1212–1221.
27. X. Wang, H. Qian, J. Xu, Y. Yang. Trap detection based tail-biting convolution code decoding algorithm. Electronics and Information Technology Journal, 2011, 33(10):2300–2305.
28. H. Pai, Y. Han, T. Wu, P. Chen and S. Shieh. Low-complexity ML Decoding for Convolutional Tail-biting Codes. IEEE Communications Letters, 2008, 12(12):883–885.
29. 3GPP TS 36.212–2007; 3rd Generation Partnership Project; Technical pecification group radio access network; Evolved Universal Terrestrial Radio Access (EUTRA); Multiplexing and Channel Coding (Release 8), Sep. 2007.
30. E. J. Candes, J. Romberg and T. Tao. Robust uncertainty principles:Exact signal reconstruction from highly incomplete frequency information. IEEE Transactions on Information Theory, 2006, 52(2): 489–509.
31. C. Zhao, W. Zhang, Y. Yang and S. Yao. Treelet-Based Clustered Compressive Data Aggregation for Wireless Sensor Networks. IEEE Transactions on Vehicular Technology, 2015, 64 (9):4257–4267.
32. EPFL LUCE SensorScope WSN. http://lcav.epfl.ch/cms/lang/en/pid/86035.
33. G. Quer, R. Masiero, G. Pillonetto and et al. Sensing, Compression and Recovery forWireless Sensor Networks: Sparse Signal Modelling and Monitoring Framework Design. IEEE Transactions on Wireless Communications, 2012, 11(10): 3447–3461
34. N. Ahmed, T. Natarajan and K. R Rao. Discrete cosine transform. IEEE Transactions on Computers, 1974, c-23(1):90–93.
35. G. Quer, R. Masiero, G. Pillonetto and et al. Sensing, Compression and Recovery forWireless Sensor Networks:Sparse Signal Modelling and Monitoring Framework Design. IEEE Transactions on Wireless Communications, 2012, 11(10): 3447–3461.
36. T. Zhou, B. Xu, T. Xu, H. Hu and L. Xiong. User-specific link adaptation scheme for device-to-device network coding multicast. IET Communications, 2015, 9(3):367–374.

37. TS 36.212: Multiplexing and channel coding.
38. SDN architectureIssue 1, OPEN NETWORKING FOUNDATION, June, 2014
39. R. Kokku, R. Mahindra, H. Zhang and S. Rangarajan. NVS:A Substrate for Virtualizing Wireless Resourcesin Cellular Networks. IEEE/ACM Transactions on Networking, 2012, 20 (5):1333–1346.
40. H. Ishii, Y. Kishiyama and H. Takahashi. A Novel Architecture for LTE-BC-plane/U-plane Split and Phantom Cell Concept. IEEE Globecom Workshops, 2012, 48(11):624–630.
41. Y. Zhang and Y. J. Zhang. User-Centric Virtual Cell Design forCloud Radio Access Networks. IEEE International Workshop on Signal Processing Advances in Wireless Communications (SPAWC), 2014:249–253.
42. G. Piro, L. A. Grieco, G. Boggia, F. Capozzi and P. Camarda. Simulating LTE Cellular Systems: an Open-Source Framework. IEEE Transactions on Vehicular Technology, 2011, 60(2):498–513.
43. J. Dongarra,I. Foster,G. Fox,W. Gropp and et al. Sourcebook of parallel computing. Computer Science Computational Sciences, 2003.
44. H. Wang, K. Li, Y. Ouyang. Shanghai Research Center for Wireless Communications. A Parallel Computing Performance Test System and Method for Testing of Communications. Application No. 20150508961.8, Application date: August 18, 2015
45. Z. Xie, Z. Hao, Q. Tao. ZTE Corporation CN101924751B, Single-Mode Service Continuity Implementation Method and Single-Mode Service Continuity System, Patent publication (announcement) date 2010–03-12.
46. H. Liu, Y. Yang, X. Yang and Z. Zhang. Semi-distributed Resource Allocation for Dense Small Cell Networks," IEICE Transactions on Fundamentals of Electronics, Communications and Computer Sciences, 2015, E98.A(5):1140–1143.
47. K. Son, S. Lee, Y. Yi and et al. REFIM: a Practical Interference Management in Heterogeneous Wireless Access Networks. IEEE Journal on Selected Areas in Communications, 2011 (29):1260–1272.
48. A. Abdelnasser, E. Hossain and D. Kim. Clustering and Resource Allocation for Dense Femtocells in a Two-Tier Cellular OFDMA Network. IEEE Transactions on Wireless Communications, 2014(13):1628–1641.
49. Y. Zhu, J. Xu, Z. Hu, J. Wang and Y. Yang. Distribution of Uplink Inter-Cell Interference in OFDMA Networks with Power Control. Proceedings of IEEE Conference on Communications (IEEE ICC 2014), 2014:5729–5734.
50. H. Hu and R. Wang. User-centric Local Mobile Cloud-Assisted D2D Communications in Heterogeneous Cloud-RANs. IEEE Wireless Communications, 2015, 22(3):59–65.
51. Q. Yu, J. Wang, X. Yang, Y. Zhu, Y. Teng and H. Kari. Inter-operator Interference Coordination for Co-primary Spectrum Sharing in UDN. China Communications, 2015, 12 (s1):104–112.

Chapter 5
Evaluation Test of Software and Hardware Co-simulation

Software simulation has high scalability and flexibility. However, its authenticity and efficiency have a certain gap to the hardware simulation. This chapter will introduce the technique of co-simulation evaluation on software and hardware, so as to make up for weakness of software simulation in authenticity and efficiency. As a systematic statement, firstly, we will introduce the requirements, forms and applications of the co-simulation evaluation test with software and hardware. Hardware-In-the-Loop (HIL) simulation evaluation is an important and effective method forthenew technology test evaluation of wireless communications. Next, we will elaborate the co-simulefation evaluation test of software and hardware forthelink-level and system-level simulations, and present the real implementations of system test cases. Finally, we summarize the co-simulation test evaluation of software and hardware.

5.1 Overview of Software and Hardware Co-Simulation Evaluation and Test

The concept of software and hardware co-simulation was proposed as early as the establishment of the Hardware Description Language (HDL) language. It offered a reliable software and hardware co-simulation technology basis through the interaction between the driver realized by the C language and various hardware devices. Valderramaet al. [1] proposeda unified model of software and hardware co-simulation, and the authors in [2–4] presented different hardware acceleration designs. The main thought of software and hardware co-simulation evaluation test is [5]: dividing a system into two main parts. One is realized by using software modules. The other is achieved by using hardware like actual equipments, devices, or actual channels. Through the integration of two parts, a minimum system which needs a technical evaluation will be achieved. It is impossible in the past that

© Springer International Publishing AG 2018
Y. Yang et al., *5G Wireless Systems*, Wireless Networks,
DOI 10.1007/978-3-319-61869-2_5

software algorithms or measured objects can be fully evaluated and tested in such a system.

As a technological means, software and hardware co-simulation evaluation and test reflect the idea of authenticity and rapidness, which is an effective testing and evaluating method to deal with the rapid development of 5G communications, dispersing error detection in the whole design process and reducing the error risks in the final product.

5.1.1   Requirements of 5G Software and Hardware Simulation Evaluation and Test

Traditionally, the design for wireless systems in mobile communications mainly uses software simulation to make algorithm verification and evaluation. However, there is a certain gap between the system environment under the software simulation and the real situation. They are mainly reflected in:

1. Authenticity

   First, the software simulation can not introduce the impact of hardware environment on communication systems. Software simulation believes that the hardware design can perfectly realize the software algorithm. But in fact, the algorithm design is often limited by the hardware conditions. And the software algorithm is often compromised in hardware implementations. For example, in the actual hardware environment, the algorithm with particularly huge computations and extremely high calculation accuracy requirements are often unable to obtain the optimal performance. The precisely designed algorithm isoften vulnerable to the outside interference. The algorithm at the front of transmitting and receiving hasto consider the power, the impact on other hardware components and many other factors. If the software simulation is only used for system verification and evaluation, it is relatively easy to neglect the above problems, which leads to a result that designed algorithms will only stay in the theoretical stage and are unable to meet the needs of the real situation. Second, the analog of the software simulation for the wireless channel has a certain gap with the real situation. The research purpose of 5G mobile communications is to realize the transmission of ubiquitous, high-quality and high data rate mobile multimedia. In high-rate data transmission, the mobile communication system is facing a very poor wireless channel environment. How to deal with the complex wireless channel has become a crucial problem in the design of mobile communications product. Especially in current software emulation conditions, the analog for wireless channel is realized by abstractly extracting it via a large number of mathematical methods, which often greatly simplifies the wireless channel. The difference from the real situations is still very obvious. Because the analog methods of the software simulation for the wireless channel are insufficient, it may result in that the designed algorithm is too simple and can not handle the real complex environment.

2. Rapidness

In general, the software simulation system is a complex computing system based on operating systems, which needs a non-strictly determined period to complete each operation. And the entire simulation is composed by tens of millions or even billions of tiny calculation units. Therefore, the complex simulation will not only take much time, but also have no definite period. For example, in the LTE system simulation, simulation operations of the analog for one Transmission Time Interval (TTI) include 798,474,240 times Fast Fourier Transform (FFT) calculations, 817,152 times matrix inversions, and 13,074,432 times matrix multiplications. If using software and hardware co-simulation and replacing the software calculation by the real channel environment and hardware emulation, we can greatly improve the efficiency of simulation, testing and evaluation.

The development of the modern technology and the market demand has caused the increasing reduction of product development cycles, while the needs for product testing and the integrity requirements are constantly growing. This requires the effective evaluation in the algorithm or code stage, which advances the original evaluation for the prototype to the technology research and the development stage, while the methods of software and hardware co-simulation evaluation test can effectively evaluate the result of technology implementation in advance in the design stage.

In the system of mobile communication verification and evaluation, it is significant to use software and hardware co-simulation test and evaluation to test the key technologies under the real environment. Users realize the designed algorithms quickly in the wireless link verification test platform based on the hardware system. Then they can test and evaluate in a real wireless link environment with many advantages. It can, provide a sufficient and effective support for universities and research institutes to develop new communication algorithms. It can alsoshortenthetechnology development cycle to a large extent and reducing R&D costs and risks of new technologies, algorithms and standard investment product. In addition, it canrealize the goal of effectively promoting mobile communication technology development and overcome the shortcomings of excessive codes and the simulation time brought by the pure software simulation.

5.1.2   Forms and Applications of Software and Hardware Co-Simulation Testing and Evaluation

Forms of Software and Hardware-in-the-Loop Simulation Testing and Evaluation

The forms of software and hardware co-simulation in this chapter are described based on bidirectional uplink and downlinkcommunication applications. The one-way communication can be used as a special case.

Figure 5.1 shows one form of software and hardware co-simulation with base station side as the test and evaluation object. It can be divided into two conduction test evaluations – direct connection and through the channel. In this application, the sample to be tested is base station physical hardware, and the test and evaluation device at the terminal side is the joint platform of hardware and software.

Figure 5.2 shows another form of software and hardware co-simulation with terminal side as the test and evaluation object. It can be divided into two conduction test evaluations – direct connection and through the channel. In this application, the test sample is terminal physical hardware, and test evaluation device measured by the base station is a joint platform of hardware and software.

Figure 5.3 does not specify the terminal side or the base station side as the test and evaluation object. Instead it adopts the form of joint software and hardware platforms on both sides, which extends the scope of test and evaluation to base station and terminal sides. In this application, the conduction test and evaluation can also be divided into two parts–through and not through the channel. This application form combines the base station with uplink and downlink terminals, as well as channels to form a closed loop by using the software and hardware combination. Thus, this kind of software and hardware co-simulation testing and evaluation is called HIL simulation testing and evaluation.

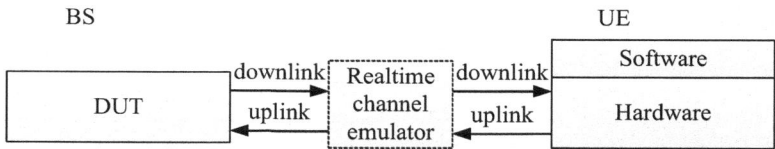

Fig. 5.1 Software and hardware co-simulation with base station side as the test and evaluation object

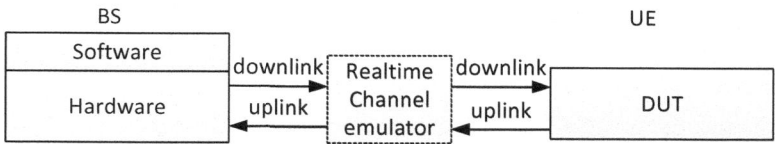

Fig. 5.2 Software and hardware co-simulation evaluation with terminal side as the test and evaluation object

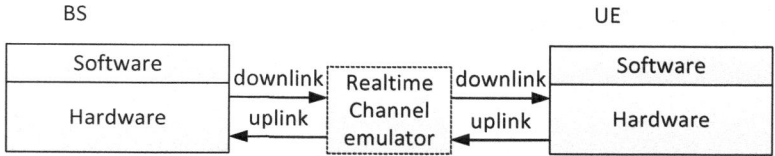

Fig. 5.3 No specification of terminal side or base station side as test and evaluation object

Superiority of HIL

HIL simulation refers that during the system test, the tested system uses a real control system while the rest parts use actual products if they can. If not, they will use a real-time digital model to simulate the external environment of the controller to test the entire system. In the HIL simulation, the actual controller and the simulation model used to replace the real environment or equipment together form a closed loop test system. The components which are difficult to establish the mathematical simulation model can be retained in closed loop, so that the test and initial matching work can be completed in the lab environment, which can greatly reduce development costs and shorten the development cycle [6].

HIL simulation test evaluation is a semi-physical simulation developed on the basis of the physical simulation and the digital simulation [7]. It is a typical semi-physical simulation method based on DUT and environment [8, 9], which realizes the function of a particular device or the external environment. In the HIL simulation test, the simulation model replaces the actual equipment or the environment. The model and real controllers constitute a closed loop test system through the interface. For the components which are difficult to establish the mathematical model (such as the inverter system), they can stay in the closed loop system. Thus, the test for the controller can be completed in the lab environment. Limit testing, fault testing and high-cost or the testing impossible in the real environment can also be carried out. The HIL simulation technology makes full use of computer's modeling convenience and simplicity and reduces the costs. It is easy to make fast and flexible changes to the system input, by which the changes of system performance can be simultaneously observed while changing the parameters. For the complex links of unessential investigations in systems, hardware can be connected directly with the simulation system. There is no need to make mathematical modeling for all details of the system [10, 11].

As a real-time simulation, since the HIL simulation technology can incorporate some physical objects into the simulation loop, it mainly has the following advantages. Real-time simulation takes the same time as the natural operation of the system. It increases the reliability of hardware. By using the HIL technology, the system function test can be carried out at the beginning of the design, which can effectively reduce errors and defects possibly existing in the process of development and design. It reduces test evaluation risks. In the process of using the simulation system environment to simulate the actual testing, the high-risk control functions, like security operations of verification systems, alarms and emergency treatment, effectively reduce the testing risks. It reduces testing costs. By using the simulation system environment to simulate the actual testing process, it can avoid the procurement configuration of various ancillary equipment in the early system design, reducing the system testing time and costs. It meets testing requirements of different application environments. Using flexible software configurations, different system environments can be simulated to meet specific test requirements.

In communication applications, the most significant advantage of HIL simulation is that it can verify the algorithm in the product R&D stage. And it can also simulate and test the hardware and software environment and work status of the future products. Meanwhile, it is able to simulate real situations of the whole equipment without producing real risks or high costs. For example, we can simulate various channel environments in the laboratory, rather than going to the realenvironment, and develop the multi-channel RF test environment more rapidly, rather than waiting for the appearance of MIMO communication products.

In addition, the HIL simulation can be used to test the communication equipment when the extreme conditions can not be achieved in a real world. Even in the strong interference environment, it can also test and simulate the ability of communication and anti-interference for communication products. HIL can help developers identify deficiencies in the equipment design, even if they only happen under certain circumstances. The HIL simulation can be used to take the output as a function of the past or current input for calculations. And the HIL simulation can also be used to test the hardware subsystem when the entire system is not well prepared, making the test become an effective component of a development process (from design to operation). The HIL simulation also helps developers make an early decision for the special design alternative scheme so as to guarantee its effective operation in the future application environment. The powerful high-fidelity real-time HIL simulation not only accelerates the time to market by shortening development cycles, but also reduces the equipment costs and the associated maintenance costs because there is no need for real hardware during the test.

5.2   Evaluation and Test of Software and Hardware Link-Level Co-Simulation

5.2.1   HIL-Based Link Simulation Composition

Currently, the relatively mature and rapid prototyping method to realize HIL function is to design and develop the system architecture based on the current Commercial Off The Shelf (COTS) module, the FPGA and the advanced multi-core microprocessor. By making use of its functions and the prototype development speed, we can accelerate the test process of the design and verification and can verify and test the new technology algorithms in the analog environment. Because the COTS system reduces the requirements for the specific customized hardware, we can get rid of difficulties of the specific customized hardware in design, maintenance and expansion.

As shown in Fig. 5.4, the HIL verification system in communication systemis similar to the real communication system. It can provide a full-duplex communication loop and usethesoftware simulation code platform to intervene the real-time hardware to simulate, verify and test the rest of system.

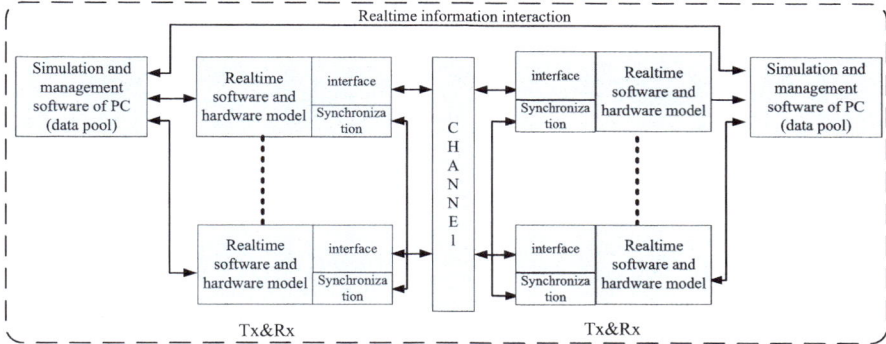

Fig. 5.4 Link simulation system configuration diagram of communication system hardware in Loop

The HIL simulation combines the physical simulation with the mathematical simulation. During simulation, it connects the computer with one part of the real system and builds a mathematical model in calculation to simulate those that are not easy to test or non-existing parts. This simulation model utilizes the computer modeling features, featuring simple modeling, low cost, easy parameters modification and flexible use. For the part that is difficult to establish the mathematical model in the system, the actual system or physical model is accessed. Thus, the operation of the entire system can be ensured, achieving the simulation of the overall system. The HIL simulation has a higher authenticity, so it is generally used to verify the validity and feasibility of the system scheme, and also to simulate the products' failure mode and have a dynamic simulation for the closed loop test of communications systems. The HIL simulation makes simulation conditions closer to the real situation. In pre-research, commissioning and testing of the product, it can reflect the performance of products more accurately and objectively.

The HIL test system uses real-time processor operation simulation models to simulate operating status of evaluated objects. It conducts the all-round and systematic test to the tested system by connecting to it via the I/O interface. Although there are HIL tests in different fields with different testing emphases, the overall simulation architecture is similar. Generally speaking, a HIL test system can be divided into three main parts, namely, the real-time processor, I/O interfaces and the operation interface.

(1) The real-time processor unit is the core of the HIL test system. Itis used to run the real-time model of tested and evaluated objects, board driver and communication exchange information of up and low position, such as hardware I/O communications, data records, stimulus generation and model execution. The real-time system is essential for non-existing physical parts in anaccurate simulation test system.
(2) The I/O interface is the analog, digital and bus signals interacting with the measured components and connecting the HIL-based communications. Signal

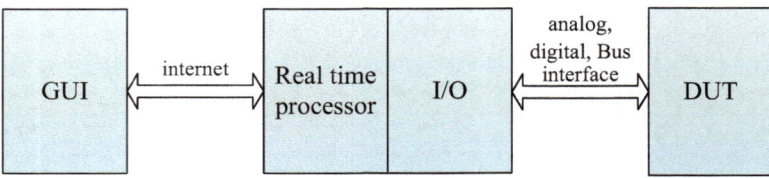

Fig. 5.5 Universal HIL test architecture

Fig. 5.6 HIL-based
Wireless link test
architecture

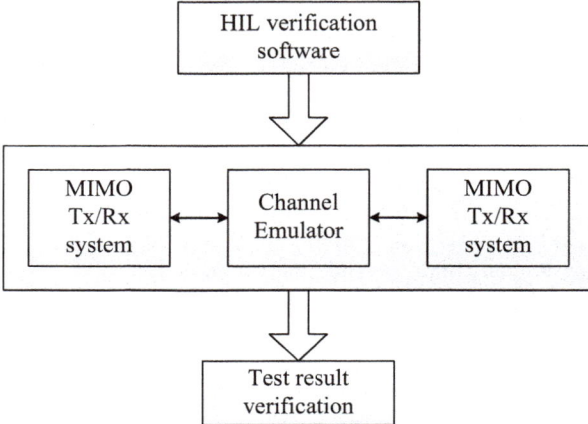

stimulus is produced through the I/O interface to obtain the data for recording
and analysis, and the sensor/actuator interaction is provided between the tested
Electronic Control Unit (ECU) and the virtual environment of the model
simulation. The operation interface can be provided for users to set parameters
and control the test platform. The HIL simulation is a powerful test method,
supporting the direct conversion from algorithms to RF signals (Fig. 5.5).

(3) The operation interface is similar to a real-time processor, which provides test
the command and visualization. It provides configuration managements, test
automation, analysis and reporting tasks. The HIL test system can simulate the
real system. It passes the hardware interface and generates physical signals to
the tested object. At the same time, the system will collect control signals from
output of the tested object and convert it into numbers to calculate, and then
combine the HIL system with the tested object to form a closed loop test
system. Figure 5.6 gives an HIL case architecture in the communication
domain. It mainly includes the signal transmitter, the signal receiver, channel
simulator between transmitter and receiver as well as the software module.
Through the real hardware, it can avoid defects of the pure software simulation
and accelerate the simulation speed.

(1) Multi-channel transmitting/receiving systems: achieving the function of the
signaltransmitter and the signal receiver, being used for testing data that the

testing system sends and receives. Transmitting module includes the embedded controller used for the remote access and control the data generation as well as the data transmission between modules, the arbitrary waveform generator used for generating modulation baseband data in the test, and the frequency converter used for up-convert a IF signal into a RF signal. Receiving module includes the embedded controller used to remote access and control data receiving and transmission between modules, the IF digital processor used to receive, synchronize and demodulates down-converted digital signals, and the frequency converter used to down-convert RF signal into IF signal or baseband signal.

(2) Channel simulator: the real channel transmission environment used to simulate different scenarios.

(3) HIL verification software module: including the HIL verification software, the test result verification module and other system software.

Users can use the operation interface in the HIL verification software to control the testing platform to make algorithm replacement test verification as well as data processing. Adopting the HIL mode can avoid virtuality and limitations of software simulation verification and increase the accuracy and validity of testing and verification. This new testing and verification method can meet requirements of complex algorithm RF in-the-loop verification in the 5G wireless communication.

5.2.2 Realization of Link Simulation Testing and Evaluation

Key Factors of Software and Hardware Simulation Testing and Evaluation

1. Hardware coordination simulation

   A software and hardware coordination simulation verification system which contains a hardware accelerator board generally includes three main parts [12]: system management tools, simulation software, and hardware platform used by hardware accelerators, that is, hardware accelerator boards. Its composition is shown in Fig. 5.7.

   System management software is responsible for coordinating the job assignment of software and hardware as well as completing the operation control of the

Fig. 5.7 Typical composition of software and hardware coordination simulation system

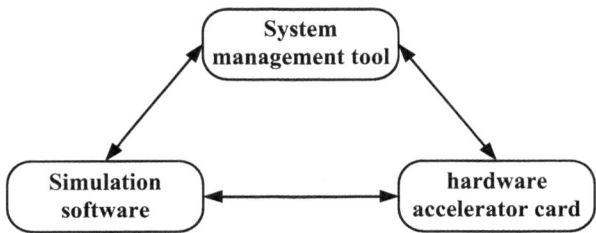

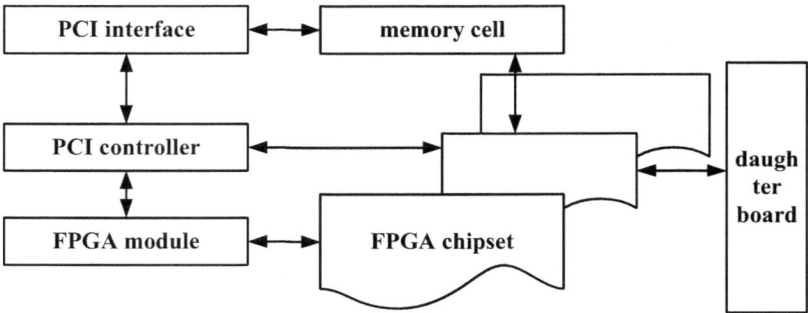

Fig. 5.8 Structureofhardware acceleration board

hardware acceleration board. Thehardware accelerator board completes the algorithm simulation verification of the specified pending acceleration module. Simulation software completes the algorithm in software, and different simulation software can be used according to different processor platforms. Based on a universal processor, C/C ++, Simulink and LabVIEW can be chosen. And based on a dedicated processor, ModelSim, VCS and NC-Verilog will be suitable. Their role is to produce the compiling data files needed by the hardware acceleration board, and to complete simulation of each module in addition to acceleration the simulation module in the hardware acceleration board.

2. Hardware acceleration board

A major part of the software and hardware coordination simulation system is the development of the hardware acceleration board. This is also the heterogeneous section in software simulation. In general, choosing PCI/PCI-E as the connection bus is a conventional manner more common and easier to achieve. The typical structure of the hardware acceleration board based on this mode is shown in Fig. 5.8.

The hardware accelerator generally comprises at least two FPGA chips, respectively for user programming and downloading and for system control. The pending acceleration module is configured in the main FPGA chip in a hardware acceleration board, while the chip which controls the FPGA chip is to control dynamic compilation of main FPGA chip. Main FPGA chip is connected with the PCI or PCI-E interface through the PCI controller to achieve data communications with the host. The hardware acceleration board also typically contains a storage module, such as Read Only Memory (ROM) and Ramdom Access Memory (RAM). Therefore, the simulation model is no longer needed as in traditional HDL simulators, which greatly improves verification ability. In order to realize the simulation of large-scale circuits on the hardware acceleration board, general hardware acceleration boards all contain sub-board connection interfaces. The sub-board can be an FPGA or a storage unit. It also can be Advanced RISC Machines (ARM) or other micro-processing chip. If necessary, adding the sub-board can be a way to increase simulation capabilities.

3. Software and hardware data exchange

Software and hardware coordination simulation verification uses the real-time hardware acceleration and the software emulation platform simultaneously. Therefore, it is a key requirement for development on how to realize communications and data exchange between the hardware acceleration board and the software platform. If the PCI/PCIe interface is used in design, we can design to use DMA or FIFO for data exchange.

The general operation principle and data exchange process of hardware acceleration verification can be expressed asfollows.

The hardware acceleration verification system first uses the system management tools to transmit HDL code compilation files to the driver of the hardware acceleration board through interface tasks, such as Programming Language Interface (PLI), FLI and VHPI, etc. Then the driver will convert these data into a specific format that the hardware acceleration board can receive, so the pending accelerated verification module will be configured to the main FPGA chip in the hardware acceleration board. Thus according to simulation requirements set in system management tools, the accelerated simulation will be realized. After the acceleration simulation, similar to the above-mentioned process, the hardware acceleration board driver reads acceleration simulation results, then transmits it to the software simulation platform through interface task for the follow-up software simulation, observation and debugging.

Since the simulation task that was originally finished in the software platform is downloaded in the real hardware, such hardware-acceleration verification system can speed up the simulation to a large extent and improve the simulation efficiency. Especially for the large-scale MIMO link simulation or multi-user simulation, acceleration effect will be particularly significant.

4. Unified development tools of software and hardware

Compared with the traditional software simulation platform, software and hardware coordinated simulation verification incorporates the real-time hardware. So it brings new development tasks and challenges todevelopers of the traditional software simulation platform Therefore, in developing the software and hardware coordinated simulation verification environment, providing a unified development tool is a demand, and also a task and challenge.

The advanced languages, such as C language, have advantages over Verilog language in the process control, so C program can be used to generate test excitation and read the signal value. Taking Windows platform as an example, users write the interface function in C language and Verilog PLI, compile codes and generate dynamic link library, and then call these functions in TestBench written by Verilog language. Thus, designers use the Verilog PLI interface to create their own system call tasks and system functions. C language can be programmed to help simulate DUT to achieve the function that Verilog grammar can not make.

In current market, there are some commercial cross-platform software and hardware unified development tools, such as System or HandleC, which can help complete the development of the software and hardware coordination simulation verification system.

Main bottleneck for system implementation

Applicationswith HIL test systems in the communications domain have become popular. But it is not difficult to find that in designing and implementing the test scheme, we need to consider technical bottlenecks faced by the closed loop in the realizing HIL test. These bottlenecks are closely associated with the increase in wireless communications requirements. New wireless communication requirements are growing and new technical standards requirements continue to increase, such as the growth of bandwidth, increase of RF channels and In-phase/Quadrature (I/Q) baseband channels, as well as the development of the digital simulation technology and the data rate improvement of digital interface, which are all technical bottlenecks faced by system implementations.

Closed loop method based on HIL link simulation

Closed loop methods based on the HIL link simulation can be divided into the following categories:

1. Signal real-time feedback based on the TCP/IP technology.
2. Signal real-time feedback based on Datasocket.
3. Signal real-time feedback based on the shared variable engine.
4. Signal real-time feedback adopting other high-speed bus (Fig. 5.9).

In a HIL verification systemforcommunicationsbased on the software defined radio platform, the design of feedback signal implementation methods should consider the following factors: software and hardware architecture of transceiver; reduction of heterogeneous components of system software and hardware; using the most reasonable, practical and efficient way to achieve feedback signal while meeting the minimum system requirements. Considering all the factorsabove, signal real-time feedback based on the TCP/IP technology and the shared variable engine may be the best scheme.

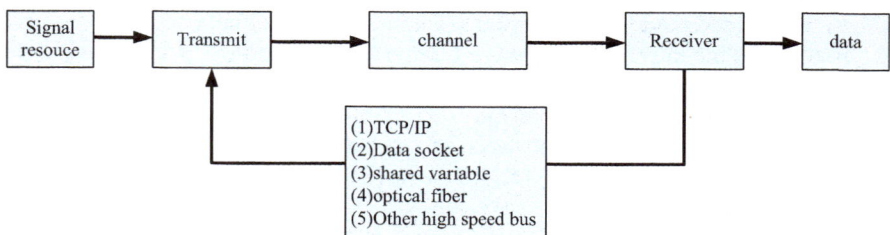

Fig. 5.9 Feedback signal implementation methods in HIL verification system of communications system

Signal real-time feedback based on the TCP/IP technology is realized by using the TCP/IP protocol. The characteristics of the TCP/IP mutual access and shared resources can be used to achieve signal's real-time feedback and variable sharing.

Signal real-time feedback based on the shared variable engine is to use shared variables to share data in different circulations of the same program block diagram or different software in the network. Three types of shared variables can be created: single-process, web publishing, and time-triggered shared variables, similar with UDP/TCP queue, real-time FIFO and other existing data sharing methods on the software platform. Wherein the network-released shared variables can be read and written on Ethernet. Network application processing is completed entirely by network-released variables. In addition to making data available on the network, the network-released shared variables also add many functions that single-process shared variables can not provide. Due to the need for a variety of additional functions, internal implementations of network-released shared variables are much more complex than the single-process shared variables. In single-process shared variables, all writing and reading operations share a single real-time FIFO. Yet for network-released shared variables, this is not the case. Every reading operation of network-released shared variables has its own real-time FIFO in a single element and multiple elements. Temporary fluctuations in a variable's reading/writing speed can be solved through the FIFO buffer in order to achieve signal real-time feedback based on the shared variable engine.

5.2.3 Case Introduction of Software and Hardware Simulation Evaluation Test

Case I: Test of Spectral Domain Channel Multiplexing Technology Based on Cyclic Delay Modulation

I. Tested technology principle and test requirements

1. Spectral domain communications technology principle

As a standard multiple-antenna diversity technology with good compatibility, Cyclic Delay Diversity (CDD) has greatly enhanced the existing standard OFDM technology to enable them to get enough space diversity gain in rich scattering wireless environment [13, 14]. Cycle delay processing can convert spatial diversity to frequency diversity, which adds redundancy in frequency domain of OFDM systems. Different from Space-Time Block Coding (STBC) [15] and Space-Time Trellis Codes (STTC) technology [16], the CDD technology can be realized only at transmitter, so the system which uses this enhancement technology maintains the compatibility for standards. Thus, CDD technology can be integrated into some existing broadcast standards (e.g., Digital Audio Broadcasting (DAB), Video Broadcasting (DVB) [17] and Digital Video Broadcasting Handheld (DVB-H) [18] and next generation of mobile communications (3GPP-LTE) [19], and also applicable to wireless Metropolitan Area Network (MAN) and LAN standards such as IEEE 802.11a [20] and HIPERLAN 2 [21]. However, the CDD technology can not provide Space Division Multiple Access (SDMA) functions and spatial multiplexing functions simultaneously.

Cyclostationarity of Cyclic Delay Diversity Orthogonal Frequency Division Multiplexing (CDD-OFDM) signals, in general, is caused by two different inherent cyclic processes, namely Cyclic Prefix (CP) and CDD process. Specifically, CP and CDD processes induce different, separate cyclostationary components separated from cyclic frequency and delay parameters on a 2D plane. In particular, the position and size of cyclostationary components induced by CDD may vary with changes of cycle delay parameters, and will be mutually linearly independent [22]. Cyclostationarity shown in the CDD-OFDM signal has been applied to the following two types of spectral domain communication system. They are spectral domain channel multiplexing transmission for singleusers based on CDD [23] and spectrum division multiple access for multi-users based on the cyclic delay channelization vector [24].

In a spectral domain channel multiplexing transmission system based on Cyclic Delay Modulation (CDM), the transmitting device of the system embeds the CDM module and the CDD-OFDM transmitter module. The CDM module is used to map the sub-information bit stream as a cyclic delay vector. The CDD-OFDM transmitter module is used to have a cyclic delay operation on the diversity OFDM symbol according to the described cyclic delay vector to achieve the multiplexing of spectral domain channel. The system embeds the CDM module in the existing CDD-OFDM standard system through the multi-antenna device and modulates the size and position of the cyclic autocorrelation function of CDD-OFDM signals. So it achieves the multiplexing of OFDM modulation channel and CDM channel and solves the problems of multiplexing spectral domain channel in multi-carrier frequency domain channel. While the system gains a CDD, it does not consume extra power and bandwidth and increases the transmission data rate of the system.

2. Implementation of the spectral domain channel multiplexing technology based on CDM

In a spectrum division multiple access system based on the cyclic delay channelization vector, the transmitting device of the system comprises the first spectrum division multiple access scheduling entity based on a cyclic delay channelization vector, and the first spectral division multiple access physical layer entity based on at least one of cyclic delay channelization vectors. Its first spectrum division multiple access scheduling entity based on the cyclic delay channelization vector comprises the first dispensing unit of cyclic delay channelization vector and the first adaptive modulation unit based on spectral division multiple access channel. The first spectral division multiple access physical layer entity based on at least one of cyclic delay channelization vectors comprises a cache unit connected in sequence, a coding and data rate matching unit, an orthogonal amplitude modulating unit, and a spectrum division multiple access processing unit. By using the multiple-antenna transmitting device, the system makes cyclic autocorrelation functions of CDD-OFDM signals become linearly independent, so as to solve the problems of ASU multiple access that the CDD-OFDM technology can not achieve.

(1) Implementation of CDD-OFDM transmitting scheme

The system block diagram of the transmitting device of a spectral domain channel multiplexing transmission system based on CDM is shown as Fig. 5.10.

As shown in Fig. 5.11, where the CDD-OFDM transmitting module generates module modification, adds a processing unit and realizes the generation of CDD-OFDM signals based on standard OFDM signals. The CDM module is used to control the signals generated by the cyclic delay diversity OFDM transmitter module. The block diagram of added a processing unit in the CDD-OFDM transmitter module is shown in Fig. 5.11.

Figure 5.12 contains a processing unit, a power normalization processing unit and a cyclic shift processing unit based on the cyclic delay vector. The processing unit is the component of the CDD-OFDM transmitter module, which is used to deal with the signal of CDM vector mapping unit. The cyclic shift processing unit based on the modulated cyclic delay vector is used to generate a shift signal.

(2) Implementation of CDD-OFDM receiving scheme

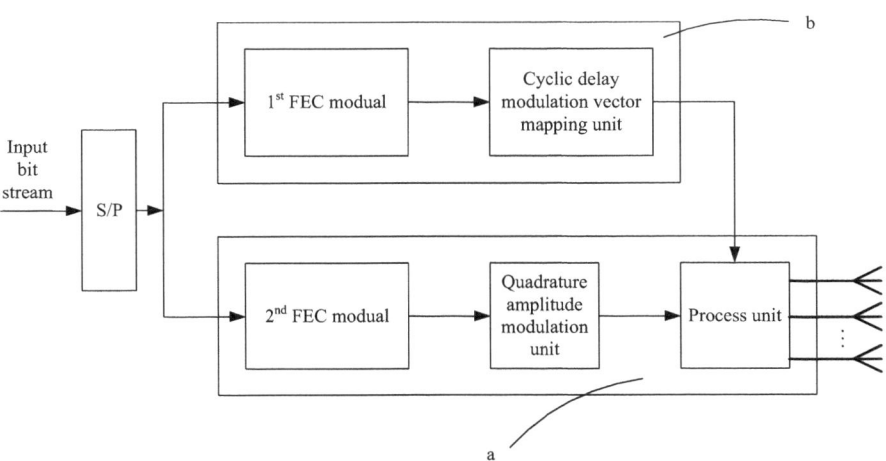

Fig. 5.10 System block diagram of transmitting device of CDD-OFDM system

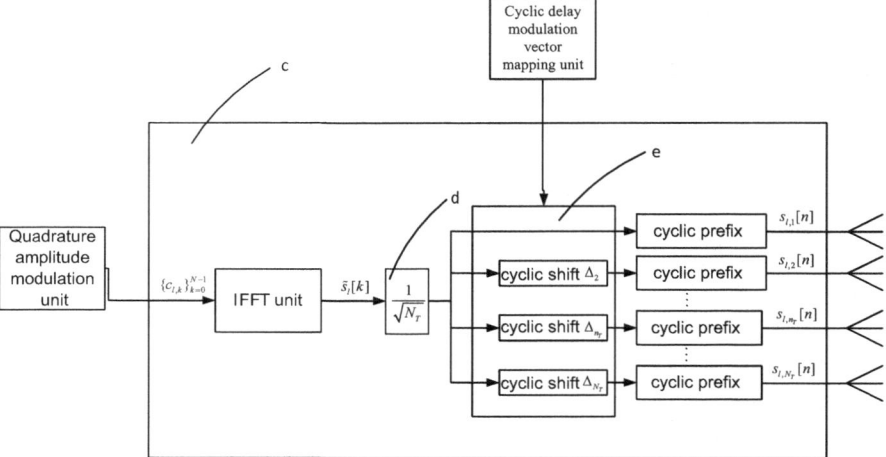

Fig. 5.11 Block diagram of processing unit in CDD-OFDM transmitter module

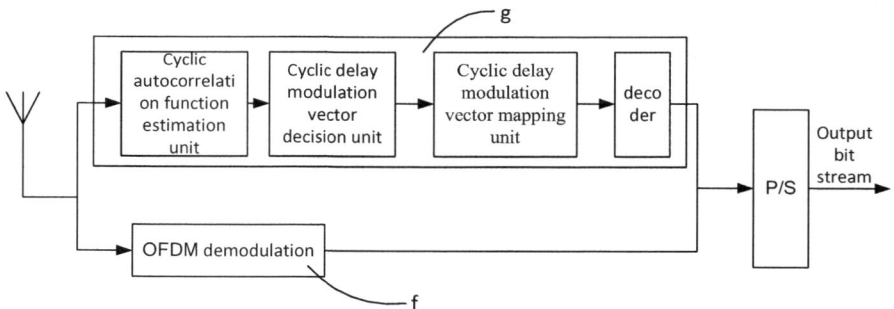

Fig. 5.12 System block diagram of transmitting device of CDD-OFDM system

The system block diagram of the transmitting device of a spectral domain channel multiplexing transmission system based on CDM is shown in Fig. 5.12.

The transmitting device of a CDD-OFDM system comprises an OFDM demodulation module and a CDM module, which are respectively used for the standard OFDM signal module demodulation and CDD-OFDM signal demodulation.

3. Testing requirements analysis

At present, the spectral domain channel multiplexing technology based on CDM is still in the theoretical research stage. The principle is usually realized by software simulation, which still has a certain gap with the actual hardware. For this, we need to use semi-physical simulation functions of the testing platform to complete the joint testing and verification of software and hardware of key technologies, providing a more objective reference for the final hardware implementation.

To complete the test verification of the CDD-OFDM technology, the test verification platform should meet the following requirements:

(1) The CDD-OFDM transmitter and receiver modules in line with the 3GPP LTE standard.
(2) Antenna configuration of transmitting and receiving devices support 2×1 and 2×2, and support 4×2 in the future.
(3) Meeting the requirements of configuring cyclic delay vectors in different OFDM symbol lengths.
(4) Able to call the baseband digital signals gained according to sampling periods in the receiving device. The minimum requirement of data storage is to support at least ten OFDM symbols once (including CP).
(5) The instrument platform can support digital signal processing modules extended by users in the transmitting and receiving devices (achieved by using MATLAB program).
(6) The instrument must support the transmission performance that two data streams can be separately analyzed.

II. Scheme design of software and hardware integration testing

1. Overview of scheme

In order to achieve the verification of the CDM spectral domain channel multiplexing technology, the semi-physical testing scheme based on SystemVue is used in this test case. On the one hand, SystemVue's powerful wireless link simulation functions can be used to achieve CCD-OFDM signal generation and reception analysis. Meanwhile the connection functions of SystemVue with instrument realize signal download and capture. On the other hand, the vector signal generator and vector signal analyzers are used to generate and receive real signals. The overall scheme is shown as in Fig. 5.13.

In order to meet the verification requirements of the CDM spectral domain channel multiplexing technology (CDD-OFDM), this scheme adopts software and hardware integration test program to achieve the HILstest. Considering that the uplink of an LTE system uses Single-Carrier Frequency-Division Multiple Access (SC-FDMA) modulation, the scheme selects a modified scheme based on the LTE downlink.

The test verification platform scheme of the CDM spectral domain channel multiplexing technology includes CDD-OFDM signal generation, CDD-OFDM signal reception and MIMO system test verification platform, which will be described respectively in the following.

(1) CDD-OFDM signal generation scheme based on SystemVue

(1) LTE OFDM signal generation Scheme

According to standard specification, the OFDM signal modulation scheme is used in 3GPP the LTE signal downlink [25]. This test

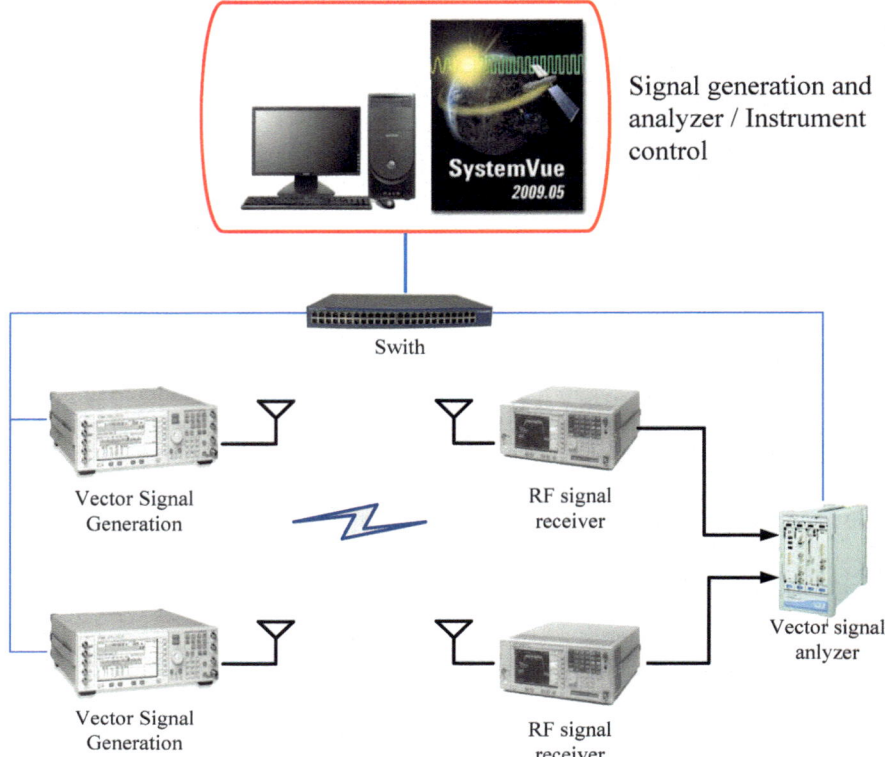

Fig. 5.13 System block diagram of semi-physical simulation testing of CDD-OFDM system

case uses SystemVue LTE OFDM signals to generate module version as shown in Table 5.1.

The schematic diagram of the 3GPP LTE downlink OFDM signal generation based on System Vue, is shown in Fig. 5.14.

(2) CDD-OFDM signal generation scheme

A CDD-OFDM signal generation scheme used in this test case is modified on the basis of the LTE downlink OFDM signal generation schematics as shown in Fig. 5.14, which mainly modifies the [LTE_DL_MIMO_2Ant_Src] module (red circle marked).The modified block diagram of CDD-OFDM generation module is shown in the blue marked circle in Fig. 5.15.

In the CDD-OFDM modulation signal generation scheme, users can set their own CDD shifting parameter. SystemVue will have a cyclic shift according to the value set by users, so as to generate CDD-OFDM signal required by users.

Table 5.1 LTE downlink signal (OFDM) generation module parameters

Name	Note		Remark
Software platform version	System Vue 2008.09		
LTE signal standards	In line with 3GPP LTE v8.6		
LTE OFDM signal settings	Duplex mode	FDD	Optional: TDD
	Signal bandwidth	5MHz	Optional: 1.4MHz, 3MHz, 10MHz, 15MHz, 20MHz, etc.
	Frame structure type	Type 1	Optional: Type 2
	Frame length	10ms	
	Sampling rate	15.36MHz	
	Oversampling	Ratio 2	

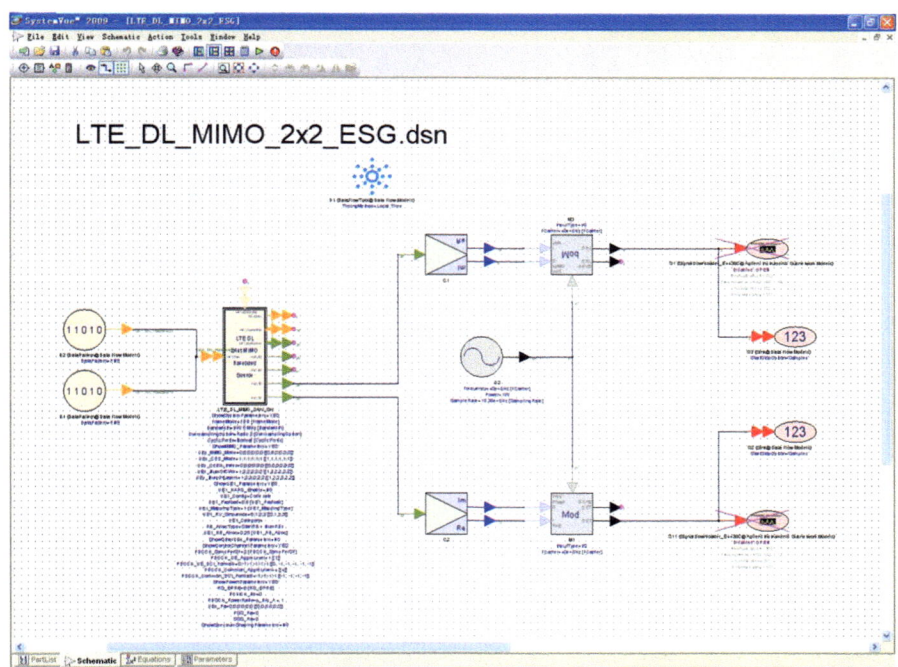

Fig. 5.14 Schematic diagram of LTE downlink OFDM signal generation based on SystemVue

(2) CDD-OFDM signal receiving scheme based on SystemVue

The CDD-OFDM signal generation scheme used in this test case is modified on the basis of 3GPP LTE downlink OFDM signal generating schematics, which mainly modifies [LTE_DL_MIMO_2Ant_Rev] module; block diagram of the modified CDD-OFDM reception module is shown in the blue marked circle in Fig. 5.16.

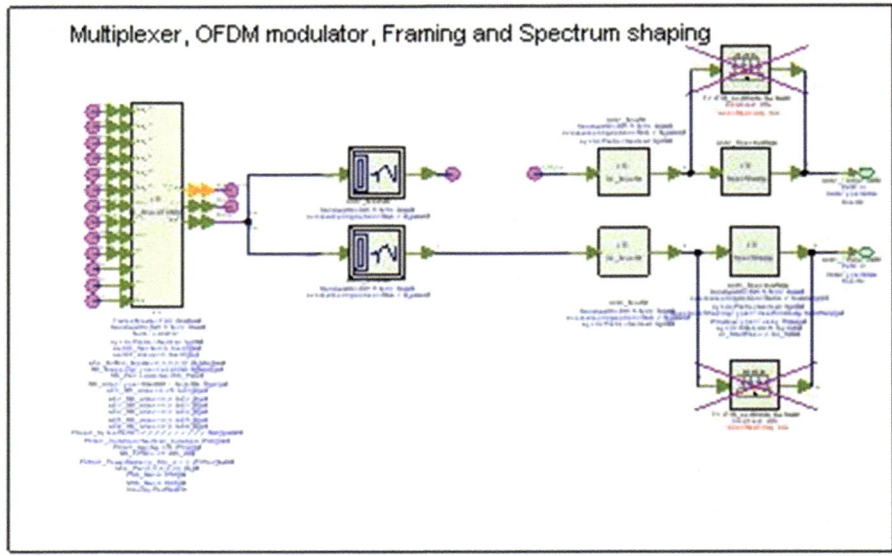

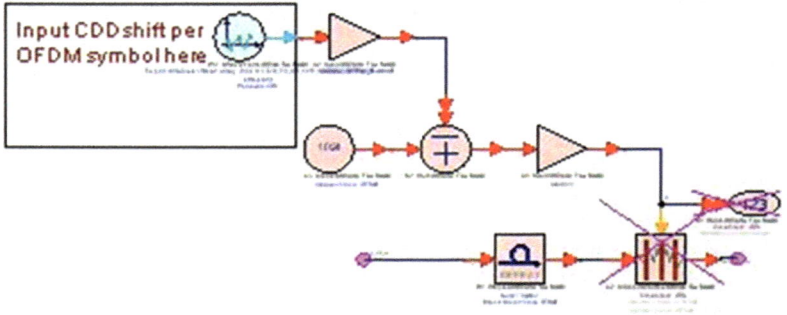

Fig. 5.15 CDD-OFDM signal generation schematics diagram based on SystemVue

In this CDD-OFDM modulation signal reception case, in order to simplify the receiver, we use the CDD shifted sequence which is assumed as known to demodulate the received CDD-OFDM signal, thus achieving CDD-OFDM signal reception analysis.

2. MIMO system experimental verification platform scheme

In order to meet the experimental verification needs, the physical map of the test verification platform that supports a 2×2 MIMO system is shown in Fig. 5.17. The test verification platform equipment consists of two ESG/MXG vector signal generators, two PSA/MXA spectrum analyzers (for down-converted to IF signal), and a one multi-channel digital oscilloscope. In addition, the synchronization wire, the RF connecting wire and the network wire and other materials are also needed.

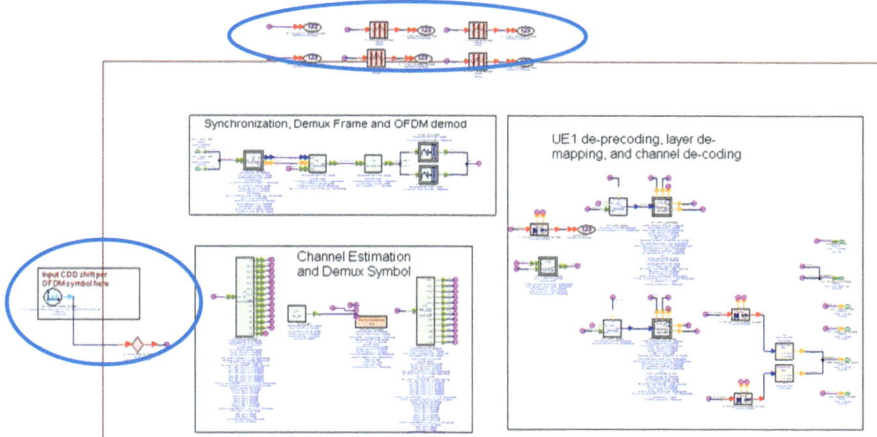

Fig. 5.16 CDD-OFDM signal reception schematic diagram based on SystemVue

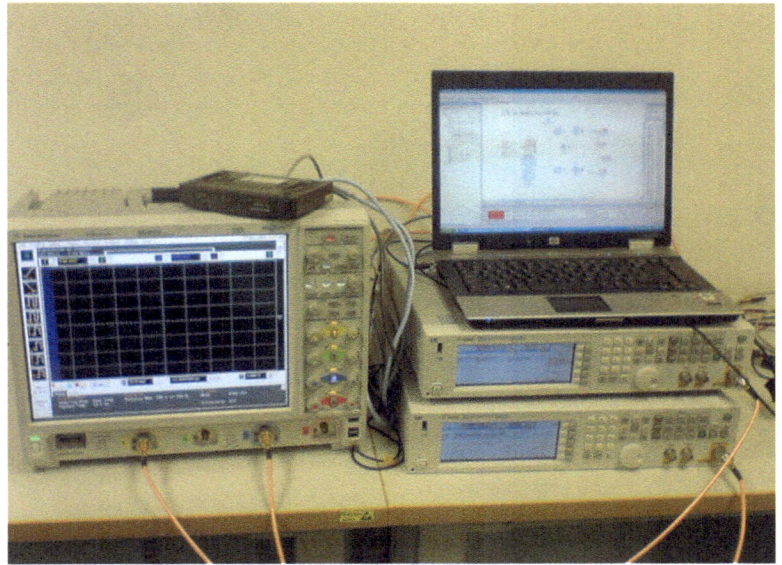

Fig. 5.17 MIMO experimental verification platform based on SystemVue

The working principle of this test platform is to use the CDD-OFDM signal generation module of SystemVue to generate user-defined waveform files, separately download them into two vector signal generators and transfer waveform files to RF modulation through the I/Q modulation function of the vector signal generatorOnthe other hand, SystemVue can send two-way time domain waveform data selected by the multi-channel digital oscilloscope to the SystemVue's CDD-OFDM signal reception module for analysis, and finally obtain test results. In this platform, the synchronization wire is used to guarantee the synchronized output of two vector signal generators, and to ensure the synchronized trigger of receiving and sending signals.

3. Results and analysis of experimental verification

(1) CDD-OFDM signal generation verification

Through the building of the experimental verification platform, we obtain the OFDM signal without CDD modulation and the OFDM signal with CDD modulation, which are respectively shown in Figs. 5.18 and 5.19. From comparison of the following two figures, we can see that CDD modulation causes shift to OFDM signals.

(1) Testing results of CDD-OFDM received signal constellation graph

The SystemVue simulation platform is used to build the CDD-OFDM system. Constellation of received signals can be obtained through simulation tests, as shown in Fig. 5.20.

(2) CDD-OFDM system BER simulation test results

The SystemVue simulation platform is used to build the CDD-OFDM system. By adding the white Gaussian noise in the system, BER and SNR relationship graphs can be obtained through the simulation test, as shown in Fig. 5.21.

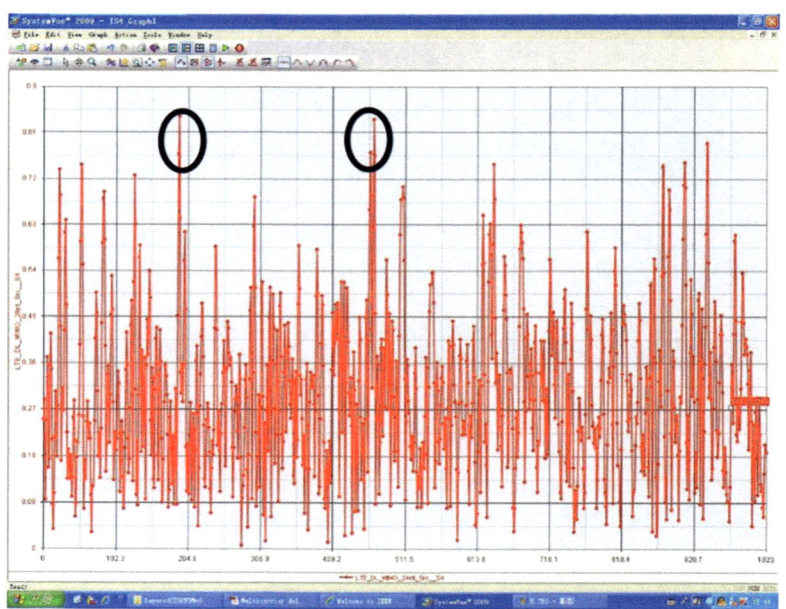

Fig. 5.18 Time domain graph of OFDM signal without CDD modulation

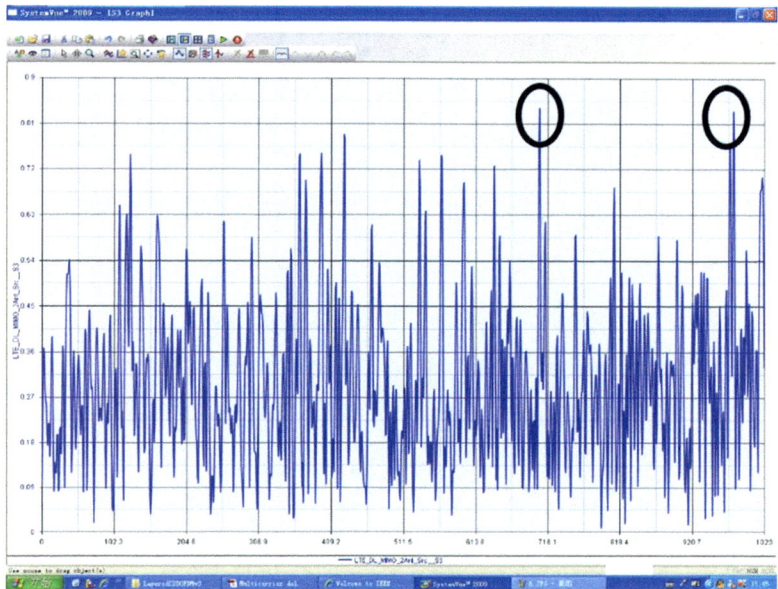

Fig. 5.19 Time domain graph of OFDM signal after CDD modulation

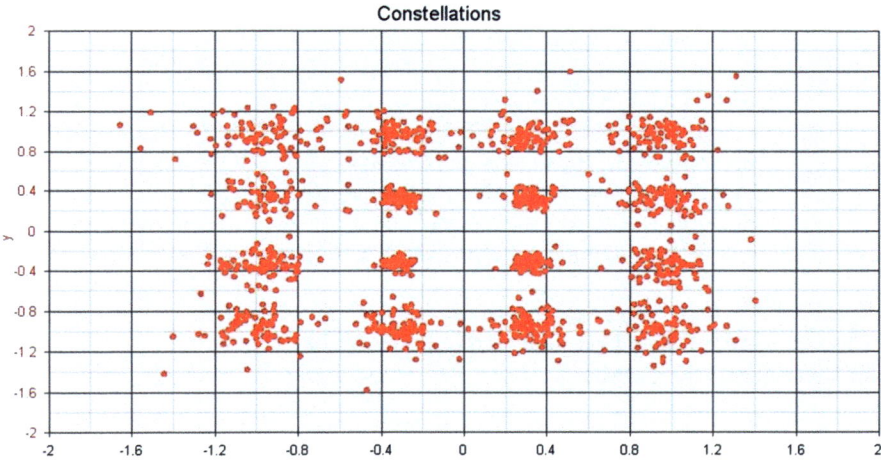

Fig. 5.20 Constellation graph of received signal from CDD-OFDM system simulation test (16QAM)

The scheme uses the software-hardware integrated test scheme to achieve the semi-physical objects simulation test, better solving the testing verification of the CDD-OFDM spectral domain multiplexing transmission technology in hardware systems. This test case uses the SystemVue's powerful simulation test function to modify the downlink

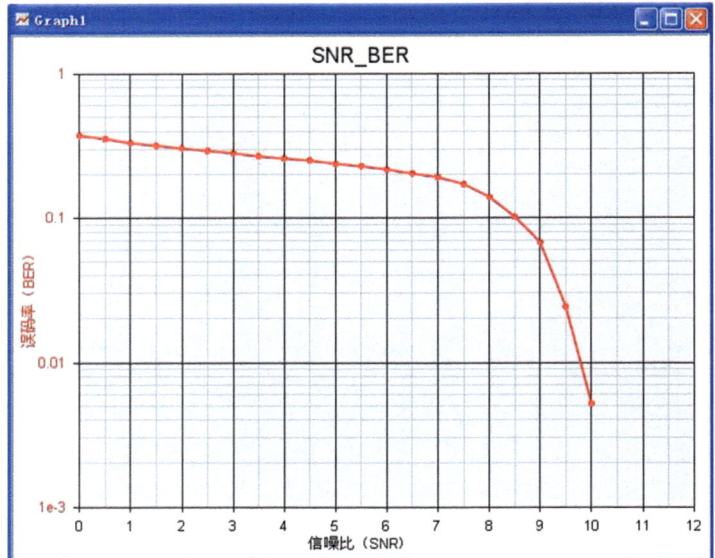

Fig. 5.21 CDD-OFDM system BER simulation test results

OFDM simulation link based on LTE MIMO configurations and quickly realize the CDD-OFDM signal generation and reception analysis by increasing the CDD software module.

On the other hand, the LTE MIMO test verification platform adopted in this scheme has good scalability, which supports the generation and reception of signals whose carrier frequency is up to 6GHz and bandwidth up to 20 MHz. It can be extended to 4×2 MIMO test platform, or even higher according to MIMO configurations. In addition, the scheme supports the addition of the channel simulator, so as to achieve key technical test verification under a typical MIMO wireless channel environment.

Case 2: Verficationof Frequency Offset Correction Algorithm in Centralized SC-FDMA

I. Overview

The LTE uplink multiple access mode is SC-FDMA, whereas the downlink mode is OFDMA. Both have higher spectral efficiency and use the frequency domain balancing technology with low complexity to suppress multipath fading. However, the SC-FDMA is a single carrier technology with even lowerpeak rate, so it is used as an uplink multiple access technology to reduce the costs of the mobile station. SC-FDMA has become an uplink transmission technology of the LTE physical layer. The technology has achieved great success in commercial and standardization activities. The reason is that SC-FDMA system has the following advantages:

(1) Good performance of resisting multipath fading.
(2) High spectrum utilization efficiency.
(3) Low implementation complexity.
(4) Flexible bandwidth configuration.
(5) High frequency diversity gain.

SC-FDMA modulation signal is in the time domain. In the frequency domain, each modulation can be extended to the entire frequency band. Therefore, we can take advantage of multipath channel frequency diversity gain to suppress narrowband interference.

(6) PAPR is lower than OFDMA system.

The SC-FDMA system is a single carrier system with lower PAPR. From another perspective, the SC-FDMA system is an OFDMA system extended from Discrete Fourier Transform (DFT). The partly offset compensation function of DFT transform and Inverse Discrete Fourier Transform (IDFT) transform of OFDMA leads to lower PAPR of SC-FDMA signals. Wherein the IFDMA system is a constant envelope system and PAPR is around 0 dB. PAPR of LFDMA signals is also lower than OFDMA signals.

In summary, compared with OFDM systems, SC-FDMA has the advantages like low PAPR, full utilization of frequency diversity of multipath channel, etc.

1. Tested algorithm

Assume that SC-FDMA uplink totally has N subcarriers and M users, with each user occupying $P = N/M$ subcarriers and all users communicating with base station through independent multipath channel, as shown in Fig. 5.22. Assume that Γ_i denotes subcarriers assigned to i, and then $\cup_{i=1}^{M} \Gamma_i = \{0, 1, \cdots, N-1\}$, $\Gamma_i \cap \Gamma_j = \varnothing$, for $i \neq j$. The subcarrier distribution scheme is as follows,

$$\Gamma_i = \{(i-1)P + r | r = 0, \cdots, P-1\} \tag{5.1}$$

Assume that the length of the cyclic prefix N_g is larger than the maximal channel delay length and the largest time shift among users.

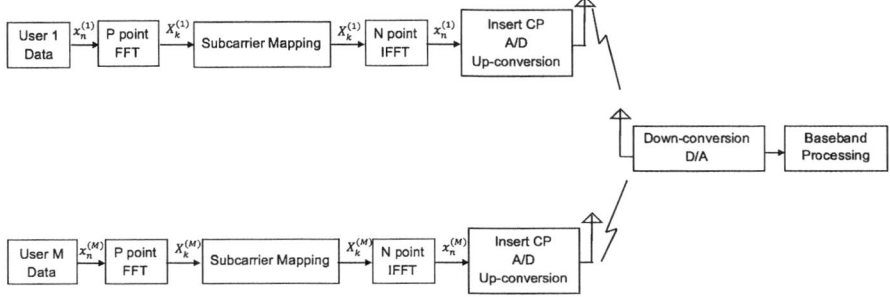

Fig. 5.22 SC-FDMA uplink model

After passing through multipath fading channel, the signal of user i is

$$y_n^{(i)} = x_n^{(i)} * h_n^{(i)} \tag{5.2}$$

where "*" presents the linear convolution. $x_n^{(i)}$ is the transmitted signal of the users i. $h_n^{(i)}$ is the unit impulse of channel between user i and base station. Only for $n = 0, \cdots, L-1, h_n^{(i)}$ is not zero, where L is the maximum channel delay spread (DS).

Taking into account of the influences of carrier frequency deviation and additive noise, the received signal of the base station can be denoted as

$$r_n = \sum_{i=1}^{M} y_n^{(i)} e^{\frac{j2\pi\varepsilon_i n}{N}} + z_n, \quad -N_g \le n \le N-1 \tag{5.3}$$

where ε_i, $i = 1, \cdots, M$ denotes user i's normalized carrier frequency offset value, and z_n denotes the additive white Gaussian noise.

(Eq. 5.3) can be rewritten into the following form:

$$\mathbf{R} = \frac{1}{\sqrt{N}} \sum_{i=1}^{M} \mathbf{Y}^{(i)} \otimes \mathbf{C}^{(i)} + \mathbf{Z} \tag{5.4}$$

where $\mathbf{R} = [R_0, R_1, \cdots, R_{N-1}]^T$, $\mathbf{Y}^{(i)} = \left[Y_0^{(i)}, Y_1^{(i)}, \cdots, Y_{N-1}^{(i)} \right]^T$,

$\mathbf{C}^{(i)} = \left[C_0^{(i)}, C_1^{(i)}, \cdots, C_{N-1}^{(i)} \right]^T$, $\mathbf{Z} = [Z_0, Z_1, \cdots, Z_{N-1}]^T$, $[\cdot]^T$ denotes the matrix transpose, $R_k, Y_k^{(i)}, C_k^{(i)}$ denote the FFT transform of $r_n, y_n^{(i)}, e^{\frac{j2\pi\varepsilon_i n}{N}}$.

The receiver detection algorithm is shown as below in Fig. 5.23.

To compensate for the carrier frequency deviation, before the FFT processing, the received sequence r_n is multiplied by the time domain sequence $e^{\frac{-j2\pi\varepsilon_0 n}{N}}$,

$$\hat{y}_n = r_n e^{\frac{-j2\pi\varepsilon_0 n}{N}}, \quad 0 \le n \le N-1$$

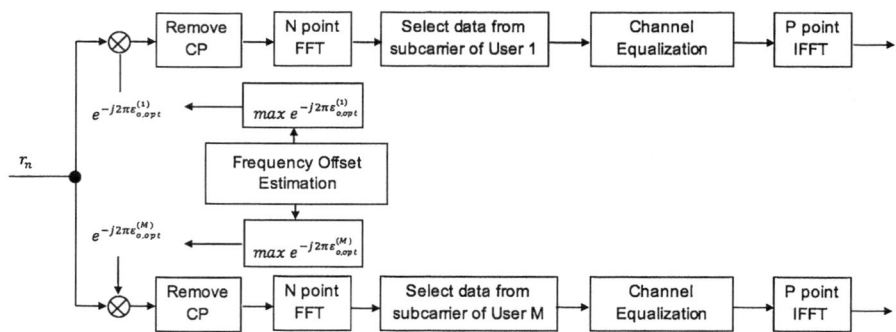

Fig. 5.23 Receiver test block diagram

The goal of algorithm is to find the optimal ε_o value, which maximizes the average Signal Interference Ratio (SIR), as shown in Fig. 5.23. It is assumed that carrier frequency deviation value has been accurately estimated.

After FFT for $\widehat{y}_n$ in expression,

$$\widehat{\mathbf{Y}} = \frac{1}{\sqrt{N}} \mathbf{R} \otimes \mathbf{C}'_o$$

$$= \frac{1}{N} \sum_{i=1}^{M} \mathbf{Y}^{(i)} \otimes \mathbf{C}^{(i)} \otimes \mathbf{C}'_o + \frac{1}{\sqrt{N}} \mathbf{Z} \otimes \mathbf{C}'_o$$

$$= \frac{1}{\sqrt{N}} \sum_{i=1}^{M} \mathbf{Y}^{(i)} \otimes \mathbf{D}_o^{(i)'} + \mathbf{Z}'$$

$$= \left(\mathbf{H}^{(m)} \mathbf{X}^{(m)} \right) \otimes \mathbf{D}_o^{(m)'} + \sum_{\substack{i=1 \\ i \neq m}}^{M} \left(\mathbf{H}^{(i)} \mathbf{X}^{(i)} \right) \otimes \mathbf{D}_o^{(i)'} + \mathbf{Z}'$$

where $\mathbf{C}'_o = [C_{o,0}, C_{o,1}, \cdots, C_{o,N-1}]^T$, $\mathbf{Y} = [Y_0, Y_1, \cdots, Y_{N-1}]^T$,
$\mathbf{D}_o^{(i)'} = \left[D_{o,0}^{(i)'}, D_{o,1}^{(i)'}, \cdots, D_{o,N-1}^{(i)'} \right]^T$, $\mathbf{X}^{(i)} = \left[X_0^{(i)}, X_1^{(i)}, \cdots, X_{N-1}^{(i)} \right]^T$,
$\mathbf{H}^{(i)} = \mathrm{diag}\left\{ H_0^{(i)}, H_1^{(i)}, \cdots, H_{N-1}^{(i)} \right\}$, $\widehat{\mathbf{Y}} = \left[\widehat{Y}_0, \widehat{Y}_1, \cdots, \widehat{Y}_{N-1} \right]^T$, $C_{o,k}$, Y_k,
$D_{o,k}^{(i)'}$, $X_k^{(i)}$, $H_k^{(i)}$, $\widehat{Y}_k$ are the FFT transform of $e^{\frac{-j2\pi\varepsilon_o n}{N}}$, y_n, $e^{\frac{j2\pi(\varepsilon_i - \varepsilon_o)n}{N}} x_n^{(i)}$, $h_n^{(i)}$, $\widehat{y}_n$.

In the expression, the first term includes the signal term of the k-th sub-carrier and Self Interference (SI); the second is MultiUser Interference (MUI); the third is additive noise. Received data on the k-th subcarrier is given by the expression.

$$\widehat{Y}_k = \frac{1}{\sqrt{N}} X_k^{(m)} H_k^{(m)} e^{\frac{j\pi(N-1)(\varepsilon_m - \varepsilon_o)}{N}} \frac{\sin \pi(\varepsilon_m - \varepsilon_o)}{\sin \dfrac{\pi(\varepsilon_m - \varepsilon_o)}{N}}$$

$$+ \frac{1}{\sqrt{N}} \sum_{\substack{q_m \in \Gamma_m \\ q_m \neq k}} X_{q_m}^{(m)} H_{q_m}^{(m)} e^{\frac{j\pi(N-1)(\varepsilon_m - \varepsilon_o + q_m - k)}{N}} \frac{\sin \pi(\varepsilon_m - \varepsilon_o + q_m - k)}{\sin \dfrac{\pi(\varepsilon_m - \varepsilon_o + q_m - k)}{N}}$$

$$+ \frac{1}{\sqrt{N}} \sum_{\substack{i=1 \\ i \neq m}}^{M} \sum_{q_i \in \Gamma_i} X_{q_i}^{(i)} H_{q_i}^{(i)} e^{\frac{j\pi(N-1)(\varepsilon_i - \varepsilon_o + q_i - k)}{N}} \frac{\sin \pi(\varepsilon_i - \varepsilon_o + q_i - k)}{\sin \dfrac{\pi(\varepsilon_i - \varepsilon_o + q_i - k)}{N}} + Noise$$

Assume $X_{q_i}^{(i)}, X_{q_m}^{(m)}, X_{q_m'}^{(m)}$ are not correlated, $E\left[X_{q_i}^{(i)} \right] = E\left[X_{q_m}^{(m)} \right] = 0$, where $q_i \in \Gamma_i, q_m, q_m' \in \Gamma_m, q_m' \neq q_m$, E[ ] represents the expectation operator. Set

$$E\left[\left|X_{q_i}^{(i)}\right|^2\right] = \sigma_X^2, E\left[\left|H_{q_i}^{(i)}\right|^2\right] = \eta_i,$$ and then the powers on received signal of

user m on the k-th sub-carrier, SI and MUI are as follows:

$$\sigma_{Signal}^2 = \frac{\eta_m \sigma_X^2}{N} \frac{\sin^2 \pi(\varepsilon_m - \varepsilon_o)}{\sin^2 \frac{\pi(\varepsilon_m - \varepsilon_o)}{N}}$$

$$\sigma_{SI,k}^2 = \frac{\eta_m \sigma_X^2}{N} \sum_{\substack{q_m \in \Gamma_m \\ q_m \neq k}} \frac{\sin^2 \pi(\varepsilon_m - \varepsilon_o + q_m - k)}{\sin^2 \frac{\pi(\varepsilon_m - \varepsilon_o + q_m - k)}{N}}$$

$$\sigma_{MUI,k}^2 = \frac{\sigma_X^2}{N} \sum_{\substack{i=1 \\ i \neq m}}^{M} \eta_i \sum_{q_i \in \Gamma_i} \frac{\sin^2 \pi(\varepsilon_i - \varepsilon_o + q_i - k)}{\sin^2 \frac{\pi(\varepsilon_i - \varepsilon_o + q_i - k)}{N}}$$

According to the expression -, SIR of user m on the k-th sub-carrier is

$$SIR_k^{(m)}(\varepsilon_o) = \frac{\eta_m \frac{\sin^2 \pi(\varepsilon_m - \varepsilon_o)}{\sin^2 \frac{\pi(\varepsilon_m - \varepsilon_o)}{N}}}{\sum_{i=1}^{M} \eta_i \sum_{q_i \in \Gamma_i, q_i \neq k} \frac{\sin^2 \pi(\varepsilon_i - \varepsilon_o + q_i - k)}{\sin^2 \frac{\pi(\varepsilon_i - \varepsilon_o + q_i - k)}{N}}}$$

It is noted that the optimal carrier frequency offset compensation value should be between the minimum and maximum values of user CFO. The optimal ε_o value of user m in SC-FDMA system is given by the following expression:

$$\begin{cases} \varepsilon_{o,opt}^{(m)} = \arg \max_{\varepsilon_o} SIR^{(m)}(\varepsilon_o) \\ s.t. \quad \min\{\varepsilon_1, \cdots, \varepsilon_M\} \leq \varepsilon_o \leq \max\{\varepsilon_1, \cdots, \varepsilon_M\} \end{cases}$$

where $SIR^{(m)}(\varepsilon_o) = \frac{1}{P} SIR_k^{(m)}(\varepsilon_o)$, $P = N/M$.

2. Test environment and setup

(1) Characteristics of input data

The input signal is the SC-FDMA signal, with the characteristics including:

(1) Total sub-carrier number (256).
(2) Cyclic prefix length is 40.
(3) Modulation mode (QPSK)
(4) Subcarrier spacing (15 kHz)
(5) Moving speed (39 km/h)
(6) Carrier frequency (400 MHz)

(2) Characteristics of radio channel

The wireless channel is assumed to be a multipath Rayleigh fading channel.

3. Test indicators

   BER: defined as the ratio of the number of received error bits and the total number of sent bits.

4. Expression forms and accuracy

   For tested data, the following expression can be used:

   (1) When other parameters are unchanged, frequency deviation is fixed. And BER performance is compared with traditional methods.

   (2) When other parameters are unchanged, frequency deviation is randomly and uniformly distributed. The range of random frequency deviation is controlled, and BER performance is compared with the value from the traditional methods.

II. Scheme design

   According to test case requirements, the design uses NI PXIe-5673E, NI PXIe-5663E and C8 to complete testing tasks (Table 5.2).

   Hardware connections in testing scheme are shown as below:

   The device connection is shown as in Fig. 5.24 According to the test algorithm, 5673E device generates the transmission signal waveform. It then passes through the channel machine C8 to receive signals, and finally enters into 5663E. The stored data is processed by PC and then BRE is analyzed.

III. Test results

   Because of equipment limitations, when in single-user situations, the user number M = 1, SC-FDMA system degenerates into a Single Carrier Frequency Domain Equalization (SC-FDE) system. Test channels are AWGN and Extended Vehicular A model (EVA) channels. BER performance of measurements and numerical simulation are indicated as in Fig. 5.25.

   In AWGN channel, the measured and numerical simulated BER performance is basically consistent. For EVA channel, the measured BER performance is worse than numerical simulation performance. When BER = 0.1, there is a loss of about 8 dB. The reasons for the poor measured BER performance in EVA channel are as follows:

Table 5.2 Instrument list used in table design

Instrument	Description
NI PXIe-5673E	Vector signal generator
NI PXIe-5663E	Vector signal analyzer
C8	Channel simulator

Fig. 5.24 Device connection diagram

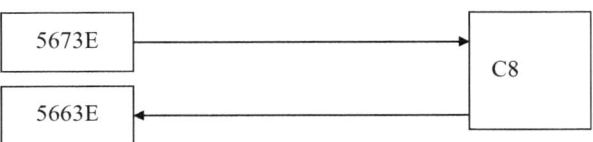

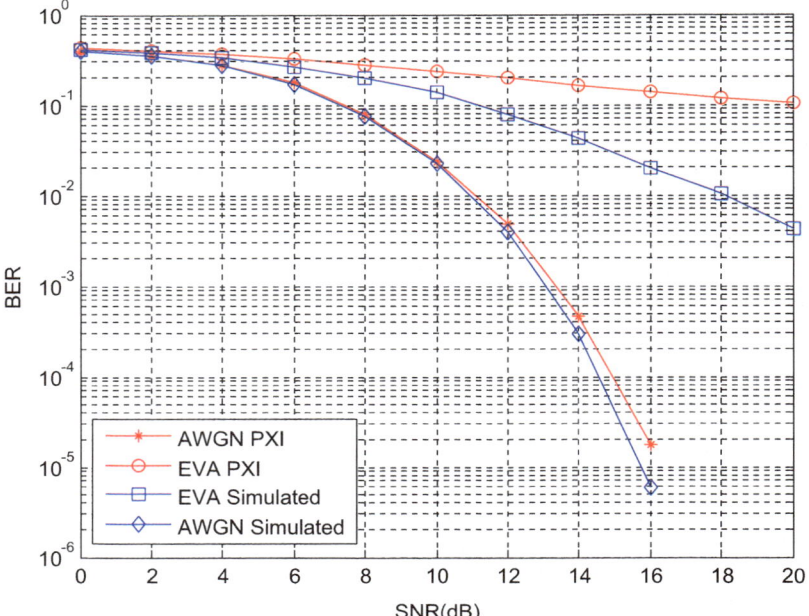

Fig. 5.25 BER performance

1. The frequency offset estimation algorithm in Labview codes does not consider the impact of multipath channel;
2. Because there is no AWGN module in the hardware platform, in the process of adding Gaussian noise in Matlab script,

It is considered that the received data does not include noise. But in fact, the received data has already contained noise caused by hardware devices. Thus, SNR in the figure is bigger than real SNR.

Therefore, in the single-user testing, the correctness of time synchronization, channel estimation algorithms and the basic system module building is verified.

Case 3: Laboratory Integrated Verification Platform Supporting Remote Use

I. The main functions and difficulties of the laboratory integrated verification platform supporting remote use

1. 4 × 4 channels wireless link platform based on modular instrument

Multichannel is also called MIMO technology. It can multiply increasing the channel capacity and spectrum efficiency of communication system without increasing the channel bandwidth and the total transmitting power. Theoretically speaking, channel capacity will increase linearly with the growth of the number of antennas. It has become the key technology and hot research topic of IMT-A mobile communications systems. Therefore, the test for multi-channel RF and wireless channels is a necessary and important part of the key IMT-A technology testing and verification platform.

The previous test solutions in the industry mostly concentrate on 2×2 channels. Their application scope is limited. Therefore, we can say that by far there is no really mature commercial solution suitable for the multi-channel (4×4–6×6, or even more number of channels) test.

German MEDAV has channel sounder products, and Finnish EleKtrobithas ittoo. They both can support the multi-channel channel tester. However, in their implementations, a single signal source transmits in different time slices from the same source through the high-speed RF switch. Therefore, they can not complete the test of diversity technology and are different from the proposed implementation method of the multi-channel adaptive and self-updating test platform. Besides, the value of each company's single product not including software is more than 5 million.

1.1 Difficulties and challenges in the system development

The multichannel adaptive and self-updating test platform will directly face the challenge of multi-channel testing. The typical challenges it brings to traditional test equipment are as follows:

1. It needs multiple signal sources to precisely synchronize generating source signals and self-correct.
2. It needs multiple analyzers to precisely and synchronically analyze the signals and self-correct.
3. Data sample transmission rate and memory depth of multi-channel testing are far beyond the support ability of the traditional test equipment.
4. It needs to develop complex channel matrix algorithms and multi-path test algorithms.
5. It supports a channel model as the foundation of performance evaluation and comparison.
6. Therefore, the challenge and complexity faced by the development are enormous. And with the increase of channels, the complexity will increase dramatically. In this case, the design object is 6×6 channel RF and channel testing system.

2. Shared inter-modulated interfaces and mechanisms with software simulation

In this case, the integrated verification platform fully uses the research results of subtasks in the case – software simulation platform. It truly fulfills the reusability of project results. This is consistent with the initial target of the verification platform design. The verification platform is a HIL verification system. By the sharing and inter-modulated interface with the software simulation platform as well as the design and implementation of mechanisms, algorithms of the simulation platform can be moved to the hardware environment for an actual verification.

In order to ensure that all algorithms can be embedded into the HIL system, the innovative concept that the algorithm can replace the interface specification is proposed in the case, and the specification is developed. Therefore, in any algorithm, as long as users follow this specification to write software algorithms and upload them to the site database in the test task, it can be replaced in the verification platform, so as to verify the HIL simulation in hardware environment. Therefore, it can truly have performance evaluation and indicator verification for the user's software algorithm. It greatly improves the efficiency of users' software development testing and verification and greatly increases the flexibility of the software algorithm verification.

The interface specification of replacable algorithm is the innovation of this case, which is the first one in China. On the one hand, if users can just follow the algorithms defined on the interface specification, they can make the HIL simulation and verification on the integrated verification platform, verifying their own software algorithms fast, flexibly and efficiently. On the other hand, users can effectively segment the needed verification software according to the interface specification, dividing chunks of codes into several small pieces for verification, so as to position errors more rapidly. It greatly reduces the workload of software development and improves its working efficiency.

3. Multiple automatic testing and verification systems of different manufacturers

The key techniques of IMT-A include various aspects, covering from the physical layer to the application layer. Such techniques without standards cannot be accomplished by using only one test equipment or one scheme. Currently in 2G and 3G test schemes, the mixed operation systems with multiple test equipment have been adopted. As the IMT-A technology is still under development, so there isn't a single equipment, instrument or scheme covering all technical standards and indicators. Test and verification of IMT-A is bound to be a mixed system that integrates the existing test instrument, effectively uses resources and complement one another, which must also be an ultra large mixed system.

In terms of multiple equipment, during mixed operations, an integrated development progress is needed. Otherwise they cannot work orderly and the resources cannot be shared. Moreover, there may exist technology and resourcescollisions. The developed platform for the IMT-A key technology test and verification is an open platform for sharing, so a lot of work such as control, coordination and maintenance should be done in the following facets: all equipment used in order and simultaneously, their using rightusage specification and operation procedure, remote openness, test data storage, test data playback, etc. Therefore, if the test equipment is not integrated with the system, it is just a group of disordered equipment, on the basis of which the system development becomes impossible, and the effective resources can not be used to finish what single equipment can't accomplish and the orderly multi-task tests. If so, it cannot be called a test and verification platform.

Therefore, it is necessary to make an overall integrated development for purchased and developed equipment, unifying a platform that specifically provides solutions and effective management for testing the key IMT-A technologies. Then via the interface of the simulation platform and the services platform, the test and verification of key IMT-A technologies are completed in a coordinated way. Meanwhile, an open website shared for log-in and use via the network is designed and developed, thus making an active remote shared lab for the long-term use.

Such integration of equipment with various manufacturers and functions is unique in China. The established platform for test and verification of key IMT-A technologies is also the first large-scale remote and combined test platform in the field of mobile communications in China.

Based on our research, at present, there isn't a similar large-scale pubic integrated test platform in the IMT-A technology domain in international mobile communications. The testing and verification platform of IMT-A key technologies is similar in concept with the integrated products group of "Next Generation Automatic test system" (NxTest) brought up by the United States Department of Defense (DoD). NxTest program is still in implementation, which is substantially at the same progress stage as our key IMT-A technology test and verification platform. NxTest program explicitly supports the system-level RF test but does not support the code-level, module-level and subsystem-level tests. That means the purpose of NxTest is merely the product-level and the system-level integrated test, rather than supporting the whole stage of R&D as the key IMT-A technology test and verification platform. Thus, the established key IMT-A technology test and verification platform will be a global-leading large-scale integrated test platform which supports IMT-A R&D.

4. Remote test configuration and operation technology that based on the Browser/Server (B/S) architecture.

On-site equipment for measurement and control undergoes the bottom-up development, with functions gradually extended to networking, openness and distributedness. However, computer networks permeate the Internet from top to bottom, till it makes a direct communications with the bottom on-site equipment. Therefore, the Internet-based remote monitoring system in this test case came into being accordingly. It connects the equipment that is distributed at different sites and can independently accomplish specific tasks through the on-site control network (or the on-site bus), enterprises the network and Internet. Thus it can be an all-round distributed equipment status monitoring and fault diagnosis system for the purpose of resources sharing, coordinated work and centralized management. This system is the product of the development of Internet, the TCP/IP network communications technology, the field bus technology, the equipment fault diagnosis technology, the browser technology and the network database technology.

The Internet remote measurement and control system includes the following functions: (1) collecting the test equipment data; (2) transmitting the test data; (3) analyzing and processing the test data; (4) controlling the equipment status. Collecting the test equipment data means to choose the signals that can best embody the test objectivesaccording to different objectives. The signals are collected and stored by the on-site test equipment. Transmitting the test data means to send the collected signals to the remote test analytic system via Internet. Analyzing and processing the test data means that the test and analysis center analyzes and processes the received data, extracts the error signals, and makes comparison with fault documents in the system database. Corresponding solutions are proposed and the on-site equipment is controlled by Internet and the on-site bus.

In this Internet-based remote monitoring and control system, the equipment and parameters on the key IMT-Advanced technology verification platform are configured by software which can also make intellectualized judgment and correction. The B/S architecture in the system can support many users and terminals for real-time and non-real-time operations via various network remote control and verification platforms.

The B/S architecture-based technology of the test server, remote testing configuration and operation of the Internet-based remote measurement and control system in this case project is the first practitionersincommunications test platform.

5. Developing test configurators, multi-test task operation managers that support remote operation

During the project development process, we focus on not only the advancement of the platform verification technology but also the platform overall architecture and future sustainability. For this purpose, test configurators, multi-task operation managers are included in the design of the platform architecture.

As a bridge for customers to realize the test through managers, the configurator also provides the test system with a configuration platform. Clients can choose the test scheme through the local general control interface or remote websites. And managers will choose configurators for equipment according to the client's scheme. As the bottom level of the whole test platform, the configurator is the most basic platform to execute the automated test.

This makes it easier for managers to configure the client scheme. By controlling test instruments, configurators collect and store the data.

Configurators are a important and indispensable parts of the IMT-Advanced test platform.

Managers integrate the deployment and management functions with the software development platform, providing the standard interface to make the data exchange. Around the core of the multitask sequence, managers develop the architecture of the whole test system, define the implementation process, collect results and report the recording database to communicate results.

Compared to the previous management module, the module in this project has a great advantage. On the one side, it supports the test module that developed by development tools of various application programs and can independently customize the test sequence in a specific tested part so as to meet the requirements of the customer's independent test. On the other side, this manager module possesses fully customizable the user interface which can offer the humanized interface for better human-computer interaction measurement and control. Moreover, it can control various instruments and equipment with the help of the software control test system so as to adapt to flexible and diverse test requirements.

The test platform of the configurator and manager applied in this case is the first practitioners in communications.

6. Developing large-scale test database that supports fast indexing with all kinds of configurator interface.

Database is the core of the system, whose design is directly related to the system implementation efficiency and stability. Only when a rational database module is designed can the difficulty of system program and maintenance be reduced, and the actual operating efficiency increased.

Established in the R&D stage of the testing and verification platform of key IMT-A technologies, the test database of IMT-A covers plenty of data parameters of mobile communications and all the frame informations of the project. It will be the first large-scale shared database in the future mobile communications R&D domain in China. This is of great significance and value for carrying out research, development, experiment, test and education on mobile communications in China in the future.

The large-scale IMT-A R&D stage test database built in this project is in accordance with the principles of naming, concurrent search, and reference of repeated table. And it minimizes redundancy to ensure the completeness and accuracy of the data. The database and configurators have various message interfaces to finish task receiving, data processing and playback for configurators. Substituting the original traditional polling mode with the message mode greatly improves the efficiency of the system. When developing and designing the database, we took every possible conditions of the messenger and designed a rational universal message passing model which greatly simplified the programming procedures of client and server as well as the maintenance of the entire large-scale test database.

The large-scale database is used to store the test instrumentation, the test task information, the test log data, etc. It also provides data service for the manager and the adaptor to accomplish all kind of interfaces of the configurator in the message mode, which realize a complete series of the process from the remote user input, configuration, testing and data playback.

II. Architecture design and implementation

The overall architecture design is shown as below and the system is divided into seven parts.

1. 4-transmitting and 6-receiving types of the wireless link platform which is based on modular instruments, have achieved synchronization among multiplex transmitters and among multiplex receivers as well as between the multiplex transmitters and receivers.
2. Common inter-modulation interface and mechanism of software simulation platform in which its algorithm can replace the interface specification.
3. Automation testing and verification systems of all the five manufacturers and seven types of instruments are supported.
4. Test server technology, remote test configuration and implementation technology based on the B/S architecture.
5. Test configurators supporting remote operation, multi-task implementation managers and task executors are developed.
6. A large-scale database that includes all kinds of test configurator interfaces and supports fast indexing is developed.
7. Websites that users can use via network and the general control interface supporting local and network use are developed (Fig. 5.26).

The verification platform of key IMT-Advanced technologies has completed the substitutable HIL key technology algorithm, 4-transmitting and 4-receiving, and the verification environment for channel real-time simulation. As an open and sharable platform, the verification platform allows different users to make remote verification configuration via network and browser and to use it both online and offline. Therefore, the verification platform has developed and accomplished the following function components:

1. Testing and verification configurator

   It is designed to input and proofread multi-users' verification parameters, analyze their validity and store the data to the database.
2. Multi-task manager

   It is designed to manage and schedule the testing and verification tasks requested by multiple users and to interact with the database.
3. Multi-task actuator

   According to manager's scheduling, it actuates orthogonal multi-tasks and manages the execution state. Composed of several standardized test modules, it can get the test results directly and store them in the database.

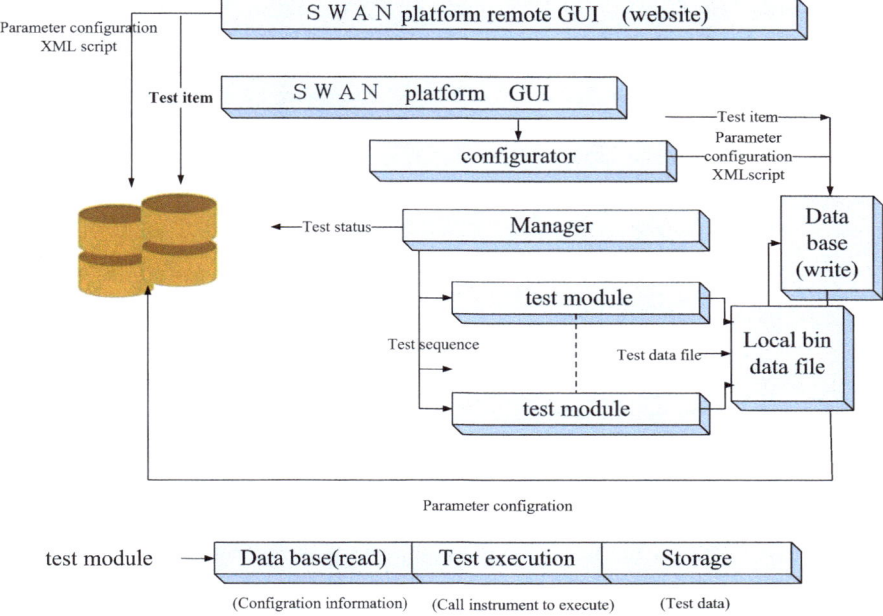

Fig. 5.26 Blueprint of overall architecture

4. Test database

It used to record the relevant user and the test data.

5. Database and configurator interface

It is used to finish the conversion from the XML-based user request to the storage data.

6. General interface

It enables users to login the verification platform through local login or remote login.

7. SWAN website

It is designed for remote users, supporting online and offline use via browser.

III. Applications of integrated test

1. Examples of standardized test scheme

(1) General parameters measurement module

The function of this module is to control the spectrum analyzer E4440A, by which to test indicators of wireless devices, mainly including channel power, power spectrum, etc. The E4440A configuration interface is shown in Fig. 5.27.

Fig. 5.27 Configuration
interface of E4440A

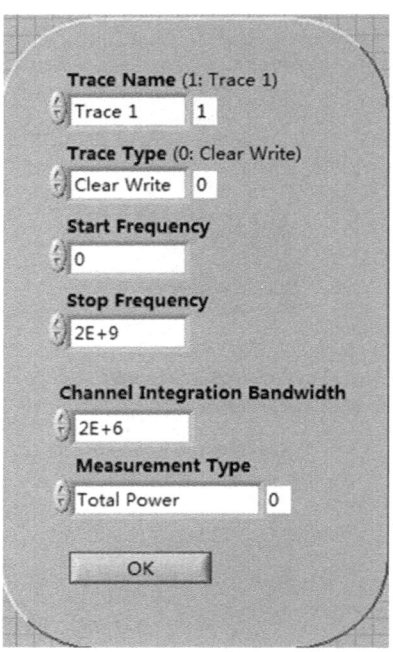

(2) Control interface of 53,132 and E4438C

The function of this module is to control the vector signal generator
E4438C and digital frequency meter 53,132. By controlling E4438C,
the vector signal is generated. By controlling the digital frequency
meter 53,132, the central frequency and other parameters are automatically measured. The configuration interface of E4438C and 53,132 is
shown in Fig. 5.28.

(3) PXI LTE RF phase noise test module

The phase noise module based on the modular instrument. It uses NI
PXI to make the PXI LTE RF phase noise test. Automatic measurement
of the phase noise at different frequency bands is performed by means
of transmitting and receiving LTE RF of PXI board of NI 5663 and NI
56732. Figure 5.29 LTE RF input.

(4) Control interface of E4440A and E4438C

This module aims to control two vector signal generators E4438C
and the spectrum analyzer E4440A. Controlling two vector signal
generators E4438C can produce signals of 3.01G and 3.00G while the
spectrum analyzer E4440A can make automatic measurements for the
third order intermodulation etc. Figure 5.30 is the parameter configuration of the third order intermodulation.

Transmitting and receiving of 1 by 1

This module aims to control PXI to verify theexperiment of
1 by 1 wireless communication. Transmitter PXI 5611 and receiver

Fig. 5.28 Configuration interface of E4438C and 53,132

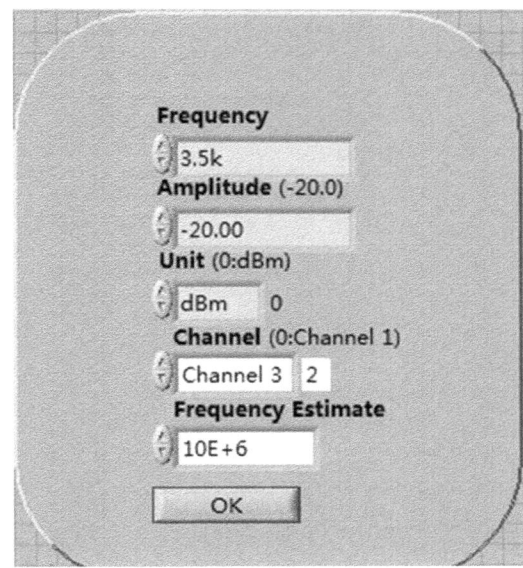

Fig. 5.29 LTE RF input

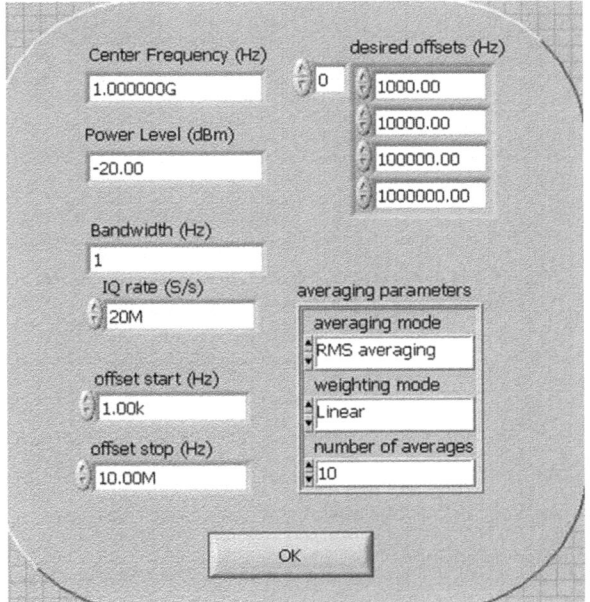

Fig. 5.30 Parameter
configuration of third order
intermodulation

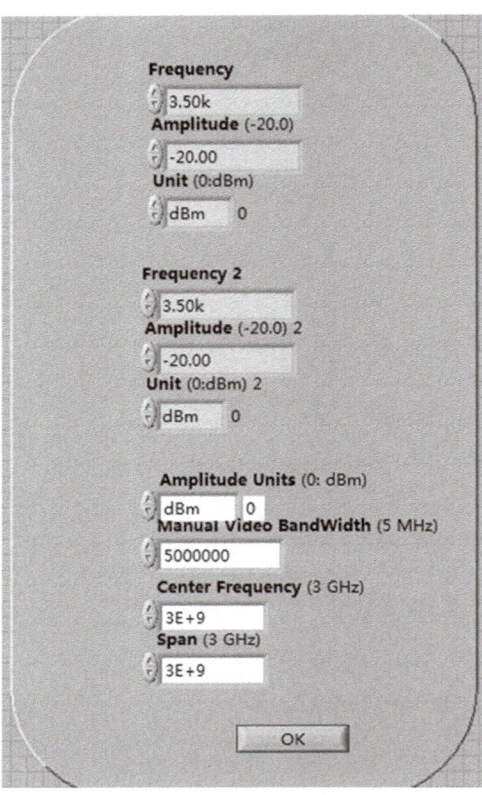

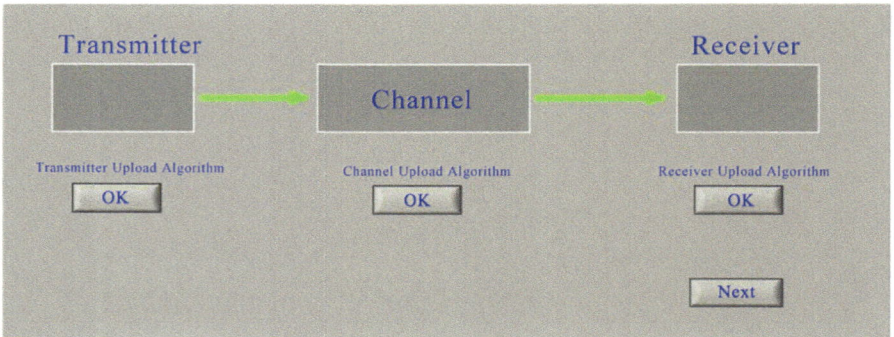

Fig. 5.31 Path configuration interface of 1×1 package

PXI 5601 are used for parameter configurations. The data of transmitter
and receiver can be seen and can be used to verify the transmitting and
receiving of algorithms withsingle antenna, as shown in Figs. 5.31,
5.32, 5.33, and 5.34.

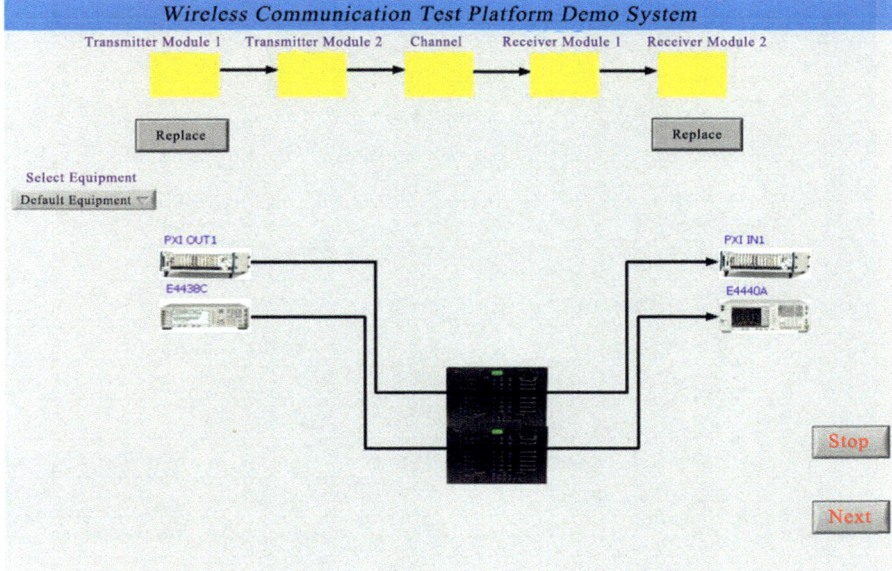

Fig. 5.32 User self-designed non-package interface

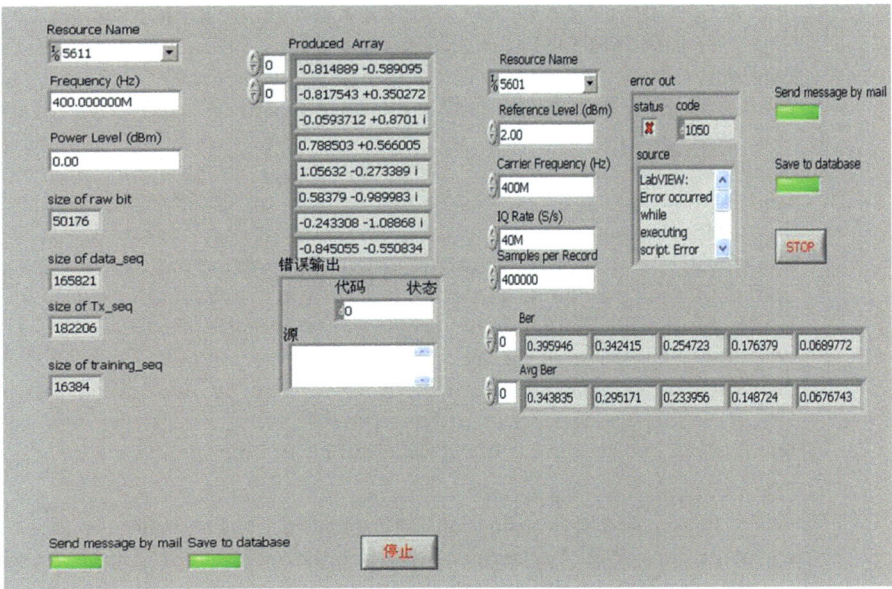

Fig. 5.33 Transmitter and receiver of 1×1 algorithm verification

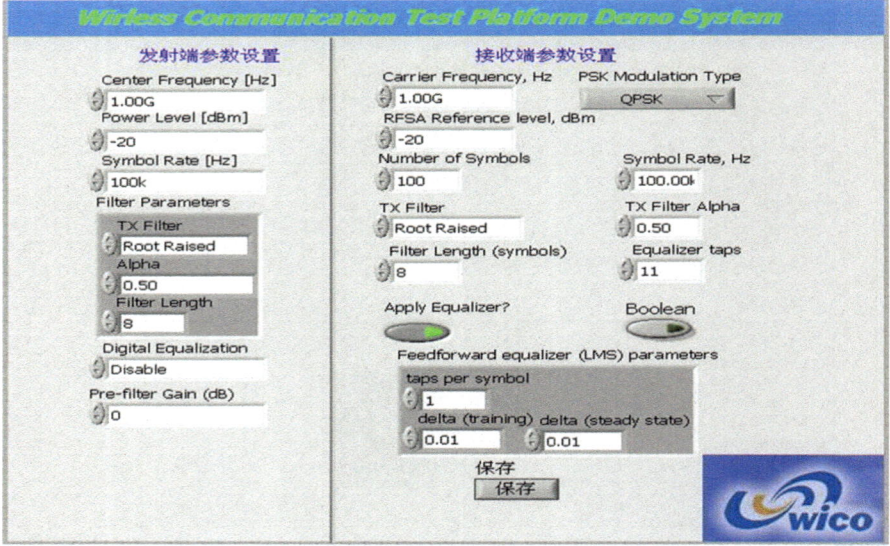

Fig. 5.34 Parameter configuration of QAM transmitter and receiver

2. Customizable test scheme

The non-package scheme is a user self-designed test scheme which allows the user to choose the instrument and the link mode while users only need to fill in basic parameters. Then the website will enter configurators according to the information that users filled in. The configurator popups the parameter interface that users need based on the sole verification code put in by the users. Users can fill the parameter according to their own needs. Then they should complete the parameter configuration and store the results that need to be stored in the database. Test order is arranged by manager on the basis of users' information. The final result is stored in the database. When choosing the test link, users can upload files via the upload button on the platform to replace the existing file, which means uploading their own algorithm to complete the test link. Users can choose apparatus of different manufacturers to complete the test. They can also choose the connection mode of instrument links. As shown in Fig. 5.32, the user chooses Agilent and PXI for 2 × 2 transmitter and receiver, and meanwhile chooses C8 channel for both routes.

(1) Algorithm verification of 1 × 1 transmitting and receiving

In this module, verification experiment of 1 × 1 algorithm is done by controlling PXI. Transmitter PXI 5611 and receiver PXI 5601 are used for parameter configurations. The data of transmitter and receiver can be seen in actuation and can be used to verify the transmitting and receiving of algorithms with single antenna, just as shown in the Fig. 5.33.

(2) Quadrature Amplitude Modulation (QAM) transmitting and receiving.
    This module is to conduct experiment on the QAM transmitter and receiver by controlling PXI. The parameter configuration is made via the transmitter PXI 5611 and the receiver PXI 5601. The data of transmitter and receiver can be seen in actuation and can be used to verify the performance of the QAM transmitter and receiver in real channels, as shown in Fig. 5.34.

5.3 Evaluation and Test on Hardware and Software System-level Co-simulation

5.3.1 Composition of Hardware and Software System-level Co-simulation (Fig. 5.35)

In this hardware and software system-level co-simulation, the physical layer of the simulation platform is partially substituted by the real physical layer and the transmission network, which can increase the reality and the instantaneity of the system.

Design and implementation of the simulation software platform should refer to Chapter 4 and the design and implementation of the real physical layer adopts the above-mentioned HIL software and hardware link system. Mapping relation from the system-level simulation software to the real physical layer should be customized in accordance with the specific requirements of the simulation evaluation. It will be introduced in detail in the real case in Sect. 5.3.3 of this chapter.

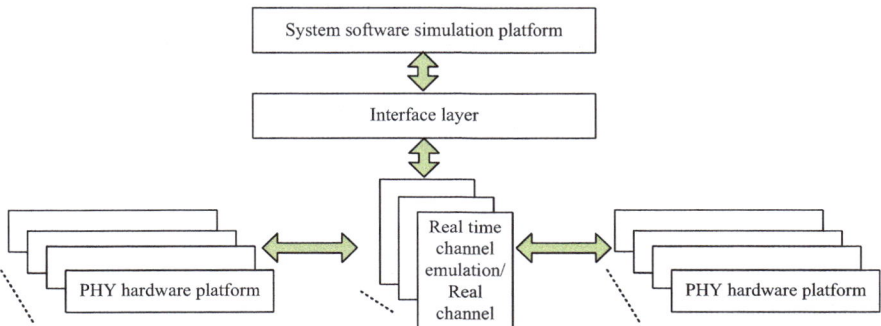

Fig. 5.35 Hardware and software system-level co-simulation

5.3.2   Key Technologies of Hardware and Software System-Level Co-simulation

Hardware acceleration based on USRP-RIO

The hardware accelerated simulation technology substitutes the software module with the hardware module to make the most of its speediness. This hardware accelerated simulation technology focuses on real-time functions to solve the bottleneck of system calculations and to accelerate the overall simulation efficiency. As a commercial collect-and-use device that can complete analog of single base station and single terminal, USRP-RIO provides a high-performance FPGA board with strong computing power and logic processing capabilities. Compared to the GPU server, the FPGA board has stronger logic capability for task management and resource scheduling. In addition, with the fast development in its development environment and compiler, the difficulty of developing FPGA has been greatly reduced. To accomplish the hardware accelerated simulation, the key link is divided into the following three aspects.

(1) Model selection of hardware accelerator board based on FPGA

With the development of the integrated circuit technology, the computing power and internal storage resources of the FPGA chip has undergone rapid growth. Meanwhile, the FPGA device has become an ideal algorithm acceleration implementation platform due to its programmability, granularity parallel ability, rich hardware resources, flexible algorithm adaptability and lower power consumption. FPGA can easily realize the distributed algorithm structure, which is very useful for acceleration scientific computing. For instance, most matrix factorization algorithms in scientific computing requireslarge quantities of multiplication and accumulation, which can be effectively realized by distributed arithmetic structures of FPGA. Today's FPGA products have achieved millions of gate-level design in 65 nm process. And all major companies are seeking to develop more logical units and more high-performance logic products.

(2) Design of interface and middleware that combines hardware accelerator board and software simulation platform.

For some computations difficult for FPGA to fulfill, C language can help. At the end of the design, the intermediate result running on hardware is imported into C for visual analysis. Through drawing the graphical waveform and the eye diagram and so no, we can check whether each part of the system design is correct. Introducing the simulation based on C/C++ to the design of FPGA has solved the FPGA debugging difficulty, as well as the frequent inconsistency with the system simulation logic. The concrete elements of the FPGA design includes a virtual adaptation mechanism, middleware design and re-design calculation.

(3) Reconfigurable FPGA logic design

Reprogrammable is an important feature of FPGA. The FPGA-based system can be changed and reconfigured easily in the development stage and in the using process, which increases the overall gain.

The implementation process of hardware accelerated simulation is as follows.

First, considering simulation design requirements, the hardware parallelism and the software system architecture, divide and design rationally hardware acceleration modules, in oder to significantly reduce the calculation time overheads of the simulation core modules.

Next, design of the FPGA-based hardware acceleration board system, involves design of adaptation based on C simulation codes, design of middleware (including board level support package) of interface layer, and design of the reconfigurable computing.

When the host receives the signal that configuration data loading is finished, the initial data will be sent to memory via PCIe interface. When it is transmitted, activating signal is sent to the accelerator through the PCI-E interface. After that, the accelerator returns the completion signal to the host via the PCIe interface. When the host receives the completion, the final results can be read in memory (Fig. 5.36).

As shown in Fig. 5.37, parts of the link in the simulation system platform adopt hardware implementation. The actual test results of physical bus transmission delay is less than 10 ms. Physical bus transmission delay is defined as

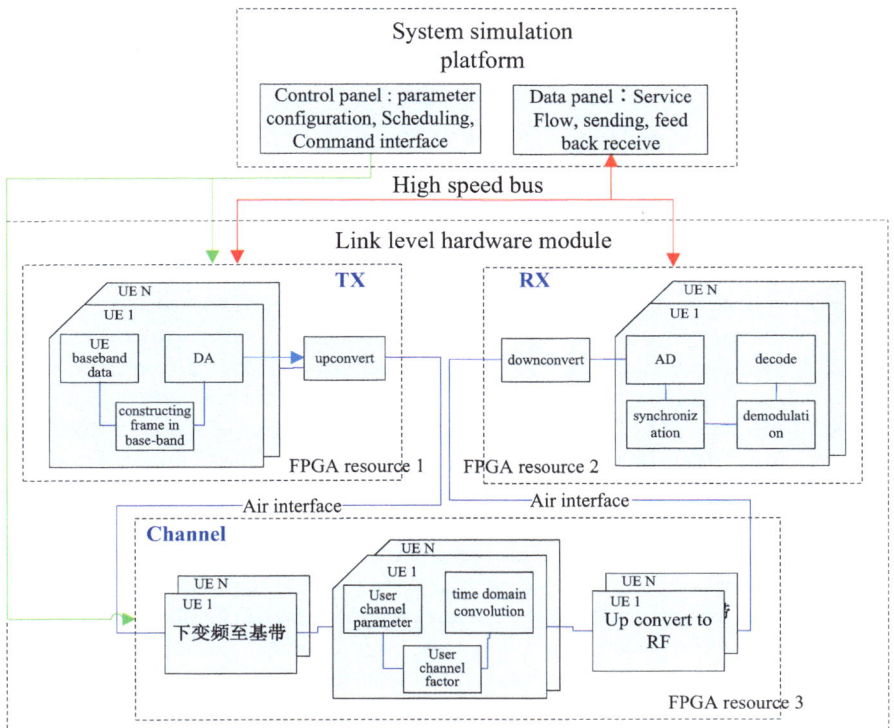

Fig. 5.36 A block diagram of system implementation of software and hardware co-simulation

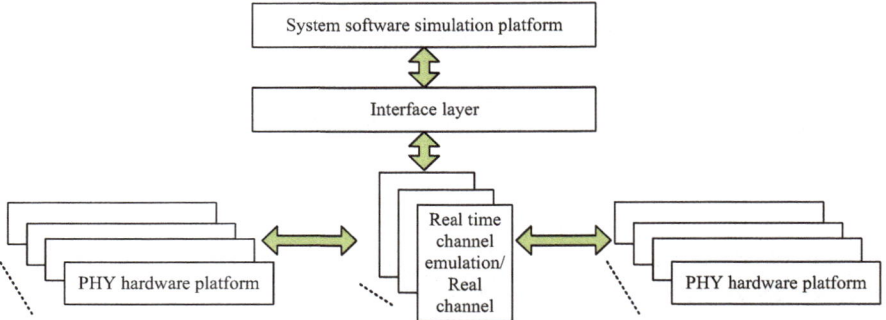

Fig. 5.37 Software implementation link

the delay of the whole process in which the data is sent from the simulation platform to the air interface via the link transmitter module, and then goes back to the platform after receiving and demodulating decode through the link receiving module.

This kind of software and hardware co-simulation methods can make full use of the hardware high-speed processing capabilities, and enables some link's system simulation performance to be close to the real-time level. Combined with the relatively perfect system function of the system simulation platform, it can better simulate the system application scenarios with high indicator requirements for system transmission delay.

Distributed data sharing technology

1. Reflective memory technology

    Reflective memory is a distributed data sharing system that enables multiple independent computers to share a universal dataset. The reflective memory network can store the independent backups of the whole shared memory in each subsystem. Each subsystem is entitled to full and unrestricted access, which allows them to write and modify the local datasets with a high-speed local memory. When data is written to the local reflective memory to backup, high-speed logic synchronization transfers it to the next node of the loop network. Each subsequent node will write new data to local backup, and then send it to the next node in the loop network. When the information goes back to the initial node, it will be removed from the network, and then, according to specific hardware and node quantities, all the computers on the network will have the same data at different addresses in a microsecond. Thus, the local processor can read the data at any time without access to the network. In this way, each computer can always have the latest local backup of the shared memory.

2. Advantages of reflective memory technology

(1) Time-delay stability

A significant advantage of the reflective memory is the stability of timedelay. As long as the network bandwidth isn't overrun unexpectedly, the timedelay won't increase significantly. The reflective memory board (node) includes a local memory, embedded interfaces and an arbitrary logic that provides the access channel for the host and the reflective memory. In addition, the reflective memory board can be installed in a physical way or be connected to a variety of computer buses which include Versa Module Eurocard (VME), PCI/PCI-X, Compact PCI, PCI Express or other standard/ special systems that can integrate PMC slots. This method enables popular workstations to be connected with one-board computer via the reflective memory.

(2) Standard-exceeding LAN

The reflective memory board provides several features beyond standard network, such as the global memory, high-speed data transmission and software transparency, all of which make the reflective memory the most attractive multi-machine communication solution. Compared with the cost produced by additional development time, testing, maintenance, documen- tation compiling and additional CPU of traditional communication means, the reflective memory is more cost-effective.

PXI multi-computing specification

In November 2009, the PXI multi-computing (PXImc) specification released by PXI Systems Alliance defined a Non-Transparent Bridge (NTB) of PCI Express with low-cost and readily available technologies, which is the technical require- ment of software and hardware for remotely connecting two or more intelligent systems based on the PCI or PCI Express interface. The PCI Express bus provides actual data throughput of multi-gigabit per second and stable microsecond delay. Therefore, it meets the requirements for applications of software and hardware co-simulation of multiple physical terminals.

1. PXImc principle

The following figure illustrates this concept. Systems A and B can fully control the resources allocation in their own domain. Meanwhile, NTB does not affect the algorithm of resource allocation of any system.

Using a NTB, two systems on the above figure are connected together via PCI Express. Response of NTB to resources from the root complex systems is similar to other PCI terminal request which is achieved through requesting a certain amount of physical address space. Then the system BIOS allocates a specific range of physical addresses to NTB. When this resource allocation occurs simultaneously on the System A and B, NTB will obtain resources in both PCI domains. As shown in Fig. 5.38, the address space obtained by NTB within PCI

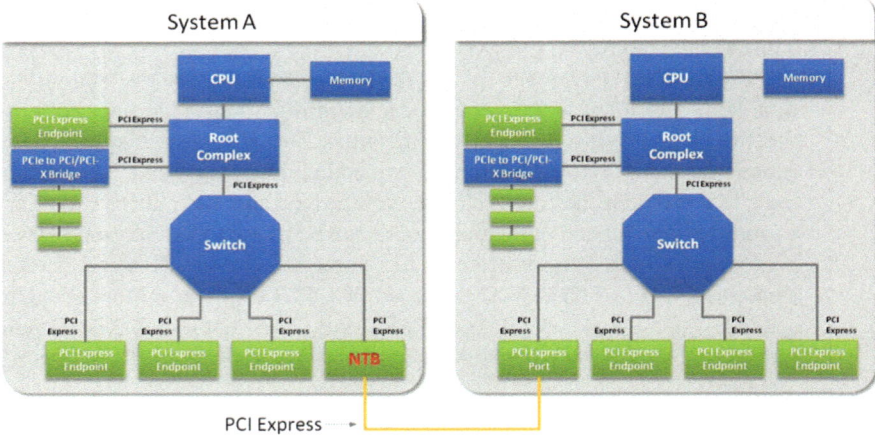

Fig. 5.38 NTB won't affect the algorithm of resource allocation of any system

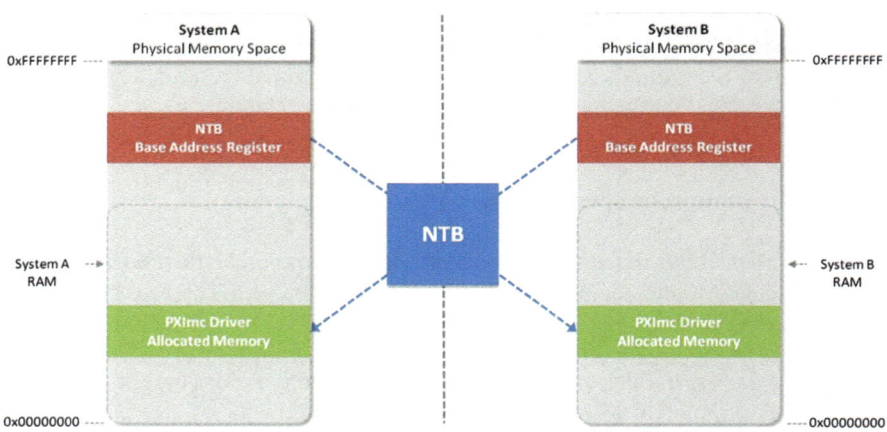

Fig. 5.39 Communication mechanism within two PCI domains using NTB

domain of System A will serve as a window into the physical address space in the PCI domain of system B while the address space of System B will be a window into the physical address space within PCI domain of System A.

After finishing resources allocation of System A and B, NTB transfers data between the two systems according to memory mechanisms. These mechanisms include memo registers for transferring data and doorbell registers for interrupt request, transferring large amount of the address space to the address space by NTB.

Figure 5.39 uses the communication mechanism within two PCI domains using NTB. The PXImc specification developed by PXISA defined specific

Table 5.3 Bandwidth value of various PCI Express links

PCI Express link	Versions	The theoretical value of unidirectional transmission rate	The theoretical value of bi-directional transmission rate
x4	1st generation	1 GB/s	1 GB/s ×2
x16	1st generation	4 GB/s	4 GB/s ×2
x4	2nd generation	2 GB/s	2 GB/s ×2
x16	2nd generation	8 GB/s	8 GB/s ×2

requirements for hardware and software components, therefore providing a standardized protocol for the communication between PCI or PCI Express systems. In term of hardware, some issues have been resolved to allow two independent systems to communicate directly via the PCI or PCI Express. What's more, in terms of software, a communications architecture has been established, allowing any system to detect and configure their own resources in order to realize communications with other systems.

2. PXImc bandwidth and delay performance

Since PXImc uses PCI Express as the physical communication layer, the performance of PXImc link depends on the type of PCI Express interface. Table 5.3 lists the theoretical bandwidth value of various PCI Express links.

In PXISA's experiment, the performance of a typical PXImc link is compared with other alternatives. In a fixed configuration, a 6 μs one-way delay and a 670 MB/s throughput were measured. Compared with Gigabit Ethernet, these indicators show a ten times increase in bandwidth and 100-fold reduction in delay. This confirms the fact that PXImc is an ideal interface to build the multicomputer test and control systems of high-performance.

3. Multi-core processor is supported for distributed process

In the hardware and software co-simulation test and evaluation, there must be a communication interface with highbandwidth and lowlatency that can assign and process tasks at different partitioned nodes. In this distributed processing system, FPGAand Digital Signal Processors (DSP) can be used to meet the requirements for acceleration of the real physical layer. However, to use the system simulation software platform, the existing ×86 software IP should be applied and computing should be accomplished in a fixed mode instead of the floating-point mode. For these cases, the PXImc technology can use the PC with the latest multi-core CPU to be the external computing node, thus to create a distributed processing system. The figure below shows an example of a feasible distributed computing system that uses the PXI Express system and the PXImc interface board (Fig. 5.40).

In this example, the master controller is responsible for collecting data from the I/O module, and subsequently allocating it to four ×86 compute nodes by the PXImc link. Depending on the requirements of processing performance, the computing nodes can be either an ordinary PC or a high-end workstation.

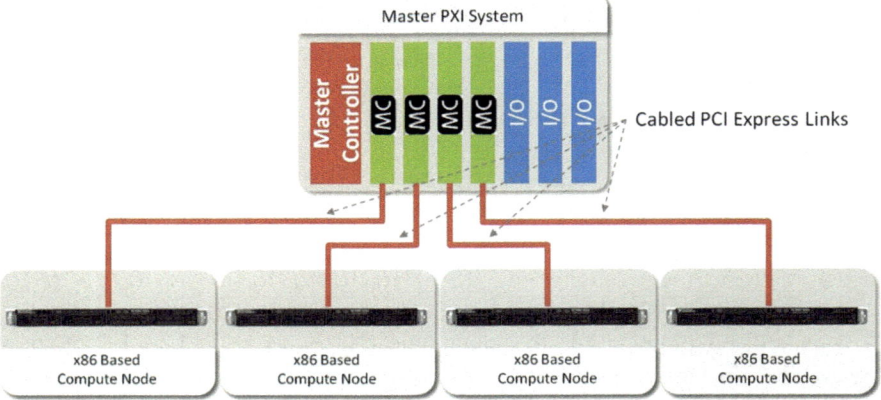

Fig. 5.40 An example of a distributed computing system that uses PXI Express system and PXImc interface board

Therefore, the PXImc specification made by PXISA expands the distributed computing function of the PXI platform, providing an interoperable scheme for the connection of multiple intelligent systems via a communication interface with high throughput and low delay, so as to meet application requirements for software and hardware co-simulation test and evaluation.

5.3.3   Case Study of Simulation

Case1: System Simulation PlatformwithPhysical Layer Acceleration Based on USRP-RIO (Fig. 5.41)

To simulate the interference of multiple users for adjacent cells to a valid user, the PXI bus architecture is used as the core, and USRP-RIO is used to simulate the real physical channel so as to complete the system configuration of upper level parameters and the downlink simulation for the LTE physical layer.

 I. System architecture

The system simulation platform architecture based on USRP-RIO physical layer acceleration is shown as Fig. 5.42. By integrating the link-level test into a complete system simulation platform, the multi-cell HIL test and simulation is realized.

While the traditional test scheme only considers the single link or a limited number of interference, this HIL scheme added the HIL performance verification of the transceiver to be tested in the scenario of real multi-cell multi-user scheduling. This scheme aims especially to verify the performance of algorithms that involves multiple cells, for example the ComP, synchronization of carrier aggregation, etc.

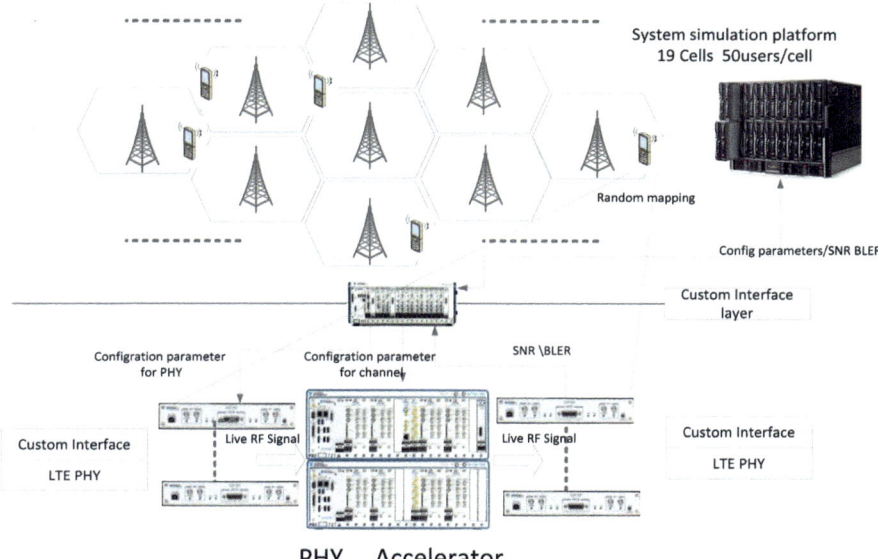

Fig. 5.41 Composition ofsystem simulation platform with physical layer accelerationbased on USRP-RIO

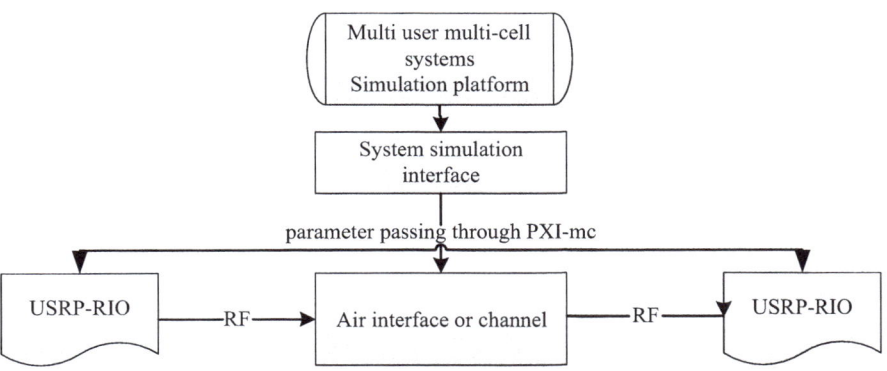

Fig. 5.42 System simulation platform architecture based on USRP-RIO physical layer acceleration

Compared with the traditional system, this scheme improves the hardware acceleration and authenticity of the system simulation platform. As for the implementation of traditional system simulation platform, the way equivalent SNR mapping into BLER will result in a loss in accuracy. Especially for complex transmission (equilibrium) scheme, the instantaneous SNR greatly varies in each time-frequency location. Besides, it is difficult to evaluate

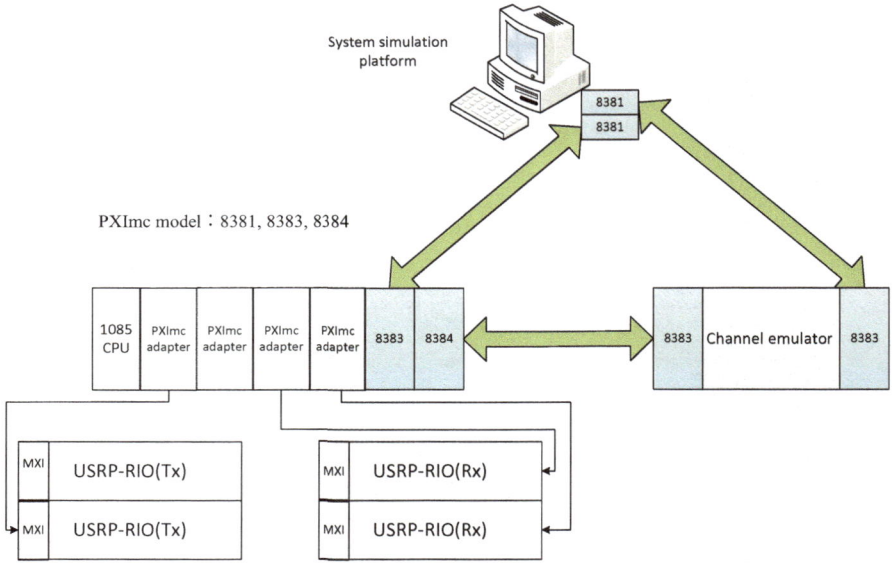

Fig. 5.43 Hardware connection frame

system performance under a non-steady-state, such as initial synchronization and frame format switching process. However the verification method of this platform can effectively simulate the above scenario.

Both the HIL and system simulation have considered the influential factors such as the actual channel environment and medium RF.

The hardware connection frame of the platform is shown below (Fig. 5.43):

Specifications of system simulation platform are as follows:

1. System simulation platform supports 19 cells.
2. Each cell supports 50 users.
3. It supports LTE DL physical layer MAC protocol (data channel).
4. SISO supported currently.
5. Standard channel model (or actual scenarios playback) generates channel factors.

A user takes the real-time test on the actual physical channel. During the process, the system simulation platform will transfer related user configuration parameters to the transmitter, the receiver and the channel simulation terminal via a high-bandwidth data board. And physical signals will be sent by the USRP-RIO (Tx) through the air interface or the channel simulator. After it is received and demodulated, the feedback information will return to the system simulation platform via the high-bandwidth data transmission board.

II. Scheme design

Firstly, setparameters of system-level cells and usersandsetthe links. The initial parameters of all users can be obtained.

Secondly, for real-time test level links, the system simulation platform will generate the MAC schedule information of the user in real time according to the scheduling algorithm. On the basis of user feedback Channel Quality Indicator (CQI) and other measurements, for real-time test level users, the observed values such as measured CQI are reported, and for other users, the simulated CQI and other measurements are used for feedback.

Thirdly, for real-time test level link, the system sends the downlink scheduling information and the channel parameter of the base station to the hardware platform, and then the platform will generate downlink signals at cell level according to the real-time scheduling information. The channel generation part in the hardware platform will generate real-time channel information on the basis of channel parameters, and then adding the actual air interface signal after channel fading to the RF signals.

Fourthly, the simulation user link calculates Estimated Signal-to-Noise Ratio (ESNR) according to each carrier signal and interference power. Based on the ESNR look-up table, the link gets BLER probability to judge the current ACKnowledgement (ACK). Real-time test level users obtain each carrier's received signal through demodulating actual transmitted data frame and estimate the signal power. Meanwhile, they will also make OFDM demodulation to downlink signals in other cells (non-target cells), and calculate actual SNR based on RB-level scheduling information from the base station, and also measure actual information such as CQI and decode them to get ACK feedback.

Fifthly, the base station calculates scheduling results of the next TTI in line with all scheduled users' feedback of ACK and CQI measurement value and makes iteration of the next scheduling period.

III. System implementation

1. Interface mode

Both of the universal processor servers and the heterogeneous platform of FPGA adopt the high bandwidth data transmission card for connection. The interactive throughput and delay can meet industrial requirements. Specific data parameter configuration is shown in the subsequent section of the interface agreement implementation.

2. Scheduling mode

(1) The total program of system platform is operated on universal CPU, and the bottom level calls the processing section of FPGA.

(2) Matlab is used to write the processing part of the system simulation. And the interface of the high bandwidth data transmission board is used to read and write in the agreed interface format.

(3) The processing section of the test link is divided into the upper computer (CPU) and the lower computer (FPGA). Wherein the upper

computer is software and is responsible for reading and writing the high-bandwidth data transmission board in accordance with the agreed interface format as well as initializing transceiver with the read user link information. The lower computer uses hardware programming language to write the FPGA program, completes the physical layer transmitting, receiving and channel passing, as well as giving implementation results feedback to the upper computer. Eventually the upper computer writes the feedback into the high bandwidth data transmission board according to the agreed interface.

3. Interface agreement (Tables 5.4 and 5.5)

The following is the actual received parameters clusters, and transmited parameters clusters.

(1) Receiving cluster of data (Fig. 5.44)
(2) Sending cluster of data (Fig. 5.45)

The mapping relation of MCS and modulation is shown in the following table (Table 5.6).

Table 5.4 MAC- > PHY (receiving)

Parameter name	Range/value/remarks	Data type/size
Device number		INT × 1(I32 bits)
Subframe ID	1 ~ 11,000 (range)	INT × 1
User ID (RNTE)	1	INT × 1
Cell ID	1	INT × 1
Base station ID	1	INT × 1
Channel type		INT × 1
RB resource allocation of target users.	4 values (1bit-1RB from MAC, represented by four INT)	INT × 4
RB resource allocation of other users	36 values (1bit-1RB from MAC, represented by four INT)	INT×(4 × 9)
MCS of target users.	One enumeration type is: MCS 0(QPSK,rate_0.12) MCS 1(QPSK,rate_0.15) MCS 2(QPSK,rate_0.19) MCS 3(QPSK,rate_0.25) MCS 4(QPSK,rate_0.30) MCS 5(QPSK,rate_0.37) MCS 6(QPSK,rate_0.44) MCS 7(QPSK,rate_0.51) MCS 8(QPSK,rate_0.59) MCS 9(QPSK,rate_0.66) MCS 10(16-QAM,rate_0.33) MCS 11(16-QAM,rate_0.37) MCS 12(16-QAM,rate_0.42) MCS 13(16-QAM,rate_0.48) MCS 14(16-QAM,rate_0.54)	INT × 1

(continued)

Table 5.4   (continued)

Parameter name	Range/value/remarks	Data type/size
	MCS 15(16-QAM,rate_0.60) MCS 16(16-QAM,rate_0.64) MCS 17(64-QAM,rate_0.43) MCS 18(64-QAM,rate_0.46) MCS 19(64-QAM,rate_0.50) MCS 20(64-QAM,rate_0.55) MCS 21(64-QAM,rate_0.60) MCS 22(64-QAM,rate_0.65) MCS 23(64-QAM,rate_0.70) MCS 24(64-QAM,rate_0.75) MCS 25(64-QAM,rate_0.80) MCS 26(64-QAM,rate_0.89) MCS 27(64-QAM,rate_0.93)	
Modulation typeof other users	The same as the above MCS of target users.	$INT \times (1 \times 9)$
Carrier Frequency		Float $\times$ 1 (32 bits)
Transmit power		Float $\times$ 1
The path loss value		Float $\times$ 1
Doppler fd		Float $\times$ 1
Target users power offset	$-10 \sim 10(0.1$ dB, U11 $< 2,9>)$	Float $\times$ 1
Other users power offset	$-10 \sim 10(0.1$ dB, U11 $< 2,9>)$	Float $\times$ 1 $\times$ 9
RB-level interference from adjacent cell 1	$\pm50(0.1$ dB, U18 $< 9,9>)$	Float $\times$ 100
RB-level interference from adjacent cell 2	$\pm50(0.1$ dB, U18 $< 9,9>)$	Float $\times$ 100
RB-level interference from adjacent cell 3	$\pm50(0.1$ dB, U18 $< 9,9>)$	Float $\times$ 100
RB-level interference from adjacent cell 4	$\pm50(0.1$ dB, U18 $< 9,9>)$	Float $\times$ 100
RB-level interference from adjacent cell 5	$\pm50(0.1$ dB, U18 $< 9,9>)$	Float $\times$ 100
RB-level interference from adjacent cell 6	$\pm50(0.1$ dB, U18 $< 9,9>)$	Float $\times$ 100

Table 5.5 PHY- > MAC (transmitting)

Parameter name	Range/value/remarks	Data type/size
Device No.		INT $\times$ 1
TTI No		INT $\times$ 1
UE ID		INT $\times$ 1
Cell ID		INT $\times$ 1
BTS ID		INT $\times$ 1
Feedback CRC		INT $\times$ 1
Feedback CQI		FLOAT $\times$ 1
Feedback ESNR		FLOAT $\times$ 1

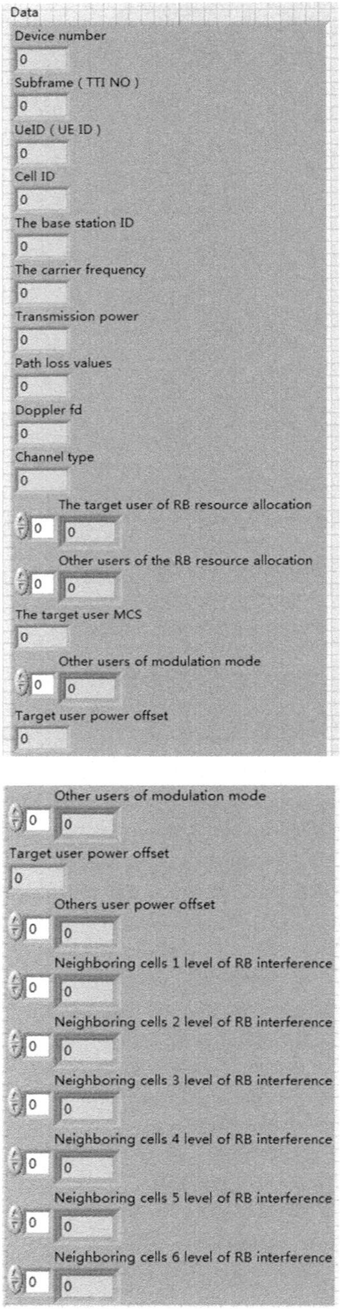

Fig. 5.44 Receiving cluster of data

Fig. 5.45 Sending cluster
of data

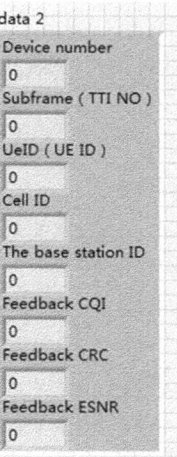

Table 5.6 The mapping
relation between MCS and
modulation

No.	MCS	Modulation mode	Code_rate
1	0	QPSK	0.12
2	1	QPSK	0.15
3	2	QPSK	0.19
4	3	QPSK	0.25
5	4	QPSK	0.30
6	5	QPSK	0.37
7	6	QPSK	0.44
8	7	QPSK	0.51
9	8	QPSK	0.59
10	9	QPSK	0.66
11	10	16-QAM	0.33
12	11	16-QAM	0.37
13	12	16-QAM	0.42
14	13	16-QAM	0.48
15	14	16-QAM	0.54
16	15	16-QAM	0.60
17	16	16-QAM	0.64
18	17	64-QAM	0.43
19	18	64-QAM	0.46
20	19	64-QAM	0.50
21	20	64-QAM	0.55
22	21	64-QAM	0.60
23	22	64-QAM	0.65
24	23	64-QAM	0.70
25	24	64-QAM	0.75
26	25	64-QAM	0.80
27	26	64-QAM	0.85
28	27	64-QAM	0.89
29	28	64-QAM	0.93

4. Implementation of real-time receiving and transmitting links

SISO implementation is considered now.

The physical layer acceleration system of software and hardware co-simulation platform is mainly used to transmit and receive the downlink data of the LTE physical layer. The operational process of the signal flow is shown as below (Fig. 5.46).

The following is the system block diagram of the implementation of transmitters and receivers of downlink in the physical layer (Fig. 5.47).

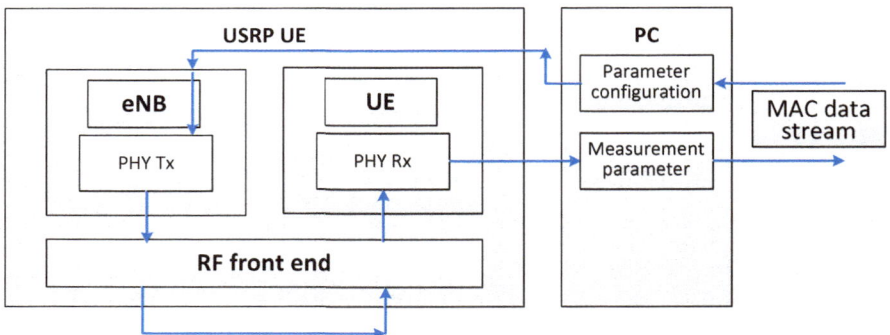

Fig. 5.46 Operational process of signal flow

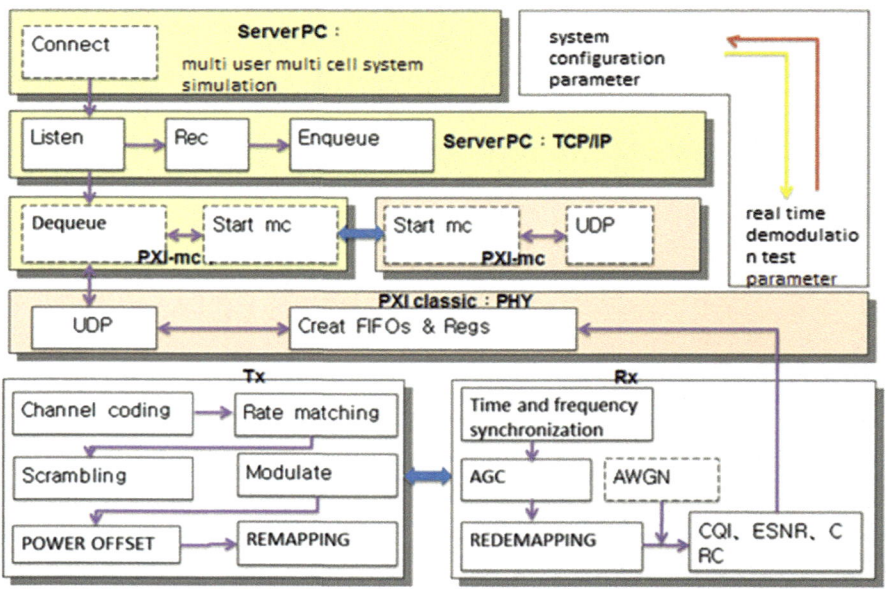

Fig. 5.47 The system block diagram of the implementation of transmitters and receivers of downlink in physical layer

(1) The transmitter:

Firstly, the transmitter generates baseband data signals according to the users' initial scheduling information input by MAC, which includes the random signal generation of data channel, coding, data rate match, QAM mapping, antenna port mapping, insertion of reference signal (synchronous signal), framing and OFDM modulation. Secondly, it converts the baseband signal to IF through the interpolation filter and the digital up-converter Direct Digital Controller (DDC), and turns to analog domain by calling DAC. Next, the IF analog signal is modulated to carrier frequency point via amplifying and frequency mixing.

(2) Channel generation section:

Channel simulator is configured according to the information input by MAC such as antenna matching, transmitting power, path loss, and channel parameters, with which increasing valid channel fading for the input RF signal. In addition, the channel can be bypass of channel emulator, instead to direct connection or via air interface or playback with the channel factor acquired.

(3) The receiver:

Firstly, the receiver transforms the attenuated RF to IF by processes like frequency mixing and the low noise amplifier.

Next, after IF ADC is adopted, through multistage downsampling, filtering and orthogonal demodulation, it is transformed to baseband signals.

Thirdly, in baseband processing, the receiver makes cell search and time-frequency passing in the frequency band of 1.4 M on the basis of the synchronization signal. After synchronization, the receiver begins to modulate the OFDM, decode frame, estimate channel, test MIMO and decode Turbo, and finally completes CRC decision output. In this process, the channel factors after channel estimation are used for related measurements including the CQI measurement.

The interference of the user can adopt the following two ways.

In the first way, the user only has one real-time link, while other carrier-level interference generated by the high bandwidth data transmission card. The hardware computes the total RB-level interference power according to the powers of each interference signal and generates the random equal-value frequency-domain variables to superpose it to the received signal. The second method is to run all the links of the target user (including user links in the cell and other base stations) on the real-time link. In this method, the target link demodulates each base station link in time division, and superposes the interference directly to the useful signal according to RB.

In receiver RB-level interference implementation algorithm, AWGN is implemented at FPGA side and Host side.

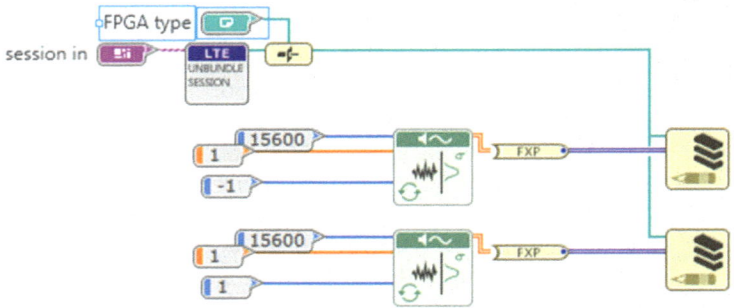

Fig. 5.48 Receiver RB-level interference

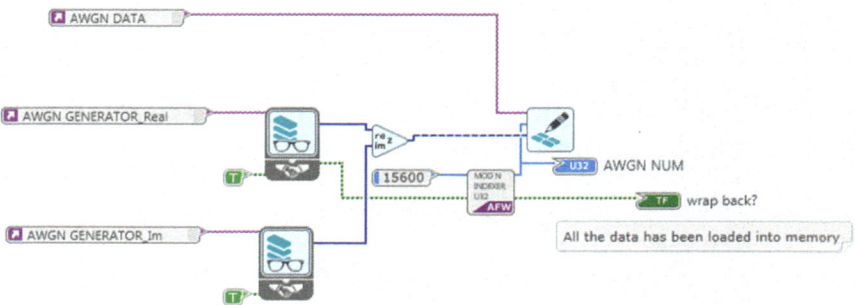

Fig. 5.49 DMA channel read

Host-side can produce the AWGN noise on each Resource Element (RE) required by ten subframes at one time and then send it to the FPGA via DMA. The noise level can be adjusted by inputting standard deviation of parameters where the upper and lower paths are I path and Q path noises. 15,600 represents RE (100 * 13 * 12) of 13 symbols (no noise is added to the first one) (Fig. 5.48).

FPGA, first in a single-cycle timing cycle, reads 15,600 data from DMA channel, and writes the data sequentially in the predefined memory for later noise superposition. The cycle will stop after writing, as shown below (Fig. 5.49):

Then AWGN Module is added between Demapper and Channel Estimation path, as shown below (Fig. 5.50):

Fourthly, lower computer transmits the feedback to the reflective memory. Currently, the CRC, ESNR and CQI are feed back.

Fig. 5.50 AWGN adding module

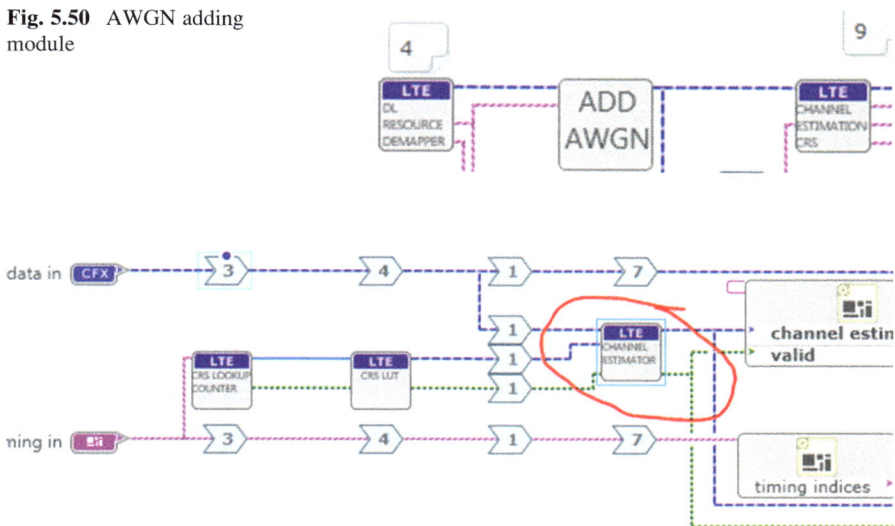

Fig. 5.51 The estimated value of original band noise channel obtained from LS algorithm

The CQI and ESNR algorithms of the downlink receiverare implemented. The algorithm is shown as below:

```
Y = S*H +N
H_est = Y/S = H +N/S
H_filt ≈ H
CQI = S*H/N = H/(N/S) ≈ H_filt/(N/S)
Whereas
  N/S = H_est - H_filt = H + N/S - H = N/S
so
  SINR = H_filt/(H_est - H_filt)
```

In the above expression, H_est refers to the value transmitted by FPGA while H_filt represents the value of upper computer after filtering.

Algorithm details are as follows:

The estimated value of original band noise channel is obtained through LS algorithm in the bottom layer of FPGA, and the value is sent to the upper computer in DMA mode, just as shown below (Fig. 5.51):

In the upper computer, a total number of 800 (100RB * 8CRS/RB) estimated channel values at CRS in the subframe of index = 5 are extracted, as shown below (Fig. 5.52).

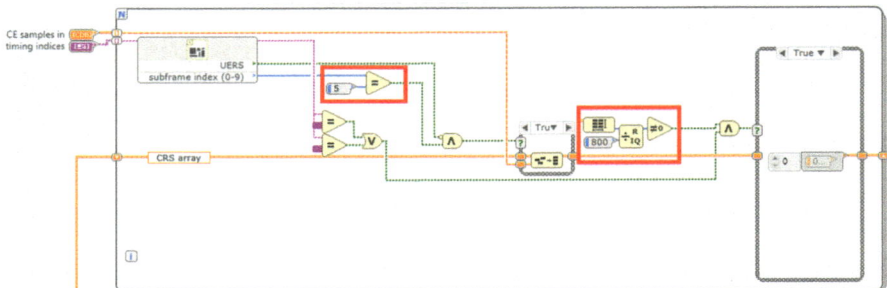

Fig. 5.52 The estimated value of CRS channel

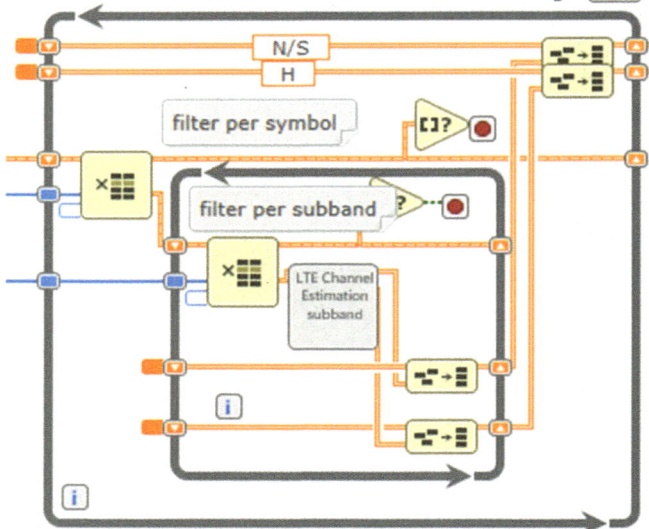

Fig. 5.53 The estimated channel values of each sub-band are extracted and then sent to filter module

From the data of a subframe, in the order of "index symbol first, and then index sub-band", the estimated channel values of each sub-band are extracted and then sent to filter module (as two steps below), as shown below (Fig. 5.53).

In the upper computer, after across a low-pass filter, an approximate value of denoising channel is obtained.

The difference of H_est obtained through the LS algorithm and H_filt after filtering serves as the original estimated value of the noise.

The above two steps are shown as below:

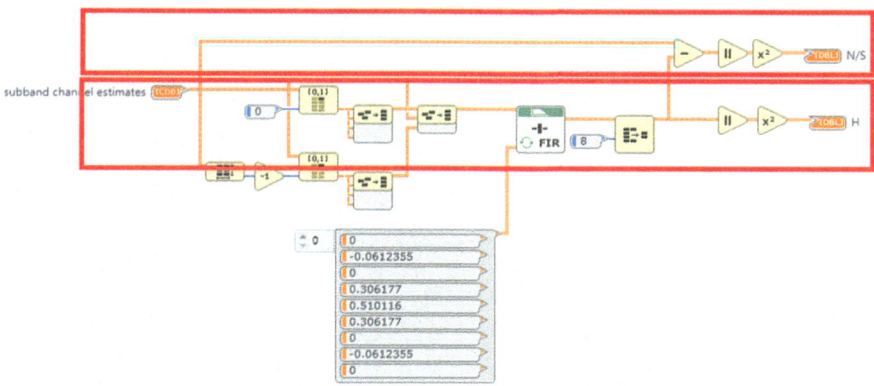

Fig. 5.54 The original estimated value of noise

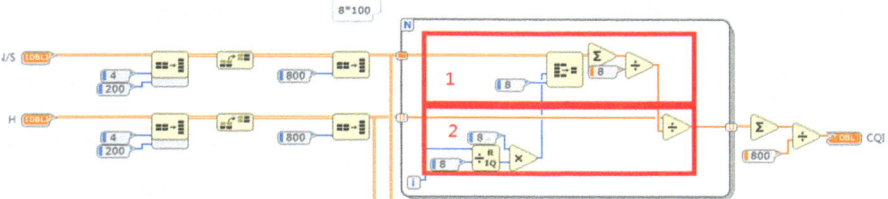

Fig. 5.55 Averaged CQI value on sub-frame

LTE Host DL.gvi- > LTE Read Channel Estimates.gvi- > LTE Calculate CQI.gvi- > LTE Calculate SubframeCQI.gvi- > LTE Channel
Estimation subband.gvi (Fig. 5.54)

N/S and H obtained in the previous stage are used to calculate CQI and
ESNR

LTE Host DL.gvi- > LTE Read Channel Estimates.gvi- > LTE Calculate CQI.gvi- > LTE Calculate SubframeCQI.gvi->

LTE Calculate SNR for CQI

For CQI, as shown below, in Part 1 the average noise of a RB is
calculated; in Part II, the CQI values in each CRS position is calculated;
finally the average is got on sub-frame (Fig. 5.55).

For ESNR, each bit of RB allocation corresponds to 4 RBs. Each RB
includes 8 CRS, so every 32 CRS is a group (Fig. 5.56).

Since there is disparity between the estimated original channel
power and original noise power, it is likely to have disparity between
the obtained CQI and ESNR. Hence, calibration is needed. The calibration equation is shown below:

$$\text{CQI/dB} = 1.8 * \text{CQIraw/dB} - 10.2 - 6 \leq \text{CQIraw/dB} < 6$$

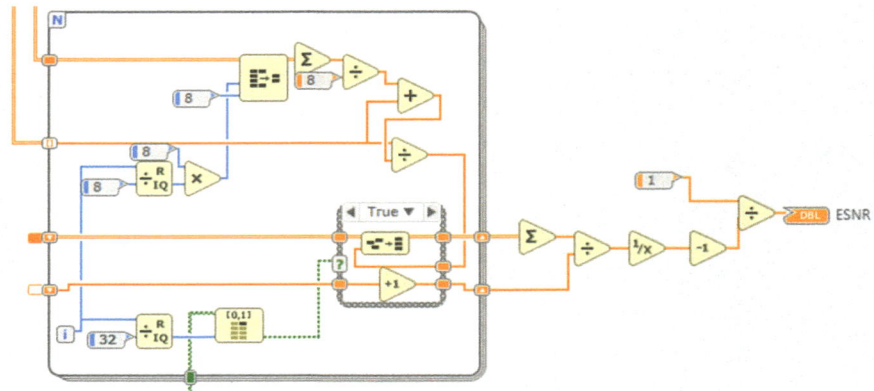

Fig. 5.56 SNR calculation

Fig. 5.57 CQI calibration

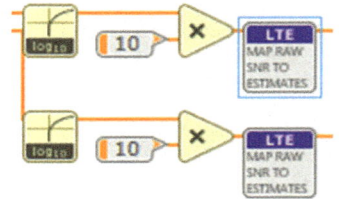

$$\text{CQI/dB} = 1.1 * \text{CQIraw/dB} - 6 \quad 6 \leq \text{CQIraw/dB} < 30$$

As shown in the below figure (Fig. 5.57):

In order to avoid initial synchronization process and solve the problem that implementation speed of the physical layer is faster than the MAC layer, the following controls are taken:

(1) Transceiver of the physical layer transmits and receives in real time and synchronization is implemented once at the beginning and tracked in follow-up steps.
(2) When MAC layer data packets of different links arrive, they are inserted into the closed downlink sub-frame.
(3) With the help of the controller, the physical layer coordinates the switch between transmitter, channel simulator and receiver.

In addition, the SNR can be changed from the parameter configration in the channel emulator.

5. Implementation of parameters transmission based on PXI-mx.

System simulation parameters are transmitted from the simulation server. The transmission parameters include channel configuration information, physical layer data configuration information, and the handshaking signals for time sequence processing. Among them, the channel configuration information is used to generate the physical layer configuration parameters to be used to transmit the LTE baseband. The details are in the above "agreed interface" table. The produced LTE baseband signal, after up-conversion, is simulated via air interface or channel. After AD sampling at the receiver and then the down-conversion, the signal is demodulated and decoded. Then the handshaking signal of ACK/Negative ACKnowledgement (NACK) is returned to the simulation server for the next data transmission.

The actual hardware need to be considered. Now is the interface verification of USRP-RIO and VST and PXI-mx.

5.4 Summary

This chapter describes the classification, methods and applications of software and hardware co-simulation test and evaluation. It puts emphasis on the HIL link simulation technology and its applications. With the HIL technology, the software and hardware co-simulation platform can be developed on the basis of existing commercial equipment. The technology evaluation and test at the algorithm stage can give us earlier test and evaluation results without increasing the product development cycle. Then this chapter further introduces the methods and applications of system-level software and hardware co-simulation evaluation and test. In the cases of link-level software and hardware co-simulation test and evaluation, we not only introduce the cases where the hardware platform and the software platform are in the same computing environment, but also introduce the cases of cross-network remote tests.

References

1. C. A. Valderrama, A. Changuel and et al. A unified model for co-simulation and co-synthesis of mixed hardware/software systems. European Conference on Design & Test, 1995:180–184.
2. R. Ruelland,G. Gateau, T. A. Meynard and J. C. Hapiot. Design of FPGA-based emulator for series multicell converters using cosimulation tools. IEEE Transactions on Power Electronics, 2003, 18(1):455–463.
3. J. Ou, V. K. Prasanna, MATLAB/ Simulink based hardware/software cosimulation for designing Using FPGA configured soft processors. International Proceeding of parallel and distributed processing symposium, 2005:148b-148b.
4. A. Hoffman, T. Kogel and H. Meyr. A framework for fast hardware-software co-simulation. Proceeding of the conference on design,utomation and test in Europe, 2001.
5. K. Wei. Software and hardware co-simulation platform. E-World, 2012:128–131.
6. W. Zheng. Research and development on new generation of hardware in the loop simulation platform. Tsinghua University, 2009.
7. B. Heiming. Hagen Haupt. Hardware-in-the-Loop Testing oflutions for Networked Electronics at Ford[C]. SAE Paper,2005.

8. R. Isermann, J. Schaffint. Hardware-in-loop simulation for design and testing of engine-control systems. Control EngineeringPractice, 1999, 7(5):643–653.
9. dSPACE GmbH, Experiment guide documents for release 4.0[K]. Germany: dSPACE Gnbh. 2003.
10. X. Liu. ECU hardware-in-the-loop real-time simulation test platform[D]. Tianjin: Tianjin University, 2009.
11. W. Tan. Research on simulation of hardware-in-the-loop and its application in wind power generation. Qingdao University of Sicence and Technology, 2014.
12. H. Jia and et al. Hardware acceleration verification technology in digital integrated circuit design[J]. Industrial technology, 2007.
13. Dammann and S. Kaiser, "Standard conformable antenna diversity techniques for OFDM and its application to the DVB-T system. Proceedings of the IEEE Globecom, 2001, 5:3100–3105.
14. J. Tan and G. L. Stuber. Multicarrier delay diversity modulation for MIMO systems. I IEEE Transactions on Wireless Communications, 2004, 3(5):1756–1763.
15. V. Tarokh, N. Seshadri, and A. R. Calderbank. Space-time codes for high data rate wireless communication: Performance criterion and code construction. IEEE Transactions on Information Theory, 1998, 44(2):744–765.
16. V. Tarokh, H. Jafarkhani, and A. R. Calderbank. Space-time block codes from orthogonal designs. IEEE Transactions on Information Theory, 1999, 45(5):1456–1467.
17. Z. Hong, L. Zhang, and L. Thibaut. Performance of cyclic delay diversity in DAB/DMB. IEEE Transactions on Broadcasting, 2006, 52(3):318–324.
18. Y. Zhang, J. Cosmas, and M. Bard, and Y.-H. Song. Diversity gain for DVB-H by using transmitter/receiver cyclic delay diversity. IEEE Transactions on Broadcasting, 2006, 52 (4):464–474.
19. 3GPP TSG RAN WG1 #46, R1–062566. Link evaluation of DL SUMIMO: Impact of generalized CDD. Seoul, Korea, Oct., 2006.
20. IEEE standard for local and metropolitan area networks part 16: Air interface for fixed broadband wireless access systems, IEEE 802.16–2004.
21. R. Van Nee, G. Awater, M. Morikura, H. Takanashi, M. Webster, and K. W. Halford. New high data rate wireless LAN standards. IEEE Communications Magazine, 1999, 37(12):82–88.
22. H. Guo, H. Hu, and Y. Yang. Cyclostationary signatures in OFDM- based cognitive radios with cyclic delay diversity. IEEE International Conference on Communications, 2009:1–6.
23. H. Guo and H. Hu. Transmission and receiving apparatus and method of spectral domain channel multiplexing transmission system (Invention patent application number: 2009 1005 4524.8).
24. H. Guo and Hu Honglin. Transmission and receiving apparatus of spectrum multiple access system and uplink and downlink access system Invention application number: 2009 1005 6374.4).
25. 3GPP TS 36.201: Evolved Universal Terrestrial Radio Access (E-UTRA); Physical Layer – General Description.

Chapter 6
5G Hardware Test Evaluation Platform

Co-simulation of software and hardware provides a more realistic testing and evaluation environment to some extent, improving the efficiency of test and evaluation. But it is still different from the real test environment. Therefore, a real hardware test platform and the environment are essential to the process of developing new technology. On July 15, 2016, the US government announced that 400 million US dollars will be invested to the Advanced Wireless Research Initiative (AWRI) led by National Science Foundation (NSF), which will build four city-scale experimental test platforms and deploy advanced future wireless research projects. The program will use two models: (1) 85 million US dollars will be invested through public-private partnership to deploy and establish an advanced wireless experimental test platform. Each platform will deploy software defined wireless networks in the urban area, allowing academics, entrepreneurs and wireless companies to test, evaluate and improve their technology and software algorithms in real-world environment. (2) The NSF will invest US$ 350 million to support advanced wireless communications theory and technology research. These wireless experimental test platforms and fundamental researches will support academic and business to develop and test advanced wireless communication technologies.

As the world's major telecommunications equipment suppliers started R&D work of 5G mobile communications system, the mainstream standardization organizations in the world all have seen the urgency of 5G technology development, and started corresponding R&D programs to promote the relative standardization work. Since 5G introduces new scenarios and new frequency bands, the requirements for 5G channel testing and modeling, as the basis for wireless communication protocol development, are becoming more and more important. In this context, National Science and Technology Major Project and 863 National High Technology Research and Development Program (HTRDP) in China also place the R&D of 5G test platform in priority. In 2013, the HTRDP, "Early Stage R&D of 5G Mobile Communication System (Phase I)", facing the requirements of mobile communications in 2020, proposed to complete the technology research of 5G mobile communications technology evaluation and testing and verification. It includes

© Springer International Publishing AG 2018   301
Y. Yang et al., *5G Wireless Systems*, Wireless Networks,
DOI 10.1007/978-3-319-61869-2_6

two aspects of works contents: (1) studying evaluation and testing methods of 5G mobile communication network and wireless transmission technology; (2) establishing the corresponding 5G technology simulation testing and evaluation platform, and finishing the evaluation and testing for 5G mobile communication network and wireless transmission technology. In 2014, National Science and Technology Major Project, in China "wireless innovative technology test platform", created public trials, verification and testing platform for wireless innovation technology of the 3GPP R12 and its follow-up standards. The project proposed to further establish the multi-scenario real channel base with the characteristics of classic Chinese environment to realize "field trial in lab", and to develop the real channel scenario system-level hardware and software co-simulation system with multi-cell multi-user to support the research and innovation of new wireless technology. It also proposed to develop and construct an open and shared, flexibly configured, new technology R&D-faced, standardized and new technology applied, and multi-cell and multi-user supported co-simulation, experiment and test platform of system-level software and hardware. In 2015, National Science and Technology Major Project in China further constructed the international standard evaluation environment of IMT-2020 for 5Gcandidate technologies and international standardization. It includes the simulation evaluation platform as well as test and verification platform for IMT-2020 candidate technologies. The two platforms can complete feasible and practical performance evaluation of potential network architecture, key technologies, algorithms, protocols of IMT-2020, etc., and support IMT-2020 candidate technology research, international standard-setting and follow-up product R&D. Based on the above platforms and facing the requirements of IMT-2020 technology, researchers can complete the evaluation and verification of massive MIMO array, UDN, HFB communications, M2 M enhancement, D2D, C-RAN, SDN, NFV, Content Delivery Network (CDN) and other IMT-2020 candidate technologies.

Overall, the current test platform presents the technology features of universalization, platform-based and software-based. This chapter will introduce a typical hardware test platform used for 5G evaluation. It includes the first parallel channel sounder platform in industry used for channel measurement and modeling and MIMO OTA platform based on a specified channel model, the first platform of hardware and software supporting software open source community of 5G terminal and base station.

6.1   Overview of Typical Hardware Platform Used for 5G Evaluation

The constitution of typical hardware test platform of 5G evaluation is shown in Fig. 6.1, which includes a parallel channel sounder platform of channel measurement and modeling, a MIMO OTA platform of designated channel model, a platform of software and hardware of open source community, and terminal and base station system based on the general purpose processor.

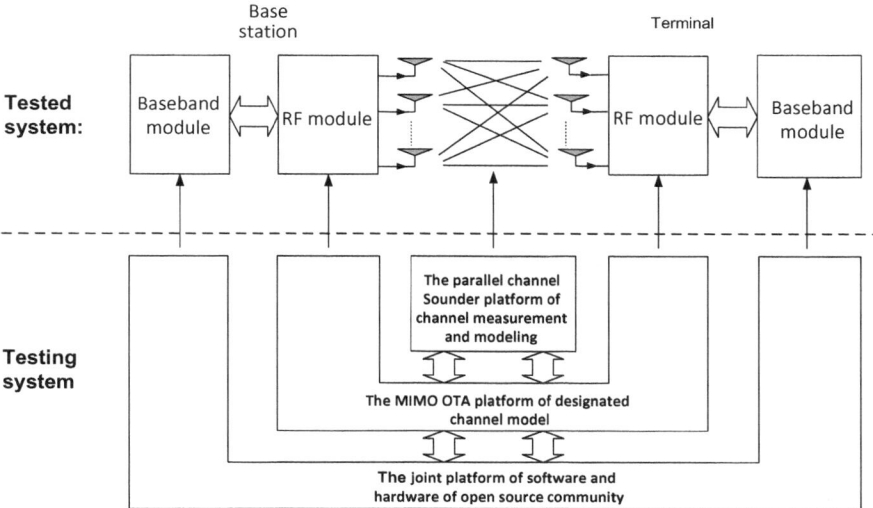

Fig. 6.1 Typical hardware platform constitution used for 5G evaluation

6.2   Test Platform of MIMO Parallel Channel

6.2.1   5G Channel Test Requirements

As mobile communication progresses towards 5G, the physical layer characteristics, such as HFB (millimeter wave) and multiple antenna (Massive MIMO), will bring the evolution of its new air interface protocols. Wireless channel measurement and modeling have always been the fundamental R&D topic of every generation of mobile communication systems. For example, the WINNER, the European Union's sixth framework project, uses the channel sounder from Elektrobit company to perform a wide range of channel test of wireless scenarios, and proposes WINNER channel model.

The main performance requirements of 5G channel test are as follows:

1. Abundant application scenarios

   IMT-2020 defines a wealth of application scenarios. These scenarios can be summarized into four characteristics: (1) seamless wide-area coverage scenario which requires the experienced rate of users can reach 100 Mbps; (2) high-capacity hot-spots scenario which satisfies the user's experience in centralized area, such as large-scale concerts, stations and other areas with high population density and traffic volume density; (3) high mobility scenario, supporting the mobility speed of over 350 km/h; (4) 3D space measurement and modeling, covering the scenarios of high buildings and dense space.

2. Coverage of a new spectrum range

5G candidate frequency band will go beyond 6GHz and is seeking the applications of a higher frequency band and millimeter wave band. The Federal Communications Commission (FCC) in the US has announced in July 2016 that the millimeter wave bands, such as 28 GHz, 37 GHz, 39 GHz, and 64–71 GHz, would be used for the future 5G mobile communication systems and applications [1]. So it is necessary to explore the propagation features of unknown electromagnetic waves of HFB and millimeter wave.

3. Multi-channel, high-precision and large bandwidth

With the introduction of new 5G spectrum and new technologies, such as massive MIMO and millimeter wave technologies, the requirements for technical parameters of channel test equipment are becoming higher and higher, which must support the real-time generation and reception of multi-channel RF data stream; signal bandwidth up to 2GHz; center frequency up to millimeter wave band; high-precision synchronization among multiple channels.

4. Raw data acquisition of channel

The large bandwidth and high mobility of 5G cause a large amount of channel measurement data captured at the time of channel measurement. Traditional channel measurement device collects channel impulse response files, while the 5G channel test needs to capture the raw data of channel, so as to make deep data mining to support the fine channel modeling requirements such as channel fingerprint.

6.2.2   Status and Shortcomings

Many universities and research institutes have made researches on channel measurement and modeling of some scenarios, but have not covered the new scenarios defined by IMT-2020. Moreover, the majority of previous channel model studies are focused on the derivation of the 2D channel model, not for 3D spatial features. It is very necessary to systematically carry out channel measurement and modeling of new scenarios. At the same time, the channel measurement equipment used in previous researches mainly adopted serial channel test equipment. For 5G specific scenarios, this equipment cannot be satisfied with 5G MIMO test and measurement requirement.

In general, classic serial channel sounder consists of a pair of transceivers and multiple antenna arrays and high-speed RF switches configured at both of transmission and reception, as shown in Fig. 6.2 . In measurement, one transmitter sends predefined waveform, such as time domain sequence. The transmitter and receiver periodically and synchronouslyswitch antenna channels to capture air interface signals. The receiver can correlate the received signal with local sequence and obtain CIR of each transmitting and receiving antenna pair. The channel parameters can be extracted by deriving CIR data using the parameter extraction processing algorithm. The processes are shown in Fig. 6.3.

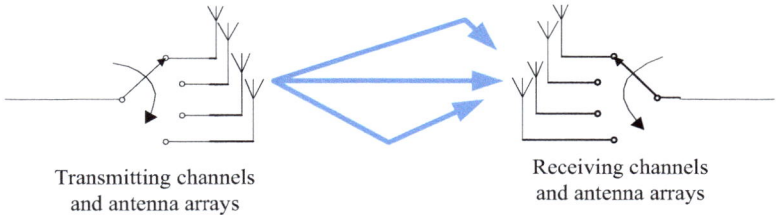

Transmitting channels
and antenna arrays

Receiving channels
and antenna arrays

Fig. 6.2 Schematic diagram of serial channel sounder (single-channel antenna array)

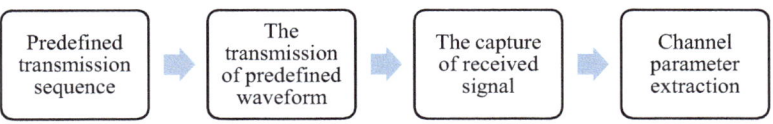

Fig. 6.3 Flow chart of channel measurement of serial channel sounder

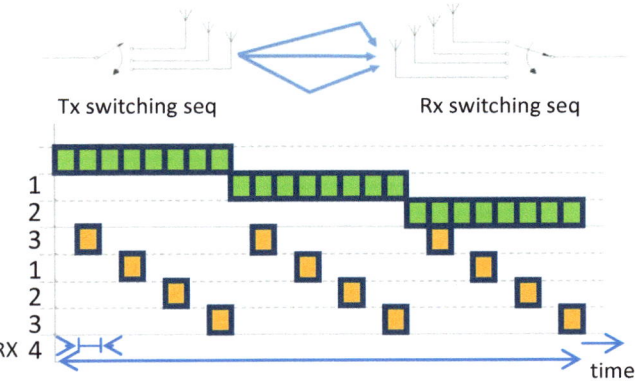

Fig. 6.4 Time-division multi-channel

As shown in Fig. 6.4, the channel sounder of this architecture has advantages that the channels are orthogonal in different ways. The calibration is convenient and the post-processing is relatively simple. This scheme, however, is restricted with the measurement period of all ergodic channels, so the channel coherence time is short. The scheme can only test channel characteristics when it is unchanged or in slowly changing static and quasi-static scenarios. This means it is not suitable for channel test under high mobility scenario. In addition, traditional serial channel sounder is limited by early equipment development conditions. It preserves CIR files, rather than raw channel test data, limiting the accuracy of channel Doppler measurement.

In other words, using the channel sounder with this architecture cannot measure time-varying channel when there are a large number of transmitting and receiving antennas. In the millimeter wave band, the coherence time of channel decreases as the carrier frequency increases. And when massive MIMO scenarios are considered, traditional serial channel sounder fails to meet the test requirements.

6.2.3 Key Technical Challenges

5G channel testing and measurement is urgently needed with the parallel channel sounder. However, the development of parallel channel sounder has to face the following difficulties and challenges.

(1) Synchronization across multiple channels

For the parallel channel sounder, picoseconds level synchronization accuracy among the MIMO channels both in Tx and Rx sides must be achieved in order to guarantee high spatial and temporal resolutions for channel parameters estimation. This goal is much challenging because of the differences between clock Phase-Locked Loop (PLL) circuits across multiple RF channels.

(2) Real-time storage of massive raw measurement data

For the parallel channel sounder, multi-channels' raw data with a high sampling rate per channel, must be simultaneously stored. Dozens of Gigabits data streaming require data transmission interface/bus with very high throughput and very efficient storage mechanisms design.

(3) Parallel channel calibration

Compared with TDM-based channel sounder which has only one Tx and Rx pair, multiple RF channels in parallel channel sounder can be considered as a multi-channel time-varying complex system. The non-ideal and different responses among every Tx/Rx pairs rooted from the multipath clock Phase-Locked Loop (PLL) circuits and RF devices. A new sophisticated parallel calibration algorithm must be designed to carefully compensate the non-ideal and difference in channel response as to guarantee the estimation accuracy of channel estimation.

(4) High-speed continuous storage of raw data

The serial channel sounder first correlates the received signal with local code to obtain the CIR, and then stores the CIR for each channel. However, this correlation yields the average impulse response over a long period. Accurate channel Doppler characteristics cannot be obtained based on this impulse response when the channel time-varying is fast. Therefore, the most effective way is to collect and store the original data. However, in the case of simultaneous storage of multi-channel with high sampling rate per channel, the archiving system must be able to achieve the high-speed transmission throughput with an efficient storage mechanism.

6.2.4   Architecture of MIMO Parallel Channel Sounder

Faced with these channel modeling requirements and technical challenges, Shanghai Research Center for Wireless Communications has developed a sophisticated parallel MIMO channel sounder. Compared with a serial channel sounder, the biggest difference is the use of multiple parallel RF channels based on code division, i.e., multiple parallel transmitter transmit simultaneously time domain sequences, and multiple parallel receiver receive the air interface signal at the same time. The use of such kind of architecture can fundamentally solve the problem of measuring fast time-varying channels in multiple antenna scenarios. Figure 6.5 is the picture of the channel sounder.

The system architecture of parallel channel sounder is shown in Fig. 6.6, mainly including the following subsystems:

- RF/baseband subsystem
- Clock / synchronization trigger subsystem

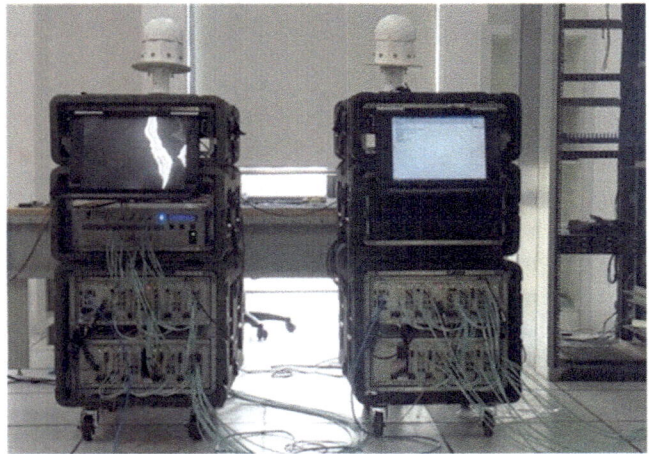

Fig. 6.5   Picture of MIMO parallel channel sounder

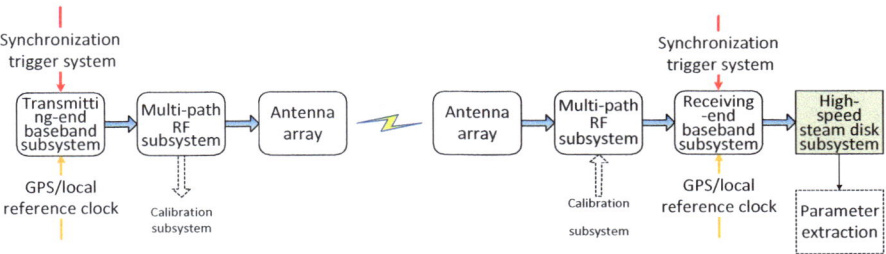

Fig. 6.6   System architecture of MIMO parallel channel sounder

- Calibration subsystem
- Antenna array
- High-speed stream disk subsystem
- Parameter extraction

1. RF/baseband subsystem

Parallel RF/baseband subsystem is the core of the entire parallel channel sounder. Its performance and stability directly determine the accuracy of measurement. The measurement system is realized by software defined radio platform based on PXI bus.

At the transmitter, the parallel channel sounder generates a set of PN with good cross-correlation characteristics as the transmission sequence of each channel. Then through up sampling and shaping filter, the signal spectrum shape is improved. After digital-to-analog conversion, the signal is converted to a high-frequency analog signal by up-conversion modules and then radiate through an antenna. In the receiver, down-conversion of parallel channel sounder and the signal after AD sampling are sent into FPGA. In FPGA, the data can be pre-processed accordingly. Then through Peer to Peer (P2P) FIFO, it is transferred to disk array for storage at a high speed (Fig. 6.7).

(1) Test signal screening

A major difficulty of MIMO parallel measurement scheme is that multiple concurrent detection signals will interfere with each other. Even with orthogonal sequences, the orthogonality cannot be guaranteed for different delay sequences, which will still produce interference. To this end, it is needed to fully consider the requirements of parallel channel to screen a set of eight signal sequences. The formula is shown in Table 6.1. These signal

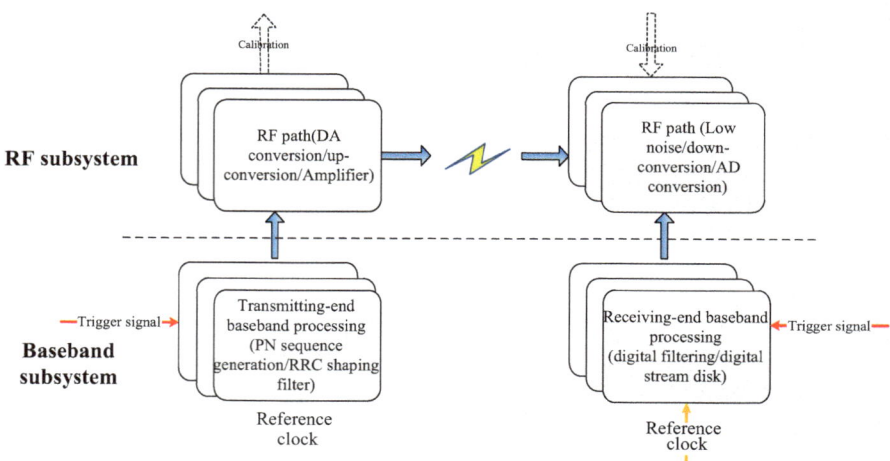

Fig. 6.7 RF and baseband subsystem

Table 6.1 PN sequence generation.

Sequence type of transmitter	PN sequence
Sequence length	2048
Primitive polynomial	$D^{11}+D^2+1$

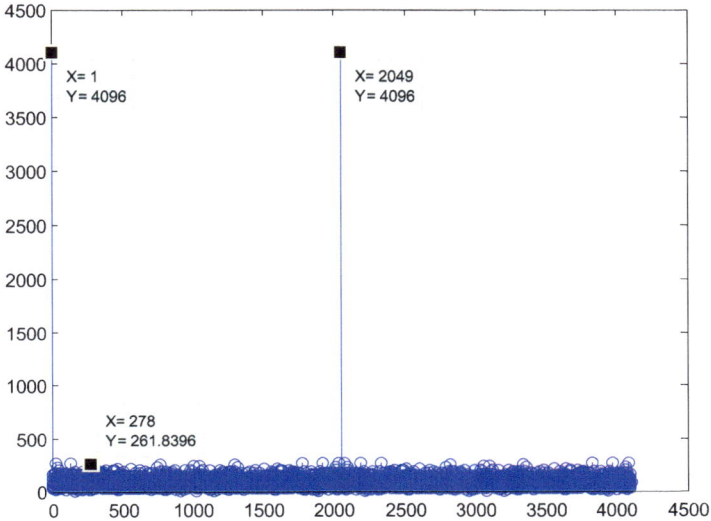

Fig. 6.8 Autocorrelation characteristic curve graph of eight test sequences

sequences are generated based on PN sequence optimization with the length of 2048. Figure. 6.8 and 6.9 respectively, show the autocorrelation and cross-correlation characteristics of the curve of eight sequences under modulation of orthogonal QPSK.

(a) Autocorrelation characteristic curve

   Figure 6.8 shows the simulation results for the autocorrelation performance of one of the test sequences, and the other seven are similar to this result. From the curve, these eight sequences all have very good autocorrelation properties.

(b) Cross-correlation curve

   Figure 6.9 shows the simulation results of seven cross-correlation curves of one test sequence, and the other seven are similar to this result. From the curve, these eight sequences all have good orthogonal characteristics, which can meet the requirements of channel measurement.

2. Clock/synchronization trigger subsystem

   Parallel channel sounder system requires strict synchronization and triggering mechanisms. It includes the following aspects: the strict synchronization of each transmitting channel, the strict synchronization of each receiving channel, and the strict synchronization between the transmitting and receiving channels.

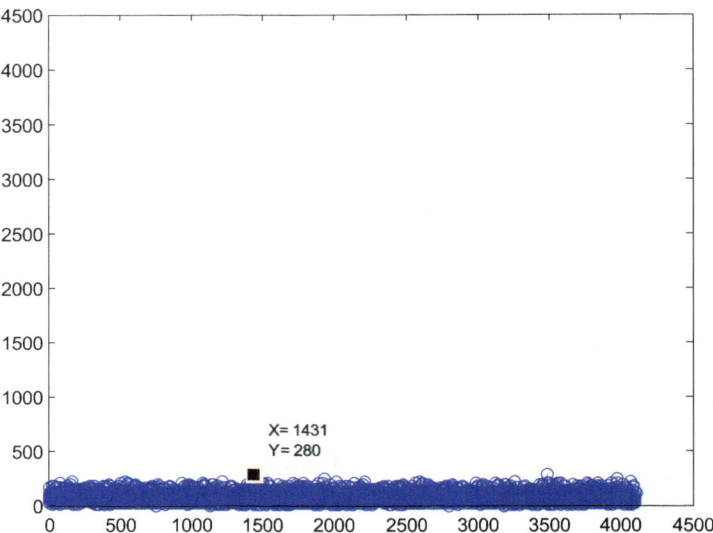

Fig. 6.9 Cross-correlation curves of eight test sequences

In order to achieve strict synchronization of the parallel channel sounder, the following synchronization mechanisms have been designed. First, each channel of receiver and transmitter of parallel channel sounder shares the same time-based reference signal and the same local oscillator. Second, the clock at both ends of transmitter and receiver must strictly implement the synchronization cycle and phase. In the system, two rubidium clocks are used at both ends of transceiver. And the jitter of two phases can be controlled to a fixed phase error after a period of recension of two rubidium clocks. Parallel channel sounder will send 10Mhz reference clock signal to both ends of transceiver, and make the reference clock of each channel at both ends of transceiver share these two 10Mhz reference signals, so as to achieve the precise clock synchronization of each transceiver channel. Both ends of transceiver adopt GPS receiver, to provide one pulse per-second signal (pps) as the initial synchronization trigger source. After triggering at the same time, the initial trigger signal, as a reference signal, is sent to both ends of transceiver to produce the periodic trigger signal. The periodic signal be send through PXI bus, sending to each channel of transmitting/receiving and triggering the FPGA transmitting and receiving signals simultaneously.

Although each channel of transmitter/receiver shares the same clock source, it is difficult to ensure that the phase of reference clock input to each channel is strictly synchronized. Also, considering more general scenarios, it is particularly desirable to have a common reference clock to align the respective sampling clocks when each channel is configured with a different sampling clock frequency. When the sampling time of each channel is aligned, synchronization trigger mechanism is adopted to achieve simultaneous transmission and reception. Parallel channel sounder increases the synchronization accuracy of channels to the picosecond level by the following strategies.

(1) Unified bus architecture of multi-channel channel test system. The same reference clock and local oscillator are used to ensure the equal clock and triggering signal path in each channel, which reduces the phase error introduced additionally.
(2) Shared trigger reference clock signal. During the initialization phase, according to different sampling clock frequencies and channel numbers of each channel, the Shared Trigger Reference Clock (STRC) signal is automatically calculated and generated, which is used for multi-channel phase-to-phase calibration.
(3) Real-time phase alignment based on FPGA. Real-time acquisition of phase difference between STRC signal and sampling clock of each channel is achieved by FPGA hardware programming for fine-tuning and aligning the phase between multi-channel sampling clocks.
(4) Regenerated trigger signal and delay sampling strategy. After the channel sampling clock completes the phase alignment, a short pulse trigger signal is regenerated. There is subtle difference in the time delay of receiving the trigger signal in each channel. After detecting the trigger signal, each channel samples at the next rising edge of STRC simultaneously. Thus, the synchronization accuracy improves further.

3. Calibrate subsystem

As described above, since parallel channel sounder can be considered as a multi-channel parallel time-varying system. Therefore, it is necessary to obtain the response of hardware system and the response of antenna system accurately before data post-processing, so as to formulate an accurate calibration scheme to deal with the time-varying characteristic.

In actual use, we design an 8-input 8-output calibration device to connect both ends of transceiver. For ease of operation and speed, an RF switch is designed at both ends of transceiver for the quick switch between "calibration" and "test".

First, the passive device in the whole system performs a system response test, getting the system response of calibration equipment and RF switch and then substituting it into parameter extraction post-processing software. At the same time, 3D pattern of antenna array is also been substituted into parameter extraction post-processing software.

Second, calibration is made. RF switch is turned to "calibration" state. The originator begins to launch, and then the receiver obtains the calibration data.

Third, the measurement starts. RF switch is turned to the "test" state. The calibrated RF path is turned off. And the receiver records measurements data at each test point.

Finally, parameter extraction is made. The system response of transmitting and receiving hardware and antenna pattern data are first removed from the measurements data, and then the extraction of channel parameters can be achieved.

4. Antenna array

As an indispensable part of channel measurement, MIMO antennas play an important role in MIMO channel measurement. Shanghai Research Center for Wireless Communications has developed an 8-antenna omni-directional antenna array as the measurement antenna.

The basic requirements for measuring antenna are:

(1) Transmitting and receiving antennas support 8-transmission and 8-reception respectively.
(2) Standing Wave Ratio (SWR) is not greater than 1.5.
(3) Antenna gain is greater than 6dBi.
(4) Antenna bandwidth is greater than 200 MHz.

To meet the technical requirements, the antenna design uses a monopole with 1/4 wavelength. Then, according to center frequency $f_c = 3.5$ GHz, $\lambda = 3 \times 10^8 / 3.5 \times 10^9 = 85.71$ mm. The antenna monopole length is 20.2 mm and diameter is 0.4 mm, with 1 pole in the center and 7 poles evenly distributing at the periphery. The distance between central monopole and any edge monopole is 40.3 mm. The antenna design is shown in Fig. 6.10. The antenna array using this scheme can effectively estimate 3D channel characteristics.

5. High-speed stream disk subsystem

The stream disk scheme of measurement system is based on the combination of zero-copy technology and asynchronous storage technology, and uses Technical Data Management Streaming (TDMS) to achieve the ultra high-speed and low-latency data storage scheme thus greatly reducing the CPU consumption.

In this measurement system, the bandwidth of PXI bus provided by case A backplane is 24 GB/s. For this high-speed bandwidth, the data transmission is achieved through the establishment of Direct Memory Access FIFO (DMA-FIFO). Another problem that affects the transmission delay is additional data copies and state transitions brought by the operating system kernel. Using

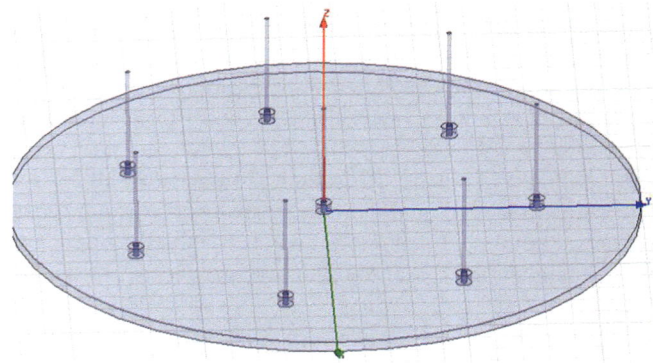

Fig. 6.10 Antenna array design

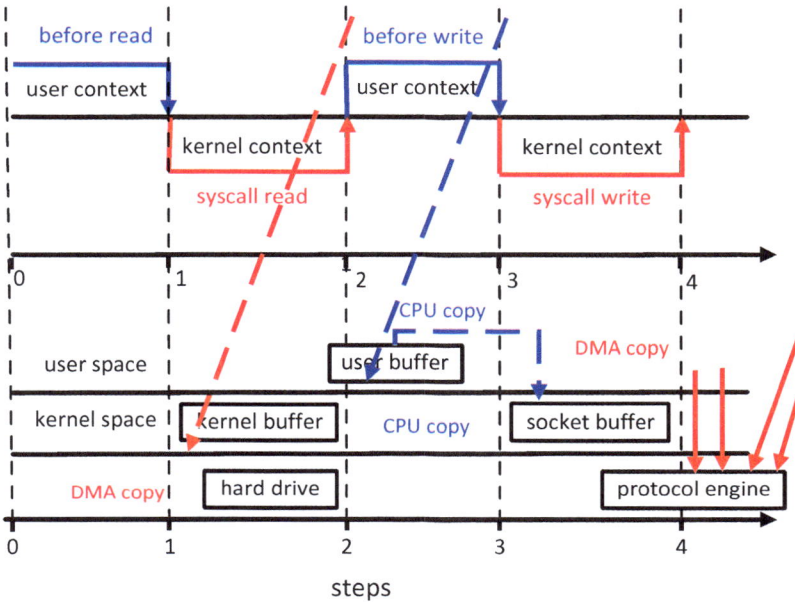

Fig. 6.11 Traditional I/O operation of a file

the zero-copy technique (Fig. 6.12) can effectively reduce the additional copy and state switch caused by the operating system kernel.

Figure 6.11 is the traditional I/O operation of file.

It can be seen that in order to complete the process from reading data in hardware to write it in disk, four copying processes are needed. First, in the system call of reading the file, the hardware FPGA transmits data to the kernel buffer in CPU through DMA. Then, CPU performs copy operation to copy the data in its user cache. In system call of writing file, CPU copies the data in user cache to socket cache, and finally transfers to disk through DMA-FIFO.

Figure 6.12 is the I/O operation based on the zero-copy technology. After the zero-copy technology is used, only the operations that hardware FPGA transmits to the kernel buffer through DMA and the socket buffer transfers to disk through DMA are retained, eliminating the need for two data copies in kernel space and reducing delay and liberating CPU. At this point, the CPU-triggered disk writing is an asynchronous operation. That is, it can execute other processes in parallel without waiting for the completion of DMA transmission.

As shown in Fig. 6.13, TDMS file storage format is a binary file storage form, which has an advantage of small space usage. In addition, it can classify stored data and abstract the composition of data into three categories: top-level file group, channel group, and channel. A file group may contain several channel groups, and a channel group can contain many channels.

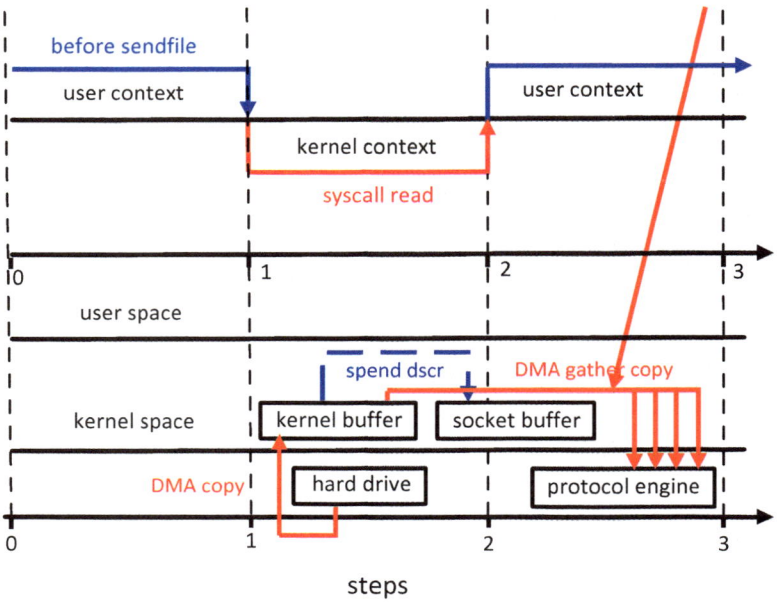

Fig. 6.12 I/O operation based on zero copy technology

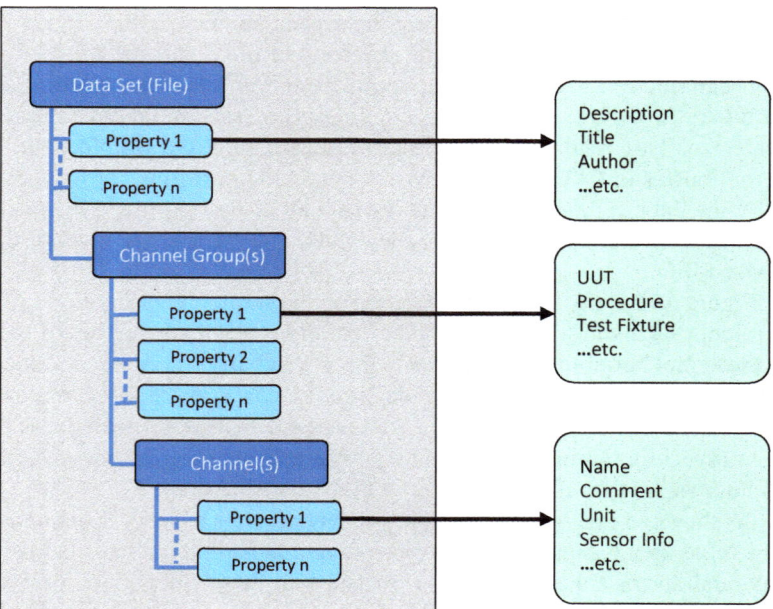

Fig. 6.13 Organization mechanism of TDMS file

Actual stream disk program includes the GPS location information, IQ data or other relevant information. These data need to be precisely synchronized. Using TMDS can easily combine these data together and package them in a file as a different data set or channel. This greatly simplifies the difficulty of different data synchronization in data post-processing.

6. Parameter Extraction

SAGE algorithm is an iterative method of maximum likelihood estimation. It can realize the optimization of multidimensional parameters by allocating the observed data in multiple subspaces and then estimating the multipath parameters in each subspace.

In order to obtain the results of the maximum likelihood estimation, the SAGE algorithm uses iterative methods to carry out the interference cancellation on the observed values in subspace, so as to ensure the monotonic rise of overall likelihood of parameter estimation with increasing iterations. The optimal estimation of output parameter set is achieved when the requirements for convergence are satisfied.

The measurement platform of parallel transmitting and receiving channel is optimized based on the multi-dimensional parameter estimation SAGE algorithm in the development of feature extraction software. The parallel sounding SAGE (P-SAGE) algorithm is proposed. This algorithm has the following functions and features:

The received signal data analysis in baseband based on parallel M × N MIMO system;

- Good orthogonality based on the baseband spread spectrum code;
- The non-uniformity among RF channels is fully considered;
- The calibration is done separately at Tx and Rx;
- The complexity of algorithm is well reduced to achieve fast estimation;
- The algorithm is optimized for the presence of phase noise and can be applied to scenarios where phase noise exists.

Figure 6.14 shows a block diagram of the parameter estimation algorithm with multi-dimensional channel characteristic based on the PS-SAGE algorithm. The innovations of PS-SAGE algorithm are:

(1) Suppression of multi-channel phase noise. Parameter extraction and data post-processing algorithm use multi-channel phase noise sample data to construct the maximum entropy statistical model. The maximum likelihood

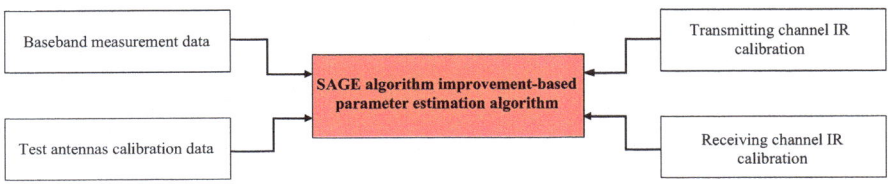

Fig. 6.14 PS-SAGE algorithm block diagram

method is used to estimate model parameters. The multi-channel phase noise model and space multipath propagation channel model are combined. Multi-channel phase noise covariance matrix and channel multipath parameter estimation perform joint iterative updating. The influence of multi-channel phase noise is suppressed by using sub-space received signal to finish phase noise bleaching processing.

(2) Multi-channel interference cancellation. The initialization parameters are set to reconstruct the interference modes, such as multi-channel multipath self-interference, orthogonal interference among transmitted signals and mutual interference among received signals. The interference is canceled in the iterative process of estimating parameter to effectively guarantee the accuracy of multi-channel multi-path parameter estimation.

(3) High-precision channel parameter estimation. The expected function iteration and maximum estimation iteration are made for multi-channel signals after canceling the interference. A more accurate channel estimation parameter is obtained and then the multi-channel interference is canceled (Step 2) to achieve rapid convergence through the cycle iteration to obtain high-precision channel estimation parameters.

6.2.5   Test Results

The test results of the parallel channel sounder in anechoic chamber are shown in Fig. 6.15:

For the static scenario in the chamber, we choose PN sequence with length of 4096 and an I/Q chip with rate of100 M/s. Each snapshot collected 100 cycles. The total length is the measurement data within 10 ns × 4096 × 100 × 2 = 8.192 ms.

We first use the 8-input 8-output connector to make calibration and measurements, and then test via air interfaces. The test results are shown as below.

Fig. 6.15 System verification test in microwave anechoic chamber

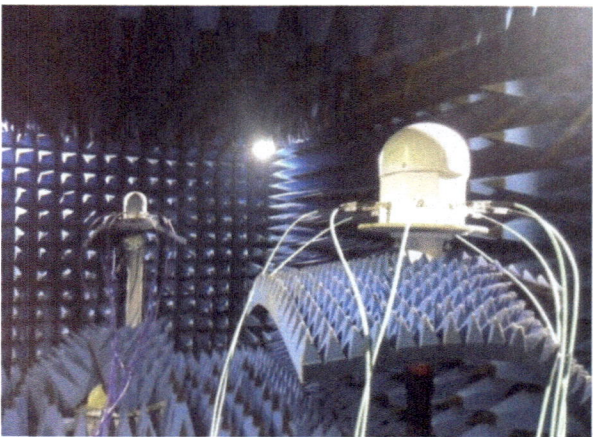

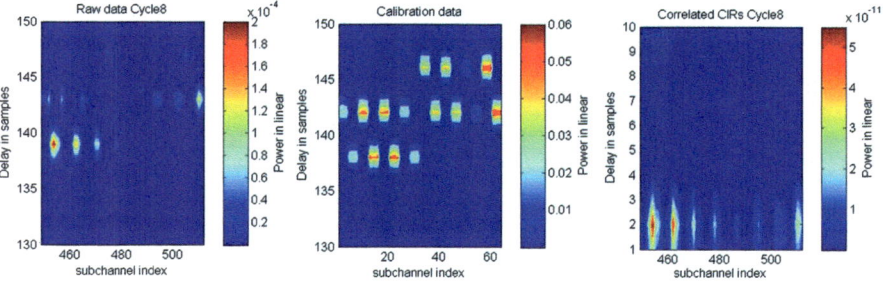

Fig. 6.16 Multi-path delay spectrum before and after calibration

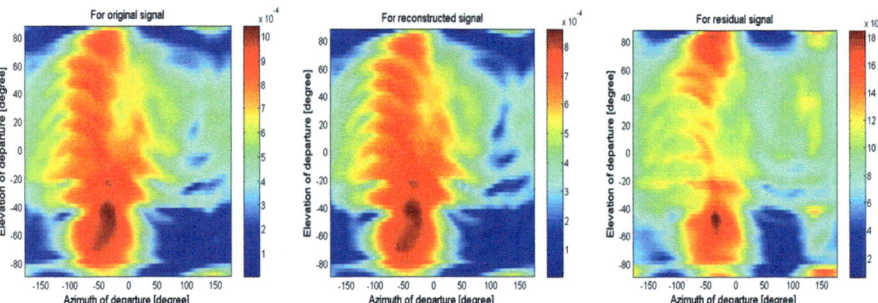

Fig. 6.17 PS-SAGE algorithm estimates the spatial parameters and reconstructs the signal spatial spectrum

Figure 6.16 is pair wise multi-path delay spectrum of 64 groups for In it, they are respectively, from left to right, air interface collection, calibration data, and the results after using calibration data compensation. It can be seen from the figure that after using 64 calibration files to calibrate original signals, the response difference of each transmitting and receiving channel can be well compensated to obtain the consistent delay as shown in the right figure.

Then here are the test results of channel parameter estimation. In Fig. 6.17, the elements from left to right, respectively are air interface data, reconstructed signal based on parameter estimation, and the power spectrum of Azimuth of Departure (AOD)/ EOD of signal after cancellation. In it, the horizontal and vertical coordinates are AOD and EOD respectively. The color represents the intensity of spectral component. It can be seen from the figure that the reconstructed signal has a strong similarity with the original signal spectrum. The energy of reconstructed signal compared to the energy the original signal is reduced by 37 dB. The chamber verification shows that the spatial parameters of wireless channels which are obtained from the post-processing algorithm estimation are very close to actual ones.

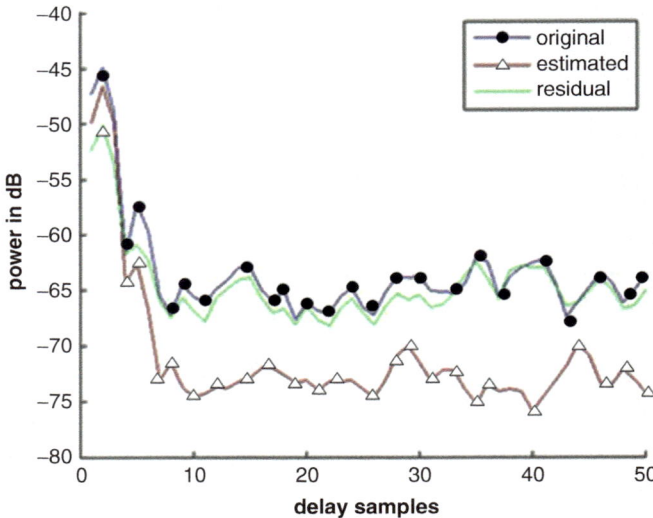

Fig. 6.18 PS-SAGE algorithm estimates the time domain parameters and reconstructs the signal time-domain spectrum

Figure 6.18 is a view of the time domain characteristic of wireless channel characteristics. The estimated delay spectrum is consistent with the morphological trend of original time-delay spectrum. And the original signal power can be reduced by more than 10 dB after cancellation.

From the above time domain and spatial parameter estimation results and the results of reconstruction and elimination, it can be seen that the channel measurement data and estimation results are reliable, which can have a more accurate correspondence with actual chamber environment.

6.2.6   Channel Measurement and Test

Scenario Test Planning and Measurement

After finishing the development and chamber verification of parallel channel sounder, Shanghai Research Center for Wireless Communications begins the planning for the channel scenario test with typical geographic characteristics defined by IMT-2020, which includes 12 scenarios with hot spots, high capacity and seamless wide-area coverage.

The parallel channel sounder has now completed the following scenario test, collecting the preserving mass data of parallel measurement channel:

- Anechoic chamber environmental verification
- Office scenarios

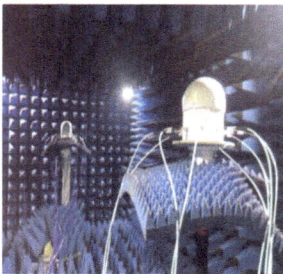  

Chamber test   Large indoor venues test   Public community test

Fig. 6.19 Main scenario graphs of progressed channel tests of parallel channel sounder test

- Large-scale indoor scenarios
- Stadium outdoor scenarios
- Rural outdoor scenarios
- Shanghai Circuit V2 V scenarios
- City hot spot regional scenarios
- High mobility scenarios (Fig. 6.19)

Channel Model Library

Based on the parallel MIMO channel sounder and the definition of 5G application scenarios in IMT-2020 (5G) promotion group white paper, Shanghai Research Center for Wireless Communications began to construct shared data channel model based on practical measured data in 2014. The designed channel model library will basically cover the following two main application scenarios.

(1) 5G mobile Internet application scenarios

The future 5G mobile communication system will meet diverse service requirements in different areas, such as residence, workplaces, leisure venues, and transportation venues. It could provide ultra-high definition video, virtual reality, real-world enhancements, cloud desktop, online gaming and other excellent service experience for customers in different scenarios. The scenarios include the places of dense residential areas, office, stadiums, open-air gatherings, subways, highways, high-speed rail, and wide-area coverage and other application scenarios with ultra-high mobile data traffic density, ultra-high mobile connection density and ultra-high mobility.

All above application scenarios can be summarized into two types: continuous wide-area coverage scenario and high-capacity scenario with hot spots.

(2) 5G IoT application scenarios

The future 5G mobile communications system will penetrate to IoT and various (vertical) industry fields, deeply integrating with industrial facilities, medical equipment, vehicles, etc., so as to effectively meet the service requirements of diversity of many vertical industries, such as industry, medicine, transportation, (ultimately) to achieve true "all things interconnected".

All of the above application scenarios can be summarized into two types: high connection scenarios with low power consumption and high reliability scenarios with low delay. We can see that these two applications are newly explored by 5G, which can be used to solve the wireless IoT applications and vertical segmentation industry applications which have not been well supported by traditional mobile communication systems.

Based on the existing channel test database of scenarios, Shanghai Research Center for Wireless Communications shares the data acquired by the parallel channel sounder with the industry peers through big data open-source platform of wireless channel under construction. And according to the requirements, the test data are developed to promote the development of channel measurement and modeling work in the industry. There are already Tongji University, Shandong University and other universities making related research and cooperation on open-source platform data.

The eight units firstly signed up for wireless channel big data open-source platform under construction are: Beijing University of Posts and Telecommunications, Tsinghua University, University of Electronic Science and Technology, Southeast University, Tongji University, Forty-first Study Institute of China Electronics Technology Group Corporation, First Study Institute of Telecommunications Science and Technology, Shanghai Advanced Studies Institute of Chinese Academy of Sciences.

6.3   OTA Test Platform

6.3.1   5G Requirements for OTA Test

The method to evaluate radiation performance of traditional SISO is now more mature. Its evaluation mainly aims for two indicators, making OTA test for Total Radiated Power (TRP) and Total Radiated Sensitivity (TRS). The two indicators respectively characterize the ability of a mobile terminal to transmit as a whole, including an antenna device and to receive signals from a base station, which can effectively evaluate the transceiver performance of a single antenna terminal in the real world. As early as in 2001, the Cellular Telecommunications and Internet Association (CTIA) began to study the performance evaluation methods of conventional SISO antennas.

Many performance parameters of passive MIMO antenna, such as efficiency, gain, are not fundamentally different from traditional SISO antennas. Since MIMO antennas contain many antenna elements, passive parameters are introduced to describe the relationships among several antenna elements. Therefore the existing SISO OTA testing techniques cannot well adapt to the requirements of MIMO application testing. Since the end of 2007, relevant organizations have begun to follow up the applications and extensions of the SISO testing technology in MIMO OTA test. Many standardization organizations and agencies including the 3GPP, COST and CTIA named the test method of evaluating the performance of multiple antennas as MIMO OTA.

The MIMO OTA test scheme is essentially a different simulation method of multipath in space propagation environment, producing wireless communication environment close to the reality to deal with the key challenge in massive antenna MIMO OTA testing techniques. The challenge is how to generate an RF channel in anechoic chamber that is closest to the real-world spatial, angular and polarization behavior. Therefore, the main requirements can be divided into three types.

1. Emulator of the real wireless propagation environment.

   This requirement is currently addressed by a channel simulator, which simulates the channel characteristics of test scenarios through a real-time channel model. These features are also reflected in time domain, frequency domain and spatial domain. Only the exact channel model can approximate the test scenarios.

2. The close-to-reality wireless propagation playback environment.

   The channel environment is played back by using different configurations of chambers and antenna arrangements to combine with the channel emulator. In particular, the spatial characteristics of signal should be considered.

3. Calibration of OTA test system.

   In MIMO systems, spatial correlation is a very important parameter, which contains the characteristics of antenna and transmission channel. In fact, the unknown antenna characteristics of known antenna model cannot get the correlation. Likewise, the unknown channel model with known antenna characteristics cannot get the correlation. Therefore, the characteristics of antenna and transmission channel to test multiple antenna terminals must be considered.

6.3.2   Status and Shortcomings of OTA Test Scheme

The Technical Report (TR) of 3GPP organization of mobile communication partnership program 37.976 summarizes seven MIMO OTA test schemes. These schemes can be divided into three categories according to chamber features: OTA test scheme based on Anechoic Chambers, OTA test scheme based on Reverberation Chambers and test scheme based on Multi-stage Method.

In the summary of 3GPP TR 37.976 candidate test scheme, OTA candidate testing schemes based on anechoic chambers include:

(1) Multi-probe method;
(2) Symmetrical ring of the probe method;
(3) Two channel method;
(4) Spatial fading emulator method.

Likewise, OTA candidate test scheme based on the reverberation chambers include:

(1) Basic or cascaded reverberation chamber;
(2) Reverberation chamber with channel emulator.

Test scheme based on multi-stage method is a two-stage method.

OTA Test Scheme Based on Anechoic Chambers

The test scheme based on anechoic chambers will connect RF channel simulator to the probe array in an anechoic chamber environment. The probe array surrounds the DUT to reproducibly simulate a wireless environment that generates complex multipath fading at the location of DUT.

(1) The multi-probe OTA test block diagram based on anechoic chamber is shown in Fig. 6.20, which mainly includes anechoic chamber, system simulator, MIMO channel simulator and test probes, etc. The multi-probe OTA test based on anechoic chamber is the most intuitive test method. The test system simulates the downlink signal of a base station with channel simulator, launches it through many test probes in anechoic chamber, and simulates the channel model with a specific AOA in chamber center, and then sets terminal under test in chamber center and tests the MIMO antenna performance in channel scenarios gotten from simulation.

Tests for multiple antenna wireless devices are typically conducted in a conduction mode, and proper fading is typically simulated using a channel simulator. The channel model used in the current test includes the signal AOD and AOA information, as well as antenna patterns on both sides of the transmitter and receiver of channels. These spatial channel models can be better adapted to environmental simulation requirements of OTA test by modifying the channel model in MIMO channel simulator.

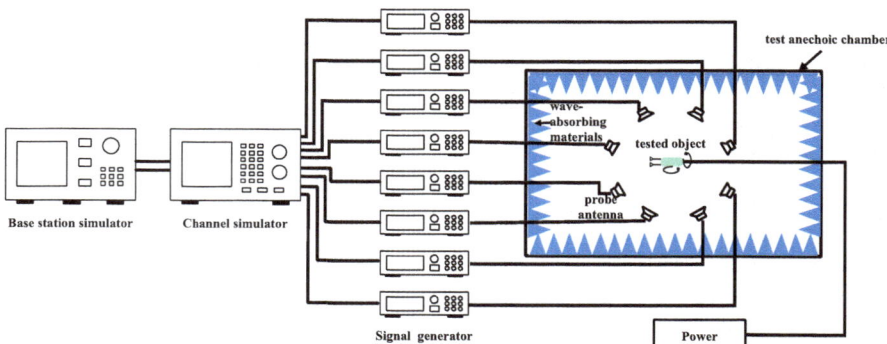

Fig. 6.20 Multi-probe test scheme based on anechoic chamber

In such a scheme, the DUT is free to use the number and position of probes to combine different test scenarios. For example, the probe can be placed in the same orientation to simulate the 2D scenario, and the antenna can be placed in different planes to simulate the 3D scenario. The system configuration can be adjusted to obtain a signal similar to that received by the device in the real world, thereby making it easier to evaluate the position, direction and impact of MIMO antenna on system. The advantages of the multi-probe method are that the concept is simple, configuration is flexible, and the testing accuracy is good. However, the disadvantage is that multi-probe may cause high cost. Another challenge of the multi-probe approach is that it can only simulate parts of the 3D channel, making it difficult to simulate a complete 3D channel.

(2) Ring-shaped probe test methods based on anechoic chamber symmetrically distribute the probe antenna around the DUT equidistantly, and place the measured object in the center of chamber, as shown in Fig. 6.21. The system consists of anechoic chamber, multidimensional fading emulator, communication tester / BS emulator, OTA chamber antenna, etc.

Similar to the multi-probe approach, each probe antenna transmits signal with a specific time domain after the processing of the channel simulator based on the ring probe test method of an anechoic chamber. Unlike the multi-probe scheme, there is no correspondence between signal AOD and the position of the probe antenna. Based on it, it is possible to simulate any 2D spatial channel model without readjusting the position of the probe antenna.

(3) The test method based on Spatial Fading Emulator (SFE) of darkroom was initially proposed by Panasonic Company in Japan. The main feature of SFE is the manufacture of spatial fading characteristics through the antenna probe and associated RF devices surrounding the DUT. The amplitude of the signal emitted by each probe is directly determined by the sampling of target PAS. Doppler frequency shift depends on the angle between the probe position and the direction of the virtual motion of DUT. Since there is no pre-fading, the

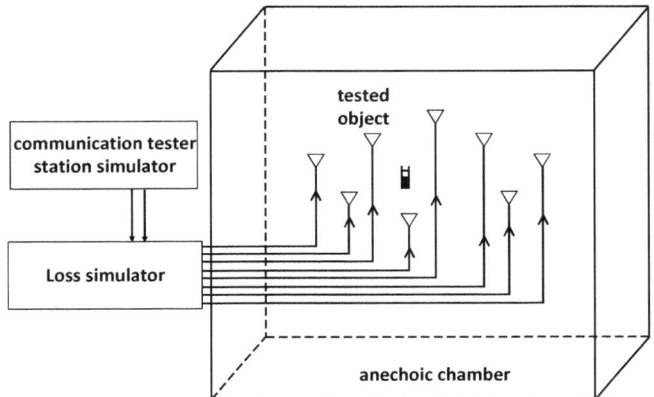

Fig. 6.21 Schematic diagram of a ring probe test method based on anechoic chamber

number of probes determines the pros and cons of SFE for the reconstruction of synthesized channel model. A theoretical study on the number of full-ring antenna probes made by NTT DoCoMo has shown that for a single-cluster, the test method of SFE based on chamber generally requires 10 probes to satisfy 1.5 times of wavelength terminal test. In the multi-cluster channel, at least 11–15 antenna probes are needed to meet the requirements of the test domain with 1.5 times of wavelength.

The SFE test method based on the anechoic chamber is essentially regarded as a variation of the ring probe test method. The test method of SFE is to use a programmable attenuator and an RF phase shift device to adjust the amplitude and phase of each antenna to replace the channel simulator in the ring probe method. The block diagram of its system is shown in Fig. 6.22. The RF signal from a base station simulator is fed into a power divider, which provides the same RF signal for each output. The number at the output end of power divider is the same as the number of probe antennas. And each output is connected to an RF phase shifter, which carries out phase offset adjustment by receiving a control signal from the digital-to-analog converter. Through the attenuator, the signal passing through the RF phase shifter is output to a horizontally or vertically polarized probe antenna. The DUT measures the signals from each antenna probe and outputs the data to a computer control terminal for further

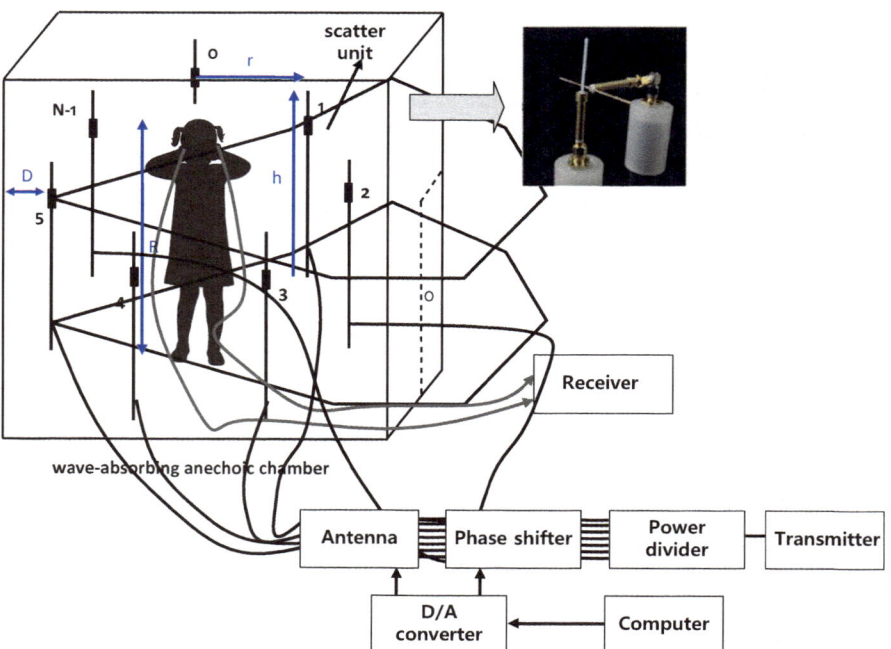

Fig. 6.22 Test method for SFE based on anechoic chamber

processing. At the same time, the computer control side also provides the configuration and interaction of simulation parameters.

The SFE test method based on the anechoic chamber can generate multipath fading of a particular distribution by controlling and adjusting the amplitude and phase of RF signal in real time, such as Rayleigh distribution. Because the scheme uses an attenuator to control the amplitude, the RF phase-shifting device adjusts the phase, which is less expensive than a commercially available channel simulator. But its flexibility has also been greatly limited. In addition, the method cannot simulate the characteristics of the transmitter.

(4) The dual-channel measurement method based on anechoic chamber is a direct and valid method to test OTA performance of MIMO equipment. Its principle is shown in Fig. 6.23. At the same distance from the UE, two polarized test antennas with a rotatable incident angle are placed, transmitting different MIMO downlink signals respectively. The overall characteristics of UE antenna are obtained by a combination of various azimuth and polarization. As shown in the figure below, the two-channel chamber contains four angular positioning devices: φ, θ1, θ2, γ angle positioners. The two test antennas A1 and A2 are distributed as 10° or 90° angle (simulate rural environment) and a communications antenna ANTUL (simulate urban environment). An optional γ angle positioner on turntable controls the tilt angle of antenna under test. All of these angle controllers can be used to implement any combination of angles for MIMO test. External devices include a base station simulator and a switching matrix. The dual-channel approach can be viewed as a special case of a multi-probe method using only two probe antennas without a channel simulator.

The advantage of the dual-channel test method is the easy updating on the basis of the original SISO test system. It only needs to add the second angular positioner control system and a second test antenna, which greatly reduces the cost. At the same time, it can also be used to verify the pattern, load and impede the smart antenna which can be adaptive with the environment changing.

Fig. 6.23 Dual-channel test method based on anechoic chamber

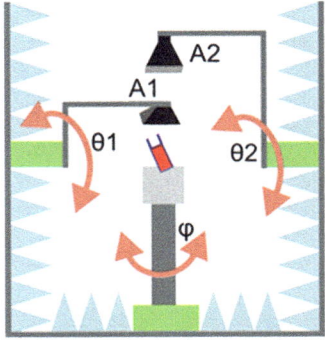

OTA Test Scheme Based on Reverberation Chamber

The basic principle based on reverberation chamber OTA test scheme is actually to provide the characteristics that the reverberation chamber is rich in reflection and the specific channel environment created by mixer for DUT testing. The purpose of the reverberation chamber is to produce a statistically uniform power distribution around DUT. And the antenna and channel simulator can be used to generate the desired delay characteristics.

Depending on whether the reverberation chamber is connected to a channel simulator, the OTA test scheme based on reverberation chamber can be divided into two types. The first is to use a separate reverberation chamber or a cascade reverberation chamber, as shown in Figs. 6.24 and 6.25. And the second is the reverberation chamber to connect the channel simulator. For example, the participation of channel simulator in the second case realizes time diversity through inputting a fading signal at different time steps, overcoming the limitations in the first case.

In general, the reverberation chamber can provide a subclass of multipath environments that can do frequency fading and time diversity simulation. The reverberation room test scheme is limited by limited analog capabilities of different fading environment, so it can only provide a limited performance evaluation for terminals.

Limited by the statistical isotropy at receiver in spatial domain of channel environment, i.e., its evenly distributed AOA, the spatial diversity cannot be simulated. In addition, the statistical isotropy in reverberation chamber also determines that vertical polarization and horizontal polarization of the channel model

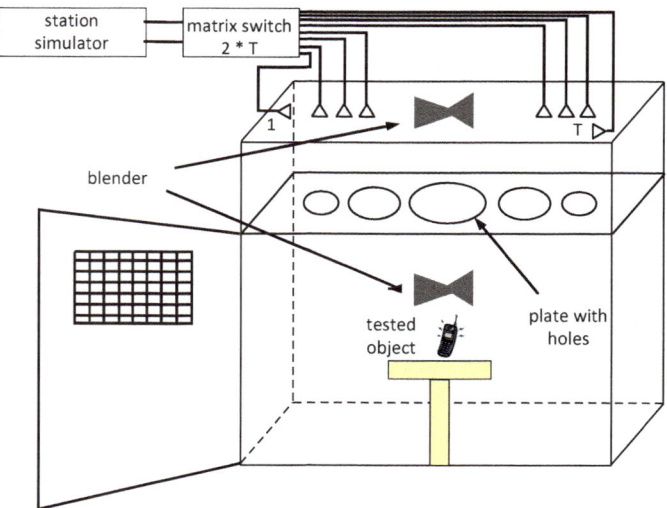

Fig. 6.24 Test method of basic reverberation chamber

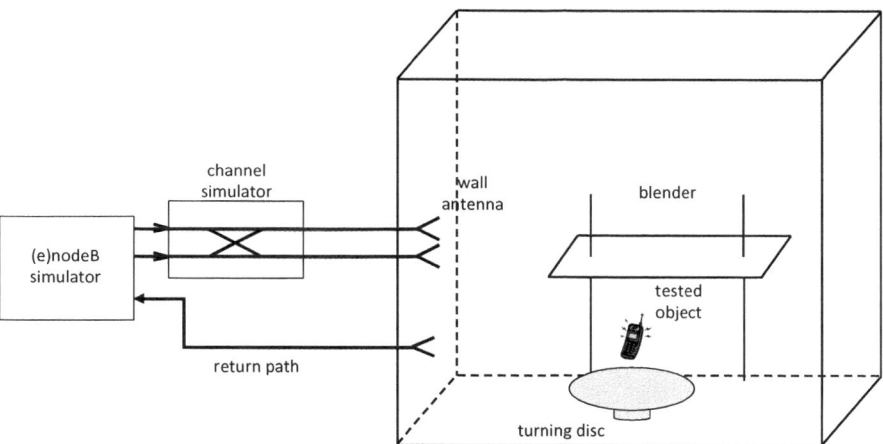

Fig. 6.25 Reverberation chamber test method with channel simulato

can only be equal. For the DUT using polarization diversity method, reverberation chamber method cannot be effectively differentiated.

The scheme based on the reverberation chamber does not specify the device. The same result can be obtained with any device. The size of test domain is not strictly required, which is very convenient for the actual test operation.

Two-Stage OTA Test Scheme Based on Anechoic Chamber

The two-stage test method based on the anechoic chamber is shown in Fig. 6.26. The two-phase test scheme consists of two test phases. The first stage, in an isotropic environment, uses a conventional anechoic chamber as a basic test system and an integrated tester to measure a complex active antenna array. The second stage combines the information of antenna array with the channel model by the following two means: using channel simulator to make conduction test or using antenna array information obtained from test to calculate a theoretical channel capacity performance by theoretical calculation. Therefore, at this point, the two-stage test method can only obtain limited data, and further research is needed to obtain accurate performance indicators.

In the two-stage OTA test method, the absolute accuracy of DUT power measurement has little effect on the accuracy of final results, since the correction is made at the second stage. The accuracy of relative phase measurement has great influence on the result. But the existing DUT, usually mobile terminal, has strong phase measurement ability, so it has limited influence on measurement result. Because 3D antenna pattern can be measured more easily, a two-stage OTA test method can simulate any 3D channel transmission condition. The quality factors

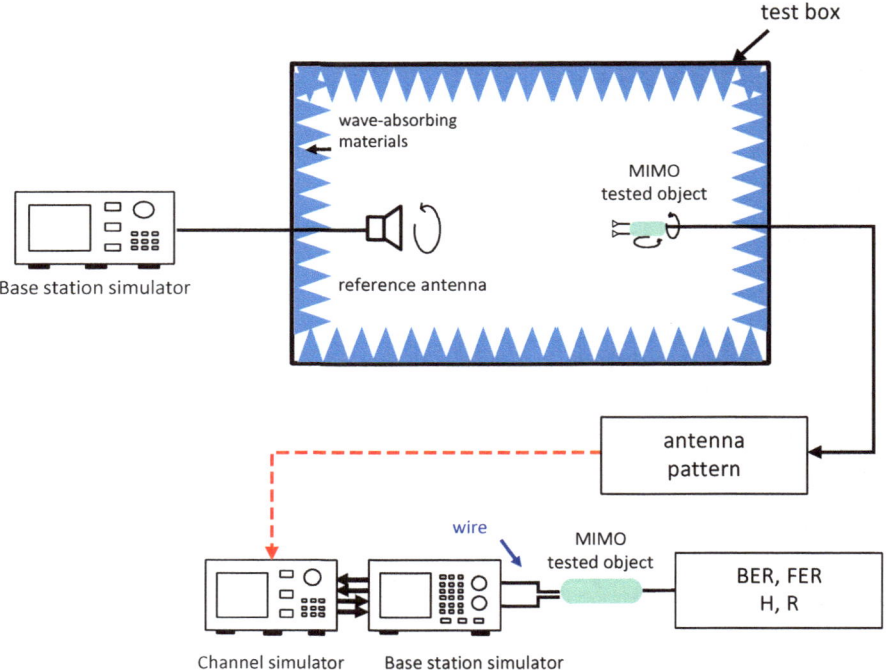

Fig. 6.26 Two-stage test method based on anechoic chamber

that the two-stage OTA test method can measure include TRF, TRS, throughput, block error rate, MIMO channel capacity, antenna correlation coefficient, etc. However, another feature of two-stage approach is that it cannot obtain the effects of self-interference.

Compared with the conventional multi-probe test scheme, two-stage OTA test only simplifies the receiving diversity performance rather than the channel-related characteristic, which is a fast, accurate, economical and efficient MIMO OTA test method. Secondly, the two-stage OTA test method can re-use the antenna pattern obtained from test to simulate a 2D or 3D channel model without re-using an anechoic chamber test, and thus improving its flexibility, and taking full advantage of test platform resources constructed in the LTE phase to rapidly expand and achieve MIMO OTA test, which is a fast and economical test solution.

In summary, the existing schemes focus more on two requirements of "building and playing back wireless propagation environments close to the reality" and "OTA test system calibration". While "wireless propagation environment close to reality obtained from emulator" all relies on ready-made commercial channel simulators, so the pertinence and applicability of channel environment have common flaws.

6.3.3 Key Technical Difficulties and Challenges

Compared with traditional single-antenna OTA testing, the MIMO OTA testing technology adds multiple antenna broadband testing. The tests must use multi-dimensional RF channel parameters, such as fading, delay, Doppler, AOA and polarization and other evaluations.

The key challenge in the development of MIMO OTA test platform is how to generate an RF channel model in anechoic chamber that is closest to the real-world spatial, angular and polarized behavior. Therefore, further study is needed for the test combining channel simulator and chamber, and antenna characteristics must be taken into account.

The key challenge in massive MIMO OTA test techniques is how to generate an RF channel model in the anechoic chamber that is closest to the real-world spatial, angular and polarized behavior. This complexity requires a lot of space and equipment investment for the R&D of MIMO OTA test platform. Its cost is too high for the majority of terminal equipment manufacturers. The following is the discussion of key technologies and difficulties of OTA test.

1. Spatial fading simulation technology

   Since wireless channels play a key role in MIMO performance, wireless channel simulator is an important part in MIMO OTA air interface test system. A test signal generated by a transmitter or a base station simulator passes through a wireless channel simulator, which simulates a wireless channel according to a predefined channel model. The signal is then separated in simulator and distributed to each probe in the chamber, each independently radiating into the chamber. Its result is that the multiplexed radiated signal is synthesized in the central space of the chamber and produces the desired wireless channel environment around DUT.

   The advanced SFE technology must be adopted to restore the real living environment in the chamber. The most typical analog parameters include path loss, multipath fading, delay spread, Doppler spread, polarization, and, of course, spatial parameters such as AOA and AS.

   In order to obtain valuable results from MIMO OTA test, wireless channel simulator must have excellent RF performance. Its Error Vector Magnitude (EVM) and internal noise level must be very low in order to minimize errors that affect the measurement results.

   Moreover, in order to obtain consistent test results in multiple measurements, the fading process must be repeatable. This is very important when benchmarking different DUT.

2. Stochastic channel model based on geometry

   The channel model for MIMO OTA test is a GSCM, in which the wireless channel is defined by the following parameters:

   - The location and array of transmitting antennas;
   - Propagation characteristics (delay, Doppler, AOD, AOA, angle spread of transmitted signal, angular spread of received signal and polarization information);

- Movement speed and travel direction;
- The location and array of receiving antenna;
- And multiple large-size parameters.

These channel models based on measurement include all parameters of wireless channel (time, frequency, space and polarization). Since both space and polarization are key parameters of spatial correlation (MIMO performance is strongly correlated with it), these two parameters are very important. Obviously, this method accurately and vividly simulates the environment required by MIMO device test. The signal received by the receiver is obtained by transmitted signals of the transmitter through a multipath environment with the spatial characteristic. Therefore, the spatial characteristic, including receiving antenna, becomes the main factor of testing receiver performance. Stochastic channel model (ie, geometric-based stochastic channel model) series must also include the channel models of Spatial Channel Model (SCM) of 3GPP, Spatial Channel Model Extension (SCME), ITU, and International Mobile Telecommunications-Advanced (IMT-Advanced).

3. MIMO OTA channel model mapping

In MIMO OTA test, the receiving antenna is integrated directly on DUT and becomes an integral part of it. Therefore, it is not necessary to load its antenna array in the channel model of simulation. The actual impact on the performance of DUT can be directly obtained through testing. In this way, the key challenge in MIMO OTA testing is to generate actual propagation characteristics in the chamber, especially AOA and its ASA. The information based on geometric parameters, such as in SCM, should create appropriate correlation parameters for the antenna of DUT. In addition, the information of the transmitting antenna array (base station) is also needed, including geometric information of the array and the array of antenna fields, as well as the speed component or Doppler frequency component of each path or cluster for terminal.

These clusters are then mapped at the same time to the corresponding OTA antenna, so that the spatial synthesis result of transmitted signals in the center of chamber is consistent with the defined model. The mapping is performed by a spatial wireless channel simulator. In order to make the angle expansion more precise, each cluster of signals is generated by the several OTA antennas. As a result, a wireless channel environment based on geometric information can be accurately generated in the chamber.

OTA systems must also support complex full-3D systems expanding from the simple single-cluster system. The high degree of correlation caused by extremely narrow AS and the requirement for the rotation of DUT will significantly affect performance results. Wider AS makes the correlation low. The channel is more easily to achieve spatial multiplexing. All 2D channel models (such as SCM, SCME, WINNER, IMT-Advanced) need to be simulated using a 2D full-circle system. A complete 3D application allows the test results to take into account not only the azimuth, but also the pitch propagation.

6.3.4   OTA Platform Example

OTA Platform Architecture

As shown in Fig. 6.27, the MIMO OTA platform developed by Shanghai Research Center for Wireless Communications is a scheme based on the anechoic chamber as a whole. The principle of the scheme has been described in Sect. 6.3.2.1, and the different points of the platform are highlighted here. The biggest difference lies in the following two aspects.

I. An on-line channel model library is added to test platform as a support for channel model. The on-line channel model not only contains the standard channel model of the traditional channel simulator, but also has a typical channel model of China, which increases the applicability of the channel model in China.

II. In addition to using the traditional channel simulator based on channel models, this platform can combine the parallel channel sounder to perform real-time playback for the captured signal under appointed locations (areas). The evaluation reliability of DUT for the suitability of appointed channel environmental capabilities is improved.

The channel model library has been described in Sect. 6.2.6 in details. The channel analog/playback device is highlighted below.

Channel Analog/Playback Device

As shown in Fig. 6.24, the channel analog/playback device of the OTA test platform developed by Shanghai Research Center for Wireless Communications uses two forms: the commercial channel simulator and the self-developed channel simulator.

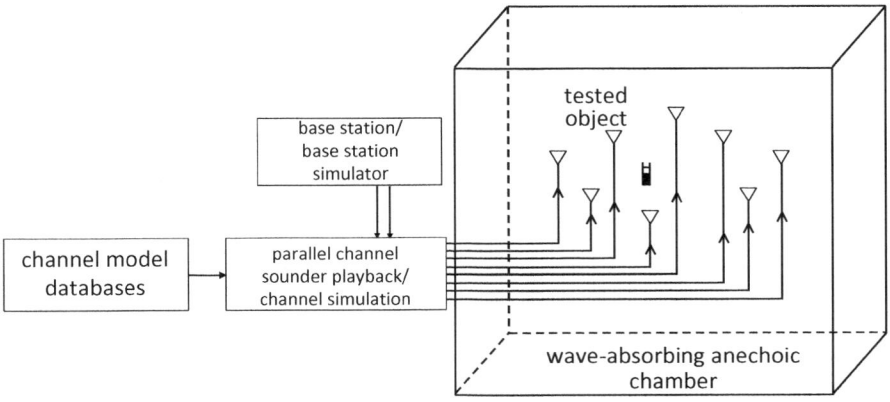

Fig. 6.27 OTA platform architecture

Channel models can be imported into commercial channel simulators to perform real-time channel simulations and then reproduce real signals through anechoic chamber based on channel model library collection, parameter extraction and modeling.

It is worth mentioning that the parallel channel test probe described in Sect. 6.2 also has the function of channel playback. As a parallel 8-channel channel detector, it can be taken as a 16-channel channel simulator in channel playback. So the same set of equipment plays the role of two sets of instruments, saving costs for the construction of OTA platform. At the same time, the system calibration scheme, as a channel detector, can also be applied to the OTA test.

6.4   5G Open Source Community

6.4.1   Introduction to Requirements

In order to meet future service requirements, mobile operators expect to be able to introduce the latest technology into existing systems in a timely manner. As a result, communications equipment is subject to ongoing system upgrades. The traditional base station will face a greater challenge. Base station is one of the most important subsystems in mobile communication systems. The mobile communication systems with different modes need the base stations with different modes for support. Traditional mobile communication base stations are designed for a single communication standard based entirely on dedicated hardware elements, such as FPGA, DSP and Application-Specific Integrated Circuit (ASIC). Hardware implementations solidify their independence of each other. There are problems of poor compatibility, cumbersome upgrades and lack of flexibility in resource between equipment. As the actual problems brought by continuous development of mobile services, the coexistence, convergence, maintenance and upgrading of many communication mode standards are becoming more and more obvious. The costs of operation and solution are rising. Therefore, designing a universal and flexible base station system which can configure various modes is the key to solve these problems.

On the other hand, the convergence of Internet and traditional telecom networks has become an irresistible trend. And standard servers have been widely used in the core network domain. As early as in 2010, in the global technology outlook released by IBM, the convergence technology of wireless and Information Technology (IT) was taken as one of the important technological trends. Intel also takes real-time signal processing of communication base stations based on CPU as a key research direction [1] For decades, semiconductor industry has been moving forward with Moore's Law. DSP and General Purpose Processor (GPP) have made great progress in architecture, performance and power consumption, providing more options for software-based mobile communication base stations. However, because the various DSP of different manufacturers or even the same manufacturers are inconsistent in backward compatibility, and the supported real-time operating systems are not the same, the industry currently lacks a unified platform and standard. In addition, the BaseBand Unit (BBU) of mobile communication base stations developed based on DSP platform system is generally a non-open proprietary platform, which still has many deficiencies in smooth upgrading and virtualization of networks. With the rapid development of relative technologies, GPP can gradually meet the requirements of high data load operations, such as digital signal processing, and provide a new choice for software to realize digital signal processing. The new technologies of these GPPs include multi-core, Single Instruction Multiple Data (SIMD) supporting fixed and floating point arithmetic, high capacity on-chip cache and low latency off-chip memory. With these new technologies, GPP can finish digital signal processing performed by a DSP, especially baseband processing functions in base station devices. The use of GPP for baseband signal processing has the following advantages:

(1) Simplified design process and shortened development cycle.

The communication system design process of traditional scheme is often based on architecture characteristics, uses a simulation platform to achieve new algorithm design, and then optimizes the corresponding codes according to the characteristics of programming. After constant optimization and correction to ensure the performance of the entire communication system, the fixed point code is reconstructed and under the premise of maintaining system performance. Finally, it is moved to the platform for corresponding programming, optimization and testing, etc. R&D process is long and inefficient, and the invested human labor costs are also rising. The migration between the platform and fixed-point design of codes all give the smooth development of the project a great deal of risks. On the GPP platform, the algorithm can be directly optimized, thus effectively improving the efficiency of programming and greatly reducing development cycle.

(2) Easy to realize multi-mode base station and achieve resource sharing.

The multi-mode base station based on the unified platform is based on a universal platform composed of modularized and standardized hardware units to realize the partial communication function of wireless equipment. It has good scalability, which can effectively extend the life cycle of base stations, save costs and seamlessly converge different communication modes. The network upgrade and smooth evolution are realized through software configuration without changing the hardware. In addition, using a multi-mode base station enables the use of a set of base station equipment to achieve multi-mode network coverage to save space, improve power efficiency and reduce power consumption. In short, the fully software-based baseband processing of base station can support multiple standards on a single system, which is conducive to resource multiplexing, improving resource utilization.

(3) Conducive to realize resource virtualization and build cloud platform.

In a centralized scenario, the operating characteristics of the GPP platform are fully compatible with the requirements of the virtualization technology. Based on universal computing platform's virtual realization of the base station system, it can maximize the advantages of hardware systems and support the protocol layer processing of high-speed communication systems. The GPP platform can perform tasks in a hardware-friendly manner, which mainly depends on computing, storage and interface resources provided by the processor. Computing resource refers to the instruction processing capacity per unit time, which is usually decided by the processor's main frequency and its arithmetic and logic computing unit. The storage resource is related to the internal and external storage capacity and cache capacity of the processor. While interface resources correspond to different levels of inter-server interaction, such as the inter-processor, the inter-processor or processor cores. By initially abstracting and equivalently calculating key capabilities of hardware as processing resource, the details of specific hardware are ignored and too-close ties with underlying hardware are eliminated, so as to implement universal and efficient virtualization. Since different processing resources of hardware processors in the centralized platform have different functions and manifestations in system implementations, they are relatively independent but also connected. Therefore, a processing resource can correspond to one dimension of virtualization to construct joint relationship of multiple processing resources on hardware platforms, and thus effectively playing the hardware virtualization capabilities.

(4) Building an open platform for easy management and maintenance.

Under the traditional communication system, a vendor usually provides a complete set of solutions, so the cost and dependence of system maintenance or upgrade are very high. Based on the centralized shared virtual base station, the multi-standard is unified, and the communications interfaces at each level are standardized, which can form an open baseband pool platform of coordinated wireless signal process, allowing access to solutions provided by any hardware vendors to achieve the smooth upgrading and the expansion of the system, so as to achieve healthy competition of industry, which is conducive to sustainable development of operators. Thus, a universal and open 5G network platform must be built. Open 5G network is conducive to work with other industries for coordinated innovation. The GPP platform has the advantage of openness in baseband signal processing of base stations, which provides a new way for base station baseband processing, and is expected to become a new generation of unified and open mobile communication systems and schemes of multi-mode baseband processing platforms, which has sustainable competition in future-oriented applications and services.

6.4.2 Development Status of Communications Platform Based on General Purpose Processor

Open Air Interface (hereinafter referred to as OAI), based on GPP, releases hardware and software version supporting Evolved Packet Core (EPC) and the base station. OAI is the most complete development platform of open source LTE. The platform uses the design idea of software defined radio. Architecture is in the Intel GPP and Linux operating system. The entire platform, including the baseband, is realized by complete software. Base station includes baseband, MAC and Radio Resource Control (RRC) signaling. The core network comprises MME, SGW, PGW and HSS. OAI's base station and core network use the standard S1 interface, which can be used alone. OAI is established and managed by EURECOM, an educational and research institution in the communication domain of French Riviera, whose partners include Agilent, China Mobile, IBM, Alcatel-Lucent, Thales and Orange.

Universal Software Radio Peripheral (USRP) [3] is a widely used software-defined radio platform based on GPP designed by the Massachusetts Institute of Technology. USRP includes hardware front end and GUN Radio of corresponding software development kit [4] USRP can connect a personal computer to RF. It is essentially equivalent to a digital baseband or IF portion of a wireless communication system. GNU Radio provides real-time signal processing software and low-cost software radio hardware. GNU radio can be used to implement software defined radio on low-cost RF hardware and GPP.

Sora system [5] developed by Microsoft Research Institute Asia, is a software defined radio system based on multi-core GPP. It uses both hardware and software technologies to address the challenges of real-time processing of wireless digital signals. A new RF control interface board is redesigned. PCIe is used to transmit broadband wireless signal sampling, which can support transmission rate of more than 10Gbps and meet the needs of most wireless technologies.

IBM Research Institute has proposed a virtual base station pool based on cloud computing, which realizes the TDD WiMAX standard on multi-core universal processing platform and a newly designed RRH, and completes the prototype design of the virtual base station pool.

Intel China Research Institute has developed the LTE mobile communication base station based on Intel architecture, and made prototype verification and reference design for wireless signal processing of physical layer in LTE standard, MAC layer and RLC layer.

The white paper of C-RAN released by China Mobile Communication Institute in October 2011 has developed centralized C-RAN prototypes based on the universal IT platform [6]

Facebook announced the new OpenCellular [2] open source platform in June 2016. It is a software-defined wireless access platform, a subsystem calculated by General purpose Baseband Computing (GBC) of integrated power supply, synchronization and other systems, and a design combined by front-end video subsystem of integrated analog to support several mobile communication standards from 2G to LTE. By providing devices and open source software needed to connect cellular networks, OpenCellular allows telecom operators, device manufacturers, researchers and entrepreneurs to develop, implement, deploy and operate wireless infrastructure based on this platform, share and develop new telecommunication hardware freely, and help normal users in remote areas with poor environmental conditions realize Internet access.

As a result, we can see that the software-defined wireless communication system has high flexibility and openness. It can realize the backward smooth upgrade of mobile communication system, which has become a hot research topic at home and abroad and achieved a lot of research results. But the realization of hardware and software architecture and composition is a complex digital signal processing system, which still lacks in-depth systematic study. 5G communications system is completed based on GPP. Achieving all-digital and softening communication signal processing is a task has not been solved well.

6.4.3 Introduction to Open 5G Universal Platform

For5G test and universal platform R&D based on GPP, Shanghai Research Center for Wireless Communications has built and completed the open LTE network platform based on universal server, which can well support the commercial terminal's access to Internet. The platform uses software defined radio design. The processor uses Intel GPP. And the operating system uses open source Ubuntu Linux. The development language uses C language. The entire platform (including baseband) is totally implemented by software. The software architecture of this platform is as shown in Fig. 6.28. The high-performance GPP is connected to optical fibers via the PCIe bus and then to Remote Radio Unit (RRU) unit to realize the complete functionality of the LTE system, including the base station, the core network and the terminal. The core network includes the realization of software, like MME, SGW, PGW and HSS and other network elements. The base station includes software implementation of physical layer, MAC layer and RRC layer. The core network may implement a software-defined full-function soft base station on the same server, or may be connected to a plurality of third-party base stations to achieve multi-user access.

The hardware architecture of Open 5G platform currently uses universal commercial server hardware architecture. The key hardware involved in wireless signal processing, transmitting and receiving includes CPU, Synchronous Dynamic Random Access Memory (SDRAM), accelerator card and interface control board. Far-end also includes Analogy Digital/Digital Analogy converter (AD/DA), RF, and FPGA chips for front-end digital signal processing and interface control.

Multi-core CPU is used to allocate the signals to be processed to the kernel which is used for baseband digital signal processing to process the transmission. And some of the large computations, such as turbo decoding, is assigned to the

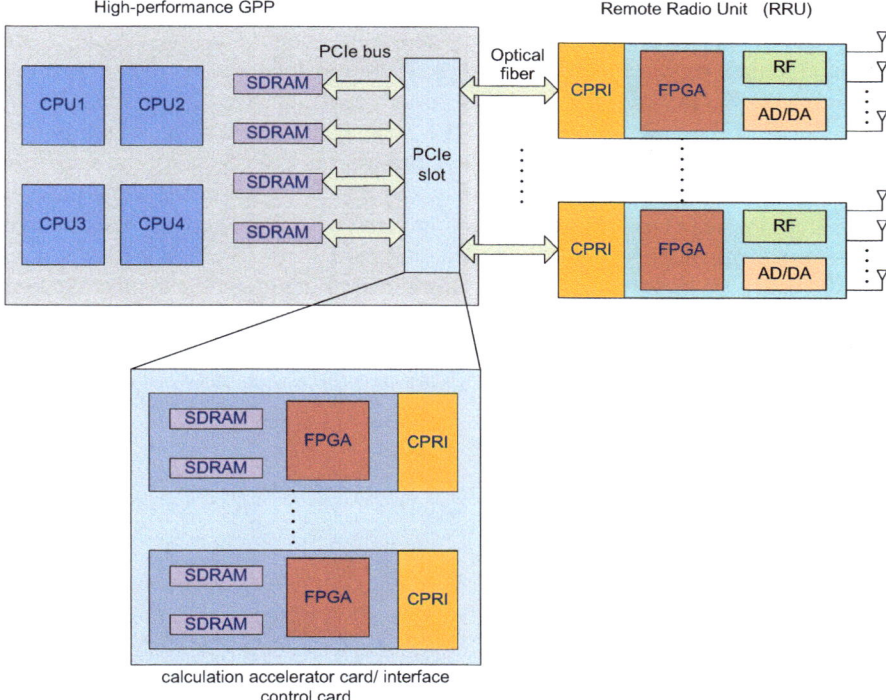

High-performance GPP

Remote Radio Unit   (RRU)

calculation accelerator card/ interface
control card

Fig. 6.28   Architecture figure of Open 5G hardware

calculation acceleration unit by task scheduler module for calculation to reduce the baseband processing load of CPU.

The interface unit is an extremely critical unit whose role is to provide high-speed data rate and low-latency data channel between BBU and RRU. One interface unit is connected to the universal server memory with the PCIe standard interface, reads and writes digital baseband signals in the universal server memory with DMA technologies. And the other end is connected to the remote RRU by optical fibers. This interface uses a Common Public Radio Interface (CPRI) protocol, including Path I and Path Q data for transmission, as well as system operation, maintenance and control signals.

The realization of real-time processing of high-speed baseband signals by universal server requires solving two technical challenges. One is real-time signal processing capability of the universal server, and the other is data interaction bandwidth of the internal interface.

With the development of multi-core GPP, the new architecture continues to emerge, so that the baseband signal processing capability has reached a high performance and has basically met the requirements of high-speed baseband signal processing. Moreover, by increasing the calculation accelerator card, CPU and FPGA/DSP form a heterogeneous computing platform, which can greatly improve platform's baseband signal processing capabilities and basically satisfy the baseband signal processing required by real-time communications. At the same time, selecting appropriate high-speed interface can increase the data interaction ability between the platform and the system and enhance the communication throughput of the universal platform. The PCIe standard interface, which is used in a universal platform, provides up to 16 Gbps transmission capacity and shares the data bus directly. The transmission delay and bus delay are in the same order of magnitude, which can ensure high-speed data rate exchange between the common server platform and the computing accelerator card.

The hardware acceleration architecture shown in Figure 6.28 can meet the real-time processing requirements of existing communication systems. As shown in Fig. 6.29, a core network architecture based on the universal server. The joint networking architecture of the core network and the access network based on the universal server is shown in Fig. 6.30.

Fig. 6.29 EPC network based on universal server

Fig. 6.30 Networking of core network and access network based on universal server

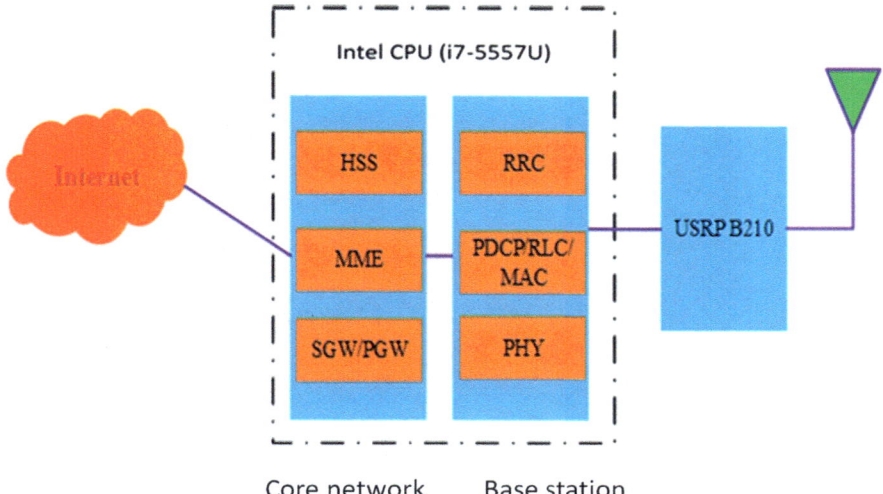

Fig. 6.31 Software-defined mobile network architecture with the key functions in core network and base stations

(1) The functions of the LTE core network and the base station are achieved based on the universal server with versatility, openness and function customization capacity of the core network equipment. Based on OAI open source EPC and through increasing the functions, such as mobility, multi-user multi-base station, stability, the open network based on universal platform EPC is constructed. As shown in Fig. 6.31, the core network of LTE is deployed in the universal server of Intel Linux. CPU uses Core i7-5557 U, 3.1GHz, dual-core and four threads, and supports 16GB of memory with 5.0GT/s of bus frequency and 4 MB L3 cache.

(2) Core network and soft base station joint network based on GPP are supported. The functions of the core network (MME, SGW, PGW, HSS, etc.) and the base station (RRC, Packet Data Convergence Protocol (PDCP)/RLC/MAC and PHY, etc.) are realized. Data transmission and reception are realized through USRP B210. Self-made SIM is used to support the commercial terminal access network, as shown in Fig. 6.32, Frequency Division Duplex (FDD) and TDD modes as well as 5/10/20 M bandwidth. Both the core network (MME, SGW, PGW, HSS, etc.) and the base station (coder and decoder and other physical layer functions, scheduling, power control, link adaptation, etc.) can be enhanced and updated.

Although some progress has been made in the GPP-based Open 5G platform, there are still some shortcomings in the existing technology solutions, which are far from practical applications. Considering all kinds of technical challenges, core technologies of baseband signal real-time processing, resource and network function virtualization, resource scheduling and task matching need to be R&D and verified.

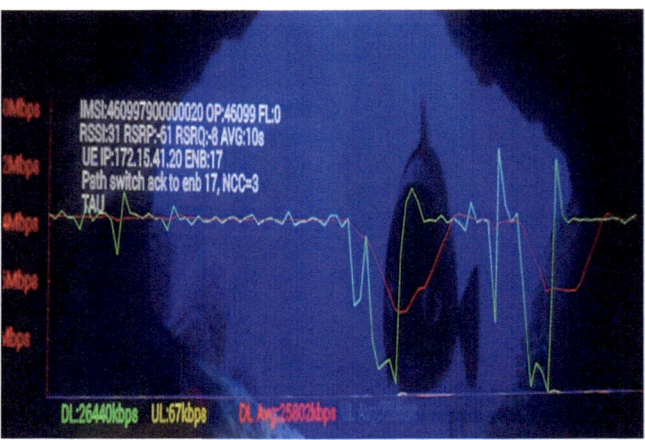

Fig. 6.32 Simultaneous access of multiple terminals and smooth switches among base stations

6.4.4   Key Technologies of Universal Platform Based on General Purpose Processor

Real-Time Processing of Baseband Signal

In wireless communication systems, the real-time processing of baseband signal possesses the characteristics of compute-intensive, and is faced with severe real-time requirements. In order to ensure the real-time processing of base stations and support the baseband data processing to guarantee the system performance, traditional communication systems often realize the baseband signal processing with special hardware, such as DSP and FPGA. However, using dedicated hardware to implement different carrier frequency bands, different coding, modulation, sampling rates and different multi-user accesses will further aggravate the independence of each other. Dedicated hardware is not conducive to the compatibility and upgrading of many heterogeneous systems with different standard modes and different networks. This not only reduces the resource utilization efficiency of base station equipment, but also directly brings operators increasing costs of maintenance and upgrading. Taking China Mobile for example, as shown in the left figure in Fig. 6.33, the expenditure of main equipment and ancillary equipment of base stations has reached 59% of CAPital EXpenditure (CAPEX) of cell site, while the figure on the right of Fig. 6.33 shows the total energy consumption constitution of China Mobile, of which 72% of energy consumption is from the base station site of the wireless network access.

Therefore, the design of a universal, flexible, low energy consumption base station system compatible with various modes is an urgent need to address the problem. Using GPP to combine with universal operating systems (such as Windows, Linux) and its convenient development environment and strong

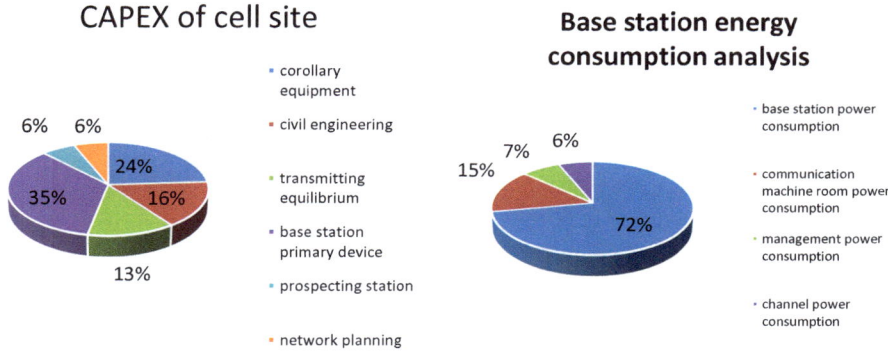

Fig. 6.33 CAPEX and energy consumption analysis of base station

reconfigurable capabilities are increasingly concerned. GPP has the advantages of flexibility and openness in baseband signal processing of a base station, which can integrate different mobile communications systems into one, providing a new realization way for wireless signal processing. Facing future applications and services, GPP real-time processing baseband signals have sustainable competitiveness, which is expected to be a solution for 5G communication systems and multimode baseband processing platforms.

Although using GPP to perform baseband signal processing has shown advantages in comparison with traditional dedicated hardware in terms of system cost, capacity and flexibility, the traditional mobile communications base stations have their own merits. Firstly, GPP is difficult to well support the computer-intensive processing, which cannot meet the real-time requirements of mobile communications. Second, limited by the architecture and processing ability of GPP, communication algorithms are inefficient in GPP with poor real-time, stable and effective features. Third, the power consumption of GPP chip largely hinders the development of a GPP for baseband signal processing. Therefore, the construction of mobile communication base station based on GPP must solve the bottleneck problem of real-time digital signal processing based on GPP under the consideration of time delay and overheads of transmission.

Software Acceleration Method

At present, the industry has not had the recognized implementation standard of wireless baseband processing based on GPP. Various architectures and algorithms are still under investigation. When dealing with network-intensive and computationally demanding baseband algorithms, the processing ability of GPP is not as good as that of a dedicated chip. Performance bottlenecks may occur, and there may be some distance from large-scale practical applications. First, because GPP usually faces many tasks, operating systems (such as Linux) mainly aim for time-division operation design. And baseband signal processing is computing intensive with high real-time requirements. Therefore, the interrupt response and delay waiting process should be avoided in the design of the signal processing algorithm, reducing the complexity of algorithm. Efficient algorithm should be used in subsequent application development as much as possible. Second, GPP has many defects in high-speed data access. However, the high-speed cache mechanism can be used to improve the storage efficiency. For example, control logic is specified in advance to determine which data and instructions are stored in the on-chip cache or memory. Finally, additional CPU resources are needed for GPP to manage multiple threads/multiple processes. However, using computer multi-threading/multiple processes to deal with and optimize big data and repetitive computations can increase the data rate exponentially. In addition, with the instruction set to optimize the CPU operation, the purpose of accelerating the baseband signal processing can be achieved.

Instruction set-based software acceleration methods include the following types: (1) Look Up Tables (LUT): LUT [7] is a compromise operation after examining computation complexity and space complexity. The LUT operation can greatly reduce on-line processing delay by replacing conventional bit operations. (2) SIMD: Intel [8] CPU has a special multi-data instructions, and SIMD instruction set can accelerate signal-level computing signal processing. SIMD repeatedly performs the same operation for symbol-level data. A single instruction of SIMD can handle several operations with low computational cost (computational resources) and fully use bit bandwidth to significantly increase CPU efficiency. (3) Sample level-Intel Integrated Performance Primitives (IPP): IPP [9] developed by Intel is a set of software library of cross-platform and cross-operating systems that can realize signal processing, image processing, multimedia and vector manipulation and other operations. In [7], IPP is used to implement Fast Fourier Transform (FFT)/Inverse Fast Fourier Transform (IFFT). The test results show that FFT/IFFT is accelerated by IPP, which can improve the performance significantly.

Actual Test of Physical Layer of OAI Open Source Platform

Based on the OAI open source platform, the bandwidth to test each 5 MHz system of FDD uplink/downlink and each module of processing delay performance of physical layer of single-base station single-user system with SISO modes are given. Each module data of physical layer on this platform is floating-point type, using SIMD instructions to achieve the acceleration of data processing.

System performance test environment of the OAI open source platform includes that EPC, base station and user equipment run on the same host. And the base station and user equipment are directly connected through the UDP. The program runs on an IBM System ×3400 M3 server. CPU uses 2.13GHz and 4-core Intel Xeon E5606 processor; 4G memory; 256G hard drive. The operating system Linux Debian 7 uses 64-bit Ubuntu 14.04 DeskTop version and installs 64-bit low-delay core. CPU uses the highest operating frequency.

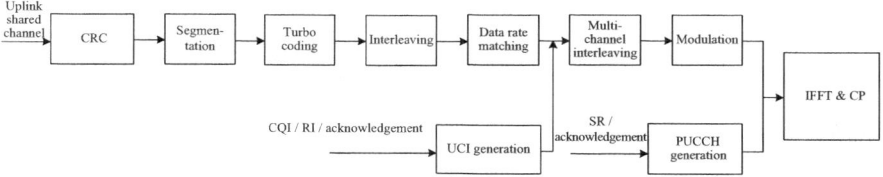

Fig. 6.34 Uplink transmitting process [8]

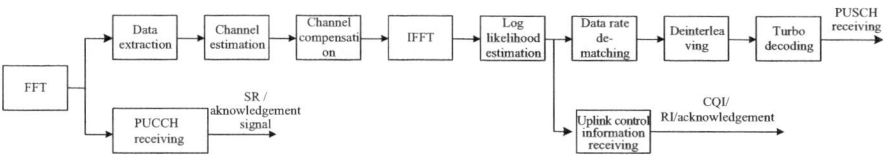

Fig. 6.35 Uplink receiving process [8]

The uplink transmitting procedure of the OAI platform is as shown in Fig. 6.34. First, the uplink shared channel information makes 24-bit CRC and segmentation. Second, turbo channel coding is performed. Then, bit interleaving and data rate matching are conducted. CQI, Rank Indication (RI) and acknowledgement information are added, and then multi-channel interleaving and modulation are carried out. Then Scheduling Request (SR) and acknowledgement information and other Physical Uplink Control CHannels (PUCCH) are combined to complete IFFT. And CP is added and finally the frame is formed. Uplink receiving flow is shown in Fig. 6.35. PUCCH can extract scheduling request and acknowledge indication information after FFT. While the acquisition of Physical Uplink Shared CHannel (PUSCH) information is subject to data extraction, channel estimation, channel compensation, IFFT, log likelihood estimation, data rate matching (at this moment, uplink control information can be solved, such as CQI, RI and acknowledgement information), de-interleaving and turbo decoding and other procedures.

Figures 6.36 and 6.37 describe the transmitting and receiving flows of downlink data respectively. The former downlink shared channel needs to complete CRC, segmentation and turbo coding first. The broadcast channel information and downlink control information respectively complete convolutional coding, interleaving and data rate matching, scrambling and modulation. Physical HARQ Indicator CHannel (PHICH) generates HARQ indication information and Physical Control Format Indicator CHannel (PCFICH) generates control format indication information. Furthermore, these channels are multiplexed with Primary Synchronization Signal (PSS), Secondary Synchronization Signal (SSS) and Reference Signal (RS). After that, FFT and CP is added, and finally physical baseband signal is generated. The latter can output HARQ indication and control format indication information respectively by FFT, channel estimation, frequency compensation and physical layer measurement. And the downlink shared channel information, broadcast channel information and Downlink Control Information (DCI) are received through

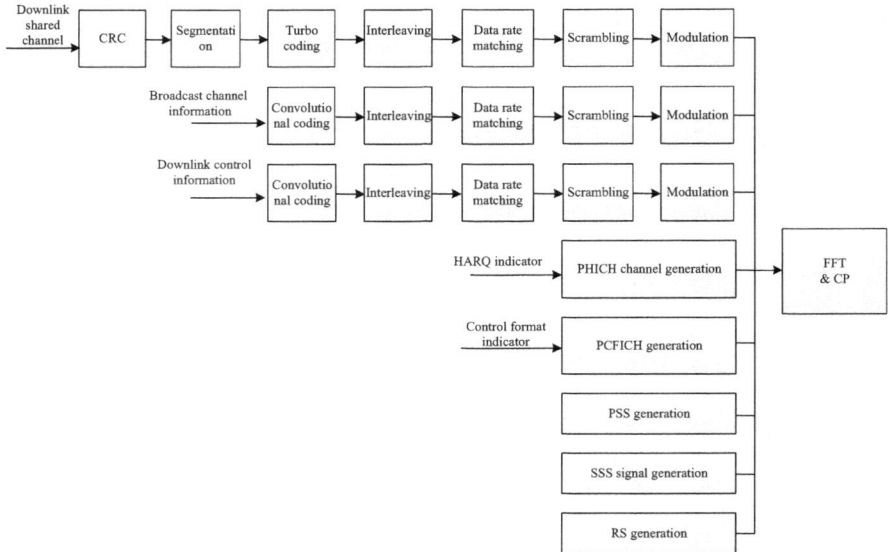

Fig. 6.36 Downlink transmitting flow [8]

data extraction, channel compensation, log likelihood estimation, descrambling and data rate de-matching, de-interleaving, etc., and then respectively through turbo decoding or Viterbi decoding.

As it can be seen from the above test results, for both the user side and the base station side, Turbo decoding module is the most time-intensive course. Although using GPP to realize real-time processing of baseband signals can reach a high performance, when GPP handles some real-time modules with high computational requirements, it may not meet the functional requirements of networks limited by the core and architecture of GPP. In order to fully develop the effectiveness and flexibility of the universal processing platform, using the hardware acceleration method, such as combining GPP with FPGA/DSP is an essential means. As a result, it can realize baseband signal real-time processing to break the bottleneck of real-time processing of baseband signals (Tables 6.2, 6.3, and 6.4).

Resource and Network Function Virtualization

In recent years, communications, as a communication tool, is undergoing significant changes. In order to bear the ever-increasing demand of users, operators are constantly optimizing their own network. At the same time, challenged by Internet services, telecommunication operators are trying to speed up the process of intelligent management to avoid becoming a pipeline. However, now the network has many problems, which hinders the network innovation and is embodied as follows.[10]

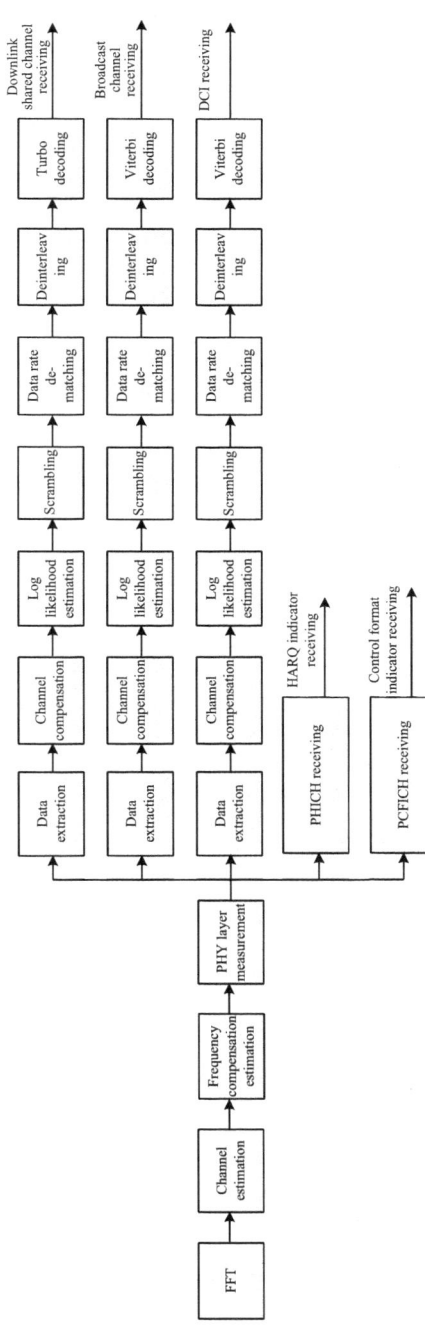

Fig. 6.37 Downlink receiving flow [8]

Table 6.2 Turbo decoding test results of user-side downlink shared channel

Data rate	2.152Mbps	4.392Mbps	8.76Mbps	11.832Mbps	13.536Mbps	17.56Mbps
Number of code blocks	1	1	2	2	3	3
Average number of iterations	2	2	4	4.08	6.25	6.54
Average value	113.44(μs)	234.68(μs)	465.01(μs)	680.27(μs)	734.86(μs)	1047.61(μs)
Variance	28.52	74.27	494.92	1796.38	3765.47	17459.83
Mean squared error	5.34	8.62	22.25	42.06	61.36	132.14
Maximum value	127(μs)	253(μs)	533(μs)	850(μs)	952(μs)	1514(μs)
Minimum value	106(μs)	221(μs)	429(μs)	644(μs)	667(μs)	923(μs)

Table 6.3 DFT module test results of base station side

Data rate	2.216Mbps	4.392 Mbps	8.76 Mbps	11.064 Mbps	14.688 Mbps
Average value	35(μs)	35.03(μs)	35.03(μs)	36.08(μs)	36.28(μs)
Variance	0	0.02	0.03	0.1	0.22
Mean squared error	0	0.16	0.17	0.31	0.47
Maximum value	35(μs)	36(μs)	36(μs)	37(μs)	37(μs)
Minimum value	35(μs)	35(μs)	35(μs)	36(μs)	36(μs)

Table 6.4 Turbo test results of uplink shared channel of base station side

Data rate	2.216Mbps	4.392 Mbps	8.76 Mbps	11.064 Mbps	14.688 Mbps
Number of code blocks	1	1	2	2	3
Average number of iterations	2.21	2.19	5.62	5.38	7.51
Average value	151.99(μs)	307.47(μs)	642.04(μs)	822.57(μs)	975.09(μs)
Variance	213.85	753.29	3122.41	4504.32	3462.69
Mean squared error	14.62	27.45	55.79	67.11	58.84
Maximum value	211(μs)	459(μs)	841(μs)	1145(μs)	1256(μs)
Minimum value	107(μs)	212(μs)	448(μs)	583(μs)	735(μs)

(1) Complex network function and poor flexibility. Mobile communication network is formed by a large number of single-function network equipment. Hardware resources are specialized and fragmented, and there are various types and large quantity in equipment. Once the network is completed, it will be difficult to change, expand and update, and resources cannot be rapidly allocated on demand within the whole network.

(2) Hardware and software of network element equipment is vertically integrated with closed architecture. In order to provide different services and multiple access modes, and meet various requirements of QoS and security, the communication network introduces a large number of control protocols and is bound to a specific forwarding protocol. The code is directly or indirectly written in hardware, constituting a closed architecture with integrated control and forwarding. In the long run, the equipment is becoming more and more bloated. The space for improving performance is small. Technical innovation and upgrading are difficult, and the scalability is limited. The service development cycle is long.

(3) The network and business form the "chimney group" phenomenon. The provision of new services often leads to proliferation of new equipment types and quantities. The division between departments is easy to form a large number of independent and closed network and business chimney groups. In addition, the cost of infrastructure construction is high, and resources cannot be shared with each other. Network and services cannot be coordinated and converged and cannot adapt quickly to new services and new models.

(4) CAPEX and OPerating EXpense (OPEX) are still high. Because the special nature of network devices results in the coexistence of equipment in different manufacturers, it requires a lot of manpower and resources for different ages and different standards, the procurement, integration, testing, deployment and maintenance of equipment. At the same time, the insufficient competition among different equipment manufacturers causes relatively high costs of equipment, operation and maintenance management and upgrade. According to 5G White Paper [16] 5G network architecture poses higher requirements in the aspects of access speed, green power saving, cost-reliance, etc. Facing with 2020 and the future, the popularity of ultra-high definition, 3D graphics and immersive video will drive a substantial increase in data rate, augmented reality, cloud desktop, online gaming and other services, which will not only make challenge for uplink and downlink data transmission rate, but also make the harsh requirements of "no perception" on delay. In the future, a large amount of personal and office data will be stored in the cloud. Massive real-time data exchange will be comparable to the transmission rate of optical fibers and make traffic pressure for mobile communication networks in the hot spot area. Therefore, the new network architecture is imperative. The new networking approach must address various problems of current networks so as to meet the changing requirements of users.

First, as people have higher bandwidth requirements of wireless networks, for mobile operators, the bulk of the increase comes from expenditures base stations construction, operation management and network infrastructure upgrades. Besides, enterprises are facing more and more intense competitive environment. However, revenue may not grow at the same rate. The traffic of mobile Internet services is on the rise due to competition, but the average revenue of single user has grown slowly and sometimes reduced rapidly, which has seriously weakened the profit of mobile operators. In order to maintain the ability to continue to make profits, mobile operators must look for ways to increase the network capacity at a low cost, thus providing better wireless service for users.

Traditional wireless base stations have the following characteristics. First, it is difficult for each independently operating base station to increase the spectral efficiency due to the capacity interference of system. Second, each base station only covers a small area and connects a fixed number of sector antennas and can only process the signal reception and transmission of the cell. Third, the base stations are usually based on a "vertical solution" developed by the proprietary platform. Compared with traditional base stations deployed by these above operators by connecting special line or optical fibers and core networks, the small base stations in the future 5G access network are causally deployed by third parties or users on the basis of requirements (such as deploying in a commercial and office area or user's home). This brings a huge challenge for operators. A large number of small base stations mean high site matching and leasing, construction investment and maintenance costs. Moreover, the greater number of small base stations means that operators will have to pay more capital and operating expenses. The average load of network of small base stations is generally much lower than the busy-time load, and the actual utilization efficiency is very low. Meanwhile, different small base stations are difficult to share the processing power, making the spectral efficiency difficult to improve. Finally, the specified platform used by small base stations means that the operator needs to maintain multiple incompatible platforms at the same time, and will require higher costs when expanding or upgrading.

At present, with the rapid development of the virtualization technology, given its characteristics like lightweight management and optimization resources, etc., building a virtual management platform for small base stations with intensive deployment is becoming more possible. Virtualization is a technology to make abstract analog for computer and communication resources. The virtualization technology, on the basis of the existing computer and communication hardware resources, simulates all or parts of virtual hardware resources, such as baseband, CPU, memory, input and output devices. These virtual hardware resources can share the same platform with local real hardware resources, which is called virtual machine. In general, from the software point of view, virtual machine and real machine have no difference. That is, the realization and operation of virtual machine are transparent for software programs.

Considering the evolution of mobile communication networks, IMT2020 has put forward new requirements in order to complete the evolution of old network architecture and adapt to 5G evolution: NFV, Cooperative Communications (CC), Automated Network Organization (ANO), Flexible Backhauling (FB) network and advanced traffic management and offloading and other key technologies [11] Among them, NFV is the next generation of network building scheme proposed and dominated by operators, with the purpose to bear more and more mobile network function software through the use of ×86 and other universal hardware and the virtualization technology, thereby reducing the high cost of network equipment. At the same time, with the hardware and software decoupling and functional abstraction, the network equipment function no longer depends on the dedicated hardware. The resources can be shared flexibly, so as to realize the rapid development and deployment of new services. The automatic deployment, flexible scalability, fault isolation and self-healing are made based on actual business requirements. The architecture is as shown in Fig. 6.38.

The nature of the virtualization technology is the division and abstraction of computing resources. The advantages, such as isolation, consolidation and migration, make it possible to integrate applications on different platforms safely and securely to the same server, and migrate an application on a server quickly to other servers, thereby improving server utilization, reducing hardware procurement and operating costs, as well as simplifying system management and maintenance. The virtualization technology has had nearly 50 years of history [12, 13] The first virtual machine was System/360 Model 40 VM developed by IBM Corporation in 1965. In recent years, with the rapid development of the computer hardware technology and continuous innovation of the computer architecture, especially in the late 1990s, desktop computer performance has been able to simultaneously support multiple systems to run. At this point, the contradiction between increasingly powerful

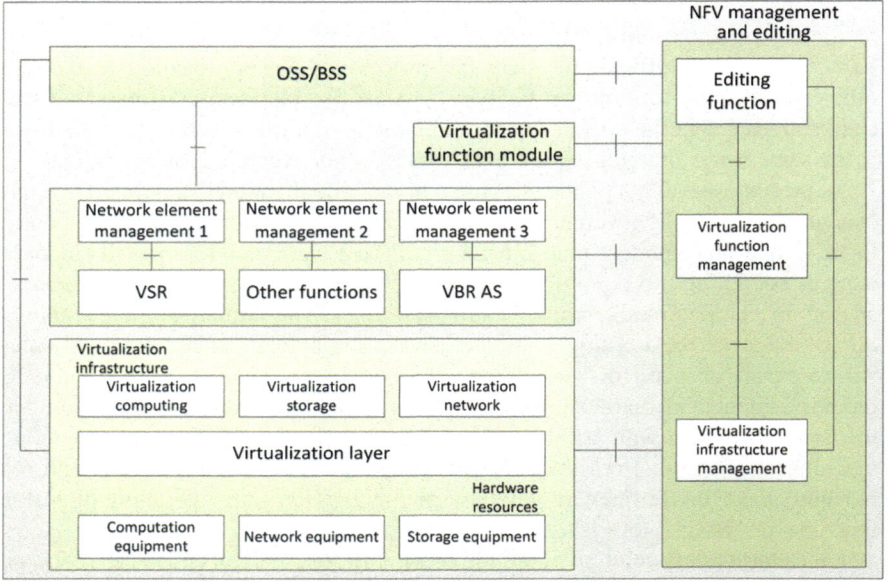

Fig. 6.38 NFV virtualization architecture [13]

computing power of the computer system and relative backwardness of the computing model has become increasingly prominent. The own characteristics of the virtualization technology can find a balance between hardware systems with rapid development and application requirements with complex changes, which provides more advantages for enterprise-class applications, such as improving resource utilization, reducing management costs, increasing the use flexibility, improving system security, enhancing the availability, scalability and maneuverability of system. Therefore, the virtualization technology has attracted the attention of academia and industry both at home and abroad, and become one of the hotspots in current research [14]

From 1990s to today, virtualization technology has made considerable progress, and a variety of technologies become mature. In addition to VMware, Denali and Xen, there are also many emerging virtualization software, such as KVM, Virtual Box, Microsoft virtualization series products (Virtual PC, Hyper-V), Paralles virtualization series products (Virtuozzo, Parallels Desktop for Mac), Citrix's XenServer, Sun's xVM, Oracle VM and VirtualIron. The main applications of virtualization today involve the following domains: (1) server domains, including data centers with high requirements for hardware platforms, cloud computing, distributed computing, Virtual servers, etc.; (2) enterprise management software, such as trusted desktop based on virtual machine, easily and effectively managing and supporting employee desktop computer; (3) individual users, including antivirus technologies based on virtual machine, program development and debugging, operating system kernel learning, server consolidation, cloud computing and data centers, virtual execution environments, sandboxes, system debug testing and quality assessment, etc. With the further development of the wireless communication domain and virtualization technology, the application space will be further expanded. In the future, 5G is promising for applications in many fields. At present, the main existed researches focus on the followings.

Peng et al. [15] present a 5G H-CRANs that is compatible with the software-defined network architecture. The architecture converges virtual cloud computing to achieve large-scale centralized coordinated processing of user data, involve the traditional base station into a new communication entity Node C, and achieve the integration of multiple RANs.

China's IMT-2020 5G Promotion Group released "White Paper On 5G Network Technology Architecture" in 2015 [16] and proposed a 5G network cloud architecture based on SDN. The architecture is mainly composed of three parts: wireless access cloud, control cloud and forward cloud. The access cloud supports the access of multiple wireless modes, including Wi-Fi, LTE and mesh networks. It can realize flexible network deployment and efficient resource management. Control cloud uses virtualization technologies to realize the centralized control of network functions logically. It can support control and open network capacity, satisfy the requirements of different services and improve the service deployment efficiency. Forward cloud realizes high reliable and low delay transmission of service data flow based on unified resource scheduling under the control of control cloud.

Agyapong et al. [17] propose a two-layer 5G network architecture consisting of the wireless network and the network cloud to support the massive MIMO system and the separation of the control plane and the user plane. The architecture consists of two layers of virtual logical networks that support wireless networks with the simplest physical layer and MAC layer functions and all virtual network cloud with higher layer functions.

Li et al. [18] propose a dual-virtualized 5G network architecture based on a small base station and user level. In this architecture, the virtualization technology is applied to small base stations with intensive deployment and users with different requirements respectively. In the virtualization of small base stations, a primary base station is responsible for the control signaling and mobility management, such as broadcasting and paging, and for communicating with and managing other base stations of a virtual cell. Virtualization of users virtualizes multiple users in one cell, reduces the communication overhead between small base station and users, and enhances the communication among different users in a virtual cell.

Through the analysis of existing research work, it can be seen that the virtualization of current 5G networks focuses on the research of the overall network architecture of the core network. The technology used mainly comes from cloud computing and SDN, emphasizing the separation of control and forwarding, thus realizing flexible network deployment and efficient resource scheduling. It is still lacking in the extensive and in-depth researches in realization of virtualization technologies of the access network and resource scheduling. In order to adapt to the low delay and high efficiency of transmission in the future 5G, deeply studying the virtualization technology of the core network and the access network can produce enormous value in the future.

Dynamic Task Allocation and Resource Matching Scheduling

Task Segmentation and Parallelization Mode

The segmentation and parallelization of tasks has two main advantages. First, it is conducive to the increase in data rate. GPP has the advantages in multi-core processing, large storage and fast cache. Using task segmentation to make parallel processing of multiple threads can increase the operating rate of processor and reduce time-consuming. Second, it improves the program structure and is easy to understand and maintain the program [19] In the case of multi-user multi-base stations, the processing of some very complex communication processes is divided into several threads, and the individual user processes can be independently run by multi-threading, which can reduce the interference of multiple users.

The parallelization improvements are made for communication procedures. First, it is needed to segment the task, dividing the large computational processes of coarse granularity into a number of fine-granularity subtasks. And then parallelism should be found according to the logic of sub-tasks. If there is a data dependency among sub-tasks, it can be vertically converged. Otherwise, the horizontal convergence will be made [20] Common methods of task segmentation of communication systems are as follows.

(1) Data segmentation. Data are divided into several independent data blocks, creating several threads to deal with respectively, so as to complete the processing of all data.
(2) Process segmentation. Complex process can be divided into a number of simple sub-processes with a certain degree of independence, using multi-thread to implement sub-processes to speed up processing speed.
(3) Problem segmentation. Complex problem is divided into several independent sub-problems. Various sub-problems are solved through multiple threads, and finally the solution of original problem is gotten.

Task parallel processing methods of communication systems, according to the amount of calculation, mutual independence, real-time requirements and other factors, can be divided into two types:

(1) Distributed parallel approach. Large tasks can be divided into several small tasks which can make parallel execution to process. As shown in Fig. 6.39 [20] the large tasks A-B-C-D are equally divided into four threads, thread 1–4. Each thread processes 1/4 of the task. This parallel approach can load in balance, which is suitable for the large tasks easy to be divided into a number of separate strong parallel subtasks [21]

Taking turbo decoding as an example (as shown in Fig. 6.40), the turbo decoder repeats the same operation for each code block. The turbo decoder has strong independence among code blocks, which is suitable for parallel processing and can effectively reduce the time consumption of processing. However, the creation and scheduling of threads all need certain extra

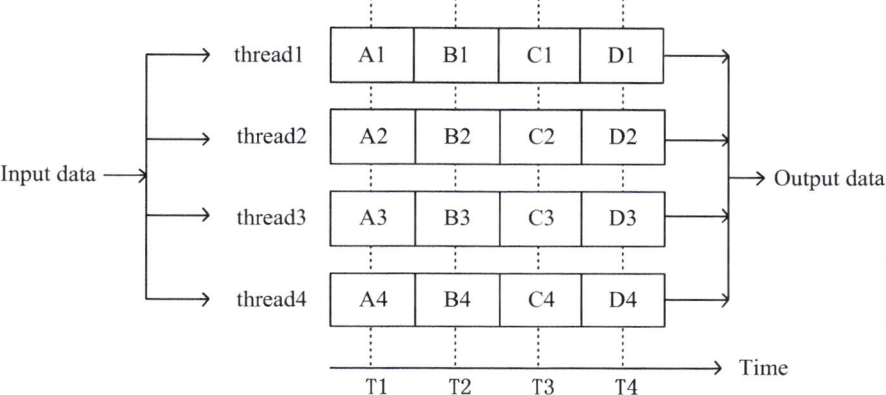

Fig. 6.39 Task distributed parallel approach [20]

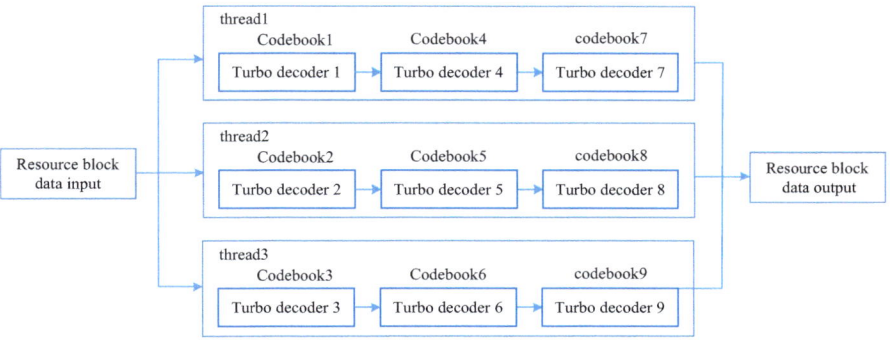

Fig. 6.40 Turbo parallel decoding structure [22]

overheads, so the number of threads is comprehensively determined depending on scheduling overhead, CPU resources, load balancing, delay and throughput requirements and other factors.

(2) Pipelined approach. A large task is divided into a series of functional modules. In parallel implementation process, each functional module is executed by independent thread, and each task module operates at the same time, inputting and outputting for each other. As shown in Fig. 6.41, thread1, thread2, thread3, thread4 are responsible for the processing at A, B, C, D stages respectively. The previous stage outputs enough data to trigger the execution at a later stage. This parallel approach is suitable for dealing with the task that data have dependence on a serial form. For data-related parts at different stages, it is needed to take synchronization measures to protect it, and then reduce tasks time consumption through the pipeline and improve the CPU utilization.

As shown in Fig. 6.42, the uplink receiving flow may be handled in a pipelined fashion. The uplink receiving flow includes symbol level signal processing, bit-level pre-decoding signal processing, turbo decoding and CRC

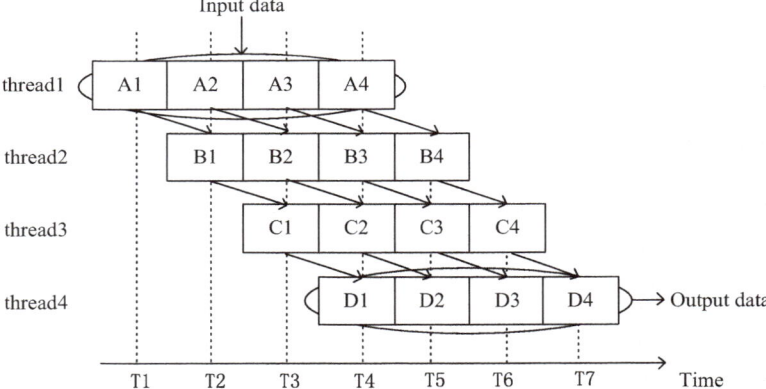

Fig. 6.41 Task parallel pipelined method [22]

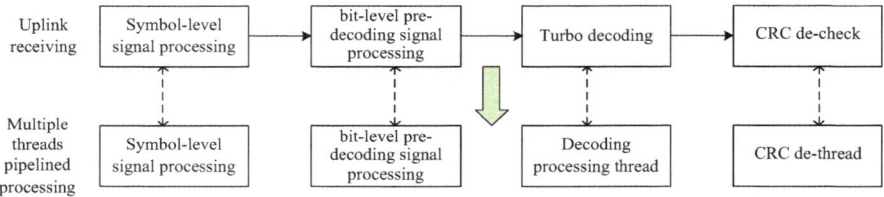

Fig. 6.42 Schematic diagram of pipeline parallel processing of uplink receiving flow [23]

de-check and other function modules. Each function module is implemented with different threads. Only when sufficient complete input data is obtained can processing operation be executed without waiting for one sub-frame to complete all the processing and execute the processing of next subframe, which greatly reduces the idle wait time of each processing module. However, in order to avoid pipeline errors, it is needed to take synchronization communication mechanism to protect the data in different modules, so that the process before and after the two steps can be connected.

Scheduling Strategy and Real-Time

Resource scheduling methods of threads have two types: dynamic and static. The former means that the scheduler is responsible for dynamically mapping the processing module needed to schedule to currently available CPU core perform task processing. Due to the flexibility of dynamic scheduling, it is not suitable for physical layer signal processing module with high data dependency. The latter divides data flow into several sub-data streams, and then statically assigns different sub-streams together with related task processing modules to different CPU core task processing. Static schedule is suitable for dealing with the physical layer data flow with a substantially constant data structure, which can improve the utilization rate of high-speed cache and reduce the overheads of data synchronization, so as to improve the integrity of communication system [23]

The existing scheduling strategies based on multi-thread programming support both real-time and non-real-time processes, and the priority of real-time processes is higher than that of any non-real-time processes. The schedule policy of real-time process is divided into the strategy of "first come first served" and the time-slice rotation strategy. The former is queued according to the priority, and the system schedules the process group with highest priority. At the same priority level, priority schedule first arrives at queue's process, and the current executing process can continue to consume system resources until a higher-priority process arrives or it exits. The latter is similar to the former. The process group with the highest priority is scheduled first. The process group with the same priority is scheduled in turn by time slice. After the time slice of current execution process is exhausted, it is placed at the end of queue.

Non-real-time process uses the time-division scheduling strategy. The system schedules each thread in a time-slicing fashion. Each process defines its priority at the time of creation. The higher the priority, the longer the time slices. After the time slice allocated by current execution process is finished, it is placed at the end of queue and waits for the next schedule. This schedule is essentially a scheduling way to share the proportion.

Hardware Resource Scheduling Allocation Method

Communications system has high real-time requirements for real-time. In order to meet the real-time requirements, the system often needs to use multi-core, multi-CPU or even multi-processor to make coordination processing. Effectively scheduling universal hardware resources to meet real-time requirements is one of the main indicators for building an Open 5G system.

A reasonable resource allocation scheduling method can make full use of the ability of hardware resources to achieve the performance limit, while unreasonable resource allocation scheduling method will seriously reduce the performance of system. Therefore, in the improvement of system performance, it is very important to optimize the allocation and schedule hardware resources. Hardware resources include computing resources, storage resources, network resources, etc. Considerations for hardware resource scheduling in setting up the Open 5G Universal platform include:

(1) It is needed to take full account of the correlation among processes when scheduler assigns the processor to process tasks. For the process with more shared data or frequent data exchange, it is needed to be allocated on the same processor, thereby increasing the hit rate of processor cache. This is because the shared data of two processes has a greater chance to be stored in cache area,

which can then be retrieved directly from the cache, and the data interaction between two processes can also be made directly through the shared cache. This acquisition speed will be much faster than memory.

(2) Scheduler should ensure the balance scheduling of load. On the one hand, it can prevent overload and shorten average response time of task. On the other hand, the full use of resources in the entire system can improve the efficiency of the resource utilization. The current scheduling methods to realize static load balance mainly are various heuristic algorithms and methods based on graph theory. The methods of implementing dynamic load balance include scheduling algorithms based on the gradient model and the probabilistic scheduling algorithm based on random selection [24] etc.

(3) Scheduler should try to minimize the migration costs in migration process. To achieve load balance, schedulers are often required to make inter-core migration of task processes in different cores. When the task is scheduled, the performance of the system is greatly influenced by the scheduling mode. Because the costs in migrating different processes among processors are different, schedulers must take full account of memory amount occupied by the process, CPU occupancy, restricted types (CPU-bound, input/output-bound), etc., and fully consider the load balance of system and the costs of dealing with load balance.

Based on software defined radio, open 5G platform can achieve a full-featured communication system with GPP. Due to the large data dependency among multiple processing modules of the communication system, it is needed to comprehensively use task segmentation and parallel processing. Allocating corresponding hardware resources to the task threads with scientific scheduling strategy, the platform can meet the system requirements of delay, timing and isolation in wireless communication protocol. Besides, by effectively scheduling the computing resources of the universal platform and designing a reasonable real-time scheduling algorithm, the utilization efficiency of computing resources can be improved, and the real-time processing of baseband signals can be realized.

6.5 Summary

This chapter mainly describes the MIMO parallel channel test platform, the channel model built on the parallel channel test platform, and the MIMO OTA platform and the 5G open source community platform developed by Shanghai Research Center for Wireless Communications. These three sets of platforms are the first platform in China currently, and are being put into the 5G R&D process. These three sets of platforms will continue to evolve, such as adding the parallel channel test function of millimeter wave, sharing millimeter wave channel test data and providing OTA test prototypes of millimeter wave equipment, etc.

References

1. https://www.fcc.gov/document/fcc-promotes-higher-frequency-spectrum-future-wireless-technology-0
2. NI TDMS format http://www.ni.com/white-paper/3727/zhs/
3. USRP[EB/OL]. http://www.ettus.com/.
4. GNU Radio[EB/OL]. http://gnuradio.org/.
5. K. Tan, H. Liu, J. Zhang, J. Fang and et al. Sora: High performance software radio using general purpose multi-core processors. Communications of the Acm, 2011, 54(1):99–107.
6. China Mobile Communications Research Institute. White Paper of C-RAN Access Network Green Evolution. Version 2.5. October 2011. (Chinese)
7. K. Niu, J. Sun, K. Chen, and K. Chai. TD-LTE eNodeB prototype using general purpose processor. Equine Veterinary Journal⟩⟩, 2004, 36(3):248–254.
8. Intel. Intel® 64 and IA-32 Architectures Optimization Reference Manual. May, 2007.
9. Integrated Performance Primitives for Intel® Architecture Reference Manual.
10. PlanetLab: An open platform for developing, deploying, and accessing planetary-scale services. http://www.planet-lab.org.
11. GENI: Global Environment for Network Innovations. http://www.geni.net/.
12. L. Peterson, Anderson T, Blumenthal D and et al. GENI design principles. IEEE Computer, 2006, 39(9): 102–105.
13. VINI: A virtual network infrastructure. http://www.vini-veritas.net/.
14. N. Feamster, L. Gao and J. Rexford. How to lease the Internet in your spare time. ACM SIGCOMM Computer Communication Review, 2007, 37(1): 61–64.
15. M. Peng, Y. Li, Z. Zhao and et al. System architecture and key technologies for 5G heterogeneous cloud radio access networks. IEEE Network, 2015, 29(2):6–14.
16. IMT-2020 (5G) Promotion Group. White Paper on 5G Concept. Beijing: Release Conference of "White Paper on 5G Concept", 2015.
17. P. Agyapong, M. Iwamura, D. Staehle and et al. Design considerations for a 5G network architecture. IEEE Communications Magazine, 2014, 52(11):65–75.
18. Y. Li, P. Hao, X. Feng and et al. Cell and user virtualization for ultra dense network. 2015 I.E. 26th Annual International Symposium on Personal, Indoor, and Mobile Radio Communications (PIMRC), 2015:2359–2363.
19. T. Wang. The implementation of sending and receiving LTE PUGG and SRS based on GPP. Beijing University of Posts and Telecommunications, 2013.
20. Q. Zhang. Research on the key technology of multi-core processor. Fudan University, 2014.
21. H. He. The Research and realization of LTE real-time communication link on GPP-based SDR platform.Beijing University of Posts and Telecommunications, 2013.
22. W. Shi. Universal multi-thread parallel processing technology of wireless signal on GPP platform. Beijing University of Posts and Telecommunications, 2014.
23. X. Zhang. Study on key algorithms and optimization of uplink receiver for LTE-A based on GPP platform. Beijing University of Posts and Telecommunications, 2014.
24. Z. Tan. Thread schedule based on multi-core system. University of Electronic Science and Technology of China, 2009.

Chapter 7
Field Trial Network

The ultimate goal of mobile communications technology research and development and evaluation is application. The field trial is the last and also the most important part of the technology from research to the application. During the R&D process of 5G key technology, the technological feasibility must also be verified by the field trial, providing experimental basis for standardization.

This chapter combines with domestic and international 5G research vision, analyzes the technical challenges of 5G system application scenarios and the field trial, introduces typical cases of the field trial and intelligent analytic methods of network testing data, and describes the feasible scheme for the R&D of the future field trial environment. In Sect. 7.1, we analyze the requirements definition and technical challenges of 5G application scenarios by EU METIS project and China's 5G Promotion Group. In Sect. 7.3, we give the corresponding test case explanation in terms of the test methods of mobile communications wireless channel, key physical layer technology verification, and network coverage performance, etc. In Sect. 7.4, based on the existing network analysis methods, we propose an intelligent network data analysis method. In the section of future test network planning and outlook, we combine the 5G field trial challenges and analyze the field R&D planning from aspects of application scenarios diversity, heterogeneous network convergence, and large-scale users, etc.

7.1 Requirements and Technical Challenges

7.1.1 Various Application Scenarios

Mobile communications technology R&D and test and verification cannot be separated from the application scenarios. The goal of R&D for key technologies is to apply to the application scenarios. With the explosive growth of

© Springer International Publishing AG 2018
Y. Yang et al., *5G Wireless Systems*, Wireless Networks,
DOI 10.1007/978-3-319-61869-2_7

communications needs, in 2020 and years to come, mobile Internet and IoT will become the main driving force of the development of mobile communications. However, with various scenarios and the extreme difference in performance requirements, it's unlikely that 5G system can form a solution of all scenarios based on single technology just like before. Therefore, at the beginning of 5G technology research, we must define the future application scenarios classifications and the corresponding technical challenges.

Back in November 2012, EU launched 5G METIS project, in which the following five characteristics of 5G application scenarios are summarized [1].

1. Fast, i.e., "amazingly fast": 5G will guarantee a higher data rate for future mobile broadband users.
2. Dense, i.e., "Great in a crowd": 5G will guarantee that the densely populated areas can get high quality mobile broadband access.
3. Complete, i.e., "Ubiquitous things communicating": 5G is committed to efficient handling of various types of terminal equipment.
4. Best, i.e., "Best experience follows you": 5G is dedicated to providing mobile users better user experience.
5. Real, i.e., "Super real-time and reliable connections": 5G will support new applications with more stringent requirements on latency and reliability.

In "White Paper on 5G Concept" (2015–02) [2] by China's 5G Promotion Group, 5G technology scenarios are described. From the main application scenarios, service requirements and challenges of the mobile Internet and IoT, four main technology scenarios can be summarized, namely seamless wide-area coverage scenario, high-capacity hot-spot scenario, low-power massive-connection scenario, and low-latency high-reliability scenario.

Seamless wide-area coverage and high-capacity hot-spot scenarios mainly target the demand for mobile Internet in 2020 and the years to come, which is also the main technical scenario of the traditional 4G.

1. Seamless wide-area coverage is the most basic coverage means of mobile communications, which provide users with seamless high-speed service experience with ensuring the user's mobility and service continuity as its goal. The main challenge for this scenario is to provide more than 100 Mbps user experienced data rate with guaranteed service continuity anytime and anywhere (including harsh environment like coverage edge and high-speed moving).
2. High-capacity hot-spot scenarios mainly target local hot-spot areas where ultra-high data rates should be provided to users and ultra-high traffic volume density needs to be handled. The main challenges include 1 Gbps user experienced data rate, tens of Gbps peak data rate, and tens of Tbps/km^2 traffic volume density.

   Low-power massive-connection and low-latency high-reliability scenarios mainly target IoT service, which is a newly expanded 5G scenario with the key focus on IoT and vertical industry applications that traditional mobile communications cannot provide good supports.

Table 7.1 5G typical scenarios classification

No.	Scenario features	Typical scenarios
1	Tens of Tbps/km^2 traffic volume density	Office
2	Gbps user experienced data rate	Dense residence
3	1 million connections /km^2	Stadium, open-air gathering
5	6 persons/m^2 super high density	Subways
6	Millisecond level end-to-end latency	Highway
7	Higher than 500 km/h mobility	High-speed train
8	100 Mbps user experienced data rate	Wide-area coverage

3. Low-power massive-connection scenarios mainly target sensor and data collecting use cases, such as intelligent city, environmental monitoring, intelligent agriculture, forest fire prevention, etc., which features small data packet, low power and massive connection. This type of terminals are widely distributed in large quantity, which should be able to not only support ultra hundreds of billion connections and meet the 100 million/km^2 connection density requirements, but also must ensure the ultra low power consumption and ultra low cost of terminals.
4. Low-latency high-reliability scenarios mainly target special application requirements of IoT and vertical industries such as Internet of Vehicles (IoV) and industrial control, which have extremely high indicator requirements on latency and reliability. The ms-level end-to-end latency and nearly 100% reliability need to be guaranteed in this scenario.

Specifically, 5G typical application scenarios include the following categories (Table 7.1).

7.1.2 Technical Challenges in Field Trial

China's 5G Promotion Group's White Paper on 5G Vision and Requirements (2014–05) [3] describes the 5G application scenarios and performance challenges.

5G typical scenario will touch many aspects of life in the future, such as residence, work, leisure, and transportation. The 5G scenarios include at least dense residential areas, office towers, stadiums, open-air gatherings, subways, highways, high-speed railways, and wide-area coverage. These scenarios, which are characterized by ultra-high traffic volume density or ultra-high connection density or ultra-high mobility, may be quite challenging for 5G.

Some typical services, such as augmented reality, virtual reality, ultra-high-definition videos, cloud storage, IOV, smart home, and Over The Top (OTT) services, will take place in these scenarios. The performance requirements for 5G are derived for each scenario, according to the predicted distribution of users, percentage of different services, and service requirements such as data rate and latency. The key performance indicators for 5G include user experienced data rate, connection density, end-to-end latency, traffic volume density, mobility, and user peak date rate.

From the mobile Internet and IoT main application scenarios, we can sum up that 5G will meet people's various service requirements in residence, work, leisure, and transportation. Even in dense residential areas, office towers, stadiums, open-air gatherings, subways, highways, high-speed railways, and wide-area coverage which are characterized by ultra-high traffic volume density or ultra-high connection density and ultra-high mobility. The extreme service experience like ultra-high-definition videos, virtual reality, augmented reality, cloud desktop, and online games can be provided to users. Meanwhile, 5G will also penetrate into and deeply converges with IoT and a variety of industries such as industrial facilities, medical equipment, and transportation tools, effectively meeting the needs of diversified services in vertical industries like industry, medical service, and transportation to realize a real "everything connected" world.

5G will respond to the challenges brought by diversified performance indicators in the various application scenarios. Different application scenarios are faced with different performance challenges in user experienced data rate, traffic volume density, latency, energy efficiency and connections, all of which may be challenging to different scenarios.

METIS project converts five characteristics of the 5G application scenarios into a series of highly challenging requirements in the form of numbers.

In the densely populated city environment, 5G will provide the rate of 10–1000 Gbps, 10–100 times of the current situation.

- The unit area or single user mobile data will increase by 1000 times, or exceed 100Gbps/km^2 or 500GB/user/month.
- The number of the interconnected terminal equipment will increase by 10–100 times.
- Battery service life of large low-power communications equipment will be extended by 10 times. And that of terminals such as sensors or pagers will reach 10 years.
- Ultra-fast response applications such as "touch Internet" will be supported. The end-to-end latency within five microseconds and the very high reliability will be realized.

7.1.3 Development Status of 5G Test Bed in Foreign Countries

In May 2014, Japanese operator NTT DoCoMo announced the start of the 5G network experiment. It was reported that the companies that participated in NTT DoCoMo 6 GHz-above spectrum 5G tests are Alcatel Lucent, Ericsson, Fujitsu, NEC, Nokia and Samsung. NTT DoCoMo announced that 5G network commercial application was expected to be launched in 2020. Compared with the 100Mbps to 150 Mbps speed in the existing LTE network, 5G network speed can be as high as 10 Gbps at maximum. NTT Docomo plans to do 5G experiments firstly in the company lab in Yokosuka, Kanagawa Prefecture, and then do the outdoor experiment. The company says that it will share its research results with peers in the industry, and strive to promote the development of the 5G network standard starting in 2016. In the 5G experimental project, NTT DoCoMo will cooperate with other communications companies to carry out research, testing and verification. The tests with Nokia mainly concentrate in millimeter wave technology and 70 GHz spectrum of beamforming. The cooperation with Samsung is mainly on 28 GHz spectrum tests. 15GHz spectrum of outdoor tests with Ericsson study new antenna technology which can support "the large-scale multi input multi output" (massive MIMO).In the tests on 3–6GHz spectrum with Fujitsu, NEC focus on "time domain transmit beam forming" in the 5GHz spectrum. The cooperation with Alcatel Lucent is mainly oriented to existing below 3GHz spectrum to mobile broadband and M2 M applications.

In October 2014, SamSung carried out the 5G technology field experiment, in which the 1.2 Gbps high-speed breakpoint-free data transmission speed in a car at the speed of 110 km/h and the 7.5 Gbps peak transmission rate in stop state were reached. The company has announced a partnership with SK Telecom in joint R&D of 5G technology and scheduled to put the 5G network in trial during the Pyeongchang Olympic Winter Games in 2018, in South Korea.

In September 2015, 5G Innovation Center (5GIC) of University of Surrey, UK was officially founded. Its goal is to carry out 5G technology test in the real world, rather than in the laboratory. It hopes to build a 5G network "testbed", which is expected to be completed in three phases. The first phase is expected to be completed around April 2015. The first phase of the testbed will focus on the construction and validation of cloud wireless access network in UDN, which will display unprecedented network capacity to users. With the continuous improvement of the testbed, a variety of 5G candidate waveforms, including Sparse Code Multiple Access (SCMA) technology will be verified in the 5GIC testbed. When the entire test platform is ready for operation, in 2016 or 2017 it can begin to be deployed on campus of University of Surrey, covering the campus composed of 17,000 faculty members and students. As one of the founder members of the 5GIC, Huawei plays an irreplaceable role in the testbed construction. All of the wireless access equipment in testbed is provided by Huawei. Huawei has also set up a 5G key technical verification experts team to provide on-site support to test bed construction. In addition, Huawei will continue to invest 5 million pounds of funds in the 5GIC joint R&D.

In September 2015, the largest U. S. mobile network operator Verizon announced that it would start 5G wireless service field trials in 2016, and would be committed to the commercial use of 5G services as soon as possible. Verizon has begun to work with Alcatel Lucent, Cisco, Ericsson, Nokia, Qualcomm and Samsung to carry out 5G tests, and constructed a 5G test site in San Francisco innovation center.

Overall, globally people have begun to prepare and plan the 5G technology test network for testing and validation. However, by far these test networks are mainly for the cellular mobile technology test verification.

7.1.4   Development and Evolution of HetNet Convergence

Different from the traditional ways, 5G will no longer feature the single multiple access technology. Instead, 5G should include cellular mobile communications technology, new-generation WLAN technology and network technology. Since the simple cell division technology is unlikely to improve the data capacity and cell edge spectrum efficiency, Heterogeneous Network (HetNet) has gradually caught attention [4]. Different communications technologies, such as cellular communications 2G/3G/4G, broadband wireless access IEEE802.16/20 and short distance communications WLAN, Bluetooth and UWB, offer a variety of services to users. Using heterogeneous structure to build a mobile network is an effective way to ease the hot-spot data traffic, which makes the HetNet will be the long-term trend of the mobile network.

HetNet, in a broad sense, refers to the integration of a variety of wireless access network technology, networking architecture, transmission mode and a variety of transmission power of the base station types, such as adding WLAN hot-spots in the mobile network. HetNet, in a narrow sense, means to add low-power nodes of the same mode under the coverage of Macro eNodeB, such as micro cell, RRH and pico cell, HeNB, relay node, etc [5].

Coordination and convergence between HetNets have become the focus of the industry attention. Through the convergence of HetNet, we can fully utilize the advantages of different types of network technologies and get various benefits. We can greatly enhance the performance of a single network, support traditional service and meanwhile create conditions for introducing new services. We can expand the overall network coverage, so that the network has better scalability. We can balance the network service load and increase system capacity. We can make full use of existing network resources and reduce the cost of network operators and service providers, thus making them more competitive. We can provide all kinds of needed services to different users, so as to better meet the diverse needs of customers, and improve customer satisfaction. We can also improve the usability and reliability of the network and enhance the system survivability.

The concept of network convergence can be traced back to the 1970s, when communications circles proposed the concept of network and service convergence, such as the famous Integrated Services Digital Network (ISDN) and Broadband Integrated Services Digital Network (B-ISDN) [6]. But limited by service and technology development, ISDN failed. Until the 90's of last century, with the development of mobile communications technology, it was proposed to develop IMT-2000 global unified mobile communications standards, but also failed. Meanwhile, with the rapid development of Internet at the end of the last century, the industry put forward the concept of the Next Generation Network (NGN). And the research idea changed from network integration to network convergence, which for the first time displayed the prospect of the convergence of the information and communications network on basis of unified IP technology. The research fruits of NGN on network convergence are embodied in IP Multimedia Subsystem (IMS) technology proposed by 3GPP, which integrates telecom network, Internet technology fixed network and mobile network technology. Network convergence integrates Internet IP technology, soft switch technology and cellular core network technology. IMS technology itself has some limitations. But as a core network convergence technology, it is widely recognized by the industry [7].

In fact, since 1990s, the convergence and development of telecom network, radio and television network and Internet have been put on the agenda, that is, what we often call "triple play". Triple play's goal is to achieve the interconnection of the three major networks, resource sharing and service convergence, to provide users with a variety of services including language, data, radio and television and multimedia services. Triple play is the product of constant information technology innovation, but also the inevitable requirement of the development of information technology. However, until now the "triple play" in its real sense has not been completed. Since triple play has entered a substantive implementation stage, this book focuses on the convergence and interconnection between heterogeneous communications networks (especially heterogeneous wireless networks).

At present, there are two main directions for the development of multi network convergence. One is based on the IP backbone network and the other is based on hoc Ad. Multi network convergence system based on Ad Hoc can extend the coverage of wireless communications, improving resource utilization, improving the system throughput, balancing the traffic, and reducing the power consumption of mobile terminals. It has become a research hot-spot in recent years. Especially a series of research fruits have been achieved in terms of the combination of wireless ad hoc network and cellular mobile communications system. And many practical network models have been proposed, such as Ad hoc assisted GSM (A2GSM), Multi-hop Cellular Network (MCN), Self-Organizing Packet Radio Ad hoc Networks with Overlay (SOPRANO), integrated Cellular and Ad hoc Relaying System (iCAR), Unified Cellular and Ad hoc Network architecture (UCAN) [8–12]. A2GSM in the traditional cellular network can support mobile terminal multi-hop relay in order to increase the system capacity, enhance cellular network coverage and solve the haunting problem of cellular coverage blind areas. MCN allows packets move via multi-hop transmission from mobile nodes to the integrated communications network model of cellular communications system base station. SOPRANO combines cellular network with wireless ad hoc network, and introduces the dynamically-distributed wireless router with routing relay function in a cellular network, so as to provide wireless Internet and multimedia services. The basic idea of iCAR is to set up a certain number of Ad hoc Relay network Stations (ARS), so as to achieve traffic load balancing between cells and to control network congestion, relieve hot spots and avoid line dropping etc.

Globally, EU is leading in the study of HetNet convergence, and has carried out a series of research projects. The BRAIN project proposes an open system with WLAN and Universal Mobile Telecommunications System (UMTS) converged. DRIVE project studies the convergence of cellular network and TV and radio broadcasting networks. MOBilitY and DIfferentiated serviCes in a future IP networK (MOBYDICK) project discusses the convergence of mobile network and WLAN in IPv6 network system [13]. My personal Adaptive Global NET (MAGNET) project provides mobile users with ubiquitous and secure personal service via design, R&D, and realization of PN in a HetNet environment. End-to-end Quality Of Service Support over Heterogeneous Networks (EuQoS) focuses on end-to-end QoS technology of HetNets [14]. WINNER project [15] hopes to use a ubiquitous wireless communications system to replace the current coexistence pattern of many systems (cellular, WLAN and short distance wireless access, etc.) and to improve the flexibility and scalability of the system, so as to adaptively provide various services in a variety of wireless environments. These projects, covering the aspects of access, network and services that are both competing and cooperating with each other, have made meaningful researches from a number of aspects and perspectives on the HetNets convergence. Although they put forward different ideas and methods on convergence of different nets, there is still a certain distance to the convergence of different HetNets. Recently, the concept of network convergence based on cognitive networks and wireless ad hoc networks has been proposed by Network Ambient. It can provide a more effective way to realize the convergence of HetNets.

7.2   5G Test Field Design

7.2.1   Network Architecture of HetNets Convergence

R&D of the 5G test field fully considers the HetNets convergence requirements of 3GPP and WLAN. And a unified control platform for HetNets convergence was designed at the beginning of the design. Through the unified control platform and the information interaction between the interfaces of the future 3GPP core network and WLAN controller, HetNets convergence can be realized in the network side. Basic architecture of the HetNets convergence network is shown in Fig. 7.1.

7.2.2   Abundant 5G Typical Scenarios

Similar to the under-construction 5G test field in U.K., our 5G test field is also chosen in the campus environment, located in the campus of ShanghaiTech University. ShanghaiTech University is located in Pudong Science and Technology

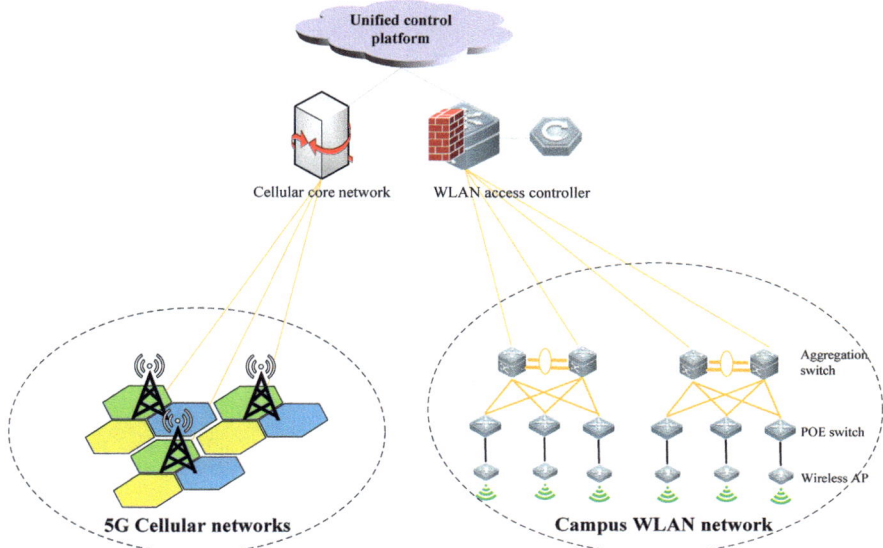

Fig. 7.1   Architecture of HetNets convergence network

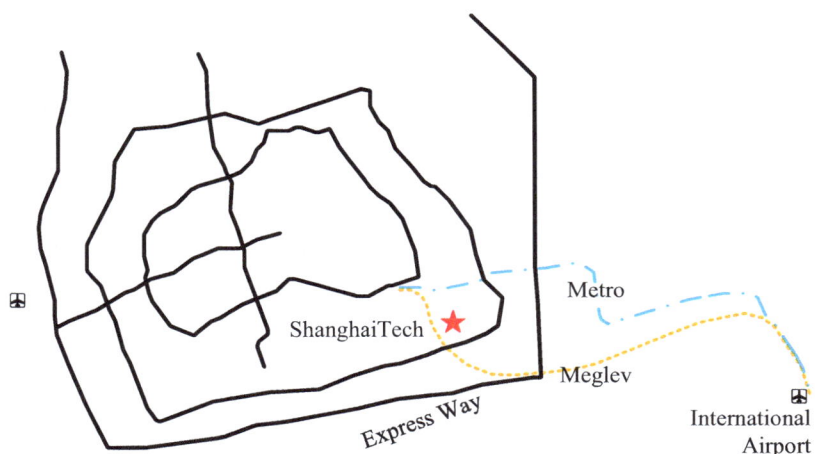

Fig. 7.2   Geographical location diagram of ShanghaiTech University

Park (central Shanghai Zhangjiang High Tech Industrial Development Zone. Its specific location is to the south of Chuanyang River, north of Middle Huaxia Road of Middle Ring, East of Luoshan Road, and west of Jinke Road) which is constructed by the Chinese Academy of Sciences and the Shanghai Municipal Government. The specific location is shown in Fig. 7.2.

Fig. 7.3 Campus aerial view of ShanghaiTech University test field

The campus covers an area of about 900 mu (about 148.26 acres). And the construction scale is 701,500 square meters of the total construction area, including 587,8000 square meters of Phase I new campus construction area (about 280,000 square meters of teaching and experiment buildings, sports facilities and auxiliary buildings; about 160,000 square meters of student dormitories and teachers apartments; about 150,000 square meters of underground construction). University's industry-university-research base (Phase II Science and Technology Park) project has a total construction area of 113,700 square meters.

Campus aerial view is shown in Fig. 7.3.

According to the above introduction, 5G mobile communications test field has the following typical scenarios.

(1) High mobility scenario

   The field is near the world's only in-commercial-operation Shanghai maglev line with the maximum speed of 430 km/h, suitable for high mobility scenario test. Meanwhile it is close to the middle ring elevated road, suitable for 120 km/h mid-to-high-speed mobility scenario test. Campus peripheral roads are suitable for below 60 km/h in slow mobility scenario test.

(2) A variety of indoor and outdoor typical scenarios

   Compared with the single function park, this test field has abundant indoor and outdoor typical scenarios. The field is divided into the area of teaching and living by function. Among them, teaching area is mainly composed of teaching buildings, research buildings, activity center, library, gymnasium and other larger volume buildings, most of which are as high as 20–30 meters. Living area is mainly in high-rises, including apartment layout—three apartments on each floor, dormitory layout--rooms on both sides of the passway, as well as a hotel. The outdoor scenarios include the ways between buildings, the open areas like square and stadium, and the areas with landscape of river, lawn, trees, etc.

(3) Large enclosed underground indoor scenarios

This field has a 150,000 square meters of underground scenario, which is not just a single underground garage, but also a planned underground space with living functions, which is very suitable for indoor coverage, D2D communications and IoT and other related technology tests.

7.2.3   3GPP Wireless Network Design

In order to realize the interconnection between LTE and non-3GPP access networks, the evolution of LTE system on the system network architecture is also studied. In December 2004 in the 26th plenary session, 3GPP was officially approved as a research project, which studies UTRA&UTRAN LTE feasibility. High data rate, low latency, IP packet-based service and becoming competitive in the next 10 years are the direction of the 3GPP system evolution research. In order to achieve this goal, in addition to considering the evolution of wireless access system, we also need to study the System Architecture Evolution (SAE) in order to support the new LTE access network system. SAE's work objectives and research direction are to achieve the full IP network architecture and provide real-time services, and realize the interconnection between the network in evolution and the existing 3GPP network or non 3GPP access networks (such as Worldwide Interoperability for Microwave Access (WiMAX), WLAN). Supporting multiple access systems is one of the basic principles of SAE, which is very important for the competitiveness of SAE.

The main features of the SAE network include:

(1) Guaranteeing to support end-to-end QoS.
(2) Overall grouping. Providing pure packet access in the true sense, while no longer providing a circuit domain service.
(3) Supporting multi-access technology. Supporting the interoperability with existing 3GPP system and the access of non 3GPP networks (e.g., WLAN, WiMAX). And supporting network roaming and switching between the 3GPP network and non 3GPP.
(4) Adding support for real time services. Simplifying network architecture and user service connection signaling process; reducing the service connection latency. The time required for connection is less than 200 ms.
(5) Flat network hierarchy. Nodes of user plane are compressed as much as possible. RNC is canceled in the access network. The user side nodes of the core network are merged into one.

SAE's objectives are consistent with LTE. The first is to improve the performance, reduce the latency, provide higher user data rates, improve system capacity and coverage, and reduce the operation cost. The second is to achieve mobility flexible configuration and implementation of an IP network-based existing or new access technology. And the third is to optimize IP transmission network. However, different from LTE, SAE tends to consider the trend and characteristics of future mobile communications from the perspective of the whole system and to determine the future direction of mobile communication from the angle of network architecture. Under the condition that the wireless network interface technology is diversified and homogeneous, the network architecture that can meet the future trend will make operators more competitive in the future. And the user's changing service needs will also be well met.

In SAE system, non-3GPP access network is defined as IP access network whose access technology is beyond 3GPP, such as WLAN, WiMAX, CDMA2000, etc. According to the trust relationship between EPC and non-3GPP access network, the access network can be divided into the non 3GPP access network and the not-trusted non-3GPP access network. The criteria of trust for access network are not decided by the characteristics of the access network, but by the operators' various strategies. The trusted non-3GPP access network indicates that the communications between UE and EPC is secure. And all the communications between the access network and the EPC is transmitted through the pre-established secure link. The non-trusted non-3GPP access network indicates that the communications between UE and EPC is not secure. And when UE joins the non-trusted non-3GPP network, Internet Protocol Security (IPSec) tunnel between the access network and EPC should be established.

3GPP Core Network Design

According to different trust relations between non-3GPP access network and EPC, and different protocols used for connection of non-3GPP access network and EPC, six communications scenarios are defined.

(1) Realize the billing information connection between WLAN and mobile cellular network.
(2) Realize user WLAN access authentication and billing functions via mobile network.
(3) Support users to get access to mobile network packet domain service via WLAN.
(4) Support users to switch between WLAN and mobile networks and achieve service continuity, though during the switch there may be brief interruptions.
(5) Support users to seamlessly switch between WLAN and mobile networks, while ensuring service continuity and non-interruption.
(6) Support users to get access to the mobile network circuit domain via WLAN.

These six kinds of connected scenarios in order provide users with more and more perfect convergent functions. At present, in WLAN and 2G/3G network, the connection of the first three access scenarios are realized. With the development of service model and technology, as well as the large-scale deployment of LTE network, users want to be able to freely switch between LTE and WLAN to achieve service switch, which requires the implementation of the interconnection between four and five scenarios. For example, when a user is in a WLAN-covered area, he hopes to get PS services via WLAN access operators. And when he leaves the WLAN-covered area, he wants to be switched to the LTE network, while in order to obtain better experience, switching between systems should be smooth. Conversely, the user in the LTE network is able to find the suitable WLAN around automatically, and be seamlessly switched to WLAN.

Core Network Design Process

Core network is the core of the entire network's control and service functions. The driving force of its development and evolution is mainly from two aspects: service requirements and wireless access network evolution requirements, which are the external causes of the evolution of the core network. The major changes in the wireless network have brought the changes in the support system. For example, the change from voice services to data services has brought the evolution from core network to circuit domain and packet domain. With the demand for high-speed data services, 2G network evolved into 3G network, and core network into soft exchange time. With the demand for differentiated data services, core network further evolved into SAE network. Network becomes flat and Policy Control and Charging (PCC) is introduced for differentiated services. On the other hand, the internal requirement for the core network, which mainly means to enhance network security, improves the resources utilization and optimizes the network structure. For example, the separation between call control and bearing has improved the networking flexibility and equipment utilization. The all-IP flat structure introduced by SAE expanded the transmission bandwidth and reduced the end-to-end latency.

Core network EPC system network planning is the upgrading and evolution of soft switching network. In principle, their network structures and soft switching networks are similar in system planning, both of which need present network data analysis, users and traffic prediction, preliminary exploration, network topology, system design (including service test) and other steps. However, since SAE is based on flat, full-IP system, in the network planning we must consider its unique system characteristics, so as to give full play to SAE's technical advantages in convergence, full-IP bearing, and high transmission efficiency. Also in the SAE network planning and construction, we also need to consider the actual deployment of 2G, 3G network, landline network and WLAN network. Not only should we learn from the planning scheme of the existing systems and cater to local conditions, but also we should take into account the system coexistence and balance operation, striving to achieve a perfect combination of excellent performance and low cost in the design stage.

Network planning is the key work before the network construction. According to factors like service demand, social and economic development, regional conditions,

and supporting facilitates, the capacity, QoS, network structure layout are configured. And the construction principles of "overall planning and step-by-step implementation" are adopted. Overall planning means, in accordance with the principle
of all-process all-network, smooth expansion, and overall stability of network
structure, to develop a plan for the SAE network (short-to-middle term rolling
planning for small-range amendments). Step-by-step implementation means to
ensure the synchronization of network construction and market demand, avoid
resource waste, match each project's investment plans, and ensure the independence of planning results as well as the sound construction and development of the
network.

Network Element Planning and Design

EPC core network includes S-GW, MME, P-GW, HSS, PCC, DNS, CG, BG. Full-
IP construction is used.

MME is the only control plane equipment of the core network. Its main functions
are access control, mobility management, session management, network elements
selection and user bearing information storage.

S-GW is located on the user plane. Each of the EPS-related UE has a S-GW
serving it at a point of time.

P-GW is located on the user plane, which is the gateway targeting Public Data
Network (PDN) and ending at Short Guard interval (SGi) interface.

HSS is a database used to store user contract information. The home network can
contain one or more HSS. HSS can also generate user security information for
authentication, integrity protection and encryption. HSS is responsible for the
connection between call control and session management entities in different
domains and subsystems.

Policy and Charging Rules Function (PCRF) is the policy and billing
control unit.

Bandwidth Requirements Bearing

For capacity allocation of the base station, we should mainly consider two aspects.
The first is the system bandwidth selection, and the second is the base station carrier
configuration.

Considering from the perspective of improving the spectrum efficiency, we
should use the large bandwidth, such as 20 MHz, 10 MHz. However, from the
perspective of improving coverage, we should choose the lower frequency bands as
much as possible. But most of the frequency bands below 2GHz in the present
3GPP-defined LTE frequency bands are currently occupied by other systems. These
bands may be released gradually in the future, but they are relatively dispersed.
While the variable bandwidth characteristic of LTE makes it possible to utilize the
dispersed bands.

When we make network plans in reality, we should consider the frequency bands division of the operators and properly choose the biggest possible bandwidth configuration.

In LTE capacity estimation, we need to consider two aspects: control channel and service channel. Among them, downlink capacity should be estimated from the maximal simultaneously scheduled users supported by downlink control channel Physical Downlink Control CHannel (PDCCH) and the average throughput of a single cell Physical Downlink Shared CHannel (PDSCH). Uplink capacity is estimated mainly from the user numbers that can be borne on one Physical Resource Block (PRB) in PDCCH and the average throughput of a single cell PUSCH.

The carrier configuration of the base station should be considered from two aspects: the number of users that can be scheduled by the control channel and the traffic that can be borne by the service channel. The overall estimation formula is shown in (Eq. 7.1).

$$\text{Number of carriers} = \max\left\{\frac{\text{Requirement of data traffic}}{\text{Througput/Carrier}}, \frac{\text{Number of users}}{\text{Scheduled users/Carrier}}\right\}$$

$$(7.1)$$

Generally speaking, sector service demand is the main determinant.

Here are two ways to convert the cell throughput rate to the maximal number of supported broadband users. One is based on the traffic, and the other is based on the single user data rate. Suppose 20 MHz bandwidth, S111 configuration, 2×2 MIMO, and the cells with the frequency efficiency of 1.74bits/s/Hz/are used.Busy time gets 15% of the whole day traffic, with the average load of 50%.

[Method One]

Single cell data service capacity is:

$$20\text{MHz} \times 1.74\text{bits/s/Hz} = 35\text{Mbit/s} \tag{7.2}$$

When the unit is converted to GByte, then the single cell per hour processable data traffic is:

$$\frac{35\text{Mbit/s}}{8192} \times 3600\text{s} = 15.38\text{GB/h} \tag{7.3}$$

The busy-time processable data traffic when the cell average load is above 50% is:

$$15.38\text{GByte/h} \times 50\% = 7.69\text{GB/h} \tag{7.4}$$

Suppose the traffic per user per month is 5GB, then the average busy-time data traffic per user is:

$$5/30 \times 5\% = 0.025 \tag{7.5}$$

Number of users can be borne by S111 base station:

$$3 \times 7.69/0.025 = 921 \tag{7.6}$$

[Method Two]

Single cell data traffic capacity: 35Mbits/s

The busy-time cell data traffic capacity when the cell average load is above 50% is: 17.5Mbits/s

Required per activated user data rate: 1Mbits/s

Busy time user activation ratio: 1/20

Average busy-time throughput per user: 50kbits/s

The number of users that the three sector base stations can accommodate:

$$3 \times 17.5 \times 1000/50 = 1050 \tag{7.7}$$

Through the above calculation, it shows that the LTE base station can accommodate a considerable number of broadband data users.

The transmission requirements of LTE base station mainly consider the transmission bandwidth requirement of S1 interface and X2 interface. At the early stage of network construction, when both the overall network traffic level and the average busy-time throughput requirement are low, in order to give full play to the advantages of LTE technology, the busy-time traffic requirements and the minimum peak rate requirements should be considered in the interface transmission bandwidth configuration.

LTE single-station transmission bandwidth requirements can be estimated based on the service throughput and transmission overhead of each sector.

$$\begin{aligned} \text{S1 interface BW} = {} & \text{Requirement of traffic} \\ & \times (1 + \text{Transmission overhead}) \end{aligned} \tag{7.8}$$

According to the test results of FDD test network, S1 interface user plane transmission overhead is between 2% and 10%. At the early stage of network construction, if we consider meeting the single-sector (20 MHz bandwidth, 2×2MIMO) peak data rate requirements, the S1 interface bandwidth configuration should be:

$$172\text{Mbit/s} \times (1 + 10\%) = 189\text{Mbit/s} \tag{7.9}$$

According to the test results of TDD test network, S1 interface user plane transmission overhead is generally less than 10%.

At the early stage of network construction, if we consider meeting the single-sector (20 MHz bandwidth) peak data rate requirements, the S1 interface bandwidth configuration should be:

$$104.54\text{Mbit/s} \times (1 + 10\%) = 115\text{Mbit/s}. \tag{7.10}$$

It should be noted that the peak data rate achieved here is the downlink peak data rate with 3:1 time slot ratio. If the time slot ratio is 2:2, the downlink peak data rate should be reduced by 1/3.

3GPP Wireless Access Network Design

Wireless network design is a systematic project covering from wireless propagation theory research to the antenna equipment indicator analysis, from the network capability prediction to detailed engineering design, from the network performance test to the system parameter adjustment and optimization. It runs through the whole process of the entire network construction, from the macro overall design ideas to the micro parameters of every cell.

Network planning means, according to the needs of network objectives and evolution and the cost requirements, to choose suitable network element equipment for planning, finally output element number and element configuration, and determine the connection between the network elements, providing basis for the next-step.

Wireless Network Design Process

The objectives of network planning and design can be summarized as "3C1Q", i.e., with the minimal Cost, to obtain the best possible Coverage; with the biggest possible system Capacity, to serve as many as possible users; meanwhile ensure certain communications Quality. In order to achieve the above objectives, we must follow a certain planning and design process to make scientific network planning and design. The whole wireless network planning is divided into four steps.

1. Preparation stage
   At this stage, the main work involves collecting the market information, geography, demographic information, customer demand, and other related information needed for the planning and design; acquiring the preliminary information of the network size; carrying out Continuous Wave (CW) test and propagation model correction, sweep frequency test, indoor penetration loss test, etc.; mastering the wireless environment of the service area, so as to provide the basis for the following coverage and capacity planning.

2. Pre-planning stage
   It mainly includes estimating and analyzing service model, network performance, network size and equipment demand, determining the network construction strategy and specific construction objectives, making budget for coverage and capacity, and determining the overall size of the network construction, making investment estimation, etc.

3. Detailed planning and design phase

It includes the specific site selection, top surface layout, equipment selection, code programming, parametric planning, etc. The effects of planning and designing can be verified via the simulation and prediction software. If the construction objectives remain unmet, the planning and designing schemes can be adjusted until they do.

4. Project implementation phase

On the basis of the previous work, the plans are implemented in reality and some necessary adjustments are made.

Propagation Model Selection

Wireless communications model plays a key role in the link budget. And the coverage radius of the base station is determined by the maximal path loss allowed in the link budget. Wireless communications model falls into two types, namely, indoor and outdoor. The difference between the two models lies in the different parameters. In the outdoor wireless propagation model, the landforms and buildings in the propagation path are the influencing factors that must be considered, because the signal fading is different in different environments. The wireless propagation in free space has the minimal signal fading. The fading in open area/suburb is greater than that in free space. The fading in general urban area is greater than that in open area/suburb. And the signal fading in dense urban areas is greater than that in general urban area. The features of the indoor propagation model are low transmitting power, small coverage, and complex surrounding environment. In the following part we are going to introduce several common communications models.

1. Free space model

Free space represents an ideal space or a space composed of isotropic media. When the electromagnetic wave propagates in this space, there is reflection, diffraction or absorption. The cause of the propagation loss is only the electromagnetic wave propagation. Satellite communications and LOS communications of microwave lines are typical examples of free space propagation. Under certain conditions, the base station and the terminal antenna can be installed at any height. In such a case, the communications between the base station and the terminal is LOS communications. If the clear LOS communications exist between the transmitting antenna and receiving antenna, the path loss follows the free space model. The propagation loss model in free space is as follows:

$$PL = 32.4 + 20\lg d + 20\lg f \qquad (7.11)$$

In it, d is the distance between the terminal and the base station, the unit is km. f represents the frequency, the unit is MHz. This model is applicable when the base station antenna and the terminal are installed at a certain height and there is LOS between the base station and the terminal.

Table 7.2 Okumura Hata Model application range

Frequency range	150～1500 MHz
Base station height	30～200 M (The height of the base station must be higher than the surrounding buildings.)
Terminal antenna height	1～10 m
Distance between transmitter and receiver	1～20 km

Table 7.3 Cost231-Hata Model application range

Frequency range	1500～2000 MHz
Base station height	30～200 M (The height of the base station must be higher than the surrounding buildings.)
Terminal antenna height	1～10 m
Distance between transmitter and receiver	1～20 km

2. Okumura-Hata model

As an evolutionary version of the Okumura model, Hata model is mainly used in urban areas, whose application range is shown in Table 7.2.

Okumura-Hata Model can be expressed as (Eq. 7.12).

$$L = 69.55 + 26.16 \lg f - 13.82 \lg(h_b) - a(h_{re}) + [44.9 - 6.55 \lg(h_b)] \times \lg d \tag{7.12}$$

For small and medium cities:

$$a(h_{re}) = (1.11 \lg f - 0.7)h_{re} - (1.56 \lg f - 0.8)\text{dB} \tag{7.13}$$

For big cities:

$$a(h_{re}) = 8.29(\lg 1.54 h_{re})^2 - 1.1\text{dB} \quad (f < 300\text{MHz}) \tag{7.14}$$

$$a(h_{re}) = 3.2(\lg 11.75 h_{re})^2 - 4.97\text{dB} \quad (f > 300\text{MHz}) \tag{7.15}$$

Among them, L is the path loss in urban areas with dB as its unit, f is the system's operating frequency with MHz as its unit, h_b represents the height of the base station antenna with m as its unit, h_{re} indicates the height of the terminal antenna with m as its unit, $a(h_{re})$ is the antenna height correction factor, d represents the distance between the terminal and the base station with km as its unit.

3. Cost231-Hata

Cost231-Hata can be used as the propagation model of the macro cell base station, and its application range is shown in Table 7.3.

Cost231-Hata Model can be expressed as (Eq. 7.16).

$$Total = L - a(h_m) + C_m \tag{7.16}$$

Where

$$L = 46.3 + 33.9 \lg f - 13.82 \lg h_b + (44.9 - 6.55 \lg h_b) \times \lg d \qquad (7.17)$$

Among them, f is the system's operating frequency with MHz as its unit, h_b represents the height of the base station antenna with m as its unit, h_m indicates the height of the terminal antenna with m as its unit, d represents the distance between the terminal and the base station with km as its unit, $a(h_m)$ is the terminal gain capability. This function is related to antenna height, terminal operating frequency and environment.

The values of C_m is determined by the landscape type. The values of C_m in the standard Cost231-Hata model are as follows:

Big cities: $C_m = 3$
Medium and small cities: $C_m = 0$
Suburb: $C_m = -2[\lg(f/28)]^2 - 5.4$
Rural open area: $C_m = -4.78(\lg f)^2 - 18.33 \lg f - 40.98$.

Since the frequency of some of the mobile communications networks is more than 2GHz, such as 2.3 GHz, 2.6 GLTE Hz and 3.5GHz, it has exceeded the application range of the standard Cost231-Hata model. Therefore, in the actual network planning and design, Cost231-Hata model must be corrected based on the results of the CW test.

4. Standard Propagation Model (SPM)

SPM is particularly suitable for predicting the model of 150 ~ 3500 MHz frequency band and long-distance communications (1 km$<d<$20 km).It is very suitable for a variety of cellular mobile communications technologies. The model is based on the terrain profile, diffraction mechanism and has considered the clutter and effective antenna height to calculate the path loss.

This model can be used in any technology, and it can be calculated in accordance with (Eq. 7.18).

$$\begin{aligned} L_{SPM} = {} & K_1 + K_2 \times \lg(d) + K_3 \times \lg\left(H_{Txeff}\right) + K_4 \times DiffractionLoss \\ & + K_5 \times \lg(d) \times \lg\left(H_{Txeff}\right) + K_6 \times H_{Rxeff} + K_{clutter} \times f(clutter) \end{aligned} \qquad (7.18)$$

The meaning of the parameters in the formula is shown in Table 7.4.

By using the simulation tool, the CW test is completed, which can be used to calibrate the standard propagation model.

Coverage Design

1. TD-LTE coverage characteristics

If the target service is data service with certain data rate, then determining the reasonable target rate is the basis of coverage planning.

Table 7.4   The meaning of parameters of SPM model

Parameter	Meaning
K_1	Offset constant (dB)
K_2	$\lg(d)$ constant factor
K_3	$\lg(H_{Txeff})$ constant factor
K_4	Constant factor of diffraction loss; K_4 must be positive
K_5	$\lg(d) \times \lg(H_{Txeff})$ constant factor
K_6	H_{Rxeff} constant factor
d	Distance from transmitter to receiver (m)
H_{Txeff}	Effective height of transmitting antenna (m)
H_{Rxeff}	Effective height of receiving antenna (m)
$DiffractionLoss$	Diffraction loss caused by obstacles. (dB)
$K_{clutter}$	$f(clutter)$ constant factor
$f(clutter)$	Average weight loss caused by ground objects

In TD-LTE, there is no circuit domain service but only PS domain service. The coverage capability of different PS data rates is different. In coverage planning, we must first determine the data rate target of the edge users, such as 500kbits/s, 1Mbits/s and 2Mbits/s. Different target data rates have different demodulation thresholds, and thus different coverage radius. So determining the reasonable target data rate is the basis of coverage planning.

(1) LTE resource scheduling is more complex, and the coverage characteristics and resource allocation are closely related.

TD-LTE network can flexibly select users' Resource Block (RB) resources and modulation coding modes and combine them so as to cope with different coverage environments and planning requirements. In the actual network, the user data rate is related with MCS and RB occupied while MCS depends on the SINR value and RB occupied will affect SINR value. So MCS, RB occupied, SINR value and user data rate will affect each other, making LTE network scheduling algorithm more complex. In coverage planning, it is difficult to simulate the complex scheduling algorithm of the real network. Therefore, how to reasonably determine the RB resources, modulation and coding modes to make the selection more suitable for the actual network conditions is a difficult problem in coverage planning.

(2) Selection of transmission mode and antenna type influences coverage planning.

Multi-antenna technology is one of the most important key technologies of LTE. The LTE network with multiple antennas has many kinds of transmission modes (there are eight transmission modes at present) and a variety of antenna types (two antennas, eight antennas, and so on at the base station side). Selection of transmission mode and antenna type has great impact on coverage performance.

(3) Inter-cell interference affects TD-LTE coverage.

OFDMA technology is introduced to TD-LTE system. Due to orthogonal frequency between different sub carriers, the interference between different users in the same cell can be ignored. But there still exists same-frequency interference between cells of TD-LTE system. With increase of network load, inter-cell interference level will increase, drawing down the user SINR value and data transmission rate accordingly, showing certain respiratory effects. In addition, different interference cancellation technology will produce different inter-cell interference suppression effects, which will affect the TD-LTE edge coverage effect. Therefore, how to evaluate the inter-cell interference is also a difficult problem in TD-LTE network coverage planning.

2. LTE wireless network coverage strategy

The main objectives of LTE system coverage planning is based on the requirements of the actual cell edge coverage. Under certain system parameter setting, we can estimate the coverage distance that the base station can achieve, so as to get the network size requirements. According to the application scenarios and the actual planning requirements, the LTE system's coverage planning strategy is generally divided into three categories.

(1) Network size estimation based on the requirements for uplink edge data rate

The first strategy is mainly used for the coverage requirement that only the uplink edge data rate is limited. Based on the uplink data rate, with the input of certain link budget parameters, the uplink coverage radius is calculated. Based on the obtained uplink coverage radius, the uplink achievable edge data rate is predicted.

(2) Network size estimation based on the requirements for downlink edge data rate

The second strategy is mainly used for the coverage requirement that only the downlink edge data rate is limited. Based on the downlink data rate, with the input of certain link budget parameters, the downlink coverage radius is calculated. Based on the obtained downlink coverage radius, the uplink achievable edge data rate is predicted.

(3) Network size estimation based on the requirements for uplink and downlink edge data rate

The third strategy is mainly used for the coverage requirement that both the uplink and downlink edge data rate is limited. Based on the uplink and downlink data rate, with the input of certain link budget parameters, both the uplink and downlink coverage radii are calculated. And by comparison the limited coverage radius can be obtained.

In the actual network planning, according to different requirements and application scenarios, we need to select the appropriate coverage planning strategy and flexibly respond to the problems in network planning.

Capacity Design

TD-LTE system capacity is determined by many factors. First of all is the fixed configuration and algorithm performance, including single sector frequency bandwidth, time slot configuration mode, antenna technology, frequency usage, intercell interference cancellation technology, resource scheduling algorithm, etc. Next, the actual network's overall channel environment and link quality will affect TD-LTE network's resource allocation and modulation and coding mode selection, so the network structure also has a crucial impact on TD-LTE capacity.

(1) Single sector frequency bandwidth. TD-LTE supports the flexible bandwidth configuration of 1.4 MHz, 3 MHz, 5 MHz, 10 MHz, 15 MHz, 20 MHz. Obviously, with greater bandwidth, network resources will be more available and the system capacity will be greater.

(2) Time slot configuration mode. TD-LTE uses TDD, which can, according to the different uplink-downlink proportions in an area, flexibly configure the uplink-downlink time slot ratio. The present protocol defined seven kinds of uplink-downlink time slot configuration modes, the special time slot of which have nine modes to choose from. Different configurations have distinct differences in uplink and downlink throughput.

(3) Antenna technology. TD-LTE uses a multi-antenna technology, so that the network can, based on the actual network requirements and antenna resources, realize single stream diversity, multi stream multiplexing, multiplexing and diversity adaptation, single stream beamforming, multi stream beamforming, etc. They have different usage scenarios, but all will affect the user capacity to certain extent.

(4) Frequency usage. Currently, the analysis shows that TD-LTE network can use the same frequency network. But the capacity performance of single cell with the same frequency network system of the same bandwidth will be poorer than that of the different frequency network systems. So in the actual operation we should comprehensively consider the frequency resource, capacity requirements and other factors to determine the frequency usage.

(5) Inter-cell interference cancellation technology. For TD-LTE system, due to the characteristics of OFDMA, the intra-system interference is mainly from other cells with the same frequency. The co-channel interference will reduce the user's SNR and thus affect the user capacity. So the effect of interference cancellation technology will affect the overall system capacity and the cell edge user's data rate.

(6) Resource scheduling algorithm. TD-LTE adopts the adaptive modulation coding method, so that the network can detect the test feedback in real time according to the channel quality, and dynamically adjust user data encoding modes and occupancy resources, achieving the optimal performance. Therefore, TD-LTE overall capacity performance and resource scheduling algorithm is closely related. Good resource scheduling algorithm can significantly improve the system capacity and user speed.

(7) Network structure. TD-LTE user throughput depends on the quality of the user's wireless channel environment. Cell throughput is determined by the overall cell channel environments. And the most critical factors affecting the overall cell channel environment are the network structure and the cell coverage radius. In TD-LTE planning, we should pay more attention to network structure than the 2G/3G systems, choose site strictly in accordance with station distance principle, and avoid high stations and the sites deviating greatly from the cellular structure.

TD-LTE capacity evaluation indicators.

The TD-LTE system capacity analysis can be separated into control plane and user plane. Control plane capacity indicators include simultaneous scheduled user number and simultaneous online user number. The simultaneous scheduled user number is the basic indicator for system evaluation, specifically including uplink and downlink control channel capacity, which is limited by the air interface resources and channel configuration. Meanwhile the simultaneous online user number is not only limited by the channel resources of the air interface control plane, but also closely related to the equipment hardware processing capacity. User face capacity indicators can be divided according to service types, including Voice over Internet Protocol (VoIP) services and non-VoIP data services. Non-VoIP data service indicators have cell peak throughput, cell average throughput and cell edge data rate, etc. The VoIP service indicator is the number of VoIP users. The basic indicators of the user plane include the average cell throughput and the number of VoIP users. The above indicators are briefly discussed in the following part.

(1) Simultaneous scheduled user number: the number of users that can be scheduled by every system TTI.
(2) Simultaneous online (activated) user number: the number of users that maintain the system connected status.
(3) Cell average throughput: When distributed in certain rules, the average throughput of the whole cell equals to the sum of all cell throughput / cell number.
(4) Cell edge throughput: throughput of the users distributed at the cell edge. In the simulation system, the edge user is defined as the user at 5% when all the users in the network are ranked in descending order according to the user throughput.
(5) The number of VoIP users: the total number of VoIP users contained in the cell. VoIP user number is related with bandwidth configuration, control channel resources and VoIP scheduling algorithm.

Site Resources Design

Base station site selection is an important link in mobile communications network planning. Reasonable site location can use the network construction funds with maximum efficiency, can avoid coverage blind area and reduce capacity insufficiency to a certain extent, greatly improving the capacity, coverage and quality of service. Generally speaking, a good site should follow the following principles.

(1) Try to choose the site with regular cell structure that can meet the coverage target of single edge user's uplink and downlink throughput requirements. Regular cellular structure can ensure the effective system coverage in the planned area, avoid weak signal area, better achieve higher edge data rate, and meanwhile facilitate future cell division. However, in the actual network construction, due to various reasons, sometimes the site cannot be built in the cellular center and the alternative suboptimal site around the ideal site has to be found. Therefore, it is necessary to provide a search circle around each site. The actual site can be found within the search coil. Generally speaking, search circle radius deviation should be less than 1/4 of the radius of the base station cell. Meanwhile, taking into account the yearly increasing requirements of the commercial network, in the area where conditions allow, it can be considered to reduce the minimum base station spacing.

(2) The site distribution should correspond to traffic distribution density. In site selection, the areas with more users like government agencies, airports, railway stations, news centers, important hotels and the main hot-spot regions, etc. should be given priority to ensure good communication and avoid the overlapping coverage. Meanwhile, the over-concentration of cell edge users should be prevented, so that the site distribution matches the traffic density distribution and the cell service users are concentrated in the cell center as much as possible.

(3) The antenna of the macro cellular base station should be higher than the average height of the surrounding objects. Exceeding values have different require-ments for different terrains, landforms and environments. The antennas of general urban site should be higher than the surrounding average by 5~10 m (about 5 m in dense urban areas), and that of suburban and rural areas can be decided according to the coverage requirements. For site planning, the site antenna height in implementation should not be too different to the require-ments of the planning, in order to avoid the emergence of the cross-area coverage. The base station site should be selected in the main coverage area with relatively high terrain or where there are high-rises or towers that can be used. If the height of high-rises cannot meet the base station antenna height requirements, there should be conditions for setting tower on roof or on ground, so as to ensure a broad vision around the base station and there is no tall buildings higher than the base station antenna, and most of the cell can be covered by base station antenna via straight line propagation.

(4) Avoid strong interference sources near the base station. The base station site should avoid the high power radio transmitting station, high power television transmitting station and high power radar station. Strong interference will greatly reduce the capacity of base stations and voice quality. Generally, in site investigation, a spectrum analyzer must be used to scan the uplink and

downlink frequency near the base station so as to ensure that there is no strong interference source. The isolation of LTE site and other wireless equipment with similar frequency should be given full consideration. The sites where the interference cannot be solved should be avoided, such as paging, microwave, and other equipment with similar frequency.

(5) The site should be selected in the locations with convenient transportation, good power supply and safe environment. Base station is the equipment that needs long-term stable operation and regular maintenance. Any failure is likely to reduce the user's trust to the network operator and reduce the user's support for the network, causing users and income reduction and even the operation failure. Transportation and city power supply must ensure the base station a long-period high-power operation and the timely maintenance by maintenance staff. Site should not be selected near the flammable and explosive buildings and stacks, or near the industrial enterprises which emit harmful gases, heavy smoke, dust, hazardous substances in the production process.

(6) Investment restrictions. In the site selection, the site with lower cost should be preferred. For those high-cost sites, other alternative lower-cost sites should be considered. In addition, while not affecting the overall layout, the existing equipment room, power supply and other facilities should be utilized as much as possible. At the early stage of network construction, in the case of insufficient funds, the coverage of important users and high density users should be guaranteed as much as possible.

Indoor Wireless Network Design

Indoor distribution is a successful solution for indoor user groups to improve the mobile communications environment in buildings. In recent years, it has been widely used in mobile communications operators all over the country.

Indoor distribution system provides a good solution for indoor signal coverage. Its principle is to use the indoor antenna distribution system to evenly distribute the signals of the mobile base station in every corner of the room, so as to ensure the indoor area an ideal signal coverage.

Construction of indoor distribution system can comprehensively improve the call quality within buildings, enhance the mobile phone connection rate, and open up a high quality indoor mobile communications area. Meanwhile, micro cellular systems can share the outdoor macro cell traffic and expand network capacity, so as to improve the overall service level of mobile network.

Indoor distribution system is to introduce the base station signal into the building, make a reasonable distribution via the power distribution device, and then transmit signals via indoor antennas.

With the increasing number of high-rise buildings in the city, the user satisfaction is also rising. These buildings are large in size and good in quality, and have a strong shielding effect on the signal of mobile telephone. Especially in the lower floors of large buildings, underground shopping malls, underground parking and other environment, mobile communications signals are very weak. And mobile phones cannot be used normally, forming the mobile communications blind areas and shadow areas. In the middle floors, due to interference from surrounding base stations, ping pong effect occurs, where mobile phones frequently switch or drop off, seriously affecting the normal use of mobile phones. In higher floors of the building, due to the height limitation of the base station antenna, normal coverage cannot be achieved. It's also a blind area of mobile communications. Besides, in some of the office buildings, although the mobile phones can be used for normal phone calls, the mobile phones are difficult to go online because of the high user density and the base station channel congestion. Under the fierce competition environment, the indoor mobile communications' network coverage, capacity, quality are key factors for operators to gain the competitive advantage. It fundamentally reflects the service level of mobile network and remains the priority of mobile operators in recent years.

7.2.4   WLAN Wireless Network Design

Analysis on New Generation WLAN Technology

The Latest Development of WLAN Technology

WLAN, featuring high throughput and low cost, can be well combined with the Internet service to provide users with convenient high-speed wireless Internet access service. With the rapid development of the Internet, WLAN equipment is rapidly becoming popular all over the world, offering great convenience for people's work and life. According to Wi-Fi Union statistics, currently there are over 1 billion WLAN users globally. Telecom operators also attach great importance to WLAN and see it as an important complement and extension of fixed network and cellular network. They deploy WLAN hot spots in large scale to provide services to the public, which further promotes the WLAN development.

In order to meet the growing market demand, WLAN technology and standards are constantly developing and improving, and data transmission capabilities continue to improve. After more than 20 years of development, the maximum information transmission rate of WLAN equipment in the current IEEE 802.11n device reach up to 600Mbits/s. In addition, the 802.11n device also uses two channel access mechanisms, static 40 MHz and dynamic 20/40, significantly improving the system throughput. IEEE has launched the next generation IEEE 802.11 ac/ad WLAN technical standards, whose data throughput will reach 7Gbits/s at maximum, better meeting the market demand for high throughput wireless data services such as wireless High Definition (HD) video transmission. IEEE 802.11 ac project has already begun as early as the first half of 2008, when it was known as "very high throughput" with the goal directly reaching 1Gbits/s. In the face of the requirements of multiple HD video and lossless audio for over 1Gbits/s code rate, IEEE 802.11 ac was helpless, especially in the indoor high speed data transmission environment. Thus, IEEE 802.11ad was proposed, which would be used to achieve the transmission of domestic wireless HD audio and video signals, and bring more complete HD video solutions for home multimedia applications.

In order to achieve higher wireless transmission rate, IEEE 802.11ad has abandoned the crowded 2.4GHz and 5GHz frequency bands, but rather uses 60GHz frequency spectrum high frequency carrier with 57 ~ 66GHz unallocated frequency band. With such a wide bandwidth, the data transmission rate can be greatly improved. Since 60GHz spectrum in most countries (including the United States) has large quantities of available frequencies, IEEE 802.11ad can realize the simultaneous multi-channel transmission with the support of MIMO technology with the bandwidth of each channel exceeding 1Gbits/s. On the basis of integration of IEEE 802.11 s and IEEE 802.11z, it totally can used to realize file transmission and data synchronization between equipment, and the speed will be faster than the second generation Bluetooth technology by more than 1000 times. Of course, its main purpose is to achieve HD signal transmission.

IEEE802.11 ac

802.11 ac is specifically designed for the 5GHz band. The unique characteristics of the new radio frequency improve the performance of existing wireless LAN until it can be comparable with the level of wired Gigabit network. 802.11 ac is a new standard of IEEE wireless technology. It draws on the advantages of 802.11n and further optimization. In addition to the most obvious characteristics of high throughput, it has a lot of improvements in many aspects.

Improvement on the 1802.11 ac standard physical layer

802.11 standards include physical layer and MAC protocols. Since the first release, the physical layer has made a number of important additions and amendments, while most of the MAC basic features remain unchanged. Here we focus on the changes in the physical layer of 802.11 ac.

(1) A wider channel bandwidth

  802.11 ac supports 80 MHz bandwidth, and choose to use continuous 160 MHz frequency band or discontinuous 80+80 MHz frequency band. Its channel distribution is shown in Fig. 7.4

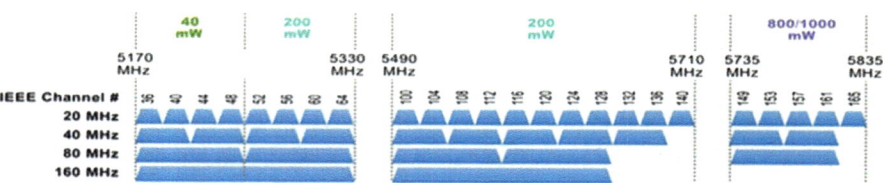

Fig. 7.4 802.11 ac channel distribution

The increase in bandwidth brings more available data subcarriers. 80 MHz has 234 available subcarriers while 40 MHz has only 108, so 80 MHz can bring about 2.16 times of the growth rate. The accompanying side effect is that the same transmission power has to be separated on the added sub carriers, resulting in a slightly reduced signal coverage.

(2) Choosing 5 GHz band

The most important reason for the significant improvement in 802.11 ac performance is technological innovation, and most crucially 5GHz band is used. Previous 802.11 devices always have to face the problem that the network equipment based on these standards have to compete for channels with other non 802.11 devices, such as baby monitors, Bluetooth earphones, or even microwave ovens, etc. These devices are also using 2.4GHz frequency band. When all the equipment is competing for the 2.4GHz band, it's just like a road crowded with cars and everyone slows down. The 802.11 ac works on the not-that-crowded or "cleaner" 5GHz band with fewer wireless devices competing for bandwidth, so the speed is also guaranteed. Of course, the 802.11 ac standard has downward compatibility, ensuring that the 802.11 ac device can be used on the existing Wi-Fi network.

(3) Higher order modulation

802.11 ac uses OFDM technology to modulate the data bits in the wireless medium transmission. According to conditions, 802.11 ac can choose 256 QAM, which makes each subcarrier's data bits increase from six to eight so that the throughput increases by 33%. But 256 QAM is only applicable in high SNR environment, that is, in good channel conditions.

(4) More spatial streams and MU-MIMO

802.11 ac supports up to eight channels of spatial streams. Supporting multiple spatial streams is optional, but the increase in spatial streams is the most effective when combined with the 802.11 ac MU-MIMO new features. 802.11 ac technology, in single user and multi-user MIMO mode, supports up to eight channels of spatial streams, up to four users. And in multi-user mode, each wireless terminal has no more than four channels of spatial streams.

(5) Carrier interception technology enhancement

802.11 ac standards also apply a number of MAC layer enhancements to further enhance the performance of RF and MU-MIMO characteristics. In 802.11n, Request to Send/Clear to Send (RTS/CTS) undertakes clearing tasks, stopping the transmission of 802.11a/g devices during the transmitting to avoid collision. In 802.11 ac, since 80 MHz uses more channels, RTS/CTS mechanism needs to be enhanced to handle communications collision on the secondary channel. The improved RTS/CTS also support the "dynamic bandwidth" mode.

In MAC layer, the 802.11n device announces its transmission intention via sending RTS/CTS frame. These frames allow 802.11a/g devices nearby to sense the channel being used in order to avoid collision.

Due to 802.11 ac's wider bandwidth and limited channels when 80 MHz bandwidth is in use, the hidden nodes on the second channel have become an important problem to be solved. The RTS/CTS mechanism has been updated to better detect whether or not a non-main channel is occupied by different transmissions.

Therefore, RTS and CTS (optional) support the "dynamic bandwidth" mode. In this mode, if the partial band is occupied, the CTS frame is transmitted only on the main channel. The client (STA) that sends RTS frame can be dropped to a lower bandwidth mode, which will help to reduce the impact of the hidden nodes. Anyway, the final transmission bandwidth always includes the main channel.

(6) Message Aggregation Enhancement

In 802.11 ac's basic MAC protocol, in order to ensure that each station can achieve fair access to media usage and avoid collision, a series of control mechanisms are used. These mechanisms can improve the system performance but also bring the fixed overhead, which limits the increase in system throughput. In order to increase the size of Aggregated MAC Protocol Data Unit (A-MPDU) and reduce the communications overhead, 802.11 ac introduces two kinds of frame aggregation methods: MAC Service Data Unit (MSDU) aggregation and Message Protocol Data Unit (MPDU) aggregation. These two frame aggregation methods reduce the overhead of the single channel RF lead code during the transmission of each aggregated frame.

IEEE802.11ad

In the face of the requirements of multiple HD video and lossless audio for over 1Gbits/s code rate, 802.11 ac was helpless. Therefore, 802.11ad came into being, which will achieve the ultra-high data rate of 7Gbits/s and be mainly used for domestic wireless HD audio and video signal transmission, bringing more complete HD video solutions for family multimedia applications .

After 60GHzVHT project got approval in December 2009, 802.11VHT working group passed the 802.11ad Draft 2.0 draft in March 2011, and the official standard is expected to be published by the end of 2012.

In order to achieve a higher rate of wireless transmission, 802.11ad abandoned the crowded 2.4GHz and 5GHz band. Instead it uses high frequency carrier 60GHz spectrum (57GHz–66GHz).Since 60 GHz spectrum in most countries have large quantities of available frequencies, 802.11ad bandwidth of each channel can reach 2.16GHz, which will be 50 times of the 802.11n channel. In addition, 802.11ad also uses adaptive beamforming, a variety of physical layer types, PBSS network architecture, mm Wave channel access, fast session migration and other enhancement technologies to improve the system throughput and coverage. However, 802.11ad is also faced with its technical limitations. For example, 60GHz carrier has very poor penetration ability, and its signal attenuation is very serious in the air, greatly affecting its transmission distance and signal coverage. So its valid connection can only be limited in a small range.

To maximize the system performance and minimize the system implementation complexity and cost and also support the interconnection with the existing 802.11 standard, 802.11ad at the physical layer and MAC layer have been enhanced via the following key technologies.

(1) Greater channel bandwidth

802.11ad uses 60GHz frequency band, which is license free and has very large available bandwidth. China's frequency allocation of this frequency band is 59G-64GHz. Compared with the 83.5 MHz available bandwidth on the 2.4GHz frequency band, 60GHz frequency band has the 5G-9GHz available bandwidth on it. This bandwidth can be divided into multiple channels, each of which has the bandwidth of 2.16GHz. With such a wide channel, 802.11ad does not need to use the channel binding technology. And it can support the application with high bandwidth requirement such as the non-compressed video stream transmission.

(2) Different physical layer types are supported for different application scenarios

802.11ad supports a variety of physical layer types, including the PHYsical Control layer (PHY Control), Single Carrier PHYsical layer (SCPHY), Low Power Single Carrier PHYsical layer (Low Power SCPHY), OFDM PHYsical layer (OFDMPHY).

Control PHY is designed for low SNR operation, in which single carrier mode is mandatory and the maximum data rate is 27.5Mbits/s. With $\pi/2$-DBIT/SK modulation, 1/2 data rate Low Density Parity Check (LDPC) code, and the maximum packet length limit of 1024 bytes, it can achieve shorter data block and get better robustness in the high noise environment.

SCPHY allows for low power, low complexity transceiver and supports the data rate as high as 4.6Gbits/s.

Low Power SCPHY reduces the processing power of the transceiver by a simplified encoding method and a shorter symbol structure. Its maximum data rate is 2.5Gbits/s.

OFDMPHY can achieve better performance in frequency selective channels. It uses 512 point FFT, including 336 data sub carriers, 16 pilot sub carriers, and three Direct Current (DC) carriers. With 64QAM modulation, in 13/16 data rate and LDPC coding, it supports the data rate as high as 7Gbits/s.

SCPHY and OFDMPHY have different application scopes. OFDMPHY, which is suitable for transmission in long distance and large scalability latency, can deal with the multi-path propagated signals. SCPHY generally can get lower power consumption and is suitable for small, low-power handheld devices.

(3) Adaptive beamforming

Due to the huge transmission loss in the 60GHz band, 802.11ad uses adaptive beamforming technology to achieve data transmission over long distance above ten meters.

Adaptive beamforming, by adaptively adjusting the antenna directions and reducing the beam width, obtains a higher antenna gain, reduces interference, and expands the signal coverage. Besides, if there is an obstacle in the line of sight of the transceiver, the transceiver can quickly evade the obstacle and rebuild a new link for communication. Beamforming can be realized by different techniques such as beam switching, phase weighted antenna array, multi antenna array, etc.

Beam shaping protocol consists of three stages: Sector Level Sweep (SLS), Beam Refinement Phase (BRP), and Tracking phase (Tracking). SLS is divided into a transmitting SLS and receiving SLS. The former is used to determine the optimal transmitting direction of the transmitting device and the latter is used to determine the optimal receiving direction of the receiving device. BRP further optimizes the beam direction through the joint adjustment of transmitting and receiving beam. Tracking stage is used to dynamically adjust the beam according to the channel changes in the process of data transmission.

(4) New network architecture PBSS

802.11ad defines a new network architecture Personal Basic Service Set (PBSS), which allows direct communications between two devices. PBSS network architecture falls into the ad hoc network, similar to Independent Basic Service Set (IBSS).But the difference between themis that in PBSS architecture, one party's STAtion (STA) bears the PBSS Control Point (PCP) function. Only the PCP can transmit beacon frames.

(5) mmWave channel access mechanism

The medium time in PBSS system is divided into Beacon Intervals (BI) structure, and the sub time interval in BI is the access time. Different channel access time follows different access rules. As shown in Fig. 7.5, one BI contains the following four types of access time. First, Beacon Time (BT): PCP sends beacon frames in different directions and discovers new STA. Second, the Associated BeamForming Training time (A-BFT): Performing the beamforming training between PCP and STA, establish beamforming link. Third, Announcement Time (AT): It is used to transfer control frames between PCP and STA. Fourth, Data Transmission Time (DTT): It is used to accomplish data frame exchange between STAs, including scheduling-based Service Period (SP) and Competition-Based access Period (CBP), or a combination of any number of the two. During the SP, the scheduled STA can access the channel. During the CBP, all STA can access the channel based on the 802.11DCF and HCF mechanisms. DTT scheduling information is transmitted from STA to PCP through beacon frame and announcement flame.

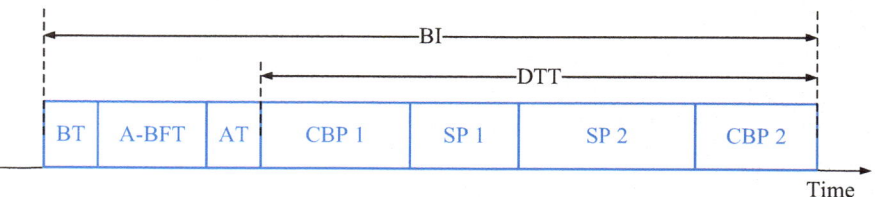

Fig. 7.5 BI structure schematic diagram

(6)   Multi-band inter-operable Fast Session Transfer (FST)

In order to achieve the interoperability of 802.11ad and 802.11a/b/g/n/ac, and support the seamless switching between Wi-Fi communications in the three frequency bands, 802.11ad uses FST technique. In such a case, as for user experience, the users with multiple frequencies device can enjoy uninterrupted continuous communication between the Wi-Fi networks of different modes. When covered by 802.11ad signals, they can experience high-speed network connection, and when 802.11ad signal is poor or does not exist, they can seamlessly switch to Wi-Fi networks.

(7)   Power saving technology

802.11ad devices can reduce power consumption by utilizing scheduling access mechanism. The two devices using directional link communications can access the channel at the scheduled time, and begin hibernation at the schedule interval so as to achieve the purpose of power saving. This technology can enable the device to manage the power overhead based on the actual operational load, which is very important for mobile phones and other handheld devices. 802.11ad defines the non-PCP power saving mechanism and the PCP power saving mechanism, allowing non-PCP STA and PCP to hibernate for several BI times.

In non-PCPSTA power save mode, after STA and STA negotiation, STA enters the hibernation mode in a specific channel access interval, and selects the wake-up interval according to STA's own service needs. PCP tracks the hibernation interval of the STA associated with it, stores MSDU cache for the STA in hibernation mode, and passes on the cache service to STA when STA is woken up.

In PCP Power Save (PPS) mode, PCP can make one or more continuous BI enter the hibernation state, so as to reduce its power consumption. PCP running in PPS mode will no longer send beacons. Before entering hibernation, PCP will notify the non-PCPSTA in PBSS about the necessary information through beacon frame or announcement frame, so that PCP is still able to communicate with STA in hibernation.

WLAN System Architecture

WLAN consists of an end STA, an AP, an Access Controller (AC), an Authentication、 Authorization、 Accounting (AAA) server and a network management unit. AAA server is an entity that provides AAA services. In the model, the AAA server supports the Radius protocol, and the portal server is the entity that is used to push the portal website. In the WEB certification, it assists to complete the certification function. Figure 7.6 is the WLAN network model.

Since IEEE802.11 standard was put forward in 1997, WLAN access speed has increased from original 1 Mb/s to today's 300 Mb/s. And IEEE802.a/b/g/n standards have all greatly promoted the expansion of WLAN.WLAN is not only a supplement to the wired network, but also gradually developing towards large-scale

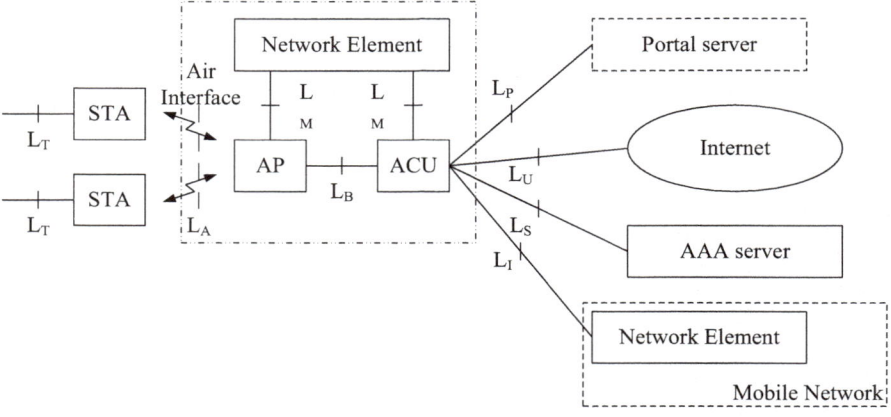

Fig. 7.6 WLAN network model

Fig. 7.7 Autonomous
WLAN network
architecture

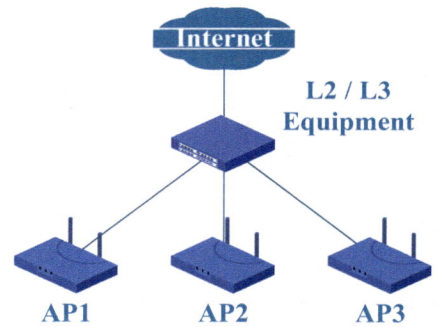

deployment and independent network, and even has replaced the wired network in some places. Traditional WLAN architecture has been unable to meet the requirements of large-scale network. Therefore, the Internet Engineering Task Force (IETF) established Control And Provision of Wireless Access Points (CAPWAP) working group, studying the large-scale WLAN solutions. After working on the current mainstream WLAN solutions, CAPWAP working group divides the WLAN system into three types: autonomous, centralized and distributed network architecture.

1. Autonomous architecture

The early WLAN structure is a kind of autonomous system structure. In this architecture, all 802.11 functions namely the 802.11 PHY Layer and MAC functions are completed by AP. In addition, some complex functions, such as the 802.11i–defined security function, 802.11-defined QoS function, and even Radius client functions are implemented by AP. As shown in Fig. 7.7 is the autonomous WLAN network architecture.

In the autonomous WLAN system structure, "fat AP" mode is mainly used. The so-called "fat AP" means that one access device has most of the network intelligence, whose function is to control most protocols including roaming, encryption, management, user authentication, etc. In this WLAN architecture, the network intelligence is put in the AP equipment, which reduces the workload of the central switch and makes the deployment more convenient. However, this system also has the following main defects.

(1) AP must be managed one by one, i.e., the control strategy can only manage one AP at a time, which increases the management costs, and thus more management resources are needed.
(2) Since the possible attack and interference cannot be seen in the whole system, the service refusal of the whole WLAN cannot be alleviated.
(3) Since the system cannot predict and analyze the enterprise activities, it is not easy to implement real-time load balancing management. Also, users cannot execute the fast switch based on real-time applications such as voice.
(4) If one AP encounters theft or destruction, security will not be guaranteed.

2. Centralized architecture

In order to solve the problems existing in the autonomous architecture, the centralized WLAN architecture was put forward, which has gradually become a hot research topic recently. The CAMP working group has developed a set of standards based on the centralized WLAN architecture to solve the problems in large scale WLAN such as network management, security, resource management and interoperability. As shown in Fig. 7.8 is the centralized WLAN architecture.

It can be seen from Fig. 7.8, compared with autonomous architecture, centralized WLAN architecture increases the factor of AC, which can be regarded as a set of logical devices to realize the management, monitoring, dynamic configuration, AAA and other functions of the whole network. Wireless Termination

Fig. 7.8 Centralized WLAN architecture

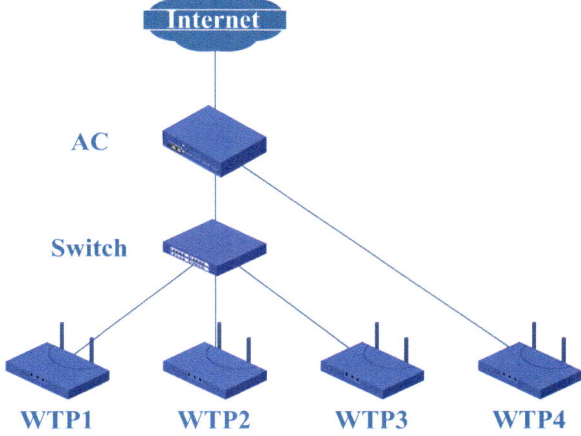

Point (WTP) is different from AP defined in IEEE802.11 standard. AP can achieve all of the functions of 802.11 while WTP only needs to achieve part of the functions. So WTP can be seen as a lightweight AP. AC and WTP have three kinds of connection modes: direct connection, layer 2switching connection and layer three routing connection. Among them, the layer three routing connection enables WTP to connect AC through IP network, which not only makes the WTP deployment more flexible, but also solves the wireless STA's seamless layer three roaming problem. It is a widely adopted connection mode. The centralized WLAN architecture has the following advantages.

(1) It reduces the required traffic filtering and strategy processing in AP and realizes the centralized control over traffic filtering and strategy execution. Any change in traffic will be announced to AP, so as to reduce the traffic processed in AP and to concentrate the limited resources in WLAN system on wireless access.
(2) For the whole WLAN system, a centralized management and traffic management are adopted so that the traffic control, authentication, encryption and execution strategies are realized (to ensure QoS and security).
(3) Through the network infrastructure or IP routing network, universal encapsulation and transferring mechanism is provided for multi-user AP interoperability, enabling the enterprise user's LWAPP console to better perform LWAPP AP storage and processing, and making it more convenient for the system to implement access strategy or service management.
(4) Functions of 802.11 are realized between AP and the centralized device, giving it good management and scalability. The centralized WLAN architecture adopts the "thin AP", which means most intelligent configuration is concentrated in the centralized equipment, thus reducing the WLAN execution management function and the total operating cost.

The above characteristics of the centralized WLAN architecture determine that it is suitable for large scale deployment.

3. Distributed WLAN architecture

The distributed WLAN architecture is a distributed network composed of wireless AP which depends on the wired or wireless medium. In this structure, wireless APs are connected through wireless links, similar to the mesh structure. Some of these APs are connected to the external network via a wired link and they can be used as gateway nodes in charge of network management. The distributed WLAN architecture is shown in Fig. 7.9

The distributed WLAN architecture has the advantages of strong robustness and easy maintenance It can also draw on the research results on mesh to optimize the deployment. Distributed WLAN architecture is currently a new research direction. IEEE 802.11 working group is developing the 802.11 mesh standard 802.11 s.Researches on its application scenarios, service management, and execution strategies are still in its infancy.

Fig. 7.9 Distributed
network architecture

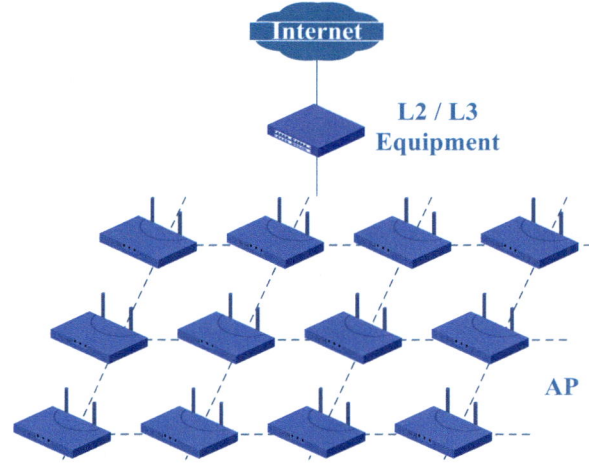

WLAN Coverage Design

WLAN Indoor Propagation Model

WLAN indoor coverage is mainly characterized by smaller coverage, and larger environmental changes. In general, we select the following two models that are suitable for WLAN analysis. Because the indoor wireless environments vary greatly, in planning the reference model and the model coefficients should be chosen according to the actual situation.

1. Devasirvatham model
    Devasirvatham model is also known as linear path attenuation model, of which the formula is shown in (Eq. 7.19).

$$PL(d,f)[dB] = PL_{FS}(d,f) + a \tag{7.19}$$

In it, $PL(d,f)$ is the indoor path loss, $PL_{FS}(d,f)$ is the free space loss, which is shown in the expression (Eq. 7.20):

$$PL_{FS}(d,f) = 32.4 + 20\lg f\,[MHz] + 20\lg d\,[km] \tag{7.20}$$

In it, d is the propagation path, f is the radio frequency, a is the model coefficient, the value is 0.47.

2. Attenuation factor model
    In terms of the spatial propagation loss of the radio wave, the electromagnetic wave in 2.4G frequency band has the approximate path propagation loss formula.

Table 7.5 Theoretical value of WLAN typical path propagation loss

Distance (m)	Propagation loss (dB)		
	n = 2.5	n = 3.0	n = 3.5
10	63.47	66.97	70.46
50	80.95	87.94	94.93
100	88.47	96.97	105.46

Table 7.6   Maximum distance of WLAN theoretical indoor propagation

Data rate (Mbps)	Sensitivity (dBm)	Indoor maximum transmission distance (m)	
		Devasirvatham model	Attenuation factor model
11	−88	53	48.9
5.5	−91	60	63.1
2	−93	64	72.4
1	−96	69	94.8

$$PL[dB] = 46 + 10 \times n \times \lg D[m] \tag{7.21}$$

In it, D is the propagation path, n is the attenuation factor. In different wireless environments, attenuation factors n have different values. In free space, the path attenuation is proportional to the squared distance, i.e., the attenuation factor is 2. Within the building, the effect of distance on path loss will be greater than that of free space. Generally speaking, the value of n in the open environment is 2.0 ~ 2.5, the value of n in the semi open environment is 2.5 ~ 3.0, and the value of n in the closed environment is 3.0 ~ 3.5. The theoretical calculated value of the typical path propagation loss is shown in Table 7.5.

3. AP signal link loss calculation

According to the model, the indoor path loss is equal to the free space loss and the additional loss factor. Moreover it grows exponentially with distance. The receiving level estimation formula is shown in (Eq. 7.22).

$$\Pr[dB] = Pt[dB] + Gt[dB] - PL[dB] + Gr[dB] \tag{7.22}$$

In it, $\Pr[dB]$ is the minimum receiving level, which is the receiving sensitivity of AP at different transmission rates. $Pt[dB]$ is the maximum transmit power. $Gt[dB]$ is the transmitting antenna gain. $Gr[dB]$ is the receiving antenna gain. $PL[dB]$ is the path loss. We can make the theoretical calculation of the limit propagation distance of AP signal as follows.

Assuming that the antenna transmitting and receiving gain is zero, when AP transmitting power is 16.2 dBm, the maximum distance of theoretical indoor propagation is shown in Table 7.6.

4. AP signal penetration loss

The experience values of the current 2.4G electromagnetic wave's penetration loss for a variety of building materials are as follows.

(1) Block by partition walls (brick wall thickness 100–300 mm): 20–40 dB
(2) Block by floors: above30dB
(3) Block by wooden furniture, doors and other wooden partition: 2–15 dB
(4) Thick glass (12 mm): 10 dB

In addition, when measuring the penetration loss of AP signals, we need to consider the incident angle of the AP signal. For a 0.5 m thick wall, when the linear connection of AP signal and the coverage area forms a 45 degrees' incident angle, it's equivalent to a 1 m thick wall. With 2 degrees' angle, it's equivalent to a wall of more than 14 m thick. Therefore, in order to get a better acceptance effect, we should try to make the AP signal through the wall or ceiling vertically (90-degree angle).

WLAN Indoor Coverage Planning Scheme

When planning WLAN network, we should firstly consider the interaction between AP and wireless network adapter signal, as well as the user's effective access to the network. Therefore, how to ensure the wireless signal coverage is a factor that must be considered in AP selection. Since WLAN has high work frequency band, low sensitivity (compared with mobile base station/mobile phones), great signal reflection and diffraction, there are several different indoor coverage solutions that the planning personnel can choose according to the actual situation on site.

On-site investigation must be carried out before designing to find out the following points.

(1) Getting to know the coverage area and signal coverage quality requirements since different locations have different coverage requirements.
(2) Investigating the existing signal distribution of the coverage area, to grasp the blind spots, hot spots and signal collision areas.
(3) Investigating the composition of the buildings in the coverage area, and the signal blocking.
(4) Signal access position and mode.
(5) Investigating the location where the equipment can be installed.

After the on-site investigation of the environment, we can choose three different plans according to the actual situation, respectively, co-cellular network indoor distribution system coverage scheme, independent AP distribution coverage scheme and cross patch coverage scheme. Here are brief introductions to the three schemes.

1. Co-cellular network indoor distribution system coverage scheme

(1) Application scope and usage requirements of the scheme
    Application scope: mid-to-large scale indoor coverage. The system structure is complex. Mainly used in medium blind area coverage or important public places to meet the coverage requirements of places such as hotels,

airports, conference centers, etc., but not suitable for the network with high capacity requirements.

Usage requirements: the system is an indoor coverage system, and the equipment is required to be installed indoors.

(2) Engineering design experience

Generally, if the antenna radiation power is 10 dBmW and a 2dBi small omnidirectional indoor antenna is used, then within 30 meters from the antenna, when in spacious conference center and without brick wall, the coverage level can reach -75 dBm/WLAN.

For the hotel room with dense wall structure, if the antenna tip radiation power is 10 dBmW, and a 2dBi small omnidirectional indoor antenna is used, then within 8 to 10 meters from the antenna, the room signal can reach $-70 \sim -85$ dBm.In general, the signal to the room door should be controlled within 4 m.

In the energy distribution, signals should be distributed as evenly as possible. In design, in the paths from the base station to each antenna, the emergence of more than two power dividers (or coupler) should be avoided, so as to ensure the effective access of uplink signals.

The use of this scheme is suitable for indoor coverage of large scale hotel lobby, airport, and conference centers with lower capacity requirements.

2. Independent AP distribution coverage scheme

(1) Application scope and usage requirements of the scheme

Application scope: It's suitable for small and medium scale indoor coverage where there is no indoor distribution system, small area coverage or important public places, such as hotels, conference centers, etc. The system has simple structure and covered by independent APs.

Usage requirements: indoor coverage, applicable to the design and installation of indoor equipment.

(2) Engineering design experience

If only one AP is installed in a hall, then the AP would be better placed in a central position in the hall. And it is better to be placed on the ceiling of the hall. If two APs are installed within the same space, they can be put on two diagonals.

The number of signals penetrating walls and ceilings should be kept minimal. 2.4G signal can penetrate walls and ceilings, but each wall and ceiling will reduce the AP signal coverage by 1–30 meters. AP and the computer should be placed in a suitable position so that the signal block path of the wall and ceiling can be kept the shortest and the loss minimum.

Linear connection between the AP and the coverage area should be considered. The location of AP should be chosen in the way that the signal can go through walls or ceilings vertically (90-degree angle).

Different building materials have different transmission effects. Building composed of metal frame or door will shorten the WLAN signal transmission distance. Therefore, AP should be placed where signals go through dry walls or open doors, instead of where signals must penetrate the metal material.

AP antenna direction is adjustable, so AP should be installed where the antenna main beam directly faces the coverage target area to ensure good coverage effect.

AP should stay far away from electronic equipment (1~2 meters), such as microwave ovens, monitors, motors, etc.

3. Cross patch coverage scheme

(1) Application scope and usage requirements of the scheme

Scope of application: for large-scale indoor coverage, the system structural design is relatively complex and the renovation project is smaller. WLAN signal source connected to the indoor distribution system, complemented by independent separate AP layout covering a few blind spots, and ultimately seamless coverage with good effect can be achieved. It can meet the coverage needs of public places and open areas like halls and airports, and can also cover the medium capacity needs of hotels and conference centers.

Usage requirements: indoor coverage, applicable to the design and installation of indoor equipment.

(2) Engineering design experience

AP signal source access is similar to co-cellular network distributed antenna system coverage scheme in terms of distribution system, design requirements and hardware transformation requirements. It needs to be noted that when the original indoor system's antenna location does not meet the WLAN coverage/range requirements, we may not consider moving the antenna or adding terminal antennas by changing the power divider. But rather, we can use independent AP planning method to cover a small amount of hot spots for coverage complement.

In cross distribution coverage planning scheme, when using the original indoor distributed system of WLAN coverage, for the problems of massive renovation project, more antenna positions' changing, and difficult power divider load matching, we can moderately use independent APs to cover blind spots and hot spots, as supplement to the original indoor distribution system, realizing good overlapping coverage effect. This scheme has the characteristics of flexible planning. Before design, we should make pre-evaluation on the renovation project scale and the expected coverage effect, and then after comprehensive consideration, we develop an ideal project plan.

WLAN Capacity Design

In WLAN network planning, the impact of the whole mesh capacity on the system performance is greater than that of the AP coverage. To avoid the situation that a single user access slows down the whole cell's transmission rate, an AP transmitting power threshold should be set up. By adjusting the AP transmission power to alter the cell size, we can make the cell smaller in design. In such a way, since there aren't many users in a cell small, each user's high transmission rate can be guaranteed.

Network Capacity Calculation

Capacity in communications can be studied from two aspects: theory and practice. Theological concept of capacity is based on the amount of information on unit bandwidth (time or area). For voice services: Erl/unit bandwidth/unit area; for data services: bit/unit bandwidth (time or area). The above theological concepts of capacity are difficult to be applied in the actual measurement and comparison of information. Practical concept of capacity is based on the amount of charges (or users) on unit bandwidth (time or area). For voice services: Erl/unit bandwidth or Erl/unit bandwidth/unit area; for data services: Channel/unit bandwidth. This practical capacity concept is applicable to engineering communications capacity measurement.

Network capacity, without a standard definition in industry, can be characterized by different parameters from different angles, and the results are also different. From the user's point of view, by using mathematical basic theories to study the network capacity, we can get the following definitions.

Definition 1: Network capacity C is defined as the sum of the communications probabilities of all users, and its expression is as shown in (Eq. 7.23).

$$C = \sum_i Q_i \tag{7.23}$$

where Q_i is the communication probability of the user i and means the average number of users communicating simultaneously at any given moment supported by the network.

Definition 2: Communications probability Q_i is the probability that the user can normally send the data when the expected user does not transmit data. Assume that user i's communications status Z_i is subject to two points, that is, $Z_i \sim (0, 1)$, in which $Z_i = 0$ means the user is not communicating and $Z_i = 1$ means that the user is in a state of communications. Assume user i's interference user number is N_i, and then the Z_i distribution rate is:

$$P(Z_i) = \begin{cases} \dfrac{N_i - 1}{N_i}, Z_i = 0 \\ \dfrac{1}{N_i}, Z_i = 1 \end{cases} \tag{7.24}$$

When $Z_i = 1$, i.e., when the user is in a state of communications, the network capacity is:

$$C = \sum_{i=1}^{M} \frac{1}{N_i} \tag{7.25}$$

Network capacity needs to be calculated for actual network. According to the definition of network capacity, as shown in (Eq. 7.23), to calculate the WLAN network capacity, we need to know the number of AP, AP coverage results etc., and also the distribution of users in the network. In reality, with constantly user leaving, coming and changing position, the actual network is also always in constant change. Therefore, the calculation of network capacity is only for the capacity of an actual network in a given state at a given time. It does not represent the evaluation results of the network capacity for a period of time.

Let $A = \{1, \ldots, n\}$ denote all APs and AP_j denote the jth AP, then $A = \underset{j}{\cup} AP_j$. Let $B = \{1, \ldots, m\}$ represent all users in the coverage and B_i denote the ith user, then $B = \underset{i}{\cup} B_i$. D_j is the set of the user covered by the jth AP, then $D_j \subseteq B$, and $\underset{j}{\cup} D_j = B$. In order to facilitate the calculation of network capacity, for A and B, we introduce a coverage matrix G. If the operator '$\rightarrow$' is defined as coverage, then $G = A \rightarrow B$. This matrix represents the coverage result of AP set A to user set B. Its mathematical expression is

$$G = \left\{ g_{ij} | i = 1, \ldots, n; j = 1, \ldots, m \right\} \tag{7.26}$$

In it,

$$g_{ij} = \begin{cases} 1, A_i \in A; B_j \in D_i \\ 0, else \end{cases} \tag{7.27}$$

Based on the coverage matrix G, the network capacity's calculation steps are as follows.

(1) Initialize parameters. Let $i = 0, j = 0$, and all elements in Array $flag[n]$ be 0.
(2) If $g_{ij} = 1$, then Array Element $flag[i]$ gets the value 1.
(3) If $i < n$, then $i = i + 1$, jump to (Eq. 7.27). Or enter the next step.
(4) $k = 1, N_i = 0$.
(5) If $flag[k] = 1$, then $N_j = N_j + \sum_{j=1}^{m} g_{kj} - 1$.
(6) If $k < m$, then $k = k + 1$, repeat Step (5). Or enter the next step.
(7) $N_j = N_j + 1; Q_j = \frac{1}{N_j}$.
(8) If $j < m$, then $j = j + 1$, return to Step (2). Otherwise, jump out of the program.

After the above process, we get each user's communications probability vector $Q = \{Q_i | i = 1, \ldots, m\}$, and then get the capacity value by the calculation of network capacity (Eq. 7.23).

Discussion on Optimal Network Capacity Based on Computational Results

According to the definition of network capacity in (Eq. 7.23), the conditions for the optimal network capacity are

$$\text{MAX}(C) = \text{MAX}\left(\sum_i Q_i\right) = \sum_i \text{MAX}(Q_i) \tag{7.28}$$

Put $Q_i = \frac{1}{N_i}$ in (Eq. 7.28) and get

$$\sum_i \text{MAX}(Q_i) = \text{MAX}\left(\frac{1}{N_i}\right) = \text{MIN}(N_i) \tag{7.29}$$

The above model shows that user i's maximum communications probability is equivalent to the number of user i's minimum interference users N_i. Since $N_i \geq 1$, $\text{MAX}(Q_i) = \text{MIN}(N_i) = 1$. Assume the number of users is n, and then

$$\text{MAX}(C) = \sum_{i=1}^{n} \text{MAX}(Q_i) = n \tag{7.30}$$

(Eq. 7.30) shows that in the network when the number of users is n, deploy n APs, and each AP covers and only covers one user. When all users are covered, the network has the largest capacity, and the maximum capacity value is n.

The Guiding Significance of Network Capacity in Actual Network Planning

The goal of the actual network construction is that in the premise of ensuring the user network throughput rate, using APs as less as possible to obtain the users' coverage as much as possible, and meanwhile pursuing good balanced coverage results. The mathematical significance of AP is

$$\text{MAX}(C) \cup \text{MIN}(AP) \cup \text{MAX}(User) \cup \text{MAX}(Q_i) \tag{7.31}$$

Obviously, the network capacity is proportional to the number of AP. After considering the cost of network construction and operation and the actual use of the environment, we cannot simply increase the amount of AP to improve network capacity. That is to say, the network deployment plan in the maximum capacity calculated in (Eq. 7.30) has no practical application value. Therefore, it is necessary to comprehensively consider different factors and find a relatively optimal network deployment scheme with a good overall cost and effectiveness.

7.2.5 Large-Scale Test User Application Scenarios

Large scale user application test is the last step to carry out the network performance evaluation of the communications system. The traditional method is to carry out the relevant work by issuing the test user number in the pre-commercial stage. For 5G systems, since technologies of various modes will be interconnected, it is absolutely necessary to carry out the relevant tests and verification at the R&D stage, which requires the test environment to provide the appropriate test conditions.

ShanghaiTech University has 6000 students. Together with faculty and staff, the overall size is more than 10 thousand people, which is very suitable for a dense crowd massively connected scenario test. In particular, the characteristics of students' needs for new services make this test field very suitable for the new service applications test.

In the preliminary experiment, it is clear that students' characteristics for network use make them very suitable for large-scale user application test. Different from traditional voice services, students' curiosity about new types of services makes them more willing to cooperate with the application test of new services. In addition, the tidal effect of the student population movement is also very suitable for carrying out load balancing technology test.

7.3 Typical Case Analysis of Field Trials

Wireless communication field trials involve all aspects of the wireless communication system, covering from radio wave propagation characteristics research to the test of air interface key technology, to network performance evaluation and validation. Every link needs field trial to confirm whether the technology meets the design requirements.

A typical field trial procedure is shown in Fig. 7.10.

This chapter gives typical test cases from three aspects, namely wireless channel measurement and modeling, wireless communications key technology test and performance test of wireless network.

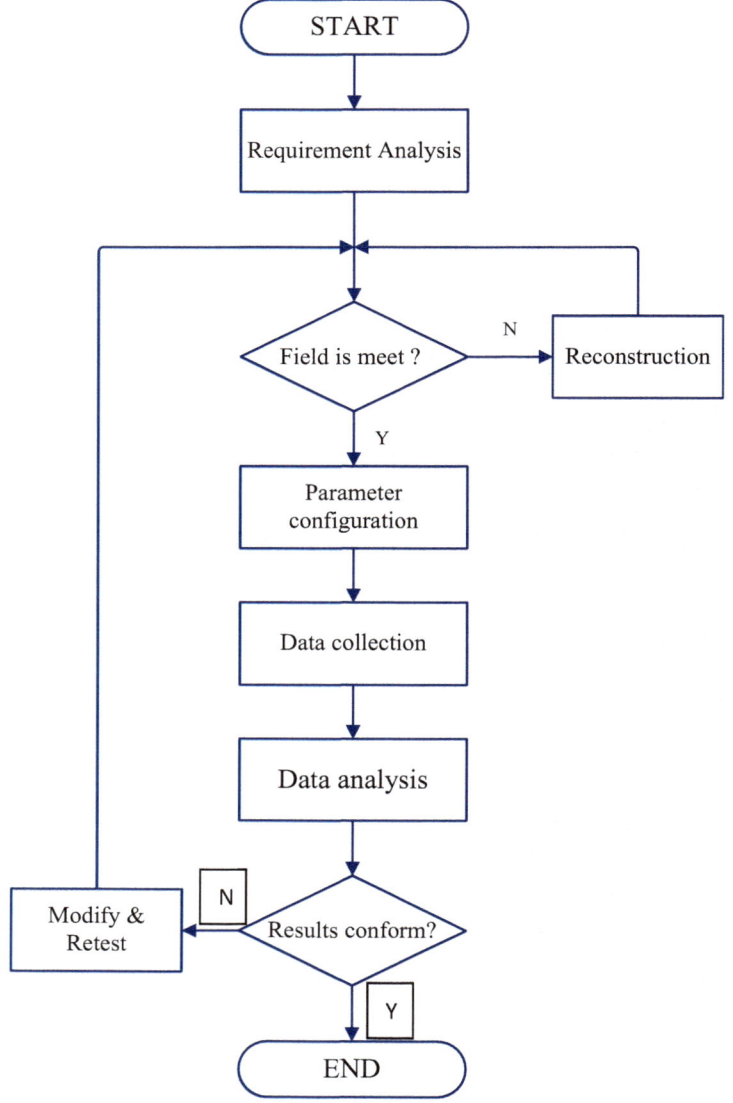

Fig. 7.10 Outfield trial flow chart

7.3.1 Case One: Research on the Irregularity of Wireless Cellular Networks

Test Purpose

In order to ensure seamless coverage of wireless network, we need to consider the coverage shape of wireless cell. Traditionally we consider that the hexagonal boundary shape is close to a circle, which is in a honeycomb shape, to achieve the continuous coverage. The research on cellular wireless coverage shape has been going on for decades. For the free space propagation environment, with a single transmitter's signals radiating evenly in all directions, the coverage of a wireless cell in two-dimensional plane should be a circle centering around the transmitter. It is assumed that the service area of the wireless cellular network can be divided into a seamless coverage service area of a plurality of normal triangles, squares, or regular hexagon, when the plurality of transmitters are deployed in an equal distance in accordance with the regulations. Taking into account the regular hexagon which is the most close to the shape of the circular shape, it is widely used as a conventional wireless cellular network coverage shape. However, in the actual network deployment process, the increase of the base station makes the cell divide continuously. Irregular terrain landforms make wireless cell coverage shape can really achieve the effect of positive hexagon, but there are irregular boundary characteristics.

An important challenge for wireless cellular network operators is to ensure the mobile users seamlessly from adjacent cells to achieve continuous coverage, especially those located in the residential area of the boundary of the user. In addition, the switching signal of the user in the adjacent cell is decided by the cell boundary in the wireless cellular network. Therefore, the cell boundary shape of wireless cellular network is a key indicator to design, deploy and optimize the wireless cellular network. How to evaluate the boundary features of a cell is a problem worthy of studying.

Test Environment

In order to achieve the purpose of studying the irregularity characteristics of cell boundary shape, we need to build the test environment in typical scenarios so as to avoid the impact of the adjacent base stations' signals on the test results. Of course, the position of the transmitter should be effective to evaluate the theory analysis results. Therefore, we have built temporary test environments in several locations in Shanghai, carried out a large number of field trials, obtained the real test data, and analyzed the test results. Figure 7.11 shows the transmitter location and test roadmap of a dense city scenario, in which the test radius reaches 1.5 km.

Fig. 7.11   Shanghai Xuhui intensive city scenario test roadmap

Test Results

Through the field data collection in three test scenarios and the analysis on the statistical fractal characteristic for the stochastic process with three methods, we find that although there are slight differences between three different typical methods in Hurst parameters estimation, the final estimations for the wireless cellular coverage boundaries are basically the same. That is, the estimated average Hurst parameters are approximately 0.9, which shows the real wireless cellular coverage boundary exhibits a statistical fractal feature.

7.3.2   Case Two: Characteristics Analysis on the Radio Wave Propagation Between Multiple Frequency Bands

Test Purpose

Spectrum, as a valuable and non-renewable resource, is the fundamental of the survival of mobile communications technology, and the basic resource of the wireless communications system. Mobile spectrum resources are currently managed by the nation, and its management mode is the fixed spectrum management mode. World Radio Conferences in 1992, 2000, and 2007 has allocated many frequency bands in succession for the wireless mobile communications system, including 450 MHz, 700 MHz, 900 MHz, 1800 MHz, 2GHz, 2.3GHz, 2.6GHz, 3.5GHz, etc., which are the main frequency bands that support the current 2G, 3G, 4G mobile communications systems of GSM, TD-SCDMA, CDMA2000, WCDMA, LTE, etc.

Wireless channel measurement, based on the feature analysis of radio wave propagation, analyzes the large scale characteristics of radio wave propagation as its basic content, that is, the path loss.

Most of the current wireless networks design has multi-mode, shared network and shared antenna. Usually there is current network data of a certain network frequency band. As for how to infer from the coverage characteristics of one network to another network's frequency band propagation coverage, we lack the experiential data of objective testing. So we can't guide the front line to use the current network data in other networks with shared network. This requires the multi-band radio wave propagation characteristic analysis. Through the actual test and the test data analysis, we can find out the rules for the differences in propagation characteristics between different frequency bands, so that we can maximize the utilization of the current network data, support the use of the coverage data from different networks in the coverage prediction of other networks, and improve the coverage prediction accuracy.

Test Environment

To analyze the propagation characteristics of different frequency bands in the scenarios with the same characteristics and in different scenarios, and to analyze the influence of different data processing methods on the propagation characteristic analysis, we have selected several test environments in Shanghai, including dense urban scenario, ordinary urban scenario and suburban scenario. And for each typical scenario, we have chosen 2~3 test areas.

Figure 7.12 shows the transmitter location and test roadmap of the suburban scenario in Zhangjiang, Shanghai.

Test Results

For this test, 11 typical frequency bands that are commonly used in wireless cellular network within 700 MHz~3500 MHz range are selected for all the test scenarios.

Through data and processing, we have obtained the comparison results of path loss indicator and shadow fading factor between same scenarios different frequency and different scenarios same frequency. In data processing, we have studied the geographical average method of the test sample points in the mobility test, and

Fig. 7.12 Test roadmap of the suburban scenario in Zhangjiang, Shanghai

compared the impact of the different processing methods on the results. The following are part of the test conclusions.

(1) Comparing the test results of the suburban, dense urban and ordinary urban environments, we find that the dense urban environment, with the most obstacles, has greater path loss attenuation value than the other two communications environments.

(2) Comparing the standard deviations of the test results and the theoretical propagation model, we can see that the standard deviation decreases with the increase of the frequency point, and the standard deviation is less than 9 dB, which meets the requirement of the network test to the standard deviation.

(3) Comparing three data processing methods of 10 meters, 10 meters grid and 17 meters grid, we find that the 17 meters grid can simultaneous meet the data processing requirements from low frequency band to high frequency band, and the smaller standard deviation can be obtained.

7.3.3   Case Three: CoMP Key Technologies Test

Test purpose

Coordinated Multiple Points (CoMP) Transmission/Reception refers to the technique that the geographically separated multiple transmission points coordinate to transmit data for one terminal (PDSCH) transmission or jointly receive the data from one terminal (PUSCH).

According to the relations between coordinated nodes, CoMP can be divided into two types, namely Intra-site CoMP and Inter-site CoMP. The coordination of IntraiteCoMP occurs within one site, so there is no limit for backhaul feedback capacity and massive data interaction can take place in one site. Inter-site CoMP occurs between multiple sites, so there are more stringent requirements for backhaul capacity and latency. That is to say, the performance of Inter-site CoMP is restricted by the current backhaul capacity and latency tolerance. According to whether the data can be obtained at multiple coordination points, it can also be divided into Joint Processing (JP) and Coordinated Scheduling/BeamForming (CS/CBF).

The systematic diagram of CoMP key technologies is shown in Fig. 7.13.

The goal of CoMP key technology verification is to carry out the system performance tests of Inter-site CoMP and Intra-site CoMP operating modes in the real indoor and outdoor scenarios, and to evaluate the performance gain with the conventional single station transmission.

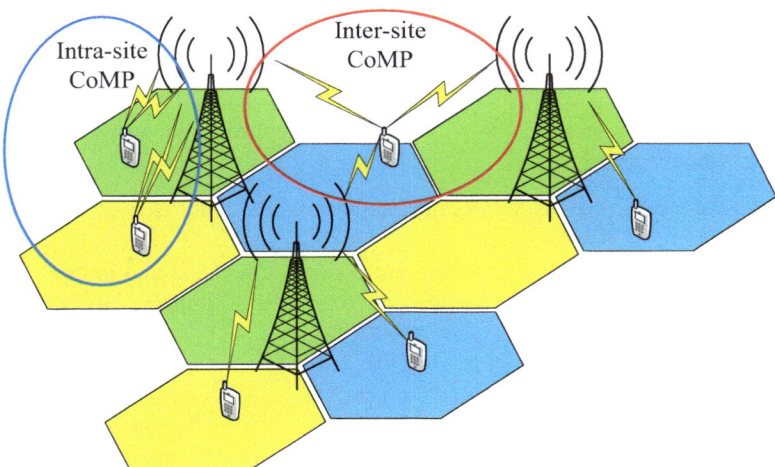

Fig. 7.13 Schematic diagram of CoMP key technologies

Fig. 7.14 Schematic diagram of CoMP key technologies verification test environment

Test Environment

For CoMP key technologies test environment, we build a multi-layer LTE wireless coverage network from outdoor to indoor according to the test requirements, which is shown in Fig. 7.14.

As shown in the above figure, macro-0 and macro-1 are two sectors (cells) of the same LTE outdoor macro cell base station (eNodeB). IA-2 is a sector of the LTE outdoor micro cellular base station. In the teaching building we also constructed more than 10 LTE indoor coverage stations, thus forming a macro cell -- micro cellular --- indoor multi-layer coverage of wireless network test environment.

In the test configuration, we can choose an appropriate position to achieve different CoMP configuration.

(1) The outdoor CoMP Inter-site test environment composed of macro-0 and macro-1;
(2) The Intra-site CoMP test environment composed of IA-2 and indoor sites;
(3) The Intra-site CoMP test environment composed of multiple indoor sites.

Test Results

This test verified the effect of CoMP in the system capacity performance improvement in multi-layer network environment. More detailed test results will not be presented here.

7.3.4 Case Four: Active Antenna System (AAS) Key Technologies Test

Test Purpose

AAS solution is to integrate the base station RF into the antenna and to use the multi-channel RF and antenna array in coordination, so as to realize the space beamforming and complete the RF signal transceiving. Active antenna is a new form of base station architecture while BBU similarly brings baseband signals to the active antenna unit. Different from BBU+RRU architecture, active antenna divides the transceiving channels into the sub antenna oscillator level with more refined particles size. Through different configurations of active antenna oscillators, we can achieve the functions like flexible beam control and MIMO, and more flexible and dynamic resource allocation and sharing, so as to achieve the goal of the optimal performance and lower cost in the whole network.

AAS technology uses the adjustable angle design and thus improves the network performance. It mainly includes the following several key technologies.

(1) Vertical sector splitting

AAS system can form multiple beams in horizontal and vertical dimensions, achieving the multiplexing of the resource with the same time frequency. For the AAS vertical sector splitting performance test, we need to evaluate the split sector performance on the same horizontal plane.

(2) 3D beamforming

Through the 3D beamforming, AAS can achieve very good spatial resolution, making MU-MIMO and spatial interference suppression field have a certain performance improvement.

(3) Proactive cell shaping

By flexibly adjusting the lower angle and the shape of the beam, AAS can be carried out in advance, in order to promote the performance balance in the macro cell and the low power node, especially in the non-equilibrium network configuration scenario.

Test Environment

The test field for AAS key technologies verification is composed of four stations (one center station, three interference stations), among which the center station has three sectors. This network topology consists of six sectors ($1 \times 3 + 3 \times 1$). The topological graph is shown in Fig. 7.15.

To facilitate the test of the network configuration, the core network EPC adopts the simulated core network and builds the core network elements with servers, including MME, SGW, and PGW.

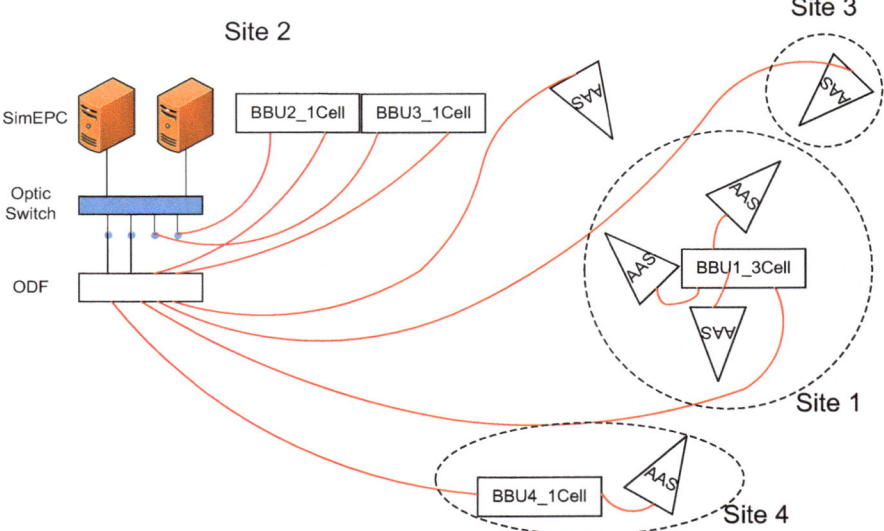

Fig. 7.15 AAS key technologies test network topology

Test Results

For AAS key technologies verification field trial, we compare the network performance in three modes, namely conventional RRU, vertical sector splitting and
virtual sector. The test results are summarized as follows.

(1) Conventional RRU mode, under the same test conditions in comparison test,
    will change with the inclination angle. And RSRP, SINR, DownLink (DL),
    UpLink (UL) and other data change regularly. In the planed road test area, there
    will be the optimal inclination angle.
(2) For vertical sector splitting mode, under the same test conditions in comparison
    test, with combination of different inclination angles, representative areas (far
    end and near end) are selected in the planed road test area. And RSRP, SINR,
    DL, UL and other data change regularly. There will be the optimal inclination
    angle combination.
       At the near end, the downlink test indicators will be worse than other modes,
    mainly because of the same frequency interference by the side lobe of the outer
    cell. At the far end, the downlink test indicators show similar performance with
    the other modes.
(3) For virtual sector mode, under the same test conditions in comparison test, with
    combination of different inclination angles, representative areas (far end and
    near end) are selected in the planed road test area. And RSRP, SINR, DL, UL
    and other data change regularly. There will be the optimal inclination angle
    combination.

At the near end, the downlink test indicator is slightly worse than the conventional RRU model but significantly better than vertical sector splitting. The main reason is that avoidance scheduling is done in the base band side according to the downlink interference detection, which significantly lowers the near-end interference. At the far end, the downlink test indicators show similar performance with the other modes.

(4) As for the uplink tests of the three modes, the conventional RRU mode is better than the other two modes, while the virtual sector and vertical sector splitting are similar. The main reason is that the same frequency interference between UEs' uplinks is at a very low level, the impact on performance is very small.

(5) In terms of network coverage evenness, vertical multi sector and virtual sector are better than conventional RRU mode, with the coverage performance improved.

7.3.5   Case Five: Inter-Cell Interference Coordination (ICIC) Key Technologies Test

Test purpose

Inter-Cell Interference (ICI) is an inherent problem of the cellular mobile communications system which severely affects the system performance. The reason is that the users of the same frequency resource in each cell can interfere with each other. Severe inter-cell interference will directly affect the system coverage and capacity. Particularly for cell edge users, it will directly cause their frequent connection lost or disability to access the network.

According to the existing research results and the research projects of standards organizations such as 3GPP LTE, IEEE 802.16 m, 3GPP2 Ultra Mobile Broadband (UMB) for inter-cell interference control processing modes, there are three types of inter-cell interference control technologies.

(1) Interference randomization technique

Interference randomization is a common technique to reduce the inter-cell interference by the interference whitening of adjacent cells. Its main advantage is that it does not affect the complexity of the receiving end scheduling and receiving process. But when the system is fully loaded, the interference randomization technology is limited in improving the system performance. The representative technologies of interference randomization are base station-based scrambling codes and various frequency hopping technologies.

(2) Interference coordination technology

Interference coordination technology mainly means to effectively coordinate the channel resources and power of space, time and frequency domain of a number of cells, and to avoid or reduce the interference to adjacent cells through limiting the use of some resources in this cell (such as frequency, power and time). The main technology includes Fractional Frequency Reuse (FFR), multi station MIMO and power control technology, etc.

(3) Interference cancellation technology

The interference cancellation technology is to decode and copy the signals from the interfering cells, and then subtract the interference signal from the cell in the received signals. The advantage of interference cancellation technology is that there is no limit to the use of cell frequency resources. But its limitation is that the target cell must also know the pilot frequency structure of interference cell to make channel estimation of the interference sources. As a result, the signaling cost and implementation complexity of the interference cancellation technology are relatively high.

Test Environment

The test field of ICIC key technology test is composed of two stations (one cell in one station, and the two cells have overlapping areas).The topological graph is shown in Fig. 7.16.

The test will take the parameters like the target cell RSRP, adjacent cell RSRP, Cell Reference Signal (CRS) SINR, CQI, etc., which are obtained based on the location of the test terminals, as the input variables of the cell interference coordination algorithm. Then it will calculate the optimized power parameters, configure the base station transmitting parameters according to the optimized parameters, and retest the validity of UE network performance verification algorithm.

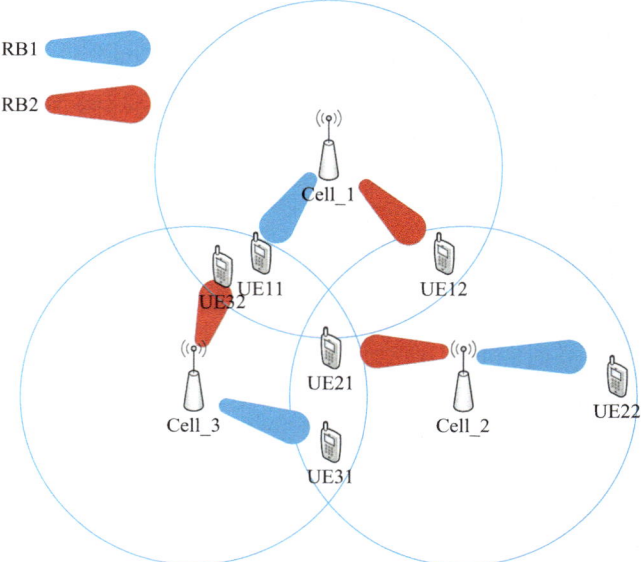

Fig. 7.16 Schematic diagram of UICIC key technologies validation test environment

The test selects two scenarios for two tests. One is that the terminal is located near the cell center, and the other is that the terminal is located in the overlapping area of the cell.

Test Results

(1) Test results of the near-station position

Both the two users are close to the base station. One user's service cell is Cell 1, while the other user's service cell is Cell 2. Table 7.7 and Table 7.8 have shown the two tests' results.

(2) Test results of the far-away-from-station position

Both the two users are far away from the base station and located in the overlapping area of the two cells. One user's service cell is Cell 1, while the other user's service cell is Cell 2. Table 7.9 and Table 7.10 have shown the two tests' results.

Table 7.7 The first test result when users are near the base station

Power configuration (W)	Cell throughput (Mbps)		Total throughput (Mbps)	Proportional fairness and value	Power efficiency (Mbps/W)
	Cell 1	Cell 2			
P1=17.31 P2=17.31	20.06	27.34	47.4	6.307	EP1=1.159 EP2=1.579
P1=6.76 P2=6.00	19.33	24.84	44.17	6.174	EP1=3.222 EP2=3.675

Table 7.8 The second test result when users are near the base station

Power configuration (W)	Cell throughput (Mbps)		Total throughput (Mbps)	Proportional fairness and value	Power efficiency (Mbps/W)
	Cell 1	Cell 2			
P1=17.31 P2=17.31	27.986	12.293	40.279	5.841	EP1=1.617 EP2=0.71
P1=4.38 P2=5.51	15.916	17.232	33.148	5.614	EP1=3.634 EP2=3.127

Table 7.9 The first test result when users are far away from the base station

Power configuration (W)	Cell throughput (Mbps)		Total throughput (Mbps)	Proportional fairness and value	Power efficiency (Mbps/W)
	Cell 1	Cell 2			
P1=17.31 P2=17.31	9.882	6.668	16.55	4.188	EP1=0.571 EP2=0.385
P1=4.38 P2=5.51	10.509	7.556	18.065	4.375	EP1=2.399 EP2=1.371

Table 7.10 The second test result when users are far away from the base station

Power configuration (W)	Cell throughput (Mbps)		Total throughput (Mbps)	Proportional fairness and value	Power efficiency (Mbps/W)
	Cell 1	Cell 2			
P1=17.31 P2=17.31	11.606	4.708	16.314	4.000	EP1=0.670 EP2=0.272
P1=4.38 P2=5.51	11.685	7.309	18.994	4.447	EP1=2.668 EP2=1.326

(3) Test result analysis

According to the test results analysis of the above data, when the user is in the cell center, the base station's power optimization ensures the network throughput and proportional fairness summation value. Its power is reduced by 70%, greatly improving the resource utilization efficiency. When the users is at the cell edge location, after the base station's power optimization, not only the network throughput and proportional fairness summation value increase slightly (network throughput by 9%, proportional fairness summation value by 4.5%), but also the base station power is reduced by 70%, greatly improving the efficiency of resource use. Meanwhile, it can be seen that the edge user scenario power optimization algorithm has more effectively improved the resources utilization.

The field trial results well verified the algorithm performance.

7.3.6   Case Six: Public WLAN Network Coverage Performance Test

Test Purpose

Because WLAN system uses the unlicensed Industrial Scientific Medical (ISM) frequency band and most of the terminals support the WLAN function, WLAN network is widely used and brings users a good wireless connection experience. However, due to the public's insufficient awareness for the unlicensed ISM band using provisions and the irrational use of network overall plan and channel, massive co-channel interference and adjacent channel interference occur, eventually making the actual user experience cannot reach the expected performance. As for how to evaluate the WLAN network coverage performance, a joint test of network transmission performance and air interface signal quality is required to avoid the incomplete evaluation for network performance by a single test.

Test Environment

This test chose the heavy-traffic public places in Shanghai, including the railway station and the airport area. Figure 7.17 shows the test roadmap schematic diagram

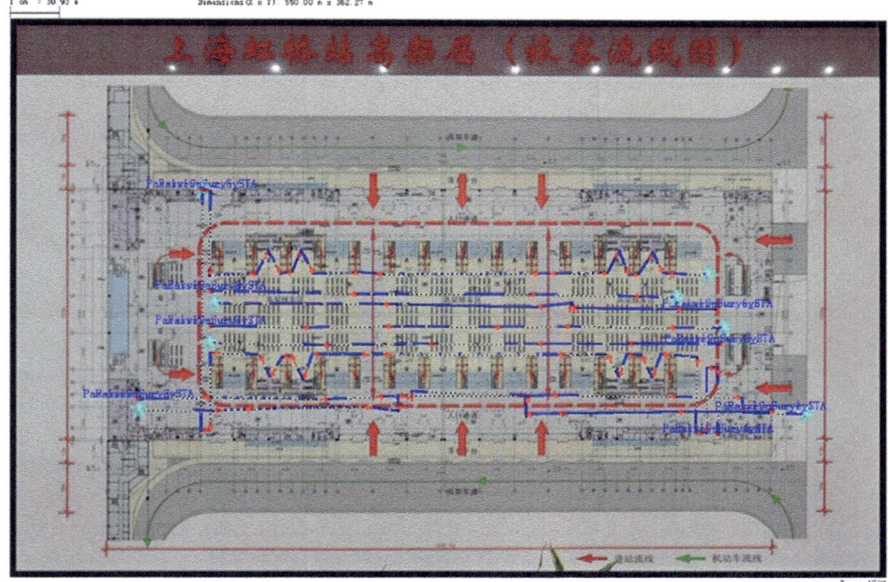

Fig. 7.17 Schematic diagram of WLAN test environment of Hongqiao Railway Station

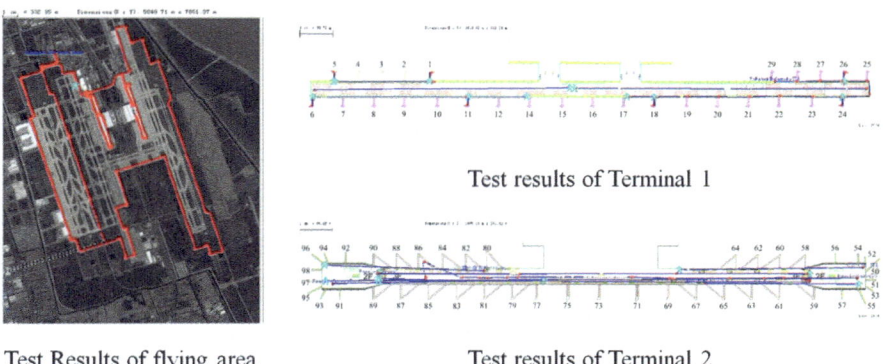

Test Results of flying area   Test results of Terminal 2

Fig. 7.18 Schematic diagram of WLAN test environment of Pudong Airport

of Hongqiao Railway Station area. Figure 7.18 shows the test roadmap schematic diagram of Pudong Airport area.

Test Results

For WLAN network, its KPI indicators system includes the following four categories.

(1) Signal optimization indicators
(2) AP performance optimization indicators

(3) AC performance optimization indicators
(4) Data side optimization indicators

For the network coverage, the signal optimization indicator is the most important indicator system, which specifically, includes the following categories.

(1) Signal coverage ratio
    In target coverage area, the proportion of the test points whose signal value >-75 dBm.Target value $\geq 95\%$.
(2) Signal coverage range
    Control the AP signal coverage to avoid the too large signal leakage range threatening user data security. Target value: within 50 m.
(3) Signal strength
    Signal strength refers to the signal value in the target coverage area. A signal strength that is greater than -75 dBm is the basic requirements for WLAN services. If the signal strength is too weak, it will affect the service access and quality of service. If there should be a signal, we should add points to patch the coverage or adjust the AP transmitting power, so as to meet the requirement for signal strength. Target value: greater than -75 dBm (key areas greater than -70 dBm).
(4) Leakage signal strength
    Leakage signal strength refers to the co-channel leakage signal intensity within the coverage area. Target value: less than -75 dBm.
(5) Signal to Noise Ratio (SNR)
    It refers to the Signal to Noise Ratio (S/N value) of the downlink signal received by the wireless network adapter of the user terminal in the 95% or more areas of the target coverage area. SNR that is greater than 20 dB is the basic requirements for WLAN services. Poor SNR will affect the service access and quality of service. If the SNR is too low, we should inspect the on-site wireless environment and re-plan the AP transmitting power and channel. Target value: greater than 20 dB.
(6) Air interface low-rate proportion
    It refers to the proportion of 1 M, 2 M low-rate air interface rate when users transmit data. Target value: $\leq 30\%$.
(7) Air interface error message proportion
    It refers to the proportion of channel CRC error message. Target value: $\leq 40\%$.
(8) Air interface packet loss rate
    It refers to the average data packet loss ratio of wireless air interface. Target value: $\leq 3\%$.

The test results of Hongqiao Railway Station and Pudong Airport show that since WLAN network operators do not make unified planning for channels and some private APs use channels irregularly, the actual public WLAN coverage performance is far below the user expectation. Reasonable planning and use of the public WLAN network is very essential.

7.4   Wireless Network Data Intelligence Analysis

7.4.1   Background and Necessity

With the gradual opening up of China' telecommunications market, mobile communications has become the main competition focus. In the context of the financial crisis, mobile communications has become a rare investment highlight. In the new competitive environment, the competition mode of the telecommunications industry has changed from "quantity" to "quality" competition. The former is often manifested as hard resource expansion like "base station expansion, telephone numbers increase", while the latter often seeks to enhance the network quality and user service satisfaction through increasing investment in software.

An efficient and stable wireless network is composed of wireless devices (wireless hardware products), network performance support system (wireless software products) and operation and maintenance personnel (people) together, all of which are indispensable. Among them, the wireless software products are playing the key value of coordinating and connecting "human" and "hardware". Restricted by network architecture, monitoring scope, control cycle, processing speed and other factors, the automatic balance capability that the wireless device itself can provide is limited. In face of complicatedly entangled technical problems, without the assistance of professional software, even the maintenance personnel with rich experience can hardly guarantee the efficient and reliable operation of the wireless networks.

After entering 3G and 4G era, mobile operators will seek more advanced and intelligent software support tools. Due to the introduction of a large number of new technologies, new mobile phone services are springing up. The complexity of a new generation of wireless network will be far beyond the traditional 2G network, and the "system theory" method is needed to update the network optimization technology system. For example, in 3G network Radio Network Controller (RNC) devices, the operating parameters of different types that impact each other has reached several thousands, which has been more than ten times of the similar Base Station Controller (BSC) devices in 2G network. Even in the telecommunications equipment developers, it is difficult to have enough technical staff that can fully grasp so much complex technical information. Frequent and fast network upgrade is pushing high the telecom enterprises' cost in learning. Telecom operators at present are in extreme lack of the operation and maintenance personnel with knowledge of the characteristics of the new-generation of network technology. The traditional model that mainly relies on personnel experience for network maintenance will be very difficult to adapt to the new situation requirements. Instead, they must rely on more professional and intelligent software solutions.

However, at the present stage, the software support is far behind the development of telecommunications equipment. In the process of wireless system development, telecom equipment manufacturer's product development focus is the most important part of the network construction investment. Wireless performance support system software is limited to verification conditions and other factors, its accumulated technology and mature products are generally later than the development of hardware devices. At this stage, in order to respond to the external policies and market situation, the immature 3G, 4G and other new mobile communications equipment has entered the large-scale network construction phase, but its supporting software technologies have failed to develop synchronously.

In fact, this imbalance has led to the low efficiency of the current network construction and operation. For example, the mobile phone user capacity and mobile Internet access rate are much lower than the network design capabilities. The signal blind area improvement is inefficient. The imbalanced development is also suggesting the important future direction of the communications technology market.

In recent years, both inside and outside of the telecommunications industry have been actively exploring the issue of self-determined choosing or even dynamically choosing the wireless network by mobile phone users. If the service model develops well, it will have very significant impact on the competition mode of the existing telecommunications industry. Telecommunication operators are no longer able to lock the users by controlling the phone numbers or phone types, and the users will be able to choose the right wireless network according to the quality of the network. This new feature will force the operators to pay more attention to the network quality assurance means to consolidate and develop the users.

Key Technologies

When the mobile communications network fully enter 3G and 4G era, network characteristics will undergo great changes. Its wireless performance optimization must also transform from 2G era's "troubleshooting" mode into "equilibrium" mode, and the analytical method of "complex giant system" in the system theory must be used to reconstruct the network optimization technology system.

The new generation wireless network intelligent analysis needs to provide the following key technological means.

(1) The technological means to make an in-depth automatic comprehensive analysis on the wireless system data
(2) The technological means to translate the wireless system data into the management information needed for the operational decision-making
(3) The intelligent correlation analysis method that can meet the requirements of system equilibrium optimization
(4) The method to carry out overall monitoring and integrated optimization for service performance and network indicators in all network protocol layers
(5) The dynamic planning and configuration management method that can adapt to the frequent adjustment characteristics of the actual operating networks
(6) The dynamic network management method that can do real-time network performance monitoring and system adjustment

(7) Collaborative performance analysis method covering both network side and terminal side
(8) Application and verification method for the mid-to-long cycle radio resources management (RRM) algorithm in the system implementation
(9) The network performance monitoring method that meet the characteristics of massive data and fast evaluation in the balanced network
(10) The systematic analysis and management method that fully cover all kinds of service performance and network indicators
(11) Data sharing and analysis method that links configuration, monitoring, planning, drive test, and customer service.

Meanwhile, the specific requirements in network operation and maintenance need to be met both in breadth and depth. Its wireless analysis capabilities must have the "6 all's" features:

(1) All targets
    Be able to meet both the targets in professionalism and management of wireless network operation and maintenance.
(2) All networks
    Be able to analyze the whole network, including special network element equipment, such as repeater, home base station.
(3) All indicators
    Be able to comprehensively analyze various types of service quality and network performance indicators, and even the equipment reliability.
(4) All links
    Be able to make balanced coordination analysis of uplink and downlink wireless links.
(5) All time
    Be able to make network performance analysis of various temporal granularity, including second, minute, day, week, month, etc.
(6) All segments
    Be able to make unified analysis on the data collected in configuration, monitoring, planning, drive test, customer service, etc.

Technical Roadmap

"Guided by the methods of system theory, combined with the advanced Business Intelligence (BI) software technologies, realize wireless professional in-depth analysis and verification means, and provide fast performance guarantee capability for the mobile communications network." Its important key technical points are as follows.

(1) Wireless network information quantitative analysis and evaluation model
    Various kinds of parameters, key events, and analytical algorithms are summarized into the tree structure. Information analysis model, independent from software platform, can realize dynamic upgrade and loading. In the model,

all kinds of data from the network side and the terminal side have unified definitions. Quantitative data association attributes are provided. The professional wireless performance analysis framework is established. It is compatible with various types of wireless networks.

(2) Multi-dimensional cube database

Multi-dimensional cube data model is designed to facilitate correlation search and association analysis, which provides a logical basis for the information analysis in the dimensions of physics, space, object, time, etc.

(3) OnLine Analysis & Process (OLAP)

The intelligent analysis on network performance is realized based on correlation search technology. The advanced parallel processing technology is used to realize the ultra-high speed real-time data correlation analysis.

(4) on-line transaction processing (OLTP)

It solves the problem of data acquisition and processing efficiency in the means of trading space for time.

(5) Real-time display technology (including dynamic display based on GIS and line graph)

It completely solves the problem of function and performance of the information release, and ensures the high concurrency in the system application, which benefits collaborative work.

Uniqueness

(1) It has the complete mobile network performance management functions.

It provides innovative automatic positioning and diagnosis ability of network performance, provides element level, network level, and service level performance intelligent analysis ability, provides the comprehensive target management of coverage, capacity, quality of service, quality of the equipment etc., and comprehensively processes network side and terminal side data, with point and surface combined and uplink and downlink balanced.

Multi-dimensional correlation search query with performance data is provided. GIS region correlation analysis function can effectively and directly reflect the true situation of network. Synchronous adjustment tracking record of key configuration parameter in wireless network is also provided.

It supports dynamic network element management, including repeater station, home base station and other special network. Network management, network measurement, instrumentation, and other performance data get unified model definition and storage. User-defined function of the evaluation model is defined to meet the needs of the development of network management.

(2) Strong quantitative management performance

Mass data association rules processing engine is embedded to adapt to large telecommunications network applications. The complete network information quantitative analysis and evaluation model can effectively reduce the complexity of the operation and maintenance. It adapts to multi service characteristics, and is close to the different service characteristics and quality level.

(3) Easy to deploy and upgrade, with extensive adaptability

Expert experience database, separated from software platform, makes it convenient for rapid upgrade and experience expansion. It's compatible with various network test tools, instruments files and data interfaces. It adapts to multiple network interfaces, data file types, and it's easy to integrate with other systems. It can be seamlessly connected and completely integrated with the vendor systems. It targets at provincial and regional multi-level mobile network application modes. And it can be integrated into the process management tools.

(4) Able to meet the needs of daily operation and maintenance of mobile network

Online linkage, and interactive management. According to different types and levels of users, provide differentiated information customization service. Provide multi-level multi-perspective network evaluation information, powerful information release and report output function, which is conducive to the integrated management of properties and configuration data related with network performance.

Provide the process management tools interfaces, so as to facilitate the work process control and management. Provide a sound data storage, management functions, in order to easily trace the original data.

Requirements to Be Achieved

(1) The unified framework of complex network analysis algorithms: providing the core theory of abstraction and integration of different types of analysis algorithms.

(2) Quantitative evaluation model of wireless network performance: providing the theoretical basis for the analysis and optimization of wireless network.

(3) Correlation analysis multi-dimensional database: providing the data model meeting the requirements of the logic association mining.

(4) Correlation association analysis engine: providing the software means for adaptive automatic analysis.

(5) Complex network visualization technology: providing the data presentation and interactive method of complex relationship information.

Through loading expert experience algorithm library, the embedded intelligent analysis engine can make automated correlation analysis and processing for system operation information and test data of various kinds of communications networks., It can provide detailed diagnostic and statistical summary information for various types of network anomaly, faults and causes in multiple forms, and provide the specific rectification measures and suggestions for operation and maintenance personnel to choose from.

It features three aspects "efficiency, expertise, and ease of use".

Efficiency is reflected in the support for the information correlation processing of multiple data sources. With massive data processing, effective information can be fully mined and utilized. Network anomaly analysis and processing can be done in a fully automated way, greatly reducing the time of manual work.

Expertise is embodied in the process that this technology fully integrates and cherishes the expertise of communications experts. The algorithm library is flexible. The analysis process is close to the requirements of the network optimization site. All the algorithms and solutions are derived from the practice of the network optimization project. The analysis results are reliable and can be used. Meanwhile, it outputs the software analysis conclusions in cutting-edge graphics, intuitively reveals the generation mechanism of the complex network anomaly, makes hard things simple, strengthens the operator's imaginal thinking, and reduces the understanding difficulty, making it easy for operators to quickly capture and memorize key information and solve complex technical problems.

Ease of use means the application and operation process is clear, convenient, instructive and intuitive. The interface is always reasonably simple, and the interactive process is comfortable, ensuring the operator's work efficiency in massive data analysis. All kinds of information in the interface are presented in a clear way with refined visual effect, improving work pleasure to the greatest extent so that the operator will not get tired easily. Meanwhile it provides a variety of online map automatic update function, and there is no need for manual installation of the map data, easy for the operator to carry out network analysis activities at any time any place.

7.4.2   Status

At present, there are many kinds of wireless mobile communications network data analysis systems in the industry. According to analysis angles (terminal side and network side) and functions (technical and management), product can be effectively categorized.

Different product functions have different focuses, roughly shown in Fig. 7.19.

As the core of network performance analysis, network optimization software and other analysis tools require very powerful expertise and comprehensiveness, and need to achieve the above-mentioned "6 all's" analysis ability and other professional requirements. However, seen from the development course of the communications system, restricted by the lack actual network verification environment, the development of network analysis technology is often lagging behind the development of hardware and equipment. And thus the development of related products lags behind the development of network technology.

The development course of the system is shown in Fig. 7.20.

The first generation analysis system provides basic data display function, while the extraction, sorting, statistical handling of the drive test recorded data need manual work and the interpretation of data is completely based on the individual experience. The second generation analysis system provides rich data display and statistical functions, and can partially replace manual work for drive test data sorting and statistics, but the data analysis process still needs manual interpretation.

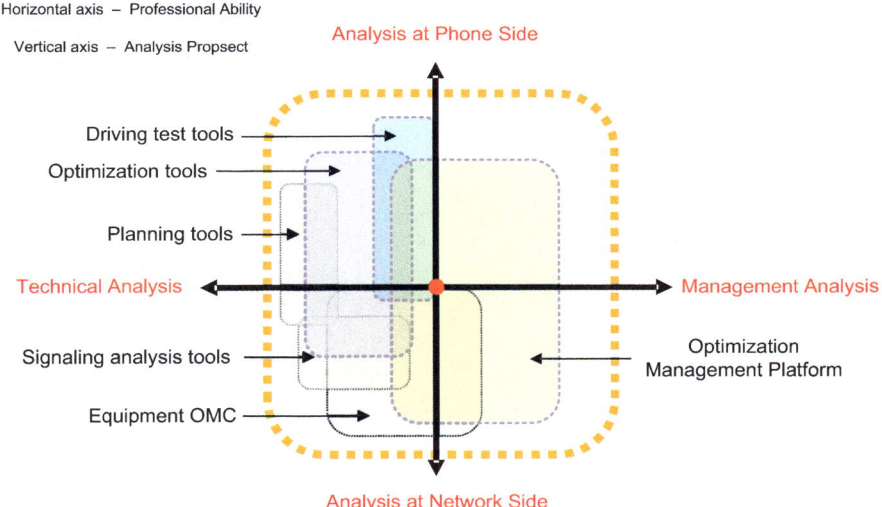

Fig. 7.19 Data analysis system for wireless mobile communication network

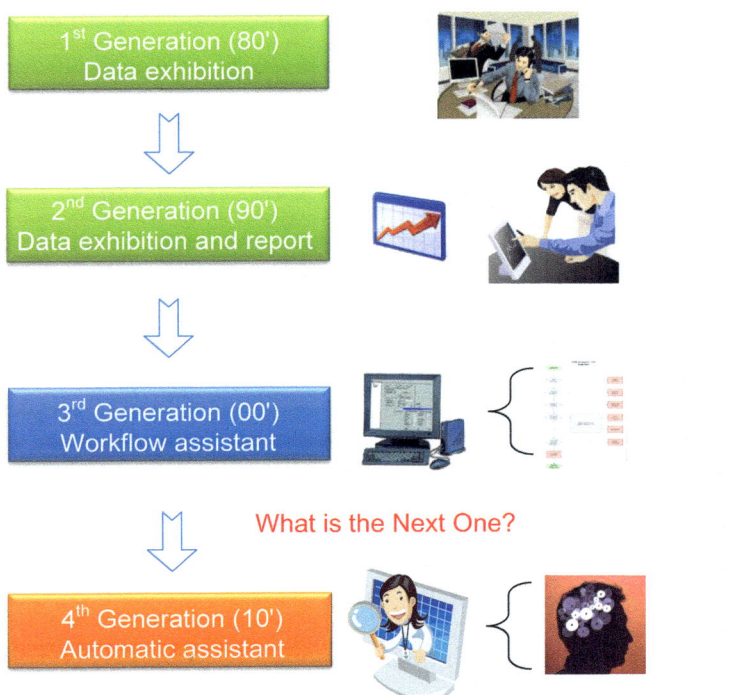

Fig. 7.20 The development course of wireless mobile communication network optimization tools

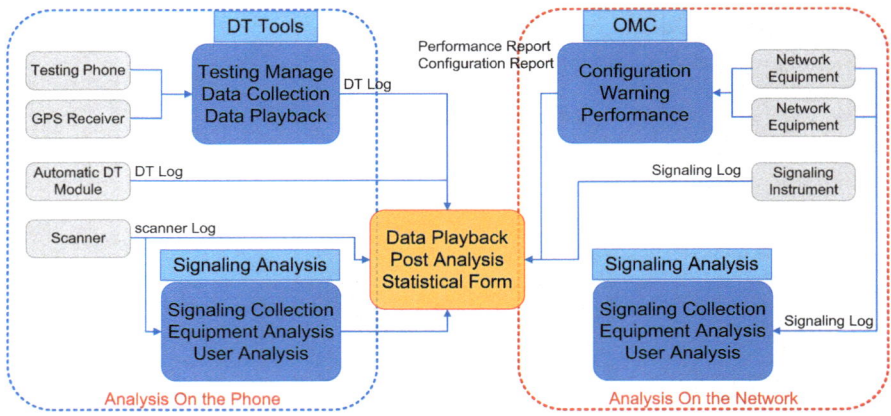

Fig. 7.21 Composition of intelligent network optimization analysis system

The third generation analysis system introduces the thematic analysis function, which while replacing human for data sorting and statistics also provides some thematic analysis process guide, but the manual interpretation remains the core of each sub process of analysis. The current industry lacks of the fourth generation analysis system, which, with wide coverage, is able to maximize the integration of expert analysis experience, and provide the whole-process automated association analysis function, while human only review and make decisions for the analysis results, realizing the intelligent software analysis system in the real sense. Wide coverage requires the system to be compatible with a variety of data sources and data formats. Integration experts need to convert the complex thinking process into machine language according to certain models. The automated process requires each link and each module of the system work together for execution in accordance with the established sequence in an orderly way. Association function requires mining the existing data to the greatest possible extent and drain data to the last drop. In such a way, we can maximize the role of machines and minimize manual work while ensuring certain degree of accuracy, and the real intelligence can be reflected.

As the core means of background analysis of the wireless network optimization department, it deals with all kinds of data in wireless network, including the test data of mobile phone test and automatic road test, sweeper test data, parameter data of planned software base station, network management system report data, signaling data, other internal data etc (Fig. 7.21).

7.4.3   Application Scenarios

The specific application scenarios of the intelligent analysis system are shown in Fig. 7.22.

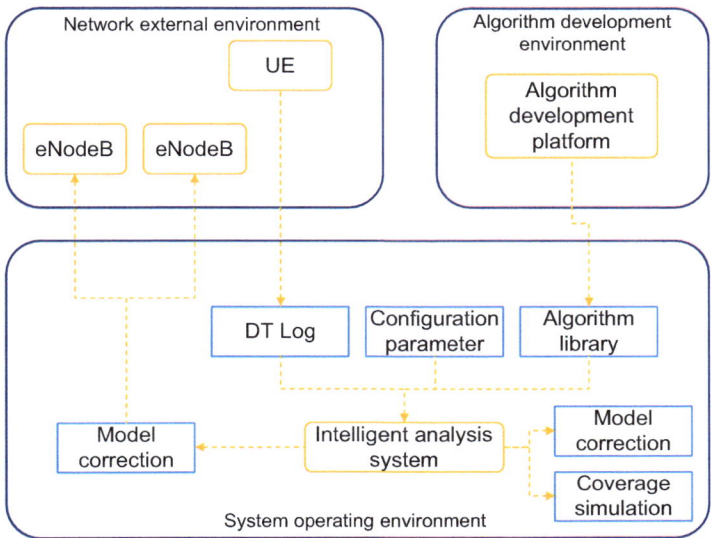

Fig. 7.22   Application scenarios of intelligent network optimization analysis system

Application process: start up field operation test → acquire network test data → import intelligent acquisition analysis system → load auxiliary analysis data → automatically detect anomalies → automatically analyze network fault → automatically analyze fault correction measures → output adjustment measures, model correction results, coverage simulation results → implement field network adjustment → restart field service test →re-acquire test data → verify network problem correction.

Innovation Points

Automatic solution generation: It fully integrates and cherishes the expertise of communications experts, and automatically generates solutions. The analysis algorithm library is flexible, and the analysis process is close to the requirements of the network optimization site. All the algorithms and solutions are derived from the practice of the network optimization project. So the analysis results are reliable and can be used.

Visualization of network problems: It outputs the software analysis conclusions in cutting-edge graphics, intuitively reveals the generation mechanism of the complex network anomaly, makes hard things simple, strengthens the operator's imaginal thinking, reduces the understanding difficulty, making it easy for operators to quickly capture and memorize key information and solve complex technical problems.

Coordinated analysis of multiple data sources: It supports the coordinated analysis of multiple networks and multiple modes, supports the information correlation processing of multiple data sources. With massive data processing, effective information can be fully mined and utilized.

Application Model: Complex Network Visualization Technology

It provides data display and interaction means of complex relational information, provides rich information view with progressive levels, so as to realize the brand-new network optimization modes – tracking, auditing, reviewing, and meet the application requirements of different types of users (Fig. 7.23).

As for the management evaluation level, generally it uses control management view and overall view, focuses on the overall situation, uses the analysis conclusion navigational function between views, and focuses on problem tracking.

As for the optimization and supervision level, generally it uses overall view and local view, uses the analysis conclusion drilling function between views, focuses on determining the overall optimization decision, and guides the optimization.

As for the audit and authentication level, it generally uses local view and manual analysis view, uses the raw data drilling function between views, focuses on the specific issues' positioning, analysis and settlement, as well as the accuracy of the audit conclusion.

In interaction, a large number of views can be summarized into "three levels and five dimensions", in which three levels refer to the three-level structure of the overall view, local view and manual analysis view, and the five dimensions refer to the management dimension, feature dimension, element dimension, geographical

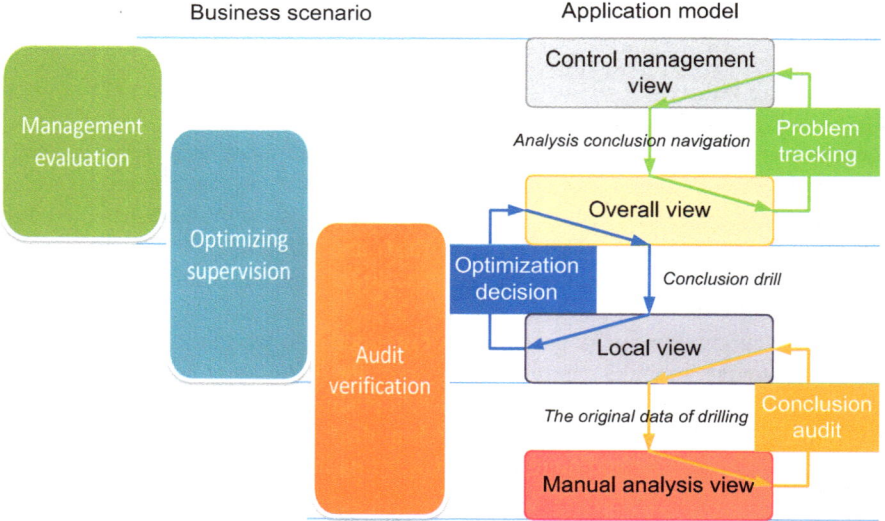

Fig. 7.23 Working mode of intelligent network optimization analysis system

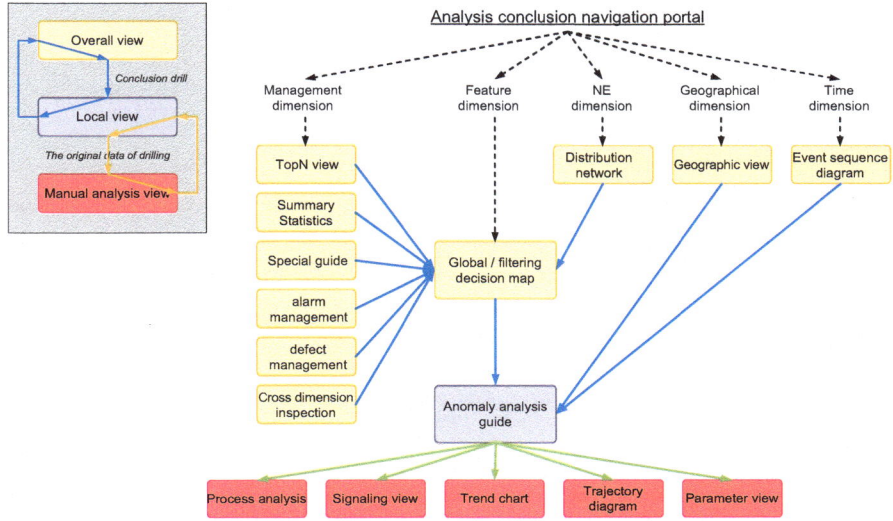

Fig. 7.24 The brand new "three levels and five dimensions" information interaction mode

dimension and time dimension. The relationship of the views is shown in the following figure (Fig. 7.24).

Application Mode: Building the Macro Optimization Intelligent System

Based on the in-depth mining of the intelligent diagnosis conclusions, it provides the connected information management and control objects of multi-level network optimization of objects, which can effectively connect the management and production tasks and meet the different levels of information needs of the network optimization operation and maintenance department (Fig. 7.25).

Risk area: It is automatically generated based on the characteristics like the distribution density of abnormal points. It is used for the active identification and analysis on the service quality high risk areas in the process of network operation.

Alarm points: According to the current network optimization goals, the abnormal points of specific types that require high attention are automatically extracted and gathered, used to determine the key rectification projects of the current network optimization.

Abnormal points: They are automatically generated in the anomalies detection process, providing detailed anomalies information, causes and rectification recommendations, used for the decisions in specific network optimization rectification program.

Based on multi-level control mode, the network optimization personnel of different functions can respectively deal with the operation and

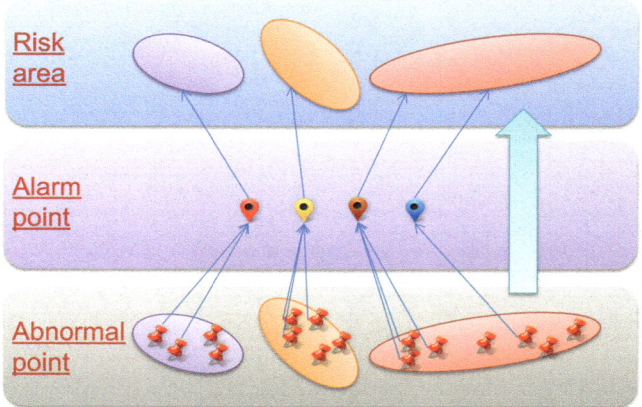

Fig. 7.25 The three-level network optimization management and control system that covers both points and surfaces

maintenance tasks with different urgency and importance levels, achieving the seamless docking of the active optimization and ensuring optimization goals between the upper and lower levels personnel, and completing the network optimization coordinately.

Application Mode: Closed Loop Intelligent Management of Network Problems

As the participants and managers of network optimization operation and maintenance, we need to grasp the overall performance of the wireless test. And the formation of abnormal events and the life cycle of the causes behind, including generation, maintaining, and elimination, which not only is necessary but also serves as a kind of effective management on the network optimization process.

The function of the closed loop intelligent management of network problems, by referring to the mature network management application model and based on intelligent analysis function, is the first in the industry to create the road test alarm/fault detection and management, effectively monitoring the tracking of problem loop.

Based on the road test alarm function, it can track the multiple batches of data and get in-depth understanding of the alarm points' status, grade, cause, distribution and other information. Flexible management can be realized on site through alarm type, road location, alarm status, date, etc. Meanwhile, for each problem, it can view the causes of the current and past problems, solutions and other relevant information (Fig. 7.26).

Analysis set automatic analysis conclusion

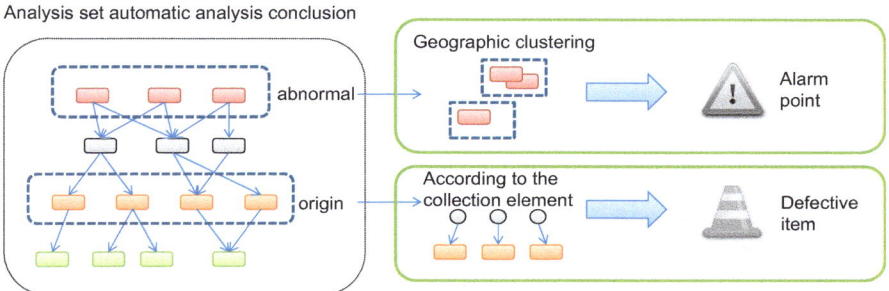

Fig. 7.26 Closed loop management of the problem points in the daily optimization stage

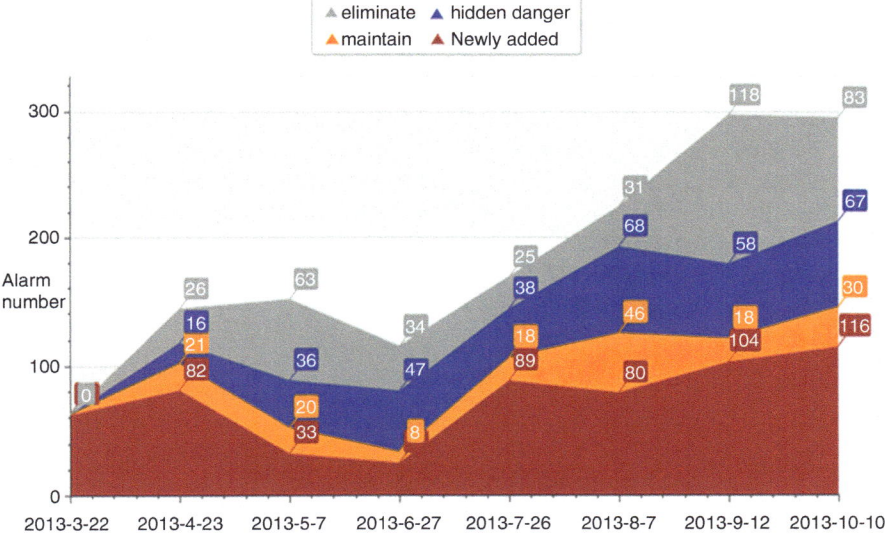

Fig. 7.27 Road test alarm history information

Clustering by Geography. Road test abnormal events, based on the configurable generation rules, alarm objects can be automatically generated. Alarm points belong to the geographic area and the tracking history can be set up according to the road test batches.

Categorizing by Network Elements. Based on the fault causes of the anomalies of the road test alarm, the defect objects are generated. Defect item falls into network elements and the tracking history can be set up according to the road test batches. Defect items support the setting of functional departments, conducive to differentiated input and coordinated rectification.

By tracking the road test alarm statistics of the multiple data, it can accurately grasp the changes in the number of network obstacles (Fig. 7.27).

Application Mode: Intelligent Single Station Batch Submission For Acceptance

The verification and optimization of single stations is the base of the whole wireless optimization of mobile network. It is the unavoidable initial stage of network optimization. The solution of single stations can help eliminate the hidden troubles in the project construction stage, and effectively reduce the pressure of the follow-up optimization work, so as to lay a solid foundation for the higher level of the whole network optimization. Network construction and maintenance department are generally in urgent need of professional and highly-efficient technical means to replace manual analysis so that they can automatically, accurately and quickly make optimization analysis and acceptance review to the new base station inspection test data, quickly master each base station's running status, performance and fault causes. In such a way, they can give effective coordinated rectification to the existing problems so that the new stations can quickly meet the acceptance criteria and enter into formal operation as soon as possible.

Intelligent single station batch submission for acceptance function can make fast, in batch and rolling judgment about whether the base station can fully meet the preset quality standards, and provide detailed information on the fault positioning. So that the various functional departments can accurately, rapidly and collaboratively make rectification and eliminate the hidden troubles of network construction to the greatest extent. Coordinated rectification can only be implemented when the network problem causes are positioned (Fig. 7.28).

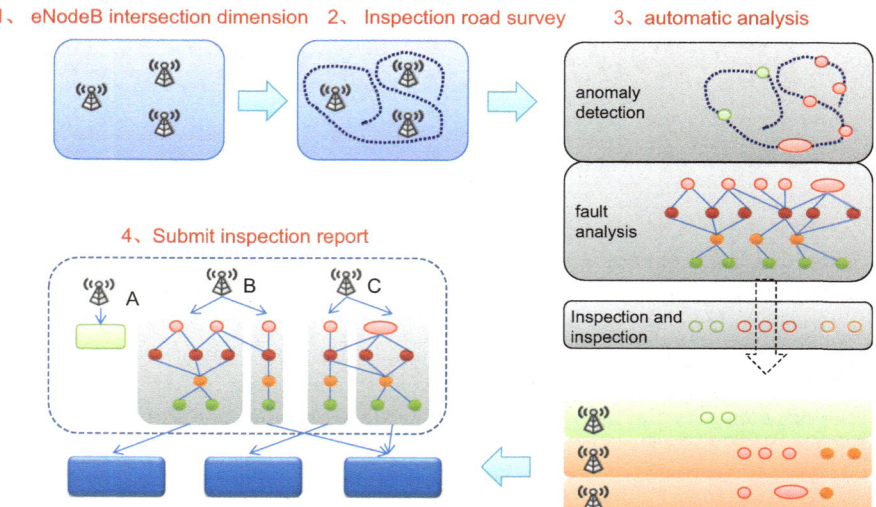

Fig. 7.28 Base station batch submission for inspection function

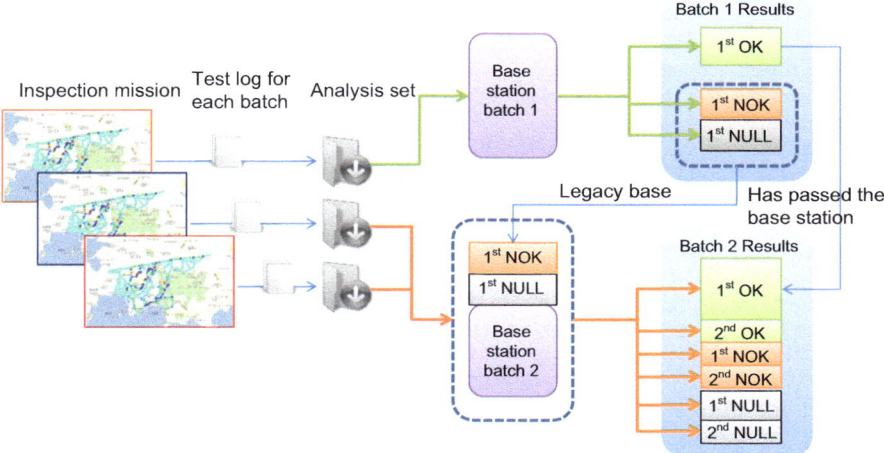

Fig. 7.29 Multi-batch rolling base station inspection

It supports the checking and analysis of multi-batch data, makes it easy to track failed base stations. The inspection results are summarized in rolling way. Based on the submission for inspection function, the inspection road test data analysis is carried out automatically. The progress and problems of the base station optimization project can be mastered synchronically (Fig. 7.29).

7.5   Summary

As stated at the beginning of this chapter, the ultimate goal of the R&D and evaluation of mobile communications technology is application. The field trial is the last and also the most important part of the process from the technology R&D to application. In particular, the acceleration of the new technology update makes it more important to carry out the field trial and verification in the R&D stage. This chapter, based on the analysis on the requirements and application scenarios of 5G mobile communications technology, gives the field trial evaluation cases in the LTE-A phase from the wireless channel measurement to the key technology validation and then to the network performance validation, providing references for the later 5G technology research. In addition, with the development of big data technology, intelligent wireless network data analysis tools will greatly enhance the network performance's self-optimization ability. At the end of this chapter, we present the architecture that is still in planning stage of the multi-scenario heterogeneous test network that supports large scale test users, providing an ideal test site for the field trial and evaluation in R&D stage for relevant research institutions.

References

1. Scenarios, Requirements and KPIs for 5G Mobile and Wireless System. METIS Deliverable D1.1, Apr. 2013, https://www.metis2020.com/wp-content/uploads/deliverables/METIS_D1.1_v1.pdf
2. White Paper on 5G Concept. IMT-2020 (5G) Promotion Group. Feb 2015. http://www.imt-2020.cn/zh/documents/download/23
3. 5G Vision and Requirements. IMT-2020 (5G) Promotion Group. May 2014. http://www.imt-2020.org.cn/zh/documents/download/1
4. A. Ghosh, N. Mangalvedhe, R. Ratasuk and et al. Heterogeneous cellular networks:From theory to practice[J]. Communications Magazine, IEEE, 2012, 50(6): 54–64.
5. B. Soret, H. Wang, K. Pedersen and et al. Multicell cooperation for LTE-advanced heterogeneous network scenarios. IEEE Wireless Communications, 2013, 20(1): 27–34.
6. W. Stallings. ISDN and broadband ISDN. Macmillan Publishing Co., Inc., 1992.
7. F. Xu, L. Zhang and Z. Zhou. Interworking of Wimax and 3GPP networks based on IMS [IP Multimedia Systems (IMS) Infrastructure and Services]. IEEE Communications Magazine, 2007, 45(3): 144–150.
8. M. He, X. Wang, T. D. Todd and et al. Ad hoc assisted handoff for real-time voice in IEEE 802.11 infrastructure WLANs. International Journal of Wireless and Mobile Computing, 2007, 2(4): 324–336.
9. R. Ananthapadmanabha, B. S. Manoj and C. Murthy. Multi-hop cellular networks:the architecture and routing protocols. Personal, IEEE International Symposium on Indoor and Mobile Radio Communications, 2001, 2:G-78-G-82 vol. 2.
10. A. N. Zadeh, B. Jabbari, R. Pickholtz R and et al. Self-organizing packet radio ad hoc networks with overlay (SOPRANO). IEEE Communications Magazine, 2002, 40(6): 149–157.
11. C. Qiao, H. Wu and O. Tonguz. Integrated cellular and ad hoc relaying system: U.S. Patent Application 10/213,058[P]. 2002-8-6.
12. H. Luo, R. Ramjee, P. Sinha and et al. UCAN: a unified cellular and ad-hoc network architecture. Proceedings of the 9th annual international conference on Mobile computing and networking. ACM, 2003:353–367.
13. M. Dick. Mobility and differentiated services in a future IP network. IST Project. http://www.ist-mobydick.org, 2003.
14. T. Braun, M. Diaz, J. E. Gabeiras and et al. End-to-end quality of service over heterogeneous networks. Springer Science & Business Media, 2008.
15. W. Mohr. The WINNER (Wireless World Initiative New Radio) project–development of a radio interface for systems beyond 3G. International Journal of Wireless Information Networks, 2007, 14(2): 67–78.

Printed by Printforce, the Netherlands